Liverpool
Community
College

PHYSICS

Other titles in the Project

Telecommunications
John Allen

Medical Physics
Martin Hollins

Energy
David Sang and Robert Hutchings

Nuclear Physics
David Sang

Biology
Martin Rowland

Applied Genetics
Geoff Hayward

Applied Ecology
Geoff Hayward

Micro-Organisms and Biotechnology
Jane Taylor

Biochemistry and Molecular Biology
Moira Sheehan

Chemistry
Ken Gadd and Steve Gurr

Liverpool
Community
College

Series Director: Prof. J. J. Thompson, CBE

PHYSICS

ROBERT HUTCHINGS

Nelson

Nelson
Delta Place
27 Bath Road
Cheltenham
GL53 7TH
United Kingdom

First published by Macmillan Education Ltd 1990
ISBN 0-333-46515-6
First edition published by Thomas Nelson and Sons Ltd 1992
Second edition published by Nelson 2000

ISBN 0-17-438731-8

00 01 02 03 04
10 9 8 7 6 5 4 3 2 1

Printed in Spain by Graficas Estella S.A.

Acknowledgements
The authors and publisher wish to thank Bill Houston for compiling the
index, Zooid Pictures for the picture research and the following for
permission to reproduce questions from past examination papers, which
are copyright material:

The Assessment and Qualifications Alliance, London Examinations, a
division of Edexcel Foundation, Northern Ireland Council for the
Curriculum Examinations and Assessment, Oxford, Cambridge and RSA
Examinations Board and the Welsh Joint Education Committee.

Every effort has been made to trace all the copyright holders, but where
this has not been possible the publisher will be pleased to make any
necessary arrangements at the first opportunity.

The author and publisher wish to acknowledge, with thanks, the
following photographic sources:

Cover and p.iii, Space Frontiers/Telegraph Colour Library **p.1**, Chris Butler/Science Photo Library;
p.4, Alex Tsiaras/Science Photo Library; **p.5**, Dagmar Schilling/Science Photo Library; **p.12**, Andrew
Lambert; **p.25 top**, Omikron/Science Photo Library; **p.25 bottom**, James L. Amos/Corbis UK Ltd;
p.27, Lowell Georgia/Corbis UK Ltd; **p.29 top**, Ed Micheals/Science Photo Library; **p.29 bottom**, Alex
Bartel/Science Photo Library; **p.34**, Zefa; **p.37**, Prof Howard Edgerton/Science Photo Library; **p.43**,
Prof Howard Edgerton/Science Photo Library; **p.44**, Wally McNamee/Corbis UK Ltd; **p.45 top left**,
Allsport (UK) Ltd; **p.45 top right**, Corbis UK Ltd; **p.45 bottom**, Jim McDonald/Corbis UK Ltd; **p.46**,
Corbis UK Ltd; **p.55**, NASA/Science Photo Library; **p.59**, Bob Child/Associated Press; **p.60** Zefa;
p.64 top, Corbis UK Ltd; **p.64 bottom**, Dimitri Iundt/Corbis UK Ltd; **p.65**, Martin Bond/Science
Photo Library; **p.69**, Peter Mezel/Science Photo Library; **p.70**, Corbis UK Ltd; **p.75 bottom left**, Corbis
UK Ltd; **p.75 bottom right**, Will & Deni McIntyre/Science Photo Library; **p.75 top**, Mere
Woods/Science Photo Library; **p.75 centre**, Ed Young/Science Photo Library; **p.76**, Corbis UK Ltd;
p.77, Corbis UK Ltd; **p.84**, Prof Harold Edgerton/Science Photo Library; **p.87**, Corbis UK Ltd; **p.88**,
Alfred Pasieka/Science Photo Library; **p.93 left**,Volvo Car Corporation (UK); **p.93 right**, Corbis UK
Ltd; **p.94**, Corbis UK Ltd; **p.97**, Pekka Parviainen/Science Photo Library; **p.98**, Andrew Lambert;
p.107, NASA/Science Photo Library; **p.110 left**, Julian Baum/Science Photo Library; **p.110 right**,
Gordon Garrardd/Science Photo Library; **p.115**, Robert Hutchings; **p.116**,Kaj R. Svensson/Science
Photo Library; **p.117**, Corbis UK Ltd; **p.121**, Royal Observatory/Science Photo Library; **p.122**,
NASA/Science Photo Library; **p.129**, NASA; **p.130**, NASA/Science Photo Library; **p.142**, Erich
Schrempp/Science Photo Library; **p.143**, John P Kelly/Image Bank; **p.150**, Gontier, Jerrican/Image
Bank; **p.15 left**, James Stevenson/Science Photo Library; **p.154**, Ford Motor Company Limited; **p.155
left**, Corbis UK Ltd; **p.155 right**, Corbis UK Ltd; **p.156**, Prof J Bories/Science Photo Library; **p.163**,
William Ervin/Science Photo Library; **p.164**, Martin Dohrn/Science Photo Library; **p.172**, Micheal
Marten/Science Photo Library; **p.176 top**, Science Photo Library; **p.176 bottom**, Peter
Aprahamian/Science Photo Library; **p.177**, Sheila Terry/Science Photo Library; **p179**, Paul
Shambroom/Science Photo Library; **p.181**, Simon Fraser/Science Photo Library; **p.183**, Digital
Art/Corbis UK Ltd; **p.196**, David Parker/Science Photo Library; **p.197**, Simon Fraser/Science Photo
Library; **p.202**, BSIP Kokel/Science Photo Library; **p.207**, Canon (UK) Ltd; **p.208**, Dr Fred
Espenak/Science Photo Library; **p.215**, Corbis UK Ltd; **p.217**, Andrew Lambert; **p.218**, Corbis UK Ltd;
p.219, Wally McNamee/Corbis UK Ltd; **p.225**, Jan Hinsch/Science Photo Library; **p.226**, NASA; **p.233
top**, Imperial College/Science Photo Library; **p.233 centre**, Imperial College/Science Photo Library;
p.233 bottom, Imperial College/Science Photo Library; **p.237**, Tim Davis/Science Photo Library;
p.240, James Stevenson/Science Photo Library; **p.248**, Peter Mezel/Science Photo Library; **p.249**,
Nelson Medina/Science Photo Library; **p.250 top left**, RS Components; **p.250 top right**, Peter
Mezel/Science Photo Library; **p.250 bottom**, Amy Trustram/Science Photo Library; **p.271**, Hank
Morgan/Science Photo Library; **p.279**, Will and Deni McIntyre/Science Photo Library; **p.282**, Peter
Mezel/Science Photo Library; **p.284**, Manfred Kage/Science Photo Library; **p.287 top**, Ray
Ellis/Science Photo Library; **p.287 bottom**, Corbis UK Ltd; **p.317**, Omikron/Science Photo Library;
p.322, Maximilian stock/Science Photo Library; **p.324**, Jerry Mason/Science Photo Library; **p.331**,
Andrew Lambert; **p.336**, Alex Bartel/Science Photo Library; **p.337**, Heini Schneebeli/Science Photo
Library; **p.343**, Charles D Winters/Science Photo Library; **p.356**, Michael Marten/Science Photo
Library; **p.357**, Robert Hutchings; **p.364**, Garry Watson/Science Photo Library; **p.379**, Philippe
Plailly/Science Photo Library; **p.385 left**, Northwestern University/Science Photo Library; **p.385 right**,
Dr Erwin Mueller/Science Photo Library; **p.389 top**, Physics of Materials, by B. Cook and D. Sang;
p.389 bottom, Physics of Materials, by B. Cook and D. Sang; **p.396**, David Parker/Science Photo
Library; **p.403**, Prof Harold Ederton/Science Photo Library; **p.406**, Klaus Guldbrandsen/Science
Photo Library; **p.411**, Prof H Hashimoto/Science Photo Library; **p.413**, Manfred Kage/Science Photo
Library; **p.417**, Magrath/Science Photo Library; **p.421**, Peter Mezel/Science Photo Library; **p.422**, John
Howard/Science Photo Library; **p.423**, Geoff Williams/Science Photo Library; **p.426**, Philip Harris;
p.432, Philip James Corwin/Corbis UK Ltd; **p.442**, Clive Freeman/Science Photo Library; **p.447**,
Science Photo Library; **p.448**, Benn Mitchell/Image Bank; **p.458**, Sam Odgen/Science Photo Library;
p.460 left, Alex Bartell/Science Photo Library; **p.460 right**, Ed Young/Science Photo Library; **p.462**,
Simon Fraser/Science Photo Library; **p.469**, David Parker/Science Photo Library; **p.470**, Jack
Finch/Science Photo Library; **p.478**, Jean-Loup Charmet/Science Photo Library; **p.480**, David
Parker/Science Photo Library; **p.504**, Powell & Fowler/Science Photo Library; **p.505 left**, Lawrence
Berkely Lab/Science Photo Library; **p.505 right**, N.Feather/Science Photo Library; **p.511**,
CNRI/Science Photo Library; **p.512 top left**, Dr. Karol Sikora/Science Photo Library; **p.512 top right**,
Dr Karol/Science Photo Library; **p.512 bottom**, Ed Young/Science Photo Library; **p.513**, Martin
Bond/Science Photo Library; **p.527**, Imperial College/Science Photo Library; **p.528 top**, BSIP
Laurent/Science Photo Library; **p.528 bottom**, J C Revy/Science Photo Library; **p.539**, Jim
Sugar/Corbis UK Ltd; **p.540**, CERN/Science Photo Library; **p.541 left**, CERN/Zooid Pictures; **p.541
right**, Philippe Plailly/Science Photo Library.

Contents

Contents

Bath Advanced Science Series: Introduction

It is now some ten years since the **University of Bath 16-19 Project** was first published. The success of the material produced by the team of talented authors has been demonstrated consistently throughout that period. At the same time there has been a great deal of change in the provision of education, not only in the UK but throughout the world. With the ongoing drive towards the broadening of post-16 education – making it *more* accessible and *more* useful to *more* people – there is greater emphasis not only on expanding knowledge, but also on learning and understanding how this knowledge can be applied, and developing the skills required to do so.

The **Bath Advanced Science** series represents an extensive review, re-organisation and update of the material produced in the original project. The author team has come together to draw on its collective expertise and experience to produce resources which reflect the demands of Post-16 education today.

The series has been written to address the Advanced Subsidiary and Advanced GCE Science specifications introduced by the unitary awarding bodies from September 2000 following the Dearing *Review of Qualifications for 16–19 Year Olds* and the subsequent consultation *Qualifying for Success*.

- The subject texts provide comprehensive coverage of the content required for the awarding body specifications in those subjects.
- The topic books each provide detailed coverage, which will support both option topics for AS and A GCE and other advanced level courses. They are also appropriate for students embarking on Foundation courses in Higher Education, or preparing for degrees in which these topics are studied.

The **Bath Advanced Science** series contains up-to-date material presented to encourage effective learning. The texts continue to be highly interactive – ensuring a degree of independent self-study and actively providing opportunities for students to acquire knowledge and understanding, to develop skills and concepts, and to appreciate the applications and implications of science.

We have designed the **Bath Advanced Science** series specifically to help all students to develop the realisation that science *is* for everyone, and to gain an appreciation and understanding of the importance and relevance of science to everyday life.

Prof. J. J. Thompson, CBE
Department of Education
University of Bath
May 2000

How to use this book

Introduction

The aim of this book is to help and encourage you in your Advanced Subsidiary (AS) and Advanced GCE physics course. Physics is not a collection of facts and equations to be memorised. It needs to be considered as a few fundamental principles that can be applied in many different situations. At the start of your course it is important to acquire habits of study and experiment that will help your understanding of the subject. Understanding is the key; memorised facts without understanding are virtually useless. The short cut to an understanding of physics is the ability to use the principles.

It has been assumed that in preparation for taking this course you have done some physics previously. This may have been as part of a GCSE science course or as a subject by itself, for example. However, much GCSE work is covered again in this core book to ensure that you have a solid grounding in the fundamentals before tackling new ideas.

The study of physics needs to be done with a notebook, pen and calculator always at hand. You cannot sit down and read a physics book like a novel.

Each chapter is divided up into sections and it is sensible to tackle one section at a time. You will find on many occasions that a topic is introduced with a numerical problem. You are advised to work through this problem before going on to see the theory done algebraically. In some of the numerical calculations you will find that, on occasions, more significant figures have been used than are reliable, but this has been done in order to avoid rounding up errors, and, where necessary, will have been corrected in the last line of the answer. This is standard procedure. Keep as many figures as possible throughout a calculation and conclude with an answer in which all the reliable figures are given together with one extra doubtful figure – and no more.

Fewer abbreviations have been used in this book than in many at this level. This has been done to limit the number of obstacles to understanding and to try and reduce the amount of jargon. At times some calculus has been used but usually it has been added so that people who have learnt calculus can see how it is used. Use of calculus is not a requirement for British awarding bodies, but you do need to be able to find a gradient of a graph and the area beneath a graph by some means and calculus is a very precise way of doing this.

I have found that physics students demonstrate their ability and knowledge better in tests if they approach the solution of a problem in a sentence or two, rather

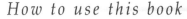

than assuming that the marker is only interested in the answer. I hope you will see the advantage of being prepared to use words where they are needed rather than relying solely on symbols and abbreviations.

I wish you success and satisfaction from the hard work you will need to put in to your course.

Robert Hutchings
May 2000

Learning objectives

Each chapter starts with a list of learning objectives, which outlines what you should gain from the chapter. Learning objectives can guide you in choosing your topics for study and help you to make notes for revision.

Examples

Each topic is introduced by defining the meaning of certain words and explaining its principles. It is then necessary to see how the ideas introduced are used in practice, so a worked example is normally included at this stage. You should work through each example carefully in order to gain greater understanding of the topic.

Questions

There are many questions in this book for you to practise using your knowledge. Do not avoid doing questions because you are a little uncertain of what to do – you will always learn something in the attempt. Frequently students say that they did not know how to start answering a question. In almost all cases these problems arise because the student does not write down what he or she is given. The following tips on how to approach questions may help:

- draw a sketch diagram;
- list, in your own words, the facts you have been given;
- make use of the units of the quantities involved;
- calculate anything obvious, even if it is not specifically asked for – when you have all the information on paper, things often fall into place;
- after making any calculation, look at all your answers to see if they are reasonable;
- look up the correct answer in the back of the book.

Questions in this book are of the following types:

1. *In-text questions* These are, hopefully, thought-provoking questions that you should be able to answer quickly.
2. *Section questions* These are questions dealing with one point at a time, which will help your understanding and ability to use principles as you progress through the course. You should do these questions as you read through each section.
3. *Assessment questions* In many physics textbooks the only questions are those taken from past GCE papers. These have their place and plenty are provided in this book at the end of the chapters for you to do at the end of a topic or for revision.
4. *Synoptic questions* Synoptic-type questions are now being set by awarding bodies. These always cover many different aspects of physics, and some have been used in this book, so you may find that some questions at the end of a chapter require knowledge from other chapters if they are to be answered fully.

Answers to questions

Answers to almost all questions, together with some notes indicating particular common mistakes, are provided in the appendix at the end of the book. Checking work is a very important aspect of physics. Everyone makes mistakes from time to time, so you need to be alert to the possibility of going wrong and to look out for careless mistakes. Powers of 10 are a frequent source of arithmetical nonsense. Cars do not travel at 40 km s^{-1} although 40 m s^{-1} is possible. Be alert to nonsensical answers and do not think that the only time to check an answer is at the end. Checking is a continuous process.

Sidelines

These are important or interesting side issues, and can be found in the margin.

Key points

These facts should, ideally, be memorised. While you can often look up such facts in reference books or data sheets, it is in practice much quicker to use them if you know them.

Investigations

The investigations in this book are designed to require thought about the experimental method being used as well as to give experience of some key principles of physics. If experiments are done thoughtlessly by slavishly following instructions, they simply use up time unprofitably.

Analysis

These extended data analysis exercises are intended to give you some insights into te way physics is used to solve the problems of everyday life and to help you learn how to apply the principles to a problem. Hints and some answers are given in the appendix C.

Summaries

Each chapter ends with a brief summary of its content. These summaries, together with the learning objectives, should give you a clear overview of the subject and allow you to check your own progress.

MEASUREMENT

Accurate measurement is central to the development of any science. The importance of measurement was apparent to ancient civilisations. Throughout history the accuracy with which measurements could be made has been improved by the use of more and more sophisticated instruments. At almost every stage, improved measuring techniques have resulted in new concepts and ideas. ■

1 PHYSICAL QUANTITIES AND UNITS

LEARNING OBJECTIVES

At the end of this chapter you should be able to:

① quote quantities in their correct SI units;

② understand the difference between base units and derived units;

③ check any equation for the homogeneity of its units.

1.1 PHYSICAL QUANTITIES

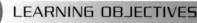

The science of physics, like all other sciences, is based upon taking measurements. All scientific theories and laws must be tested experimentally, and all experiments necessitate making measurements. One of the reasons for the development of physics as a science is that the measurements that are made in physics experiments are usually more reliable than the measurements made in some other sciences. It is very difficult to make accurate measurements on, for example, aspects of human behaviour. Sociology is therefore a less precise science than physics, although the scientific procedures that it uses are basically the same.

Physical quantity is a term that is used to include measurable features of many different items. The area of a playing field, the mass of a bag of sugar and the speed of an aeroplane are all physical quantities. Abstract quantities, such as love, fear and hope, are not measurable in the same way. In quoting any measurement of a physical quantity, two items need to be stated. The first is the **numerical value** of the quantity, sometimes called the **magnitude**; and the second is its **unit**. *It is important at all times to think of and write the value and the unit of any quantity together*. Apart from the technical accuracy of writing in this way, it will also help you to acquire a mental appreciation of the size of the quantity you are considering. For instance, if the diameter of a wire is found to be 1.46 metres, then something is wrong. A value stated as simply 1.46 does not give this check and is meaningless.

Realising whether a particular value is possible or not is something that comes with experience, but it is only by thinking about the sizes of physical quantities that this experience is gained.

1.2 THE INTERNATIONAL SYSTEM OF UNITS (SI)

History has provided for us far more units than are necessary. If units of length only are considered, we have miles, furlongs, yards, feet, light-years, inches, fathoms, nautical miles, metres, microns and ångstrøms, to say nothing about units of length that are not now used, such as chains, rods and leagues. One unit of length can, of course, be converted into another by using a conversion factor. There are a huge number of conversion factors, however, so that few people know more than a few of them. The main problem with having so many units is that familiarity with one unit under one circumstance cannot easily be transferred to a different circumstance. Measuring distance up a mountain in feet, and distance along a road in yards or miles, makes it difficult to relate one to the other when the gradient of a road is needed.

The SI system

The **metric system** of units was introduced at the time of the French Revolution to rationalise the chaos of units that then existed. It has been modified over the last two hundred years and, in part at least, all countries now use the metric system of units called the **Système International (SI)**. The advantage of the SI system of units is that any quantity has only one unit in which it can be measured. In the case of length, the metre is the only unit of length used, together with multiple units such as the kilometre, and submultiple units such as the millimetre. Table 1.1 gives the value of each of these multiple and submultiple units. The prefixes may be added on to any SI unit.

When changing from one multiple to another, you should be particularly careful to get the arithmetic the right way round. If it is not done correctly, then the answer is usually nonsensical. For example, if you change 5.09 kilometres into metres and accidentally divide by 1000 instead of multiplying by 1000, then you get the answer 0.005 09 m. A mental image of these two lengths should immediately tell you that something has gone wrong. In this particular case your equation is incorrect by a factor of a million. In your study of physics you need always to be alert to nonsensical results.

SI base units

The SI system of units uses seven **base units** from which all other units can be defined. One of the seven base units, the candela, is a unit of luminous intensity, and it will not be used in this book. The other six are listed in Table 1.2.

The value of the base units was chosen arbitrarily. This means that there is no particular reason for the size of the unit chosen. In practice, there are historical reasons for the approximate size of the base units, but, as it has become possible to measure more accurately, earlier definitions have ceased to be used. The metre, for example, was intended to be one ten-millionth of the distance from the equator to

TABLE 1.1 Prefixes that can be used with any SI unit*

PREFIX	MULTIPLYING FACTOR	SYMBOL
atto	10^{-18}	a
femto	10^{-15}	f
pico	10^{-12}	p
nano	10^{-9}	n
micro	10^{-6}	µ
milli	10^{-3}	m
centi	10^{-2}	c
deci	10^{-1}	d
deca	10	da
hecto	10^{2}	h
kilo	10^{3}	k
mega	10^{6}	M
giga	10^{9}	G
tera	10^{12}	T
peta	10^{15}	P
exa	10^{18}	E

*Several of these terms are used only infrequently.

TABLE 1.2 The six SI base units that are used in this book

QUANTITY	UNIT	SYMBOL
mass	kilogram	kg
time	second	s
length	metre	m
electric current	ampere	A
thermodynamic temperature	kelvin	K
amount of substance	mole	mol

Units versus quantities

With all measurements it is important to distinguish between the quantity being measured and the unit in which it is being measured. Later, when definitions are being given, it will be important to define units in terms of units, and quantities in terms of quantities. Expressions such as distance per second should not be used, as this is a mixture of a quantity and a unit. The quantity *distance per unit time* or the unit *metre per second* should be used.

the North Pole, and while this is approximately true, it is now realised that the Earth is slightly irregular in shape so that all distances from the equator to the North Pole are not the same. Added to this is the fact that it is necessary to have the standard of length within a laboratory so that comparisons can be made with it. Similarly, the second was originally based on the time the Earth takes to rotate once, the day. Accurate measurements now show that the Earth wobbles appreciably and is slowing down gradually, so that all days are not the same length.

Units should be defined in a definite order. The kilogram is arbitrary and does not depend on anything else. The same is true of the second, which is defined the way it is because extreme accuracy can be achieved with clocks using caesium beam tubes. Such a clock (see **Fig 1.1**) using the precise frequency of the oscillation within the atoms of caesium would have an uncertainty in a time measurement of less than 1 part in 10^{13} (1 second in 300 000 years).

The metre, on the other hand, is dependent on the second having been defined first. The metre is now defined in terms of the speed of light. It was first defined in this way in 1983 when accurate measurements of the speed of light, together with the accuracy of time measurements, made previous length measurements a little uncertain. The effect of this definition is that the speed of light in a vacuum is now, by definition, 299 792 458 metres per second *exactly*. This also means that experiments that were once experiments to measure the speed of light have become experiments to measure distance.

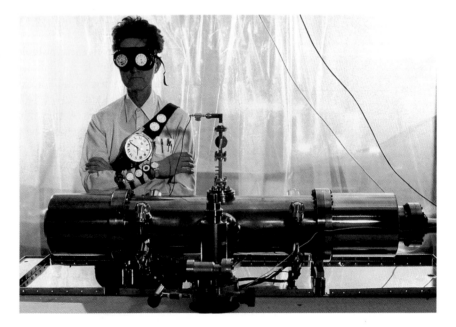

FIG 1.1 An atomic clock at the National Physical Laboratory, near London. Although it looks nothing like a normal clock, it is capable of measuring time to 1 second in 300 000 years and is used as the standard in the United Kingdom.

1.3 DERIVED UNITS

U sing the base units it is possible to construct a system of units that can be used to measure all the quantities we shall need to measure. A list of quantities and their units is given in appendix B. Units additional to the base units are called **derived units** and there are very many of them. Some of these derived units have specific names such as newton, joule and watt; others are clearly derived units as they do not have names and just use combinations of the base units, such as metres per second and metres cubed. Many quantities can be measured in two or more equivalent units. Often the measurement of two quantities can be used to calculate a third (**Fig 1.2**). Note that, although units such as the newton are correctly abbreviated to N as in 6.7 N, the capital letter is not used when the word is written in full, so this will become 6.7 newtons. The plural form of the abbreviation, Ns, is never used because N s (with a space) means newton second.

Table 1.3 shows 20 of the key quantities upon which physics is based. It is essential that you know how to use all of them and their units by the end of your course. They are given in a particular order so that, as each quantity is introduced, its definition depends only on quantities already defined. The quantities in **bold** type are SI base units.

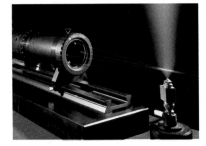

FIG 1.2 Multiple beams from an argon laser are being used to measure the velocity of a stream of gas. The Doppler effect (see p. 536) causes a change in the frequency of the argon laser light which can be used to determine the velocity of the gas.

TABLE 1.3 A possible sequence for defining certain quantities and their SI units. Thorough knowledge of these 20 quantities and their units is absolutely essential by the time you finish your course

QUANTITY	UNIT	ABBREVIATION	DEFINITION (PAGE)
mass	**kilogram**	**kg**	4
time	**second**	**s**	4
frequency	hertz	Hz	144
length	**metre**	**m**	4
angle	radian	rad	108
area	metre squared	m^2	–
volume	metre cubed	m^3	–
density	kilogram per metre cubed	$kg\ m^{-3}$	380
velocity	metre per second	$m\ s^{-1}$	36
acceleration	metre per second squared	$m\ s^{-2}$	38
momentum	kilogram metre per second	$kg\ m\ s^{-1}$	46
force	newton	N	55
pressure	pascal	Pa	393
energy	joule	J	85, 90
power	watt	W	89
electric current	**ampere**	**A**	4, 323
electric charge	coulomb	C	239
potential difference	volt	V	242
resistance	ohm	Ω	244
capacitance	farad	F	294

1.4 HOMOGENEITY OF EQUATIONS

t is possible to express any unit in terms of base units only. Because of the multitude of names of units, and because the same unit can be expressed in different ways, a comparison of units can best be made by using this form. Use may also be made of units to check an equation for correctness, or even to suggest the form an equation may take. Any physical equation must have the same units on each side of the equation; it is then said to be a **homogeneous equation**. Also any addition or subtraction of physical quantities can only be done if the quantities have the same unit. It is, for example, meaningless to add a mass of 6 kilograms to an electric current of 4 amperes. These points can best be illustrated by examples.

EXAMPLE 1.1

Express the joule in terms of base units.

From the definitions on the pages given in Table 1.3, we have:

joule = newton × metre = N m

newton = kilogram × metre per second squared
$$= \text{kg m s}^{-2}$$

So combining these, we get

joule = $\text{kg m s}^{-2} \times \text{m} = \text{kg m}^2\,\text{s}^{-2}$

Therefore the joule in base units is $\text{kg m}^2\,\text{s}^{-2}$.

This result shows why the name *joule* (J) is needed. It would be very clumsy if an energy of 25 joules always had to be referred to as 25 kilogram metres squared per second squared.

EXAMPLE 1.2

Bernoulli's equation relates the pressure difference p between two points along a pipe with their height difference y, the density ρ of the fluid flowing along the pipe, its velocity v and the acceleration of free fall g:

$$p + \rho g y + \tfrac{1}{2}\rho v^2 = k$$

Show that each term in the equation has the same units and find the base units of the constant k.

This is done in the table below.

The base units of each term are the same and the equation is therefore homogeneous. The constant k must have the same units as well. Constants do often have units; they can also be unitless, as the $\frac{1}{2}$ was in this example. The unit of $\frac{1}{2}\rho v^2$ is the same as the unit of ρv^2. The analysis of this equation only shows that it is a *possible* equation; it does not and cannot check whether or not the equation is correct. There is no way in which a unit check can determine constants.

QUANTITY	DEFINING EQUATION	BASE UNIT	
height	distance	m	
velocity	distance/time	m/s	$= \text{m s}^{-1}$
acceleration	(change in velocity)/time	$(\text{m s}^{-1})/\text{s}$	$= \text{m s}^{-2}$
density	mass/volume	kg/m^3	$= \text{kg m}^{-3}$
pressure	force/area = (mass × acceleration)/area	$(\text{kg} \times \text{m s}^{-2})/\text{m}^2 = \text{kg m}^{-1}\,\text{s}^{-2}$	

TERM IN EQUATION	BASE UNIT
1st term p	$\text{kg m}^{-1}\,\text{s}^{-2}$
2nd term $\rho g y$	$\text{kg m}^{-3} \times \text{m s}^{-2} \times \text{m} = \text{kg m}^{-1}\,\text{s}^{-2}$
3rd term $\frac{1}{2}\rho v^2$	$\text{kg m}^{-3} \times (\text{m s}^{-1})^2 = \text{kg m}^{-3}\,\text{m}^2\,\text{s}^{-2} = \text{kg m}^{-1}\,\text{s}^{-2}$

QUESTIONS

1.1 Find the base unit of each of the following quantities (look up any of these that you have not yet met): frequency, force, pressure, work, power, potential difference, resistance, capacitance.

1.2 When a body moves through a fluid there is a resisting force acting on the body. In the case of a sphere of radius r moving with constant velocity v through a liquid of density ρ, the force F is given by the equation

$$F = k\rho r^2 v^2$$

Show that k does not have a unit.

1.3 Which of the following equations are homogeneous? If an equation is not homogeneous, it cannot be the correct equation.

a $T = \dfrac{1}{2\pi}\sqrt{\dfrac{l}{g}}$

where T is a time, l a length and g an acceleration.

b $F = mv^2 r$

where F is a force, m a mass, v a velocity and r a distance.

c $E = mv^2$

where E is an energy, m a mass and v a velocity.

d $I = \dfrac{ML}{12} + \dfrac{ML^2}{4}$

where I is a moment of inertia, L a length and M a mass.

e $c = \sqrt{\dfrac{p}{d}}$

where c is a velocity, p a pressure and d a density.

1.4 Show that the unit of momentum given in Table 1.3 as kg m s^{-1} is the same unit as the N s.

1.5 Express the unit of force and the unit of charge in terms of the SI base units. Using the equation

$$F = \dfrac{q_1 q_2}{4\pi\, \varepsilon_0\, r^2}$$

where F is the force acting between two charges q_1 and q_2 when placed a distance r apart, express the unit of ε_0 in terms of the base units.

1.6 If velocity v, measured in zipps Z (an imaginary unit!), energy E, measured in joules J, and force F, measured in newtons N, had been chosen as the base quantities, what would be the units of mass, length and time when expressed in base units?

SUMMARY

- All physical quantities should be quoted with their numerical value and their unit.
- The SI system of units will be used exclusively throughout the course.
- The six base units used are kilogram, second, metre, ampere, kelvin and mole.
- Table 1.3 gives 20 key quantities and their units in a correct sequence.

ASSESSMENT QUESTIONS

1.7 With the aid of an example, explain the statement 'The magnitude of a physical quantity is written as the product of a number and a unit'.

a Explain why an equation must be homogeneous with respect to the units if it is to be correct.

b Write down an equation which is homogeneous, but still incorrect.

London 1996

1.8 The list gives some quantities and units. Which are base quantities of the International (SI) System of units?

coulomb   force   length   mole
newton   temperature interval

a Define the volt.

b Use your definition to express the volt in terms of base units.

c Explain the difference between scalar and vector quantities.

d Is potential difference a scalar or vector quantity?

London 1995

1.9 Copy the table below. In your table, classify each of the terms in the left-hand column by placing a tick in the relevant box.

	Base unit	Derived unit	Base quantity	Derived quantity
Length				
Kilogram				
Current				
Power				
Coulomb				
Joule				

London 1998

1.10 Copy and complete each of the following statements with a single word:

a The rate of change of displacement is called

b The rate of flow of charge is called

c The rate of doing work is called

d The rate of change of momentum is called

London 1998

1.11 When a body moves with speed v through a liquid of density ρ, it experiences a force F known as the drag force. Under certain circumstances, this force is related to ρ and v by the expression

$$F = k\rho v^2$$

where k is a constant. Given that the unit of ρ is kg m^{-3}, determine the unit of the constant k in terms of SI base units.

OCR 1999

1.12 Kepler discovered that the orbital periods T of the planets about the Sun are related to their distances r from the Sun. From Newton's laws, the following relationship may be derived:

$$T^2 = \left[\frac{4\pi^2}{GM}\right] r^3$$

where M is the mass of the Sun.

a Use the equation to find units for G in terms of base units.

b For a planet orbiting the Sun, draw two free-body force diagrams, one for the Sun and one for the planet.

c Explain how the forces on your diagrams are consistent with Newton's third law of motion.

London 1996

1.13 The volume flow rate of a fluid through a pipe of cross-sectional area A and length l is given by

$$\frac{V}{t} = \frac{cA^2 \Delta p}{\eta l}$$

where c is a constant having no unit.

a **i** What quantity does the term η represent?

ii The SI unit of pressure in base units is $kg\ m^{-1}\ s^{-2}$. State the SI unit for each of the following quantities:

$V \quad t \quad A \quad l$

iii Determine the unit of η in terms of SI base units.

b The figure shows a system used to provide a patient with a blood transfusion. The required fluid pressure to enable blood to flow through the hollow needle, and into the patient, can be achieved by raising or lowering the container.

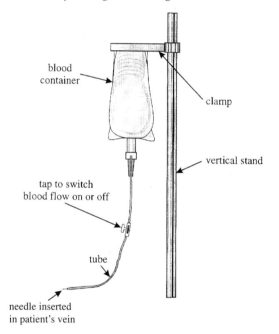

blood container

clamp

vertical stand

tap to switch blood flow on or off

tube

needle inserted in patient's vein

In one case a needle, 0.035 m long and of internal cross-sectional area $3.0 \times 10^{-7}\ m^2$, is used. The required volume flow rate into the patient is $2.5 \times 10^{-7}\ m^3\ s^{-1}$.

The magnitude of η in SI units $= 4.0 \times 10^{-3}$

The value of $c = 0.040$

i Use the equation given to determine the pressure difference between the ends of the needle that will produce the required blood flow rate.

AEB 1998

1.14 a An A Level student suggests a number of equations, not all of which are correct. The equations are given below. Indicate whether a consideration of the units involved makes each equation possible or impossible.

i surface area of a sphere $= \frac{4}{3}\pi r^3$

ii speed of a wave $= \dfrac{\lambda}{T}$

iii period of an oscillating pendulum $= 2\pi\sqrt{\dfrac{g}{l}}$

iv pressure of a gas $= \left. \frac{1}{3}\rho \langle c^2 \rangle \right)$

Here r, l, λ are distances, c is a speed, g is an acceleration, T is a time and ρ is a density.

b Give an SI unit and an estimate of the magnitude of each of the following physical quantities. (Marks will be awarded for the correct order of magnitude of each estimate, not for its accuracy.)

i the weight of an adult

ii the power of a hair drier

iii the energy required to bring to the boil a kettleful of water

iv the resistance of a domestic filament lamp

v the wavelength of visible light

OCR 1999

1.15 The figure shows the junction between two pipes. Water flows from the wide pipe W into the narrow pipe N.

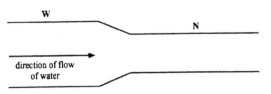

W

N

direction of flow of water

a The continuity equation for the mass flow of water at the junction between the pipes may be written as:

$$\rho v A = k$$

Determine an appropriate unit for k.

b The internal diameter of pipe W is 10 mm and that of pipe N is 7 mm. Calculate the speed of the water in pipe N when the water flows through pipe W with a speed of 24 cm s⁻¹.

AEB 1996

1.16 The permittivity of free space ε_0 has units $F\ m^{-1}$. The permeability of free space μ_0 has units $N\ A^{-2}$.

a Show that the units of $\dfrac{1}{\sqrt{\varepsilon_0 \mu_0}}$ are $m\ s^{-1}$.

b Calculate the magnitude of $\dfrac{1}{\sqrt{\varepsilon_0 \mu_0}}$

c Comment on your answers.

London 1997

2 MEASUREMENT TECHNIQUES

LEARNING OBJECTIVES

At the end of this chapter you should be able to:

1. find systematic and random uncertainties;
2. calculate the uncertainty in an experiment;
3. plan an experiment to give reliable readings;
4. use graphs accurately.

2.1 MEASUREMENT UNCERTAINTY

 henever a measurement of a physical quantity is made, some measuring instrument has to be used to make that measurement. The instrument may be as ordinary as a ruler or as sophisticated as a modern mass spectrometer for chemical analysis. In using the instrument an experimenter has to make use of his or her own skill to obtain as accurate a reading as possible. Built in to the instrument however is a limit of accuracy within which the experimenter is working. The result of this is that the readings that the experimenter takes have a degree of **uncertainty**. Physical quantities cannot be measured exactly with any instrument. An accurate clock might measure a time interval to a millionth of a second, but even this is not quite exact. If a person is using a ruler, the reading will probably be taken to the nearest millimetre. The reading might then be stated as (208 ± 1) mm. This implies that the person taking the reading thinks that the best value is 208 mm, and that the value will not fall outside the range from 207 mm to 209 mm. The ±1 mm is called the uncertainty of the reading. Sometimes people refer to this as the **error** in the reading, but the word 'error' seems to imply that a mistake has been made. This is not the case. Any reading *always* has an uncertainty. Many modern instruments hide this fact by giving digital displays of their readings.

Significant figures

A digital ammeter stating that the current is 568 mA seems to be stating that the current is exactly 568 mA. Correctly, the current in this case cannot be stated with any greater accuracy than to the nearest milliamp. It may be the case however that the uncertainty is greater than 1 mA. The meter itself may be fluctuating around the 568 mA figure; the meter may have been badly calibrated in the first place; the meter may have been misused; or the current itself may be changing. All of these factors may make it more suitable to quote this reading as, say, (568 ± 20) mA. This raises another point about readings. If the uncertainty in this reading is ± 20 mA, the figure '8' in the 568 mA is so uncertain that it is not worth while recording it. The reading then becomes (570 ± 20) mA.

This way of writing the value of the current introduces doubt about the number of **significant figures** being used. It is preferable to state the current in amperes as

$$(0.57 \pm 0.02) \text{ A} \quad \text{or} \quad (5.7 \pm 0.2) \times 10^2 \text{ mA}$$

Either of these statements makes it quite clear that the uncertainty is in the final figure of the reading, the '7' in this example. Since the readings are themselves uncertain, the size of the uncertainty is even more difficult to be sure about. There is therefore almost never any need to give uncertainties to more than one significant figure.

Absolute, fractional and percentage uncertainties

The size of an uncertainty needs to be considered together with the size of the quantity being measured. The example given above of the ruler measuring a distance of 208 mm with an uncertainty of 1 mm may be an acceptable uncertainty for a particular experiment. However, an uncertainty of 1 mm will not be acceptable if the actual distance you are trying to measure is itself only 2 mm. In other words:

- (208 ± 1) mm is a fairly accurate measurement
- (2 ± 1) mm is highly inaccurate

In order to compare uncertainties, use is made of **absolute**, **fractional** and **percentage** uncertainties. For the reading (208 ± 1) mm:

- 1 mm is the absolute uncertainty
- $1/208$ is the fractional uncertainty $(= 0.0048)$
- 0.48% is the percentage uncertainty

Note that the absolute uncertainty has units, mm in this case, but that the fractional and percentage uncertainties are ratios and therefore do not have units. As we usually require uncertainty to only one significant figure, the two values given above would be used as 0.005 and 0.5%.

In general, if a reading a is measured with absolute uncertainty δa, i.e. the reading is $a \pm \delta a$, then $\delta a/a$ is the fractional uncertainty and $100\delta a/a$ is the percentage uncertainty. (The Greek letter delta δ is read as 'a small change in'.)

QUESTIONS

2.1 State both the fractional uncertainty and the percentage uncertainty if:
- **a** a distance of 7.84 m is measured to the nearest centimetre;
- **b** a time of 10.03 s is measured to the nearest 0.02 s;
- **c** a mass of 6000 kg is measured to the nearest 5 kg.

Approximations

Note that, when we make approximations like those in question 2.2, they are errors rather than uncertainties because there is no need to make such assumptions. It can, nevertheless, be useful to use approximations such as these when checking calculations, or when only approximate answers are required.

2.2 What percentage error is introduced by taking:
  a π to be 22/7;
  b π^2 to be 10;
  c 1 year to be $\pi \times 10^7$ s;
  d g to be 10 m s^{-2} when its value is 9.807 m s^{-2};
  e the speed of light to be 3×10^8 m s^{-1}, when its actual value is 2.998×10^8 m s^{-1}.

SYSTEMATIC AND RANDOM UNCERTAINTIES

Fig 2.1 The man taking this reading will obtain a correct value. The girl does not have her eye on the correct level so will have a parallax error in the value she records.

Uncertainty can be of two types: **systematic** or **random**. A systematic uncertainty will result in all the readings taken being faulty in one direction. Using a stopclock that is running fast will result in all time readings being too big; using an ammeter with a zero reading of −0.2 A will result in all readings taken being 0.2 A too small. Calibration errors result in systematic uncertainty, and experimenters are somewhat too fond of assuming that any systematic uncertainty must be due to the apparatus they are using. This is not necessarily so. Systematic uncertainty can be introduced into an experiment by poor experimental technique. A parallax error (**Fig 2.1**) can affect all of an experimenter's readings in the same way.

Systematic uncertainty is difficult to estimate and eliminate, but there are some simple procedures that can reveal it. Two ammeters placed in series must have the same current flowing through them. If their readings are not the same, then there must be a systematic error in one of them. Use a third meter if you are not sure which one is in error. Two thermometers must read the same value in the same environment. Reducing systematic error can be done by using good experimental technique and by varying the instrumentation being used. Systematic uncertainty can never be eliminated by taking repeated readings with the same built-in faults.

Random uncertainties, as their name implies, result in a scatter of readings about a mean value. They have an equal chance of being positive or negative. Random uncertainty results from the inability of the observer to repeat his or her actions precisely. If the period of oscillation of a pendulum is being measured, an experimenter might be timing 50 swings. There are several things that cannot be reproduced exactly each time:

■ The reaction time in using the stopclock might vary a little.
■ The start of the first swing and the end of the fiftieth swing may not be noted exactly.
■ The same starting amplitude may not be used each time so that, although the period of oscillation is nearly independent of the amplitude, there is a small influence of amplitude on period.
■ The hand of the stopclock may not be viewed from quite the same angle, so introducing random parallax uncertainty.

A good visual way of comparing systematic and random uncertainties is shown in **Fig 2.2**. The diagram shows how a target appears if an archer has small and large, systematic and random uncertainty in shooting.

There are standard statistical methods for handling random uncertainties. These can give standard deviations for a series of readings. But when the number of readings is not large, it is useful to have a method to obtain the approximate value of the uncertainty without doing a formal statistical analysis. Work through example 2.1 to see one method of doing this.

QUESTION

2.3 The width of a table is measured at six different places. The readings, taken from a metre rule, are:

634 mm, 634 mm, 636 mm, 637 mm, 639 mm, 634 mm

Calculate the average width of the table. How many significant figures is it sensible to give in your answer?

FIG 2.2 The appearance of a target when an archer has the following uncertainties in aiming: (a) small random uncertainty, small systematic uncertainty; (b) small random uncertainty, large systematic uncertainty; (c) large random uncertainty, small systematic uncertainty; and (d) large random uncertainty, large systematic uncertainty.

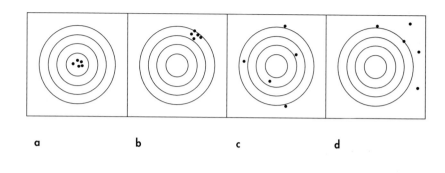

a　　　　b　　　　c　　　　d

EXAMPLE 2.1

From the set of data given in Table 2.1, obtain values for d, the diameter of a wire, and T, the period of swing of a pendulum.

TABLE 2.1 Readings of the diameter d of a wire and of the time for 50 swings, $50T$

d/mm	$50T$/s
0.83	104.3
0.83	104.5
0.85	104.5
0.83	104.5
0.85	104.2
0.86	104.7
0.85	104.4
average　0.843	104.44

First obtain the differences between the average value and the individual values. Ignore positive and negative signs. This gives the results of Table 2.2, which is normally the only table drawn up.

The average difference is a measure of the uncertainty of any reading, so giving these values:

diameter $= (0.84 \pm 0.01)$ mm

$50T = (104.44 \pm 0.12)$ s

$T = (2.089 \pm 0.002)$ s

TABLE 2.2 The readings of Table 2.1 with the differences added

d/mm	DIFFERENCE /mm	$50T$/s	DIFFERENCE /s
0.83	0.013	104.3	0.14
0.83	0.013	104.5	0.06
0.85	0.007	104.5	0.06
0.83	0.013	104.5	0.06
0.85	0.007	104.2	0.24
0.86	0.017	104.7	0.26
0.85	0.007	104.4	0.04
average 0.843	0.011	104.44	0.12

Note that in each case one significant figure has been dropped when quoting the final values. The uncertainty in the diameter measurement is much larger than the uncertainty in the period. This becomes apparent if percentage uncertainties, rather than absolute uncertainties, are given:

- fractional uncertainty in diameter reading $= 0.01/0.84$ $= 0.012$
- percentage uncertainty in diameter reading $= 1.2$
- percentage uncertainty in $50T = 0.1$
- percentage uncertainty in $T = 0.1$

The percentage uncertainty is not affected by dividing by 50. This is why a large number of swings need to be used in such a timing experiment. The absolute uncertainty in measuring the time for 50 swings is not much different from the absolute uncertainty for measuring the time for one swing. This gives only a small percentage uncertainty for $50T$ and corresponding accuracy for T.

QUESTION

2.4 The following readings of time interval were taken for a capacitor to discharge to 30% of its fully charged value:

67.3 s, 71.6 s, 68.4 s, 68.9 s, 63.6 s, 68.0 s,

70.2 s, 69.5 s, 70.6 s, 69.4 s, 69.0 s, 70.8 s

What is the time interval for this discharge and what is its fractional uncertainty?

2.3   ACCURACY AND PRECISION

These two words are often taken to mean the same thing. It is possible, however, to have readings taken with great precision that are not accurate. This will happen if there is a systematic error. Similarly, it is possible to have readings that are accurate but not very precise. This will occur if there is only a random error in the readings taken.

A precise reading will be taken to a large number of significant figures, but do be careful to use instruments of appropriate precision. It would be quite unsuitable to measure out food for a recipe using a balance capable of measuring to a milligram. It would take hours to get 100.000 grams of butter into a mixing bowl – and then some of it would stick on the wooden spoon! Similar absurdities can happen in a physics laboratory, resulting in students spending a long time taking unsuitable readings. While the aim in doing any experiment is to be as accurate as possible, there is no point in taking some readings to a very much higher precision than other readings in the same experiment. Ideally for each of the readings taken, the fractional uncertainty of each reading should be the same. In practice, things do not work out ideally, but the fractional uncertainty for each reading should be known, at least approximately.

Once a judgement has been made about the uncertainty of each reading, it is necessary to calculate the uncertainty in the final numerical value quoted. Uncertainties can be combined using the rules shown in the box. For example, if δb and δc are the absolute uncertainties in readings b and c, respectively, we find that

Combining uncertainties

- For *addition* and *subtraction*, **add** absolute uncertainties.
- For *multiplication* and *division*, **add** percentage uncertainties.
- When using *powers*, **multiply** the percentage uncertainty by the power.

$$\text{if}\quad y = b - c \quad \text{then}\quad y \pm \delta y = (b - c) \pm (\delta b + \delta c)$$

$$\text{if}\quad x = b \times c \quad \text{then}\quad \frac{\delta x}{x} = \frac{\delta b}{b} + \frac{\delta c}{c}$$

$$\text{if}\quad z = b^n \quad \text{then}\quad \frac{\delta z}{z} = n\frac{\delta b}{b}$$

There are some circumstances where the uncertainty in the final value is best found by working the problem through twice, once with the readings as taken and once with the limiting values, which will give the maximum result. Equations

EXAMPLE   2.2

Find v and the uncertainty in v if

$$v = \frac{(a - b)\,d^2}{q\sqrt{C}}$$

and $a = (1.83 \pm 0.01)$ m, $b = (1.65 \pm 0.01)$ m, $d = (0.001\,06 \pm 0.000\,03)$ m, $q = (4.28 \pm 0.05)$ s and $C = 920 \pm 30$.

First calculate the percentage uncertainties in each of the four terms:

$(a - b) = (0.18 \pm 0.02)$ m   uncertainty 11%

(note that here absolute uncertainties have been added)

$d = (0.001\,06 \pm 0.000\,03)$ m	uncertainty 3%
$q = (4.28 \pm 0.05)$ s	uncertainty 1.2%
$C = (920 \pm 30)$	uncertainty 3%

The uncertainty in $(a - b)$ is now very large, although the readings themselves have been taken carefully. This is always the effect when subtracting two nearly equal numbers. The uncertainty in d^2 will be twice the uncertainty in d; the uncertainty in $\sqrt{C}$ will be half the uncertainty in C because a square root is a power of $\frac{1}{2}$. This gives:

uncertainty in $v = 11 + (2 \times 3) + 1.2 + (\frac{1}{2} \times 3)$
$= 19.7\% \approx 20\%$

which gives the final value of v as

$$v = (1.6 \pm 0.3) \times 10^{-9}\ \text{m}^3\ \text{s}^{-1}$$

containing trigonometrical ratios, or exponentials, or equations in which some of the terms appear both on the top and on the bottom of the expression, such as

$$f = \frac{uv}{u + v}$$

are best dealt with this way.

At the conclusion of the experimental calculation, or when giving the answer to a numerical question, the correct number of significant figures must be given. If you are in any doubt about the number of figures to use, then you will not usually be very far out if you use three significant figures. In practice it is essential to give all the figures about which there is no doubt and then give one figure for which there is some uncertainty. These points are illustrated in example 2.2.

QUESTIONS

2.5 Two lengths are recorded as (1.873 ± 0.005) mm and (1.580 ± 0.005) mm.

    a What is the maximum possible value of the sum of the two lengths?

    b What is the maximum possible value of the difference between the two lengths?

    c What is the fractional uncertainty in both the sum of the two lengths and the difference between the two lengths?

2.6 In a simple pendulum experiment to determine g the equation used is

$$T = 2\pi \sqrt{\frac{l}{g}}$$

where T, the period, is found to be (2.16 ± 0.01) s when the length l of the pendulum is (1.150 ± 0.005) m. Find the value of g and its uncertainty.

2.7 The volume of liquid V flowing through a pipe of radius r in time t is given by

$$\frac{V}{t} = \frac{\pi r^4 (p_1 - p_2)}{8\eta L}$$

where p_1 and p_2 are the pressures at each end of the pipe, L is its length, and η is a physical constant called the viscosity of the liquid. Use the following readings to determine η together with its uncertainty:

$r = (0.43 \pm 0.01)$ mm
$p_1 = (1.150 \pm 0.005) \times 10^5$ Pa
$p_2 = (1.000 \pm 0.005) \times 10^5$ Pa
$L = (5.5 \pm 0.1)$ cm
$V = (10.0 \pm 0.1)$ cm^3
$t = (4.0 \pm 0.1)$ s

(These figures are based on typical figures for the flow of water through a hypodermic needle. The effect of the r^4 term means that, for a nurse administering an injection, needle size is much more important than thumb force in controlling the rate of flow of liquid through the needle.)

2.4 EXPERIMENT PLANNING

areful design of an experiment can also play a large part in increasing the reliability of results. Many physical experiments are of the 'cause-and-effect' type. They are designed to answer questions such as: 'What will happen to the length of a wire if the tension in the wire is increased?' Here the tension is the cause and the extension is the effect of that cause. It is of the greatest importance in such experiments to ensure that all other factors are kept constant. In this case the temperature of the wire must remain constant throughout the experiment.

Q **What other factors might change during the course of the experiment? How would you keep them constant?**

The design of individual experiments varies from one topic to another. Many experimental details are given in different places throughout this book. But there are some techniques that are common to many experiments, and it is sensible to be aware of them and to work to establish good experimental practice. Some of these, together with some common faults, are listed below:

1 Support apparatus firmly. It is impossible to obtain reliable readings of force, for instance, if the piece of apparatus is wobbling. A firm support will also make it more difficult for apparatus to be damaged by falling over.

2 Make certain that the measuring instruments that you are using are suitable for the readings you are trying to make. An ammeter reading up to 30 mA will not be suitable to measure a current of 1 mA. Similarly a metre rule might be suitable for measuring the length of a wire but unsuitable for measuring its extension, when a vernier scale will probably be needed.

3 Repeat readings where possible. Random uncertainty can be reduced by taking many readings using the same apparatus, and systematic uncertainty can be reduced by using different apparatus and procedures. It is often possible to increase and then to decrease a quantity, taking readings in both directions. A particular example of this would be in examining how the resistance of a resistor varies with temperature. The temperature can first be raised, while values of resistance are measured, and then lowered, again taking measurements of resistance. If the two sets of readings are markedly different, then perhaps the resistor has been damaged by the high temperature. If they are very similar to one another, then an average can be taken. It is likely that the temperature of the resistor on the way up is recorded too low, while on the way down it may well have been recorded too high.

Q **Why might the temperature of the resistor have been recorded too low on the way up and too high on the way down?**

4 Avoid parallax error when reading from a scale (see **Fig 2.1**). A mirror placed behind a scale can be used to ensure that the person taking the reading is directly in front of it. When viewed from the correct position the person will see an image of his or her eye directly behind the scale, and if there is a pointer over the scale he or she will not be able to see the image of the pointer.

5 Be careful to read scales correctly. Make certain you know what the smallest division on the scale is, and check that the reading you have just taken is sensible. For example, 2.7 is *not* sensible for the scale reading in **Fig 2.3**.

6 Be honest in taking readings. Do not decide in advance what you are going to find for a particular reading. By all means check that the reading is approximately what you expect. That is a good way of avoiding mistakes, but if you think that there is an established pattern in the figures you are recording and

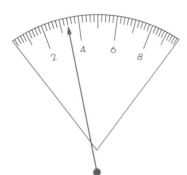

FIG 2.3 A common mistake with scales leads to this scale reading being given as 2.7. A quick check that it must be between 3 and 4 enables the correct value of 3.4 to be obtained.

then slavishly follow it, it is likely that you will have to make only small early corrections to the actual figures but that the size of these corrections will be of an ever-increasing size. You will also miss any real change that does take place in the pattern. There are examples in scientific history where small deviations from expected results have been absolutely crucial in establishing new theories. Einstein's theory of relativity is one such example.

7 Check for mistakes. Do not go blindly on when it should be clear that there is something drastically wrong. Try to find the cause of the mistake. If the mistake occurs in only a few, out of many, readings, then eliminate those readings from your calculations.

8 When performing counting experiments of any kind, make certain that you start counting from zero and not from one. If you are counting dots on a ticker-tape from 1 to 10, then although you have counted up to 10 you will only have included *nine* time intervals.

9 When measuring length, as, for instance, in measuring the difference in height between the two levels of liquid surfaces in a manometer, measure both levels from the same fixed horizontal surface (see **Fig 2.4**). This gives much greater accuracy than trying to hold a ruler in mid-air and the zero alongside one of the liquid levels. Remember also that the difference in height between the levels of the liquid surfaces in a manometer gives only a difference in pressure. It does not give the total pressure in a system. Whether an absolute value of a quantity is required or the difference between two values of the quantity is often of crucial importance.

10 Take readings promptly. You will be more likely to keep constant those features which you want to keep constant if you do not delay taking readings unnecessarily: a battery will not have as much time to go flat; room temperature will not have as much time to alter; a resistor will not have as much time to heat up; etc. There can be some contradictory demands here, so each experiment must be taken on its merits, and you will have to use some personal judgement. Experiments involving heating are often the exception to this advice, as it is often necessary to wait until the whole system under test has reached the same temperature.

11 Record readings actually taken on a prepared table. Do not do any arithmetic before recording these values. Do not use odd scraps of paper. Leave plenty of room for any extra columns that you may need, and allow plenty of lines for more readings. Your table should look like Table 2.3 as you start to take readings.

12 Use a graphical presentation of your results where possible, as explained in the next section.

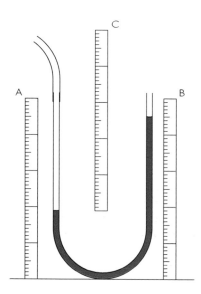

FIG 2.4 If the height difference between the liquid levels is required, take two readings, one from A and one from B, and subtract. A single reading from C will have a large uncertainty.

TABLE 2.3 A table should have plenty of space for extra columns and extra lines

M/kg	x/m	y/m	(y − x)/m	
0.100				
0.200				
0.300				
0.400				
0.500				
0.600				
0.700				

A graph can be used to show very clearly the dependence of one quantity upon another. A single line graph can only be used to relate two quantities to one another, so again everything else must remain constant. Normally the cause is plotted on the *x*-axis. This is called the **independent variable**. The effect of that cause is plotted on the *y*-axis. Because the effect is dependent upon the cause, this is called the **dependent variable**. If time is one of the variables, it is usually plotted along the *x*-axis.

If one quantity varies with another in a complex way, then the graph will show this complexity. Often, however, a graph is used to establish a pattern in the variation; that is, to establish a law that can be used to predict what will happen on another occasion. In order to do this, it is usually necessary to plot a straight-line graph. The reason for this is that one curve can look very like another, even though the algebraic equations that they are obeying may be different. A parabola can appear like a circle if only portions of both are drawn.

Straight-line graphs

Straight-line graphs

The generalised equation of a straight-line graph is

$$y = mx + c$$

where m is the gradient of the graph and c is the intercept on the *y*-axis. Note that, for this general form, the first term contains only a single variable y, the second term contains a constant m and a variable x, while the third term contains only a constant c.

If the graph from a series of readings is a straight line, then it is possible to find the gradient and the intercept directly (see box). Consider the graph drawn in **Fig 2.5**, which has been plotted to show distance against time. The gradient of this graph will give the speed, and the intercept will give the distance from a fixed point at the start of the measurements. This gradient is given by

$$\frac{68.2 - 27.0}{20 - 0} = 2.06$$

so the speed is 2.06 m s^{-1}. The intercept on this graph is 27, so the distance from the fixed point is 27 m at the start. This cannot be found to any greater certainty than to the nearest 0.5 m.

Note that this way of calculating the gradient makes it quite clear which points have been used to find the gradient and that the values on the *y*-axis have been estimated to a fifth of a small square. The gradient is quoted to three significant figures. A common mistake when calculating gradients is to quote inappropriate numbers of significant figures. Often this is because the calculation is done with too small a figure on the bottom of the division. It is not good practice to hunt for apparently convenient points on the graph. In this case it would not be correct to use the values of d when $t = 14$ and $t = 11$ to get a gradient of $(56 - 50)/3 = 2$. This has ignored the actual position of the line, and the accuracy of such a method is low. (It helps when plotting and reading from graphs to think decimal rather than fractional. Using $\frac{1}{4}$, $\frac{1}{2}$ and $\frac{3}{4}$ is awkward because the whole number has no decimal places, $\frac{1}{2}$ has one decimal place, whereas $\frac{1}{4}$ and $\frac{3}{4}$ have two decimal places. If instead you think in terms of reading 0.0, 0.2, 0.4, 0.6 and 0.8, then you are consistent in the number of significant figures that you use.)

If the relationship between two quantities being plotted is known, but is not a linear relationship, then it is still possible to plot a straight-line graph by carefully choosing what to plot. Consider the gas law equation $pV = nRT$. If an experiment is being done to measure how the volume V of a fixed mass of gas varies with the pressure p at a constant temperature, then a straight-line graph can be obtained by plotting $1/V$ against p. This graph will have the gradient $1/nRT$ and will pass through the origin if the gas is obeying this gas law.

Consider as an example an experiment on an amplifier connected to a series of different loudspeakers. The output resistance of the amplifier is r and the resistance of a loudspeaker is R. The power output P is measured for each different

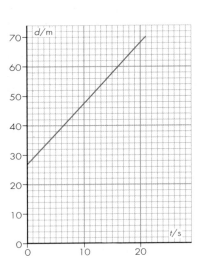

FIG 2.5 A straight-line graph.

loudspeaker for a standard setting on the amplifier volume control. A straight-line graph can be used to find the e.m.f. E of the amplifier's output and its output resistance, since the relationship between these terms is known:

$$P = \frac{E^2 R}{(R + r)^2}$$

The equation has first to be rewritten in the form $y = mx + c$ (see box on page 18). So, for our amplifier equation, we can have any combination of R and P in the first term, anything in the second term, and only a combination of r and E in the third term. It is necessary to try various possibilities until one is found that satisfies these requirements, e.g.

$$P(R + r)^2 = E^2 R$$

$$(R + r)^2 = \frac{E^2 R}{P}$$

$$(R + r) = E\sqrt{\frac{R}{P}}$$

$$R = E\sqrt{\frac{R}{P}} - r$$

If therefore R is plotted against $\sqrt{(R/P)}$, the gradient of the graph will be E and the intercept will be $-r$.

One other feature of the advantages of using graphical analysis is that, if several of the readings taken produce points that are rather too close together, additional work could be done to increase the number of points where they are too far apart.

Logarithmic graphs

Often the equation relating the two quantities being plotted is not known. In this case a log graph can be plotted. Suppose you are trying to find a relationship between the diameter of a piece of fuse wire and the current needed to cause it to melt. First of all you must have the fuse wire under controlled conditions, probably in a fuse holder. Then you will be able to measure the fusing current by increasing the current slowly until the fuse cuts off the current, and record the maximum ammeter reading. A table can be drawn up of current I against diameter d.

Assume that I is proportional to d to a power n, i.e.

$I = kd^n$ where k is a constant

Taking logs of this equation gives

$\log I = \log k + n \log d$

More details about logs can be found in appendix B.

The equation is now of the form required to plot a straight-line graph. Log I can be plotted on the y-axis against log d on the x-axis. The slope of the graph will be n, and the intercept will be log k.

Table 2.4 gives a series of values obtained for this experiment. When these points are plotted, the graph is as shown in **Fig 2.6**. Note here how the positions of the points do not lie very well along the line. This shows that the readings themselves were not very reliable. Whenever this happens, the graph plotted must be the best straight line through the points. If the points on a graph lie on a curve, then the line drawn should be a smooth line rather than a zig-zag line joining the points. The spread of points around the line indicates the random nature of the readings themselves.

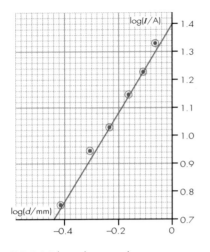

FIG 2.6 A logarithmic graph.

TABLE 2.4 The data for a logarithmic graph

d/mm	I/A	LOG(d/mm)	LOG(I/A)
0.4	5.5	−0.398	0.740
0.5	8.4	−0.301	0.924
0.6	10.3	−0.222	1.013
0.7	13.5	−0.155	1.130
0.8	16.2	−0.097	1.210
0.9	21.0	−0.046	1.322

For this graph the gradient is 1.6, showing that

$$I = d^{1.6}$$

An estimate of the uncertainty can be obtained from any graph by considering the limits between which the line may possibly have been drawn. In this case the uncertainty was estimated to be 1.6 ± 0.1. This is a difficult experiment in which to get reliable results, and the high uncertainty reflects this.

TABLE 2.5 Readings for question 2.9

t/s	v/m s⁻¹
0	3.7
0.5	5.7
1.0	7.5
1.5	9.1
2.0	11.1
2.5	13.0
3.0	14.8
3.5	16.8

TABLE 2.6 Data for question 2.10

f/Hz	Z/Ω
100	931
200	1020
300	1150
400	1310
500	1500
600	1690
700	1900

TABLE 2.7 Data for question 2.11

Separation, d/m	Force of repulsion, F/N
0.050	2.03
0.055	1.53
0.060	1.17
0.065	0.92
0.070	0.74
0.075	0.60
0.080	0.49

QUESTIONS

2.8 What would you plot on the *x*-axis and the *y*-axis if you wanted to obtain a straight-line graph from the equation

$$\frac{1}{u} + \frac{1}{v} = \frac{1}{f}$$

if values are known for the variables *u* and *v*, and the graph is to be used to find the constant *f*?

2.9 The readings in Table 2.5 were obtained from an experiment to determine the acceleration of a vehicle. The velocity *v* of the vehicle at times *t* was recorded. Plot a graph and use it to find:

a the acceleration;

b the velocity when $t = 0.8$ s;

c the velocity when $t = 4.0$ s.

d Why is interpolation, that is, using the graph to find an intermediate value as in (b), more reliable than extrapolation, that is, using the graph to find a reading beyond the end of the graph as in (c)?

2.10 The impedance Z of an electrical circuit is given by

$$Z = \sqrt{R^2 + 4\pi^2 f^2 L^2}$$

where *R* and *L* are constants and the frequency *f* is variable. Use the data of Table 2.6 to plot a suitable graph that will enable you to find values of *R* and *L*.

2.11 The force of repulsion *F* between two magnets placed in line with one another, N pole to N pole, varies with the separation *d* of their centres in the way shown in Table 2.7. Find the law relating the repulsive force to the distance, assuming it is of the form $F = kd^n$.

SUMMARY

- For a reading $a \pm \delta a$, δa is the absolute uncertainty, $\delta a/a$ is the fractional uncertainty and $100\delta a/a$ is the percentage uncertainty. δ (the Greek delta) is read as 'a small change in'.

- When quantities are added or subtracted, the uncertainty in their total is the sum of the absolute uncertainty in each.

- When quantities are multiplied or divided, the fractional uncertainty in their total is the sum of the fractional uncertainties in each.

- When two values $b \pm \delta b$ and $c \pm \delta c$ are to be combined then:

  if $y = b - c$ then $y \pm \delta y = (b - c) \pm (\delta b + \delta c)$

  if $x = b \times c$ then $\dfrac{\delta x}{x} = \dfrac{\delta b}{b} + \dfrac{\delta c}{c}$

  if $z = b^n$ then $\dfrac{\delta z}{z} = n\dfrac{\delta b}{b}$

ASSESSMENT QUESTIONS

2.12 In an experiment involving the stretching of a rubber cord of unstretched length 90 mm, the stretching force F and the extension ratio λ were measured. The extension ratio is the stretched length divided by the unstretched length. A graph of the result of the experiment is shown below.

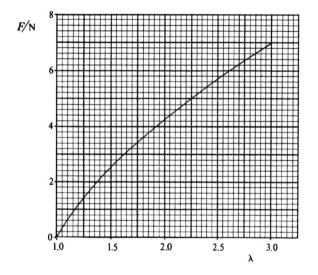

a What was the maximum stretched length of the rubber cord in this experiment?

b For values of λ less than 1.1, $F = kx$, where k is a constant and x is the extension (the increase in length) of the cord. Use the graph to find a value for k.

c It is suggested that

$$F = G\left[\lambda - \frac{1}{\lambda^2}\right]$$

where G is a property of the rubber cord. Read off values of F and λ from the graph and prepare a suitable table of values for

$$F \text{ and } \left[\lambda - \frac{1}{\lambda^2}\right].$$

Deduce a value for G.

London 1996

2.13 A large heavy spherical mass was suspended from a rigid support on a thin rope. The distance, l, from the support to the top of the mass was measured. The mass was allowed to swing through a small angle and the time for 20 oscillations measured. Then l was reduced and the measurements repeated. The readings obtained are shown in the table.

distance l/m	2.48	2.24	2.02	1.83	1.70
time for 20 oscillations/s	63.4	60.4	57.4	54.6	52.7

a Assume that the arrangement behaves as a simple pendulum of length $(l + r)$, with a time period

$$T = 2\pi\sqrt{\frac{(l+r)}{g}}$$

where r is a constant.

i Rearrange the equation into a form suitable for drawing a straight-line graph from the results given.

ii Tabulate readings suitable for plotting such a graph.

iii Plot your tabulated readings on graph paper.

iv From your graph, obtain a value for the acceleration due to gravity.

b Does the fact that the measurements were made from the support to the top of the mass affect the accuracy of the value obtained for the acceleration due to gravity? Explain the reason for your answer.

NEAB 1996

2.14 A heavy laboratory clampstand is laid on its side as shown in the diagram. It is supported with the rod horizontal by a vertical spring balance. The centre of gravity of the clampstand is at G.

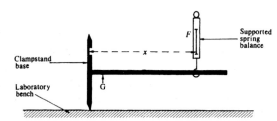

a Explain briefly how you would have found the position of the centre of gravity of the clampstand before arranging it in this way.

b Draw a free-body force diagram showing all the forces acting on the clampstand. Label each force with a capital letter.

c The spring balance can be moved to different places along the rod of the clampstand, each time holding the rod horizontal.

i By considering the physical principles associated with the mechanics of this situation, derive a relationship which predicts how the force F will vary for different values of the distance x. Take the distance of the centre of gravity of the clampstand to be a distance z from its base.

ii The table shows some results for F and x:

x/cm	34.5	41.0	49.5	60.0	68.5	77.0	88.0
F/N	10.0	8.3	7.0	5.8	5.1	4.5	3.9

Explain how you would use these results to test the prediction you made above.

iii Draw up a suitable table of values and plot a graph from which you could deduce the mass m_c of the clampstand. Use your graph to determine m_c given that $z = 7.6$ cm.

d The rod of the clampstand is removed from the base and used to drive a vertical hole in a patch of damp soil. To do this it is raised several times to exactly 0.50 m above the ground and released. The depth d below the level of the ground at which the bottom of the rod comes to rest after each drop is measured. It is suspected that d is related to the number of drops N by the equation

$$d = d_0(1 - e^{-aN})$$

where d_0 and a are constants.

i Describe briefly how you would measure d for values of N from 1 to 6. You should explain how you would ensure that the rod is moving vertically as it strikes the soil.

ii The equation can be rearranged as

$$\ln(d_0 - d) = \ln d_0 - aN$$

The graph shows $\ln[(d_0 - d)/\text{cm}]$ plotted against N for one student's set of results. Use the graph to find the values of d_0 and a.

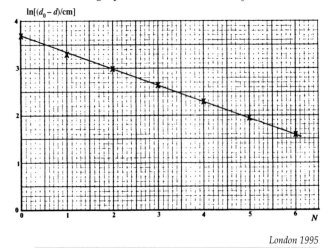

London 1995

3

SCALARS AND VECTORS

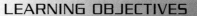
3.1   SCALARS AND VECTORS

When dealing with physical quantities in Chapter 1, it was stated that all such quantities have a numerical size and a unit. Some physical quantities also have *direction*. Any physical quantity that has direction is called a **vector**. A physical quantity that does not have direction is called a **scalar**. To specify a scalar, only its 'size' and its 'unit' need be given. To specify a vector, the three terms 'size', 'unit' and 'direction' must all be given. The numerical size of a vector is often called its **magnitude**. This gives:

Scalar   *magnitude + unit*
Vector   *magnitude + unit + direction*

    The way in which the direction of a vector is given varies from one situation to another, but should at all times be stated explicitly even when it seems obvious. It might be given as in the direction of motion, or as upwards or downwards, or as due north, or (as in **Fig 3.1**) as N 67° E. It should not be given as vertical, or horizontal, or along the axis, or parallel to the surface, because in each of these cases there is more than one possibility: 'vertically' may be vertically upwards or vertically downwards. The important point is that the direction must be clear. Often the simplest way of clarifying the direction of a vector is to draw a diagram.

Writing vectors

In books, vectors are normally labelled on diagrams or written in equations in ***bold italic*** type (as in this book), e.g. ***F*** or ***v***. When *you* write them, put a bar below or above the letter, e.g. F̠ or v̄.

    The magnitude of a vector is shown in books in *italic* type, e.g. *F* or *v*.

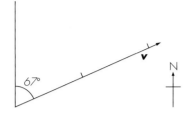

FIG 3.1 A vector showing the velocity *v* of 21.8 m s⁻¹ in a direction of N 67° E. Each unit of length represents 10 m s⁻¹.

Scale diagrams are often useful to show vectors, in which case the length of the line drawn represents the magnitude of the vector, and the direction of the line shows the direction of the vector. **Fig 3.1** shows a typical vector scale drawing.

Table 3.1 gives a list of many of the physical quantities you will be using, and divides them up into scalars and vectors. There are some fine distinctions that need to be made in deciding what words to use and hence into which category a quantity is placed.

TABLE 3.1 Scalar and vector quantities

SCALAR	VECTOR
length	displacement
speed	velocity
time	acceleration
volume	force
mass	weight
energy	momentum
frequency	torque
density	moment
pressure	electric current
power	magnetic flux density
charge	electric field
capacitance	
potential difference	
temperature	

VECTOR ADDITION

The addition of scalars is no problem; the two quantities, which must have the same unit, are added together by the normal rules of addition. A mass of 6.4 kg added to a mass of 8.3 kg gives a total mass of 14.7 kg.

Adding two vectors together is a different matter because of their directions. If they happen to be in the same direction, then direct addition is possible. But usually they are *not* in the same direction, and it is possible to add a vector of magnitude 5.3 to a vector of magnitude 7.0 to give a total vector of anything between 1.7 and 12.3.

In order to add vectors, use is made of the **parallelogram law**. This is illustrated in **Fig 3.2**, where vector *A* is being added to vector *B* to give vector *R*. It is frequently convenient to draw only half of the parallelogram, in which case the second vector is added on to the first nose-to-tail, as in **Fig 3.3**. It can be seen from **Fig 3.4** that *A* + *B* produces the same result as *B* + *A*. The result of adding two vectors together, *R* in this case, is called the **resultant** vector. Since so many physical quantities are vectors, it is essential that you can find a resultant either by scale drawing or by calculation using the cosine rule (see box and Appendix). You should now be able to work out what forces are involved when trees are bent by a strong wind (**Fig 3.5**).

This process of addition can be extended to finding the sum of as many vectors as are wanted. **Fig 3.6a** shows the various forces acting on a tractor when it is being transported on a lorry. The forces act at many different points on the tractor.

Cosine rule

Calculations involving the triangle of forces often require the use of the cosine rule for a triangle (see Appendix). This is

$$a^2 = b^2 + c^2 - 2bc \cos A$$

where *a*, *b* and *c* are the lengths of the three sides of the triangle, and *A* is the angle between the sides of length *b* and *c*.

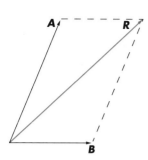

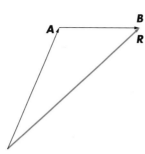

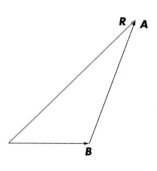

FIG 3.2 The parallelogram of forces.

FIG 3.3 A triangle of forces showing the same addition as Fig 3.2.

FIG 3.4 A triangle of forces showing that *B* + *A* gives the same result as *A* + *B*.

FIG 3.5 In a high wind a tree moves out of the vertical. Gravity still acts vertically downwards, but the force the ground exerts on the tree has a component upwards and horizontally into the wind. For equilibrium the upwards component equals the weight of the tree and the horizontal component into the wind equals the force the wind exerts on the tree.

Nevertheless, the resultant force on it can be found by adding all the forces together using a nose-to-tail method as shown in **Fig 3.6b**, where the resultant *R* is given by

$$A + B + C + D + E + F = R$$

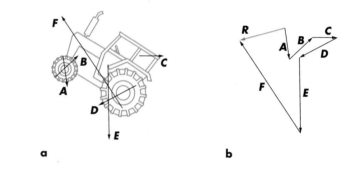

a **b**

FIG 3.6 The sum of many vectors can be obtained by drawing the vectors to scale in this nose-to-tail way. All the forces on the tractor give a resultant force *R*.

FIG 3.7 There will be a large number of forces acting on each of these cars, but they will be fixed to the trailer. So the resultant force on a car must give it the same acceleration as the lorry.

The same resultant is obtained whatever the order in which the forces are added. This procedure can be extended to three dimensions if required, but the geometry of the diagram can then be quite difficult to work with.

If the result of adding three vectors together is zero, then a **triangle of vectors** is obtained. If more than three vectors are added together and give a total that is zero, then the figure obtained is called a **polygon of vectors**.

3.3 VECTOR SUBTRACTION

The process of vector subtraction is, in principle, similar to vector addition. The process does, however, produce some surprising results. It is worth while first considering ordinary subtraction (a simple scalar problem), before we look at vector subtraction. We do this in the following examples.

EXAMPLE 3.1

A ball at one moment has a kinetic energy of 11 J. Later its kinetic energy has increased to 18 J. Find the increase in its kinetic energy.

The problem is almost too trivial; the answer is

$$18 J - 11 J = 7 J$$

Note that the problem could have been done by finding what needed to be added to the first value to obtain the final value, i.e.

$$11 J + 7 J = 18 J$$

This illustrates the way that can be used to subtract vectors.

EXAMPLE 3.2

A car goes round a right-angled corner so that its velocity changes from 12 m s^{-1} due east to 16 m s^{-1} due south (see Fig 3.8). What is its change in velocity?

The first point to make is that here we are *not* dealing with scalar quantities and that therefore the answer is *not* 4 m s^{-1}. Just subtracting 12 m s^{-1} from 16 m s^{-1} finds the

change in the *speed* of the car, but the change in its *velocity* is affected very much by the change in direction. This is illustrated by Fig 3.9. The required change of velocity is the green vector, which when added to the start velocity gives the final velocity. The change in velocity is 20 m s^{-1} in the direction of the green arrow.

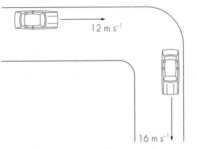

FIG 3.8 A car that changes its *speed* by 4 m s^{-1} as it goes round a corner changes its *velocity* by a very different amount.

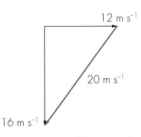

FIG 3.9 The change in the *velocity* of the car is shown to be 20 m s^{-1} in the direction of the green arrow.

QUESTIONS

3.1 Show that, for any two vectors **A** and **B**, the vector (**B** – **A**) is equal to –(**A** – **B**).

3.2 Find the change in velocity of a yacht if it changes its velocity from 5 m s^{-1} due north to 3 m s^{-1} due west.

3.3 Find the change in velocity of a tennis ball if it approaches a tennis racket at 30 m s^{-1} and leaves the racket in the opposite direction at 40 m s^{-1}.

continued

3.4 An aircraft travels through the air with a velocity of 300 m s^{-1} due east. It is found to be travelling over the ground with a velocity of 220 m s^{-1} in a direction 20° south of east. The difference between these two velocities is due to the wind velocity. What is the wind velocity?

3.5 The Moon is travelling around the Earth in a circle at a constant speed of 1000 m s^{-1}.

a What angle does it turn through in its orbit in 1 s if it takes 28 days for one complete revolution?

b What change in velocity takes place in 1 s?

c In what direction does this change in velocity occur?

The solution to question 3.5 is referred to in Chapter 8 on circular motion (see section 8.2). Similar problems to these can be constructed in terms of any vector and not just velocity.

3.4   COMPONENTS OF VECTORS

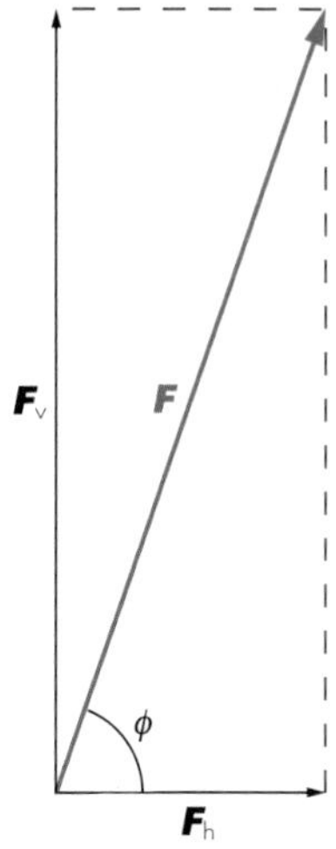

n some circumstances it is useful to split a vector up into two parts. These parts are called the **components** of the vector. Very often the components are taken horizontally and vertically, in which case the two parts are called the horizontal component of the vector and the vertical component of the vector respectively. (If working in three dimensions, three components are often used.) **Fig 3.10** shows how in two dimensions vector F is split into its two components F_h and F_v. The two components, taken together, have exactly the same effect as the single vector F. The magnitudes of the two components are given by

$$F_h = F \cos \phi \qquad F_v = F \sin \phi$$

There is a real danger when using diagrams like **Fig 3.10** that too many vectors will be used. You should use *either* the components *or* the vector itself. All three must not be used together.

FIG 3.10 Force F is shown resolved into two components: F_h is the horizontal component and F_v is the vertical component.

FIG 3.11 All the forces acting on a yacht determine its angle to the vertical.

It is sometimes convenient, especially in a complex problem like the yacht in **Fig 3.11**, to resolve all the forces. In **Fig 3.12** four forces have been resolved into their vertical and horizontal components. Example 3.3 shows how this enables the resultant force to be calculated.

EXAMPLE 3.3

Find the resultant of the four forces shown in **Fig 3.12.**

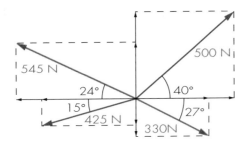

FIG 3.12 *Forces in many different directions can all be resolved into two directions at right-angles to each other.*

In the horizontal direction to the left, the total force is given by:

$$(545 \text{ N} \times \cos 24°) + (425 \text{ N} \times \cos 15°)$$
$$- (500 \text{ N} \times \cos 40°) - (330 \text{ N} \times \cos 27°)$$

$$= 498 \text{ N} + 411 \text{ N} - 383 \text{ N} - 294 \text{ N} = 232 \text{ N}$$

In the vertically upwards direction, the total force is given by:

$$(545 \text{ N} \times \sin 24°) + (500 \text{ N} \times \sin 40°)$$
$$- (425 \text{ N} \times \sin 15°) - (330 \text{ N} \times \sin 27°)$$

$$= 222 \text{ N} + 321 \text{ N} - 110 \text{ N} - 150 \text{ N} = 283 \text{ N}$$

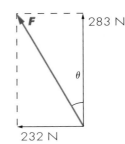

FIG 3.13 *The resultant is found from the triangle of forces. As here, such diagrams do not need to be drawn exactly to scale.*

Using Pythagoras's theorem enables the resultant force F to be found. From the triangle in **Fig 3.13**, we get both the magnitude F and direction θ:

$$F^2 = 232^2 + 283^2 = 133\ 913$$

$$F = 366 \text{ N}$$

and

$$\tan \theta = 232/283 = 0.8198$$

$$\theta = 39.3°$$

The resultant force F therefore has a magnitude of 366 N inclined in a direction 39° to the left of vertically upwards.

ANALYSIS

Forces in frameworks

Figs 3.14 and **3.15** are photographs of two very common types of structure. In civil engineering terms they are called pin-jointed structures. The design of such structures must depend on the use to which the structure is to be put and the loading forces that are applied to it. The strength, and therefore the size, of each of the struts in such a structure needs to be calculated so that the most economical design can be constructed. This exercise deals with a similar, but simplified, structure of a roof,

which is constructed out of triangles with sides of length 3 m, 4 m and 5 m. **Fig 3.16** shows *all* the forces acting on each of the joints in the framework of the roof. The forces in the struts are either forces of compression or forces of extension, and they always act in line with the strut. Other external forces are applied to the structure as shown. At each joint the total force has to be zero if the structure is to be in equilibrium. (Further details concerning equilibrium are given in Chapter 6. This exercise is basically concerned with the addition of force vectors.)

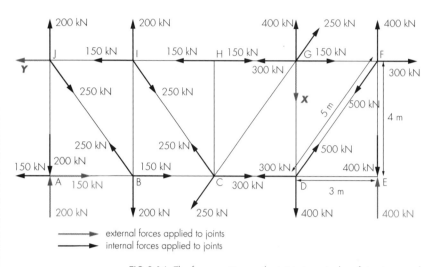

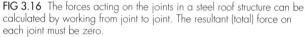
external forces applied to joints
internal forces applied to joints

FIG 3.16 The forces acting on the joints in a steel roof structure can be calculated by working from joint to joint. The resultant (total) force on each joint must be zero.

FIG 3.14 The construction of a steel pylon can be seen to be a series of triangles.

Using the information given in **Fig 3.16**, answer the following questions.

a Explain how the diagram shows that joints E and H are in equilibrium.

b Explain similarly why joint A is in equilibrium.

c Draw the triangles of forces to show that the total force acting on each of joints B, C, D, F and I is zero.

d Draw the polygon of forces for joint A.

e Draw the triangle of forces for joint J. Use the fact that the total of all the forces at the joint must be zero and hence find the value of Y.

f Draw the polygon of forces for joint G and use it to find X.

g Which of the struts are in tension and which are in compression?

h Which four struts are, for this loading, redundant? That is, which struts are neither in compression nor in tension?

This exercise could have been reversed. You could have been given all the external loadings, drawn the force diagrams one by one around the structure and hence found all the internal forces applied to the joints. This reverse process is carried out by a structural engineer (or a computer) in the design process to determine the particular steel struts to use.

FIG 3.15 A roof under construction shows the extensive steel framework consisting of a series of triangular frames.

There is much more to multiplying vectors than is being stated here. Further detail can be found in books on vector analysis.

To multiply a vector by a scalar is straightforward. If vector **F** is multiplied by scalar *t*, then the product is a vector of magnitude *Ft* in the same direction as **F**.

Multiplying a vector by a vector can give either a scalar product or a vector product. Examples 3.5 and 3.6 give one of each.

EXAMPLE 3.4

Multiply a force of 6.0 N acting in a direction due north by a time of 15 s.

You will see in Chapter 7 that this alters the momentum of a body by an amount of 6.0 N × 15 s = 90 N s in a direction due north.

EXAMPLE 3.5

Multiply a force of 650 N in a direction due east by a velocity of 30 m s^{-1} in the same direction.

You will see, also in Chapter 7, that this calculation is needed to find the power output of the machine producing the force. Power is a scalar quantity, and the power output has no direction; in this case it has a value of 19 500 W.

If there had been an angle θ between the two vectors of force **F** and velocity *v*, then the resulting power would have been $Fv \cos \theta$.

EXAMPLE 3.6

A current of 3.2 A flows vertically upwards through a wire of length 0.045 m. The wire is at an angle of 30° to a magnetic field of flux density 0.089 T (tesla, see Chapter 18). Find the force on the wire.

Clearly two vectors have to be multiplied together (see box). The equation relating these quantities is given in Chapter 18 as

$$F = BIl = BIl \sin \theta$$

where **F** is the force, **B** is the magnetic flux density, **I** is the current, θ is the angle between **B** and **I**, and *l* is the length of the wire. The result in this case is

$$F = 0.089 \text{ T} \times 3.2 \text{ A} \times 0.045 \text{ m} \times \sin 30° = 0.0064 \text{ N}$$

This force acts in a direction at right-angles to the plane containing **B** and **I**. You are probably familiar with this idea when you have used Fleming's left-hand motor rule for the direction in which a wire moves when carrying an electric current in a magnetic field.

Multiplying vectors

The rules for multiplying two vectors **A** and **B** together are
- to get a *scalar* product:
  the magnitude of the scalar is $AB \cos \theta$
- to get a *vector* product:
  the magnitude of the total vector is $AB \sin \theta$ in a direction at right-angles to the plane containing **A** and **B**

where θ is the angle between the vectors.

> ### QUESTIONS
>
> **3.6**   How much work is done by a force of 3.0 kN when it moves a distance of 0.36 m
>
> **a** in the direction of the force,
>
> **b** in a direction at right-angles to the direction of the force,
>
> **c** in a direction at 40° to the direction of the force,
>
> **d** in a direction opposite to the direction of the force?
>
> Draw a diagram of a situation that gives rise to (d) and explain what is happening.

3.7 Using the equation given in Example 3.6, find the force that a magnetic flux density of 0.085 T exerts on a wire of length 0.063 m when it is carrying a current of 4.3 A and

   a the wire is at right-angles to the field,

   b the wire is in the same direction as the field,

   c the wire is at an angle of 25° to the field.

SUMMARY

- Scalars have magnitude only. They can be added, subtracted, multiplied and divided by normal arithmetic.

- Vectors have direction as well as magnitude. They must be added or subtracted by the parallelogram rule (see **Fig 3.17**).

- Vectors can be multiplied, in which case they can give
  – either a *scalar* as in
     force × displacement = work
  – or a *vector* as in
     magnetic flux density × current × length = force

- Vectors can be resolved into two vectors at right-angles (see **Fig 3.18**).

FIG 3.17 **a** Vector addition is shown by *A + B*; **b** vector subtraction is shown by *A − B*.

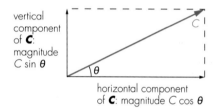

Fig 3.18 Vector *C* is resolved into two components.

ASSESSMENT QUESTIONS

3.8 a Physical quantities may be classified as *scalars* or as *vectors*. Three physical quantities are listed below. Copy and complete the table by placing a tick in the appropriate space to indicate whether the quantity is a scalar or a vector.

quantity	scalar	vector
speed		
momentum		
kinetic energy		

   b The figure shows a sectional diagram of a car bonnet ABC raised at an angle of 30° to the horizontal.

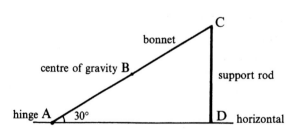

The bonnet, of mass 14 kg, is hinged at A. Its centre of gravity is at B, the mid-point of AC. The bonnet is held in the open position by the vertical support rod DC.

   i Calculate the upwards force of the rod DC supporting the bonnet.

ii The bonnet is now held open at an angle of 60° to the horizontal by a new, longer, support rod, again placed vertically at C. What is the upwards force of the new rod supporting the bonnet?

CCEA 1998

3.9 This question is about resolution of forces.
  a i Define the terms *vector* and *scalar*.
   ii Give one example of each.

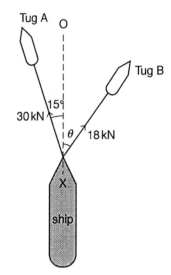

b Two tugs A and B pull a ship along the direction XO. See figure. Tug A exerts a force on the ship of 3.0×10^4 N at an angle of 15° to XO. Tug B pulls with a force of 1.8×10^4 N at an angle θ to XO.
  i Find the angle θ for the resultant force on the boat to be along XO.
  ii Find the value of this resultant force.

O & C 1997

3.10 State the difference between scalar and vector quantities.
  a A lamp is suspended from two wires as shown in the diagram. The tension in each wire is 4.5 N.

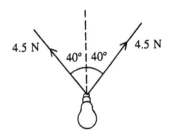

Calculate the magnitude of the resultant force exerted on the lamp by the wires.
  b What is the weight of the lamp? Explain your answer.

London 1999

3.11 a Distinguish between a scalar quantity and a vector quantity.
  b A car travels one complete lap around a circular track at an average speed of 100 km h^{-1}.
   i If the lap takes 3.0 minutes, show that the length of the track is 5.0 km.
   ii What is the total displacement of the car at the end of the lap?

NEAB 1996

3.12 The shot-putter shown in the figure throws the shot forwards with a velocity of 12 m s^{-1} with respect to his hand, in a direction 54° to the horizontal. At the same time, the shot-putter's body is moving forward horizontally, with a velocity of 3.0 m s^{-1}.

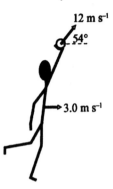

  a Draw a vector diagram to show the addition of the two velocities of the shot at the moment of release. Hence, or otherwise, show that the vector sum of the two velocities has a magnitude of approximately 14 m s^{-1}.
  b At the moment of release, the shot is 2.3 m above the ground. The shot has a mass of 7.3 kg.
   i Calculate the kinetic energy of the shot at the moment of release.
   ii Calculate the potential energy, relative to the ground, of the shot at the moment of release. The acceleration of free fall, g, is 9.8 m s^{-2}.
   iii Use the law of conservation of energy to calculate the speed of the shot as it hits the ground. (Neglect air resistance.)
   iv Copy the axes below and on it sketch the variation of the shot's speed with time, up to the moment when it touches the ground.

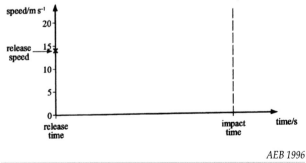

AEB 1996

3.13 a When a person is standing upright and still, the load on the head of one femur (thigh bone) is vertical. The neck of the femur makes an angle of 50° to the vertical as shown in the figure.

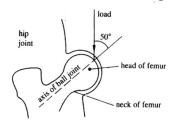

The weight, *W*, of the person is 700 N and the load on the head of one femur is 0.35*W*. This load will produce a compressive force along the axis of the ball joint and a shear force normal to the axis.

  i Copy the figure and on it indicate how the compressive and shear components of the load act on the head of the femur.

  ii Calculate the values of these two components.

b The figure shows the forces acting on the lumbosacral disc of the same person still standing upright. The vertical downwards load, 0.60*W*, and the upwards support force, *S*, are equal in magnitude but produce compressive and shear forces on the disc.

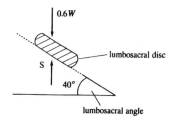

  i If the effective area of the disc which supports the compressive load is 3.0×10^{-4} m², calculate the compressive pressure on the disc.

  ii A person with a poor posture may curve the spine and change the angles of the vertebrae relative to each other. Describe the consequences of such poor posture over many years for the lumbosacral disc and for the person.

NEAB 1996

NEWTONIAN MECHANICS

Sir Isaac Newton published his classic work *Philosophiae Naturalis Principia Mathematica* over three hundred years ago in 1687. It has had a profound effect on the development of science since then, and the principles put forward in it are still used a great deal today. ■

4 MOTION

LEARNING OBJECTIVES

At the end of this chapter you should be able to:

① relate distance travelled, velocity, time and acceleration for any uniformly accelerated motion;

② use distance–time and velocity–time graphs;

③ calculate the momentum of an object.

4.1 SPEED AND VELOCITY

The speed of a car is often referred to as being, say, 120 kilometres per hour. This does not mean that the car must travel 120 kilometres or that it is on a journey that takes one hour. The expression implies a rate of movement of, in this case, 33.3 metres per second, or 33.3 millimetres per millisecond, or even 0.0333 millimetres per microsecond. All of these figures, when put into SI units, will be 33.3 m s^{-1}. This illustrates the point that the speed of an object can be found at a particular moment, and this is called the **instantaneous speed** of the object. Instantaneous speed can be defined by the equation:

K $$\text{instantaneous speed} = \frac{\text{small distance travelled by the object}}{\text{small time taken}}$$

when the small time taken gets closer and closer to zero.

Those of you who are familiar with calculus will see that this is similar to one of the basic calculus ideas, differentiation, in which the gradient of a graph is found (see **Fig 4.1**). Indeed, the instantaneous speed of an object can be defined as the gradient of the distance-time graph for the object.

The **average speed** of an object is defined by the equation:

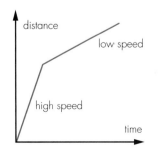

FIG 4.1 The gradient of a distance–time graph gives the speed.

average speed = $\dfrac{\text{total distance travelled}}{\text{total time taken}}$

If an object travels at a constant speed for a whole journey, then the instantaneous speed will be constant and equal to the average speed. In practice, such journeys are unusual, so it is much more reliable to write carefully which speed is being referred to. For example, 'the speed at the start is …' or 'the speed after 20 seconds is …'. The statement 'speed is …' should be avoided; it is much too vague. Always try to give an adjective and a noun whenever you start a new line of working, e.g. *total time* = …, *average speed* = …. It will help to eliminate many careless mistakes.

In many cases not only is the speed of an object known but also the direction in which it is travelling. This means that the quantity being dealt with is a vector. To make it quite clear when a vector is being used, a different term is applied, **velocity**. Velocity is always a vector, and the **instantaneous velocity** of an object is defined by the equation:

 instantaneous velocity = $\dfrac{\text{small distance travelled in a stated direction}}{\text{small time taken}}$

as the small time approaches zero. This is a rather clumsy definition, and so if **displacement** is the word used to mean the distance moved in a stated direction, then we can define instantaneous velocity by the rather briefer statement:

 Instantaneous velocity **is the rate of change of displacement.**

As with speed, we can also use the term **average velocity**. This is defined by the equations:

 average velocity = $\dfrac{\text{total distance travelled in a stated direction}}{\text{total time taken}}$

$$= \dfrac{\text{displacement}}{\text{total time}}$$

$$= \dfrac{\Delta x}{\Delta t}$$

Delta notation

The symbol Δ, the Greek letter capital delta, is read as 'the change in'. So average velocity can be read as 'delta x divided by delta t' or as 'the change in x divided by the change in t'. The symbol δ, the small Greek letter delta, is used to mean 'the small change in'.

The delta notation is described in the box. Instantaneous velocity can therefore be written as

$$v = \left(\dfrac{\delta x}{\delta t}\right) \text{ as } \delta t \text{ approaches zero}$$

Using calculus terminology this is written as

$$v = \dfrac{\mathrm{d}x}{\mathrm{d}t}$$

Measurement of velocity

Measurement of speed and/or velocity is often carried out by measuring a distance and dividing by the corresponding time. Standard methods for doing this in the laboratory may only involve the use of a metre rule and a stopwatch.

Another method involves the use of a ticker-timer, which makes a mark every 1/50th of a second on a piece of paper moving at the required speed. Electronic methods for measuring the time interval can also be used. These involve the object, whose speed is required, interrupting a beam of light to a photocell. The time that the photocell is not working is the time it takes for the object to pass through the beam (see **Fig 4.3**). There are several data-logging systems available that can link sensors to computer programs. If you have access to one of these systems, make certain that you know what it is recording and what calculations the computer is making for you. Multi-flash photographs, too, can be used, provided there is a scale for measuring the distance (see **Fig 4.4**).

EXAMPLE **4.1**

Find the average speed and the maximum speed shown on the ticker-tape in **Fig 4.2**. The dots are printed every 1/50 s and the tape can be considered to be shown full size.

The average speed is (total distance)/(total time):

total distance = 10.0 cm
total time = (23/50) s = 0.46 s
average speed = (10.0 cm)/(0.46 s)
= 21.7 cm s^{-1} = 0.217 m s^{-1}

The maximum speed can be obtained less accurately; the maximum distance between any two dots is 1.3 cm. This gives the maximum speed as (1.3 cm)/(0.02 s) = 65 cm s^{-1}. The maximum speed is an instantaneous value of the speed. Here we are calculating for a time interval of 0.02 s. In order to work with any smaller time interval, it would be necessary to plot a graph of distance against time and find its maximum gradient.

FIG 4.2 Dots on a ticker-tape for Example 4.1.

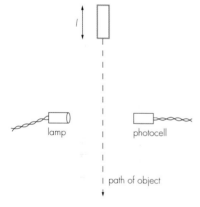

FIG 4.3 The time taken for an object of length *l* to pass between the lamp and the photocell can be used to calculate the speed of the object.

QUESTION

4.1 Find the speed of (a) the golf ball and (b) the golf club head in **Fig 4.4**. The scale is 1 to 3 and the multi-flash photographs were taken at the rate of 600 per second.

FIG 4.4 Make measurements on this photograph to find the speed of the golf ball and the golf club (question 4.1).

4.2 ACCELERATION

An object whose velocity is changing is said to have an acceleration. **Acceleration** is defined as the rate of change of velocity, and since velocity is a vector, then acceleration must also be a vector. This can be written using equations corresponding to those given above for velocity:

K   average acceleration $= \dfrac{\Delta v}{\Delta t}$

or

K   instantaneous acceleration $= \left(\dfrac{\delta v}{\delta t}\right)$ as δt approaches zero $= \dfrac{dv}{dt}$

The number of equations here can confuse, but in practice both velocity and acceleration can be taken to mean instantaneous velocity and instantaneous acceleration unless it is stated clearly that this is not so. This means that the two important definitions are similar:

K   *Velocity* **is the rate of change of displacement (unit m s^{-1}).**
Acceleration **is the rate of change of velocity (unit m s^{-2}).**

The unit of acceleration is metre per second in a second. This is written metre per second squared (m s^{-2}). Determination of acceleration practically is usually done by the same methods as were given above for velocity. In this case, however, it is always necessary to find two velocities, one at the beginning and one at the end of a time interval Δt.

4.3 GRAPHICAL REPRESENTATION OF MOTION

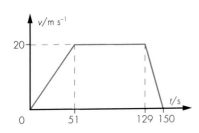

FIG 4.5 The velocity–time graph for the underground train.

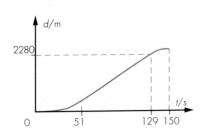

FIG 4.6 The distance–time graph for the underground train.

Many graphs can be drawn to show how a body moves. They are particularly useful for indicating motion, and in very many cases time will be on the *x*-axis of the graph. This is one way of overcoming the difficulty of presenting movement on a static piece of paper. The same problem is encountered with showing moving waves on paper, and again graphs are resorted to. Whenever a graph is drawn or used, it is essential to be certain what is being plotted, as the same information plotted on two different graphs can appear to be very different.

Consider the movement of an underground train between two stations. The motion of the train consists of an acceleration, then a period of constant velocity, followed by slowing down to a stop. The slowing down is sometimes called a **deceleration**, which is a negative acceleration. If a velocity–time graph is plotted for a movement, it appears to be very different from the corresponding distance–time graph, although they are giving exactly the same information. **Figs 4.5** and **4.6** show two such graphs, plotted for idealised circumstances. It has also been assumed that the track is straight, so that we are correctly dealing with velocity and not speed.

With many ways of presenting the same information, the question arises as to how to obtain information from any of the graphs. Two features of graphical work should be clear already from the basic definitions of velocity and acceleration. Because velocity is the rate of change of displacement, velocity is the gradient of a displacement–time graph. Similarly, because acceleration is the rate of change of velocity, acceleration is the gradient of a velocity–time graph.

Displacement can also be obtained from a velocity–time graph. Consider the graph shown in **Fig 4.5.** When the underground train is travelling at a constant velocity of 20 m s⁻¹, it is travelling for 129 – 51 = 78 seconds. During this time, therefore, it has a displacement of 20 m s⁻¹ × 78 s = 1560 m. This value, 1560, is also the area of the rectangle beneath the flat portion of the graph. This can be extended to any shape of graph.

> **K** The area beneath any velocity–time graph always represents the *displacement.*

QUESTIONS

4.2 Draw a third graph showing the same information as is contained in the two graphs for the underground train (Figs 4.5 and 4.6), but plotting acceleration against time.

4.3 What acceleration is shown by the graph in Fig 4.7?

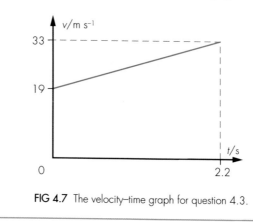

FIG 4.7 The velocity–time graph for question 4.3.

4.4   UNIFORMLY ACCELERATED MOTION

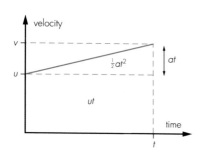

FIG 4.8 Graph showing uniform acceleration from velocity *u* to velocity *v* in time *t*. The area beneath the graph is the distance. The gradient of the graph is the acceleration.

In many practical applications the acceleration of an object is constant. That is, its velocity is increasing at a steady rate. This motion is called **uniformly accelerated motion** and is shown in **Fig 4.8** for an object moving in a straight line. Here an object at time $t = 0$ has a velocity u. After moving for time t its velocity has increased to v. Assume during this time that its acceleration is a and its displacement is s.

Since the acceleration is constant, its value is given by $a = \dfrac{v - u}{t}$

Therefore the increase in velocity is at as marked. The area of the bottom rectangle is shown as ut, and the area of the triangle is $\frac{1}{2} \times at \times t$. This gives the area beneath the graph and hence the displacement s in three different ways:

Constant acceleration

These equations of motion apply *only* to a system undergoing a constant acceleration. Mistakes are frequently made by applying them to situations where this is *not* the case. Of the five quantities, u, v, a, s and t, one is omitted from each of the equations.

$$s = ut + \frac{1}{2}at^2$$

$$s = vt - \frac{1}{2}at^2$$

$$s = \frac{(u + v)t}{2}$$

The last equation is (average velocity) × time.

If t is eliminated from any two of the above equations, the relationship between u, v, a and s can be found as

$$v^2 = u^2 + 2as$$

Car stopping distances

The distance a car travels during an emergency stop can be calculated using the equations just quoted. The distances are quoted in the *Highway Code* and car drivers taking a driving test are expected to understand the ideas involved.

When an emergency situation is noticed, it takes the driver a brief time to react to the emergency and to apply the brakes. During this time the car's speed is constant at the value it had before the emergency was noticed, and the car travels a distance called the *thinking distance*. Once the brakes are applied, the car rapidly decelerates and travels a distance called the *braking distance*. The sum of the thinking distance and the braking distance is the *stopping distance*. Any stopping distances calculated will increase if the road surface is wet or slippery and if the car is heavily loaded. A minibus is considerably more difficult to stop when full of passengers than when empty.

EXAMPLE 4.2

Calculate the stopping distances from speeds of 20 m s^{-1} and 40 m s^{-1}. (These speeds are roughly 40 and 80 mph.)

Make the assumptions that the driver has a thinking time of 0.20 s and that the car can have a deceleration of 10 m s^{-2}.

We have

initial speed	20 m s^{-1}	40 m s^{-1}
thinking distance	20 m s^{-1} × 0.20 s	40 m s^{-1} × 0.20 s
	= 4.0 m	= 8.0 m

Using $v^2 = u^2 + 2as$ gives $0 = u^2 - 2as$, and hence

$$s = u^2/2a.$$

Therefore we get

braking distance	$\dfrac{20^2}{2 \times 10}$	$\dfrac{40^2}{2 \times 10}$
	= 20 m	= 80 m
stopping distance	24 m	88 m

EXAMPLE 4.3

Using the details from the underground train example above, show that the total displacement is 2280 m.

For the three sections of the graph in **Fig 4.5**, we get the following displacements.

■ First section, acceleration:

$$s_1 = \frac{(u + v)t}{2} = \frac{20 \, \text{m s}^{-1} \times 51 \, \text{s}}{2} = 510 \, \text{m}$$

■ Second section, constant velocity:

$$s_2 = 20 \, \text{m s}^{-1} \times 78 \, \text{s} = 1560 \, \text{m}$$

■ Third section, deceleration:

$$s_3 = \frac{(u + v)t}{2} = \frac{20 \, \text{m s}^{-1} \times 21 \, \text{s}}{2} = 210 \, \text{m}$$

So

total displacement = 510 m + 1560 m + 210 m = 2280 m

ANALYSIS

Grand Prix facts and figures

During the course of one circuit of the Japanese Grand Prix motor race, the speed of one car varied in the way shown in **Fig 4.9b**. It took 96.293 seconds to complete the lap from a flying start. Also in **Fig 4.9** are given the axes for a distance–time graph **a**, and an acceleration–time graph **c**.

On a copy of **Fig 4.9**, and using approximate numerical values, complete the following:

a Plot the acceleration–time graph from the speed–time graph.
b Plot the distance–time graph from the speed–time graph.
c Find the distance for one lap.
d Find the average speed for this lap.
e Comment on the effect that tyres and the state of the track have on the shape of the acceleration–time graph.

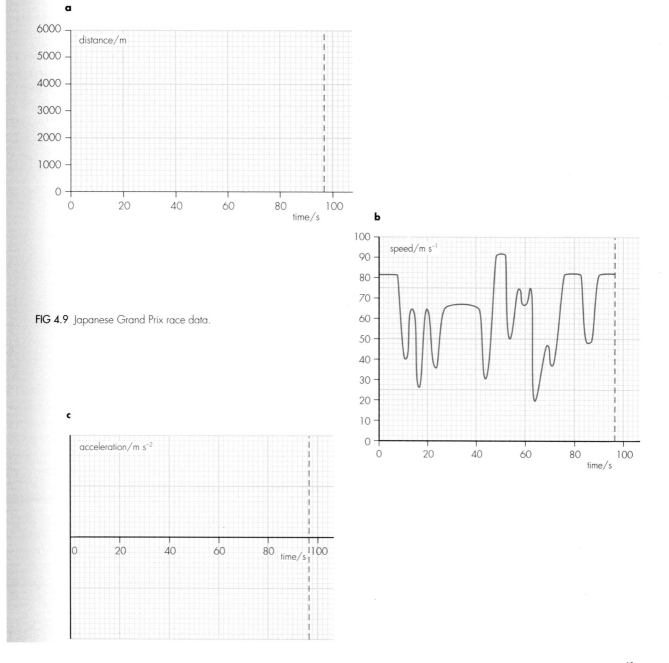

FIG 4.9 Japanese Grand Prix race data.

QUESTIONS

4.4 Find the distance travelled by a train in the first minute of its journey if it has a constant acceleration of 0.32 m s^{-2}.

4.5 What acceleration does a car require if it is to accelerate uniformly from 10 m s^{-1} to 35 m s^{-1} while it travels a distance of 300 m along a slip road of a motorway?

4.6 A ball approaches a tennis racket with a velocity of 21 m s^{-1}. The tennis racket gives it an average acceleration of 3500 m s^{-2} for 0.020 s in the opposite direction to its initial velocity.

a What is the velocity of the ball after leaving the racket?

b How far does the ball travel while undergoing this acceleration?

4.7 Repeat Example 4.2 for speeds of 5, 10, 15, 25, 30 and 35 m s^{-1}. Plot graphs, using the same axes, to show how the thinking distance and the stopping distance vary with the initial speed of the car.

4.8 A speeding motorist is travelling at a constant speed of 40 m s^{-1} on a motorway. He passes a stationary police car, which immediately accelerates at a rate of 2.5 m s^{-2}, to a constant speed of 50 m s^{-1}. Draw a sketch graph showing the variation of speed against time for each car and find

a the time it takes for the police car to catch the motorist;

b the distance each will have travelled by that time.

4.5 FALLING BODIES

Free fall

An object is said to be in **free fall** if the only force acting on it is the gravitational pull of the Earth. It then has an acceleration downwards relative to the surface of the Earth of approximately 9.81 m s^{-2}. This value varies from place to place on the Earth. It decreases as the height above sea-level increases, and it also decreases as you get closer to the equator. Uneven distribution of matter within the Earth also affects the value. As changes caused by these factors are small, the downward acceleration caused by free fall is often taken to be a constant and given the symbol g. True free fall will only occur in a vacuum, where air resistance will be zero.

Consider a ball being hit so that it starts with an upward velocity of 26.1 m s^{-1}. If wind resistance is neglected, the ball is in free fall from the moment it ceases to be in contact with the bat. Its acceleration will have a value of 9.8 m s^{-2} (to two significant figures) in a *downward* direction. This will have the effect of reducing its *upward* velocity by 9.8 m s^{-1} during each second of its movement. Values of velocity and time are therefore:

Time/s	0	1	2	3	4	5
Velocity/m s^{-1}	26.1	16.3	6.5	−3.3	−13.1	−22.9

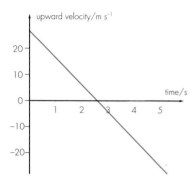

FIG 4.10 An object in free fall has its upward velocity reduced by a fixed amount each second. The acceleration is −9.8 m s^{-2}, so the gradient of the graph is −9.8.

A velocity–time graph will therefore be a straight line going from positive to negative values. There is no discontinuity when the ball reaches the top of its flight and stops momentarily after 2.66 seconds. The acceleration throughout is equal to 9.8 m s^{-2}, even when the ball is stationary. The velocity–time graph will look like **Fig 4.10**. The area of the triangle is

$$\tfrac{1}{2} \times 2.66 \times 26.1 = 34.7$$

so the distance the ball will travel upwards is 34.7 m. The ball will fall back to the level it started from when the negative area beneath the graph equals this positive area. This will be after a further 2.66 seconds, making the total time 5.32 seconds.

FIG 4.11 Before the tennis ball is hit by the racket, it can be seen to have a path that is parabolic. After being hit, the path is still a parabola, but the much larger forward velocity makes this less apparent.

If, during the time the ball is in mid-air, the ball also has a horizontal velocity, its vertical velocity is totally independent of its horizontal velocity. In the absence of air resistance, its horizontal velocity will be constant. In the photograph in **Fig 4.11**, the tennis ball, before being hit by the racket, is not only accelerating under gravity but also has a small forward velocity. This independence of vertical and horizontal motion is important in determining the path of any projectile.

EXAMPLE **4.4**

Assuming that the maximum speed at which a man can run is 11.0 m s⁻¹, and that for maximum distance a long-jumper should take off at an angle of 45°, find the maximum theoretical value of the men's world long-jump record in terms of g.

Assuming that a long-jumper can change his velocity to an angle of 45° without changing his speed, a vector diagram of his take-off is given in Fig 4.12.

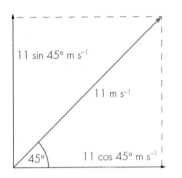

FIG 4.12 The horizontal and vertical components of the velocity of a long-jumper on take-off at 45°.

For the vertical motion, the velocity upwards has to change from $+11 \sin 45°$ m s⁻¹ to $-11 \sin 45°$ m s⁻¹. Therefore

$$\Delta v = 2 \times 11 \sin 45° \text{ m s}^{-1} = 15.6 \text{ m s}^{-1}$$

So the time the long-jumper is in the air is $(15.6 \text{ m s}^{-1})/g$. The horizontal distance d travelled in this time is

$$d = (15.6 \text{ m s}^{-1}/g) \times 11 \cos 45° \text{ m s}^{-1} = (121 \text{ m}^2 \text{ s}^{-2})/g$$

In Helsinki, Finland, where g is 9.827 m s⁻², this gives $d = 12.31$ m. At sea-level in Mexico, where g is 9.794 m s⁻², this gives $d = 12.35$ m. In Mexico City, at an altitude of 3000 m, where g is 9.787 m s⁻², the distance becomes $d = 12.36$ m.

FIG 4.13 Mike Powell long jumping in the Olympic Trials.

The other, and larger, effect of jumping at high altitude is that the air resistance is lower in the less dense air. As a result of these effects, the world long-jump record was increased dramatically in October 1968 to 8.90 metres. However, athletes still have some way to go before they reach the theoretical maximum. The world record was increased to 8.95 metres by Mike Powell in August 1991 (**Fig 4.13**). In practice, it would not be possible to take off at an angle of 45° without some springboard assistance. Gymnasts are able to travel considerably further than long-jumpers. This is not only because they have something to vault over but also because they do use springboards. Even the thickness of the rubber in the soles of the shoes of a long-jump athlete makes a considerable difference and causes difficulties for the authorities laying down the rules of the competition.

Air resistance

In practice, totally free fall does not happen because of air resistance. The effects on the diver or the bunjee jumper shown in **Figs 4.14** and **4.15** will be small because, at the maximum speed at which they travel, air resistance will only be a small fraction of their weight. This clearly cannot be true for a parachutist where, deliberately, air resistance is used to limit the maximum downward velocity of the parachutist. A free-fall parachutist (**Fig 4.16**), during the time before the parachute is opened, usually reaches a maximum speed of about 60 m s⁻¹. At this speed the weight of the parachutist (a downward force) equals the air resistance (an upward force). The parachutist is therefore in equilibrium and has a constant velocity; this constant velocity is called the **terminal velocity**. **Fig 4.17a** shows how the force of air resistance on a falling body increases with time, and **Fig 4.17b** shows the corresponding increase in the velocity of the body. The gradient of this graph is the acceleration. At the start of the fall, the acceleration is 9.8 m s^{-2}; at the terminal velocity, the acceleration is zero.

FIG 4.14 For heights up to 10 m, air resistance makes very little difference to the movement of the diver.

FIG 4.16 A parachutist in free fall will be travelling much faster than after the parachute has opened.

FIG 4.15 Air resistance also has very little effect on the bunjee jumper.

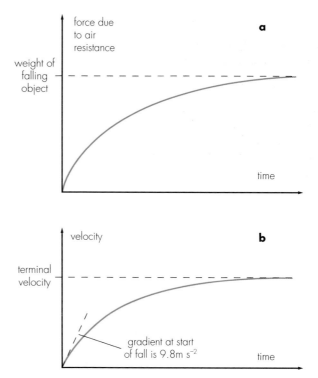

FIG 4.17 Graphs showing how **a** the force due to air resistance and **b** the velocity change with time for a falling body. For low velocities, just after a falling body has been released, the acceleration is high. After a long period of falling, the force of air resistance has increased to be equal to the weight. The body will then have zero acceleration and will fall with constant velocity.

FIG 4.18 A competitor at the Ski Jumping World Championship in 1989.

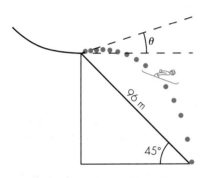

FIG 4.19 The path followed by a ski-jumper is here shown to be parabolic. The effects of wind resistance are being ignored.

QUESTIONS

4.9 How long does a high-diver take to reach the water from a 10.0 m diving stage? With what speed will she reach the water?

4.10 A football is kicked at a speed of 20 m s^{-1} at a launch angle of 45°. Find its range, assuming negligible air resistance.

4.11 A stunt rider leaves a ramp at an angle of 30° and lands at the same height after travelling a horizontal distance of 36 m. What was his or her take-off speed?

4.12 A rounders ball is thrown with a velocity of 30 m s^{-1} at an angle of 42° to the horizontal. Calculate

    a the time it takes to reach its maximum height;

    b the total time taken to return to its original level;

    c the maximum height reached;

    d the horizontal distance travelled.

4.13 A ski-jumper (**Fig 4.18**) lands 96 m from his or her take-off point (**Fig 4.19**). The slope is at an angle of 45° and the jumper is in the air for 4.3 s. Making the rather sweeping assumption that air resistance is negligible, find

    a the horizontal distance travelled;

    b the horizontal component of the velocity at take-off;

    c the vertical distance from take-off to landing;

    d the vertical component u of the take-off velocity, using the equation $s = ut + \frac{1}{2}at^2$ [be careful to use signs, so that you are consistent with directions: s is down, u is up and a is down];

    e the angle of take-off, θ;

    f the speed of take-off.

4.6 MOMENTUM

The **momentum** of a body is defined as the product of its mass and its velocity:

K   **momentum = mass × velocity**

It is a vector quantity, being the product of a scalar (mass) and a vector (velocity). The direction of the momentum is always the same as the direction of the velocity. The unit of momentum does not have a special name. The obvious derived unit to use is kg m s^{-1}, namely the unit of mass multiplied by the unit of velocity. It can be shown that this unit is exactly the same as the N s, and this is often used. The importance of the momentum of a body will be discussed in Chapters 5 and 7. At this stage it is useful to get some idea of the momentum of different objects.

EXAMPLE 4.5

Find the momentum of the Earth in its orbit around the Sun. The mass of the Earth is 6.0×10^{24} kg and the radius of its orbit is 1.5×10^{11} m.

The direction of the momentum will be in the direction of the Earth's velocity; this is at a tangent to its orbit. The Earth takes a year to rotate around the Sun, so the velocity v of the Earth in this direction is given by

$$v = \frac{\text{circumference of orbit}}{\text{period}} = \frac{2\pi \times 1.5 \times 10^{11}\,\text{m}}{365 \times 24 \times 60 \times 60\,\text{s}}$$

$$= 3.0 \times 10^{4}\ \text{m s}^{-1} = 30\ \text{km s}^{-1}$$

Therefore

momentum of Earth $= (6.0 \times 10^{24}\,\text{kg}) \times (3.0 \times 10^{4}\,\text{m s}^{-1})$
$= 1.8 \times 10^{29}\,\text{N s}$

QUESTIONS

4.14 Find the momentum of a car of mass 1200 kg travelling with a velocity of 40 m s^{-1} due east.

4.15 An ice-hockey puck has a mass of 0.40 kg and at one instant has a velocity of 30 m s^{-1}. After a time interval of 0.0030 s its velocity is 45 m s^{-1} in the opposite direction. Find the change in the momentum of the puck and its average acceleration during the time interval.

4.16 Find the momentum of an ion of mass 7.3×10^{-26} kg, travelling with a velocity of 4.2×10^{5} m s^{-1} in a direction parallel to a magnetic field.

4.17 By how much does the momentum of the Earth change over a period of

a six months and

b a year?

4.18 An animal of mass 50 kg changes its velocity from 3.0 m s^{-1} due north to 15 m s^{-1} due west in order to escape from a predator. What is the change in the momentum of the animal? [The answer is *not* 600 N s.]

SUMMARY

■ Velocity is the rate of change of displacement (unit m s^{-1}). It is a vector. The magnitude of the velocity is the speed. Speed is a scalar.

■ Acceleration is the rate of change of velocity (unit m s^{-2}). It is a vector.

■ Equations of motion:

– for constant velocity $v = x/t$

– for constant acceleration $v = u + at$

$$s = ut + \frac{1}{2}at^2$$

$$s = vt - \frac{1}{2}at^2$$

$$s = \frac{(u+v)t}{2}$$

$$v^2 = u^2 + 2as$$

■ The terminal velocity of a falling object is the constant velocity of fall when the object's weight is equal and opposite to the air resistance on it. It then continues to fall with zero acceleration.

■ Momentum is the product of the mass and the velocity of a body. It is a vector.

ASSESSMENT QUESTIONS

4.19 Select a method for measuring g, the gravitational field strength at the surface of the Earth. For your chosen method:
a sketch the apparatus you would use;
b state the measurements that you would need to make;
c state how you would make your measurements;
d show how you would use your measurements to determine the gravitational field strength.

AEB 1997

4.20 The graph shows the horizontal speed v of a long jumper from the start of his run to the time when he reached the take-off board.

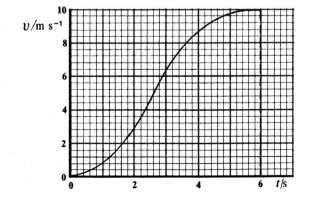

a Use the graph to estimate his maximum acceleration.
b Use the graph to estimate the distance of the 'run-up'.

London 1997

4.21 The diagram shows a mass attached by a piece of string to a glider which is free to glide along an air track.

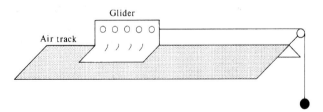

A student finds that the glider takes 1.13 s to move a distance of 90 cm starting from rest.
a Calculate the speed of the glider after 1.13 s.
b Calculate its average acceleration during this time.
c How would you test whether or not the acceleration of the glider is constant?

London 1997

4.22 Demonstrate that the following equation is homogeneous with respect to units. The symbols have their usual meanings.

$$x = ut - \tfrac{1}{2}at^2$$

The following graph shows the speed v of a body during a time interval of just over 3 seconds.

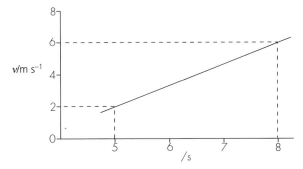

a Use the graph to determine the magnitude of the acceleration a.

b Find the distance travelled by the body between $t = 6$ s and $t = 8$ s.

London 1998

4.23 a The equations

$$s = \tfrac{1}{2}(u + v)t \quad \text{and}$$
$$v = u + at$$

frequently appear in the study of uniformly accelerated motion.

   i State the meanings of the symbols s and u.

   ii Use these equations to obtain an expression for a in terms of u, v and s only.

b A spanner of mass 0.40 kg is accidentally dropped by a worker from a bridge 80 m high. The spanner falls freely from rest. Calculate, neglecting air resistance,

   i the speed of the spanner as it reaches the ground,

   ii the time taken for the spanner to reach the ground,

   iii the maximum kinetic energy of the spanner.

c i On a copy of the figure, sketch a line to represent the variation with time of the displacement of an object falling freely from rest near the Earth's surface in a vacuum. Label this line F.

   ii Also on your figure, sketch a line to represent the variation with time of the displacement of the object when it is subject to air resistance. Label this line R.

Cambridge 1997

4.24 This question is about the bounce of a ball.

a A ball is released from rest at a height of 0.90 m above a horizontal surface. Find its speed as it reaches the surface.

b The effect of the bounce is to reduce the speed of the ball to two-thirds of the value in part **a**. Find

   i the change in speed in the impact,

   ii the change in *velocity* in the impact.

c Plot a graph of the *velocity* of the ball against time from the moment of its release until it reaches the maximum height after the first bounce. The ball takes 0.43 s to reach the surface. Assume that the bounce takes a negligible time. Label your axes fully and carefully. Show all your calculations.

O & C 1997

4.25 The diagram shows a toy truck, about 30 cm long, accelerating freely down a gentle incline.

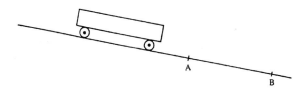

a Explain carefully how you would measure the average speed with which the truck passes the point A.

b You find that the measured average speed of the truck is 1.52 m s^{-1} when it passes the point A and 1.64 m s^{-1} when it passes the point B. The distance from A to B is 1.20 m. Calculate the acceleration of the truck.

London 1999

4.26 A steel ball bearing is released from just beneath the surface of a clear liquid in a tube. As the ball bearing passes certain marks, the time since release is recorded. The results obtained are shown in the table.

distance fallen/m	0.00	0.10	0.20	0.40	0.60	0.80	1.00
time/s	0.00	1.80	2.60	3.90	5.10	6.30	7.50

a Plot a graph of distance fallen (vertical axis) against time (horizontal axis).

b Using information from the graph which you have plotted, calculate the terminal speed of the ball bearing.

c With reference to the forces acting, explain why the ball bearing achieves a terminal speed.

NEAB 1998

4.27 An aeroplane is flying horizontally at a steady speed of 67 m s^{-1}. A parachutist falls from the aeroplane and falls freely for 80 m before the parachute opens. For the purposes of calculation, you may assume that air resistance is negligible before the parachute opens.

a i Show that the vertical component of the velocity is approximately 40 m s^{-1} when the parachutist has fallen 80 m. The acceleration of free fall g is 9.8 m s^{-2}.

ii Determine the magnitude and direction of the resultant velocity of the parachutist at this point.

iii State and explain how the magnitudes of the horizontal and vertical components of the parachutist's velocity are actually affected by air resistance before the parachute opens.

b The parachutist lands with a vertical velocity of 7.0 m s^{-1} and no horizontal velocity. The parachutist, of mass 85 kg, lands in one single movement, taking 0.25 s to come to rest.

i Calculate the average retarding force on the parachutist during the landing.

ii Explain how the parachutist's loss of momentum on landing is consistent with the principle of conservation of momentum.

AEB 1998

4.28 a The table of results is taken from an experiment where the time taken for a ball to fall vertically through different distances is measured.

distance fallen/m	1.0	2.0	4.0	6.0	8.0	10.0
time/s		0.45	0.63	0.90	1.10	1.30	1.40

i Plot a graph of distance fallen on the y-axis against time on the x-axis.

ii Explain why your graph is not a straight line.

iii Calculate the gradient of the graph at 0.70 s.

iv State what your gradient represents and use it to determine a value for the acceleration of free fall, assuming that the ball started from rest.

b A falling body experiences air resistance and this can result in it reaching a terminal speed. Explain, using Newton's laws of motion, why a terminal speed is reached.

NEAB 1997

4.29 The diagram shows a water-skier being pulled at a steady speed in a straight line. Her mass plus the mass of the ski is 65 kg. The pull of the tow-rope on her is 520 N.

a i What is the vertical component Y of the push of the water on the ski?

ii What is the horizontal component X of the push of the water on the ski? (Ignore air resistance.)

iii Component X and the 520 N towing force form a clockwise couple acting on the water skier. Explain how she can remain in equilibrium as she is towed along.

b She suddenly lets go of the tow-rope. Calculate her initial deceleration. Why does her deceleration reduce as she slows down?

c On another occasion while being towed, she moves in a curved path from behind the boat to approach a ramp from which she makes a jump, remaining in the air for over two seconds.

i Explain why the pull of the tow-rope on her is greater as she moves in the curved path than when she is being towed in a straight line.

ii Explain why she feels 'weightless' while in the air during her jump.

d The speedboat pulling the water skier produces waves which travel away from the boat. Those with a wavelength of over a metre travel faster than those with a wavelength of less than a quarter of a metre.

The waves reach and pass through a gap of two metres leading into a boatyard. Draw a diagram to show their appearance soon after the speedboat passes. Label your diagram carefully.

London 1996

4.30 The diagram shows a velocity-time graph for a ball bouncing vertically on a hard surface.

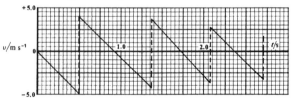

a At what instant does the graph show the ball to be in contact with the ground for the third time?

b The downwards-sloping lines on the graph are straight. Why are they straight?

c Calculate the height from which the ball is dropped.

d Sketch a displacement–time curve for the first second of the motion.

e What is the displacement of the ball when it finally comes to rest?

London 1996

4.31 A rubber ball of mass 0.120 kg is dropped from a height of 2.00 m (measured from the bottom of the ball) on to a flat horizontal patch of hard soil.

   a Calculate the speed of the ball when it hits the ground.

   b The rubber ball loses speed each time it bounces. The following graph shows how the height of the bottom of the ball varies with time during the first second of its motion.

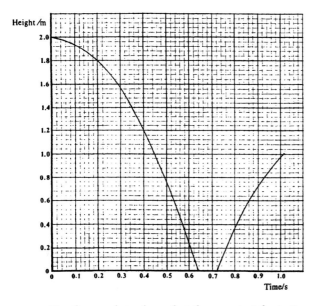

Height /m

Time/s

   Use the graph to show that the speed of the ball as it leaves the ground is approximately 4.7 m s⁻¹.

   c Calculate the average force exerted by the ground on the ball while it is in contact with the ground.

   London 1995

4.32 **a** Water flows from a nozzle with an initial velocity of 5.8 m s⁻¹ at an angle of 45° to the horizontal, as shown in the diagram.

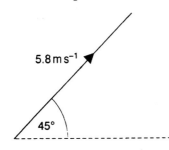

5.8 m s⁻¹

45°

   For this velocity, use a vector triangle or calculation to

   i show that the horizontal component is 4.1 m s⁻¹,

   ii explain why the vertical component has the same magnitude as the horizontal component.

   b The nozzle is part of a sprinkler used to water a lawn. The sprinkler is at ground level and when the water leaves the nozzle at an angle of 45° to the horizontal, the path of the jet of water is as shown in the diagram (not to scale).

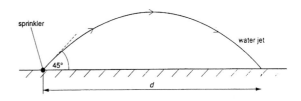

sprinkler

water jet

45°

d

   The initial speed of the water is 5.8 m s⁻¹ and air resistance may be assumed to be negligible. Using the information in **a**, calculate, for one drop of water in the jet,

   i the time taken to reach its maximum height,

   ii the total time between leaving the nozzle and hitting the ground,

   iii the horizontal distance *d* from the nozzle to the point of impact with the ground.

   c Copy the diagram in part **b**. On it, sketch paths of the jet of water when the water emerges from the nozzle at a velocity of 5.8 m s⁻¹ and at an angle to the horizontal of

   i 60° (label this path H),

   ii 30° (label this path L).

   d Suggest the angle to the horizontal at which the nozzle on the sprinkler will give the maximum value of *d* (see diagram).

   Cambridge 1999

4.33 A catapult fires an 80 g stone horizontally. The graph shows how the force on the stone varies with distance through which the stone is being accelerated horizontally from rest.

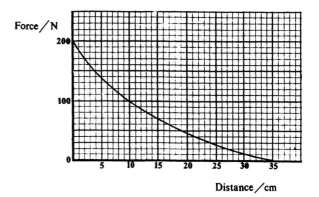

Force / N

Distance / cm

   a Use the graph to estimate the work done on the stone by the catapult.

   b Calculate the speed with which the stone leaves the catapult.

   London 1995

5
NEWTON'S LAWS
OF MOTION

LEARNING OBJECTIVES

At the end of this chapter you should be able to:

① quote and use Newton's three laws of motion;

② draw and use free-body force diagrams;

③ distinguish between mass and weight.

5.1 INTRODUCTION

Newton's laws form a principal part of the foundation stones of classical physics. An understanding of their meaning is vitally important for all students of physics, and you are well advised to spend time giving considerable thought to this chapter. Gaining the necessary understanding will come only with effort and with the application of the laws to practical examples.

Newton published his laws in 1687, and they have transformed science and engineering in the three hundred years since then. Einstein's theory of relativity has modified the laws so that they are now regarded as an approximation to the theory of relativity. But that approximation is so precise that in practice it is only necessary to consider differences from Newton's laws when speeds approaching the speed of light are being used.

To simplify things at the start, assume that any object being considered

■ has a constant mass,
■ is not rotating, and
■ is in surroundings that are not accelerating.

5.2 NEWTON'S FIRST LAW

I n formal language this law, translated from Newton's original Latin, states that:

> **K** Every object continues in its state of rest or uniform motion in a straight line unless it is compelled to change that state by external forces acting upon it.

Another way of putting this would be to say that any object has a constant velocity unless it is acted upon by a resultant external force. This effectively defines **force**: a force is necessary to accelerate an object; or, a force is that influence which, acting alone, can cause an object to be accelerated.

There are practical situations where it is easy to apply Newton's first law, but there are other occasions where the law seems to contradict common sense and these often cause difficulty.

Consider the horizontal forces of **drive** and **drag** acting on a car travelling at a constant velocity of 30 m s^{-1} (about 70 miles per hour) – see **Fig 5.1a**. The total force on the car really is zero. That is, the force that drives the car forward exactly equals the drag force caused by friction and wind resistance. If the two forces are not equal in size and opposite in direction, say the drive force is larger than the drag force, then the car will accelerate – see **Fig 5.1b**. If drag is greater than drive, then the car will decelerate – see **Fig 5.1c**. Use is made of this in braking, when the drag force is increased considerably. It is only when drive and drag are equal in size that the car has a constant velocity.

a   drive   drag   **b**   drive   drag   **c**   drive   drag

FIG 5.1 **a** A car will have a constant velocity if the drive force on the car is equal and opposite to the total drag force on the car – the drag force includes road drag and wind resistance. **b** A car will accelerate if the drive force is larger than the drag force. **c** A car will decelerate if the drag force is larger than the drive force.

The same analysis can be carried out for a person travelling inside a vehicle. Consider a woman travelling at a constant velocity of 300 m s^{-1} in an aeroplane. The seat on which she is sitting supports her with a force equal and opposite to her weight. She needs **nothing** to move her forward. There is **no** forward force: just two forces balancing to give **zero** total force and therefore no acceleration. She did need a forward force on her when accelerating on the runway. That was when she was conscious of the seat pushing her, but in mid-flight it is just like sitting on a chair in the garden (apart from any vibration from the engines shaking the aeroplane, vibration that designers of vehicles of all sorts go to great lengths to try to reduce).

That is why, if a steward pours her a drink, there is no need to hold the glass some way in front of the bottle to allow for the movement of the aeroplane. It is also one reason why drinks are not served during take-off!

Newton realised that there is no difference at all between the forces acting on an object travelling at a constant velocity and those on a stationary object. This was developed by Einstein as one principle of his theory of relativity. He stated that the laws of physics take exactly the same form for all systems that have a constant velocity relative to one another. **Fig 5.2** shows the application of Newton's first law to some practical situations.

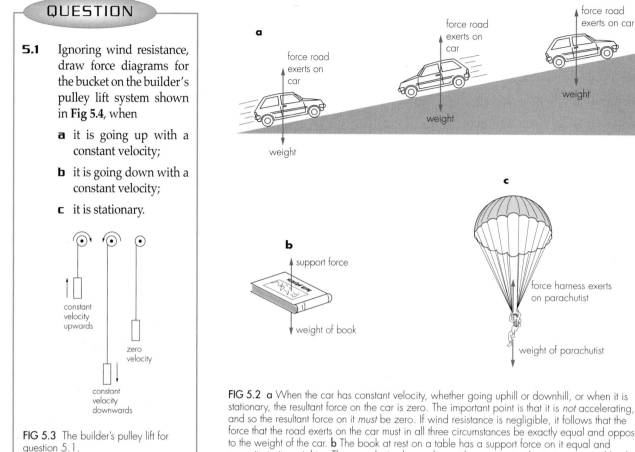

5.1 Ignoring wind resistance, draw force diagrams for the bucket on the builder's pulley lift system shown in **Fig 5.4**, when

a it is going up with a constant velocity;

b it is going down with a constant velocity;

c it is stationary.

constant
velocity
upwards

zero
velocity

constant
velocity
downwards

FIG 5.3 The builder's pulley lift for question 5.1.

FIG 5.2 a When the car has constant velocity, whether going uphill or downhill, or when it is stationary, the resultant force on the car is zero. The important point is that it is *not* accelerating, and so the resultant force on it *must* be zero. If wind resistance is negligible, it follows that the force that the road exerts on the car must in all three circumstances be exactly equal and opposite to the weight of the car. **b** The book at rest on a table has a support force on it equal and opposite to its weight. **c** The parachutist descending with a constant velocity is supported by the harness. This support force is equal and opposite to the parachutist's weight.

5.3 NEWTON'S SECOND LAW

The first law defined force. If a fixed force is applied to two different masses, one large and one small, they will not have the same acceleration. The second law deals with the size of the effect that a force has on an object. In formal language the second law states that:

K **The rate of change of momentum of a body is proportional to the total force acting on it and occurs in the direction of the force.**

Consider an object of constant mass m being pushed by a constant force F so that its velocity increases from u to v in time t. Newton's second law gives

$$\frac{\text{change in momentum}}{\text{time}} \propto \text{force}$$

$$\frac{mv \pm mu}{t} \propto F$$

$$\frac{m(v \pm u)}{t} \propto F$$

$$m \times a \propto F$$

where a is the acceleration. As an equation, this becomes

$$F = k \times m \times a$$

where k is a constant, which has no unit but whose size depends on the unit to be chosen for force. This now needs to be done:

K One *newton* (1 N) is the force that causes a mass of one kilogram (1 kg) to have an acceleration of one metre per second squared (1 m s⁻²).

Substitution into the above equation gives $1 \text{ N} = k \times 1 \text{ kg} \times 1 \text{ m s}^{-2}$, and so we must have $k = 1$ for this system of units. An equation that must be thoroughly learnt therefore is

K force in newtons $=$ mass in kilograms $\times$ acceleration in metres per second squared

As a direct result of the way the newton has been defined, k has been made equal to 1. This makes use of the equation straightforward, but it must be remembered that the simplified equation is valid *only* if the correct units are used. It is a frequent cause of mistakes to substitute values with incorrect units into the above equation. The equation must be used with care in several respects:

- The units used must always be newtons (N), kilograms (kg) and metres per second squared (m s⁻²).
- The force must always be the total force acting on the object. If there are several separate forces acting on an object, then they must be added together to find the total force. The total force is the *vector* sum of all the forces acting on an object, and is often referred to as the **resultant** force.
- Both the total force and the acceleration are vectors, and these are *always and without exception* in the same direction. This is not necessarily the direction of the velocities. Velocity and force are often in different directions, but acceleration and force are never in different directions. If you have the solution to a problem in which the force and the acceleration are not in the same direction, then something has gone wrong!
- It is important to realise that acceleration is caused by force (**Fig 5.4**). Accelerations do not cause forces. Cause and effect frequently occur in physics. In this case force is the cause and acceleration is the effect.

The equation given above is a mathematical summary of both of Newton's first and second laws of motion for a body of constant mass. Besides giving the acceleration of an object for a known force, it also shows that, if the force is zero, then the acceleration will also be zero.

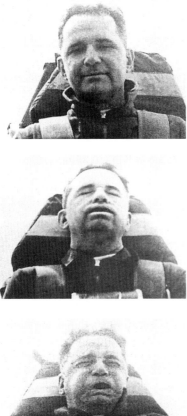

FIG 5.4 An astronaut undergoing increasingly high acceleration. All parts of the astronaut's face must have large forces on them to cause their acceleration. Note that it is not the acceleration that causes these forces – rather, the forces cause the acceleration.

INVESTIGATION

The arrangement shown in **Fig 5.5** is often suggested as an experiment to verify Newton's second law. The acceleration is usually found by using a ticker-timer and ticker-tape attached to the trolley.

1 Set up the apparatus so that the acceleration a of the trolley may be found.

2 Vary the mass m of the hanging mass so that a graph may be plotted of a against m. In doing this m should vary from being much less than M to a value greater than M.

3 Explain why your graph is not a straight line.

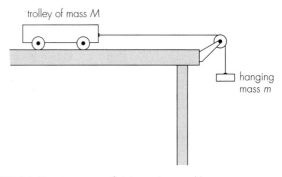

FIG 5.5 Experiment to verify Newton's second law.

5.4 NEWTON'S THIRD LAW

This law is a measure of the genius of Newton. Since the days of Newton, no one has ever found any situation in which the law does not apply. Much of Einstein's theory of relativity is based upon it, and the law of conservation of momentum can be proved from it. In recent years it has become usual to write the law in a different form from that which Newton himself used, because in its original form it was often misunderstood. Formally it is now written as:

 If body A exerts a force on body B, then body B exerts an opposite force of the same size on body A.

The two forces must also be of the same type. That is, if one is an electrical force, then the other must be electrical too. Note that the two forces are acting on two different bodies: here, one of the forces acts on A and the other on B.

Newton's third law applies in *all* circumstances. The two bodies may be enormous stars or minute parts of atoms; one may be huge and the other may be tiny. The bodies may be stationary or moving or accelerating. In particular, it is important to realise that it is not necessary to have equilibrium in order for the law to be true. Newton's third law is *always* true, not just when things are nicely balanced. When a dart enters a dart-board, then at all instants, the force the dart exerts on the dart-board is equal and opposite to the force the dart-board exerts on the dart.

Statements like this can be made whenever forces exist. Newton realised that force is a mutual influence between two objects, and that it is impossible to have a single force. Forces always exist in pairs. In order to become used to the ideas associated with this law, it is recommended that clear force diagrams are always made and that the forces are clearly labelled to show what object is exerting the force. Care needs to be taken to draw on any diagram the correct force out of the pair of forces being considered. If the wrong one is used, then it will be pointing in exactly the wrong direction. For instance, in the case of the dart hitting the dart-board, the force the dart exerts on the dart-board is a forward force; this force would only be shown on a diagram showing the forces on the dart-board. The force the dart-board exerts on the dart is a backward force, which causes the deceleration of the dart; this force would only be shown on a force diagram for the dart.

In **Fig 5.6** one pair of forces is shown. One force acts on the Earth and one force acts on the Moon. The gravitational force that the Moon exerts on the Earth is shown on the force diagram of the Earth; and the equal and opposite force that the Earth exerts on the Moon is shown on the force diagram of the Moon.

FIG 5.6 A pair of gravitational forces, which, according to Newton's third law of motion, must always be equal and opposite.

5.5 TYPES OF FORCE

Some words used in physics have a meaning that is more closely defined than the word's meaning in everyday use. There is a multitude of words in the English language that represent force. Some examples of these words are: push, pull, hit, tension, knock, shove, effort, load, pressure, strength, power, and vigour. In science it is essential to be careful in the use of words, so that when a word is used its meaning is clear. For instance, in the above list, some of the terms are scientifically inaccurate. Power means work done per unit time; it does not mean force. Pressure means force per unit area; this is different from force. Other words in the list are simply descriptions of particular situations where forces occur; tension, effort and load come into this category.

When dealing with types of force, however, we find, surprisingly, that outside the nucleus of atoms there are only two possible types of force, which are:

- electromagnetic force,
- gravitational force.

Electromagnetic forces exist between moving or stationary charges. Since all atoms have charged particles within them, it is electromagnetic forces that bind atoms together in solids and liquids. On some occasions the electrical nature of a force is important, sometimes the magnetic nature is important. In these cases there is not usually any problem in pin-pointing where the force exists. In the vast majority of mechanics problems, however, it is the electromagnetic forces between atoms that are of prime importance. Whenever the atoms of one object are close to the atoms of another object, there will be a contact force between them. *All* forces of contact are electromagnetic forces. In the list given above, push, pull, hit, knock, shove, effort and load are all examples of contact forces. Tension is also a contact force, but is used in a rather special way involving internal electromagnetic forces between atoms in a string as well as the contact force between the string and the object to which it is attached.

Gravitational forces exist between any two masses and can usually be neglected unless one of those masses is very large. The gravitational force that a car exerts on a caravan is negligible; the electromagnetic force of contact that the car exerts on the caravan is the force that pulls the caravan along. In practice, the only gravitational force that usually concerns us is the gravitational attraction of the Earth.

5.6 WEIGHT

One of the results of the work of Newton was the introduction of the concept of gravity. The application of Newton's third law to gravitation is dealt with in more detail later, but the link between gravity and weight needs to be considered here. Gravity is a very mysterious force. How one mass exerts a force on another when there is no contact between them and nothing in the space between them is the mystery. Because gravity is such a familiar force, its strangeness is often overlooked. What can be stated here by the application of the third law is that, as always, forces exist in pairs, so that the gravitational force the Earth exerts on you is equal and opposite to the gravitational force you exert on the Earth. In other words, because both you and the Earth have mass, you and the Earth are pulled together by gravitational attraction.

Weight is the word that is often used for the gravitational force that the Earth exerts on an object. Further consideration will be given to the term in section 9.5. The relationship between the weight of an object and its mass becomes much clearer if it is realised that weight is always a force and will therefore always be measured in newtons. Mass is always measured in kilograms.

Consider a ball of mass m falling freely towards the Earth, being pulled down by its weight w but with no other force acting on it (**Fig 5.7**). Using Newton's second law in this situation gives

FIG 5.7 A ball in free fall has only its own weight acting on it.

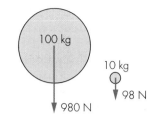

FIG 5.8 A large force of gravity on an object of large mass gives rise to the same acceleration as a small force of gravity on an object of small mass.

force (N) = mass (kg) × acceleration (m s^{-2})

$$w = m \times g$$

where g is the acceleration of free fall of the ball. (Note that g is *not* gravity.) If g is measured experimentally, its value is found to be approximately 9.8 m s^{-2}. The above equation can therefore also be written as

$$\frac{w}{m} = 9.8 \text{ newtons per kilogram} = 9.8 \text{ N kg}^{-1}$$

Fig 5.8 shows two masses in free fall; each has 9.8 N of gravitational force acting on it per kilogram of mass. The large mass has a large force acting on it, and this gives it the same acceleration as the smaller force acting on the smaller mass.

EXAMPLE   5.1

The gravitational force acting on a golf ball on the Earth is 0.430 N. If it were on the Moon, the same golf ball would have a gravitational force on it of only 0.0725 N. Find the mass of the golf ball and the acceleration of free fall on the Moon.

Consider the ball falling freely on the Earth:

force (N) = mass (kg) × acceleration (m s^{-2})
0.430 N = m × 9.8 m s^{-2}
m = (0.430 N)/(9.8 m s^{-2}) = 0.0439 kg

When the ball is on the Moon its mass is still 0.0439 kg, so now:

force (N) = mass (kg) × acceleration (m s^{-2})
0.0725 N = 0.439 kg × a
a = (0.0725 N)/(0.0439 kg) = 1.65 m s^{-2}

So the acceleration of free fall on the Moon is 1.65 m s^{-2}.

5.7   FRICTION

Consider a stone at rest on a road. If the contact between the road and the stone is examined closely, it can be seen that the two rough surfaces make close contact only at relatively few places. Where contact is made, the road will exert a force on the stone as shown in **Fig 5.9a**. The sum of all these forces is shown using a different scale in **Fig 5.9b**, and this single force is the contact force that the road exerts on the stone.

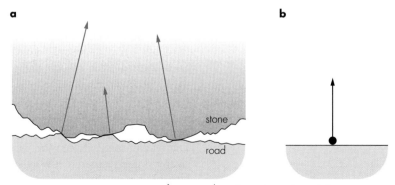

a

b

stone

road

forces road exerts on stone: stone stationary

FIG 5.9 a A stationary stone on a flat road has various forces exerted on it by the road at points of contact. **b** The resultant of these forces is vertically upwards.

The word 'reaction'

Note that the 'normal contact force' (**Fig 5.10**) is sometimes called the 'normal reaction', but this latter term is avoided here because in the old form of Newton's third law the word 'reaction' was also used. Sometimes these two terms were the same thing, and sometimes they were not. Much confusion was caused.

Problem-solving

In solving problems concerned with Newton's laws, the following routine is strongly recommended:

1 State clearly which object is being considered.
2 Draw a free-body sketch of that object only.
3 Mark on the sketch the gravitational pull on the object, its weight.
4 Mark on the sketch all the points where the object touches anything else, and draw in the contact forces at these points. Label all forces clearly.
5 Decide which direction to call positive for the total force and acceleration.
6 Apply Newton's second law equation.

If this routine is followed, you will find that it can be used to solve all problems. You will not need to use a different approach when more complicated situations arise, and you will be far less likely to make unnecessary mistakes. In use, the routine is direct, reliable and quick. Supposed short-cuts in solving such problems lead to many mistakes being made.

The essential need is for complete free-body force diagrams rather than composite pictures of more than one object.

If, however, the stone happens to be sliding across the road, to the left say, **Fig 5.9** changes to **Fig 5.10**. Now there are rather more forces in the direction opposite to the direction of travel than in the forward direction – **Fig 5.10a**. This results in the contact force that the road exerts on the stone being tilted as shown in **Fig 5.10b**. The tilted force can be considered as being the vector sum of a horizontal and a vertical component. The horizontal component of this force is called the **frictional force**, and the vertical component is called the **normal contact force**. For two surfaces where friction can be neglected, the only force will be the normal contact force.

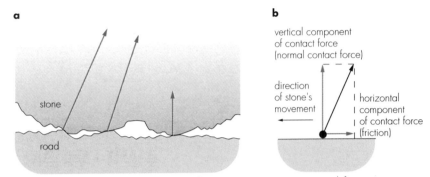

forces road exerts on stone: stone moving to left

FIG 5.10 **a** When the stone moves to the left, the forces that the road exerts on it are likely to be both upwards and to the right. **b** The resultant of all these forces is tilted to the right. It has components vertically upwards and horizontally to the right.

Friction itself is usually a component of a contact force. The distinguishing feature of friction is that it is in the opposite direction to the direction of motion. Friction is often regarded as a nuisance, but it is a force essential to everyday life. There are times when friction needs to be minimised, but without friction everything would collapse – literally. If two surfaces could exert forces on each other only at right-angles to the surface, i.e. if there was no friction, then there would be no nails, no nuts and bolts, no glue, no screws, no sewing, no knitting, no fabrics, no buildings, no cars, no people. The list is virtually endless. Friction is an extremely useful force; walking would be impossible without it. Chaos usually follows even the reduction in friction that takes place on roads during a snowfall (**Fig 5.11**)

FIG 5.11 Less friction than usual causes many problems!

EXAMPLE 5.2

A digger is exerting an upward force of 10 000 N on a load of mass 800 kg in its bucket. Find the acceleration of the load.

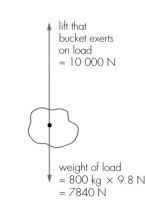

lift that
bucket exerts
on load
= 10 000 N

weight of load
= 800 kg × 9.8 N
= 7840 N

FIG 5.12 The free-body diagram for the digger's load.

Following the numbered sequence suggested on p. 59:

1 Consider the forces acting on the load.
2 Sketch the load, as in **Fig 5.12**.
3 Add the weight of the load to the free-body diagram.
4 Add the force the digger exerts on the load.
5 Consider the upward direction as positive.
6 Use the equation

force (N) = mass (kg) × acceleration (m s^{-2})

So

$10\ 000\ \text{N} - 7840\ \text{N} = 800\ \text{kg} \times a$
$2160\ \text{N} = 800\ \text{kg} \times a$
$a = (2160\ \text{N})/(800\ \text{kg}) = 2.7\ \text{m s}^{-2}$

This problem is rather trivial, but it is amazing how often the working is done in such a way as to give, for this question, an answer of 12.5 m s^{-2}.

FIG 5.13 A helicopter has to be able to control its vertical height and its horizontal velocity and acceleration. It does this by altering its 'attitude'. If it is stopping, its nose is lifted up so that the rotor forces air downwards and forwards. The force this air exerts on the rotor is therefore upwards and backwards.

Q **What method is used to obtain the incorrect answer of 12.5 m s^{-2}? Why are the method and answer incorrect?**

QUESTIONS

5.2 A train has an acceleration of 0.53 m s^{-2} and a mass of 250 t (1 t = 1 tonne = 1000 kg). Calculate the force required to cause this acceleration.

5.3 An electron, mass 9.11 × 10^{-31} kg, has a force of 2.80 × 10^{-13} N acting on it when in a cathode ray tube. Calculate its acceleration.

5.4 A helicopter of mass 5000 kg (**Fig 5.13**) is rising with an acceleration of 2.4 m s^{-2}. What thrust must be exerted on it by its rotor?

5.5 An aeroplane of total mass 30 000 kg needs to be able to take off within a distance of 1200 m on a runway that is 2200 m long. The extra distance is necessary for safety reasons. If the take-off and landing speeds of the plane are both 60 m s^{-1}, find:

   a the constant acceleration needed to reach take-off speed in 1200 m;

   b the resultant force on the aeroplane during this acceleration;

   c the thrust provided by the engines if the drag due to ground and wind resistance is taken to have a constant value of 10 000 N.

   d On landing, the mass of the aeroplane has decreased to 25 000 kg. (Why?) What minimum force is necessary to stop it on the runway?

5.6 A lift has a mass of 1800 kg and the rope supporting it exerts an upward force on it of 15 000 N. What is the acceleration of the lift? If this lift is travelling *upwards* at a velocity of 3.8 m s^{-1}, how long will it take to stop?

[Note that, when the acceleration in question 5.6 is found to be downwards, there is a tendency to assume that the lift is travelling downwards. Although this might be the case, it is also possible, as here, for the lift to be decelerating while moving upwards. The total force on the lift and the acceleration of the lift *must* be in the same direction but the acceleration is *not* necessarily in the direction of the velocity.]

EXAMPLE 5.3

A man of mass 100 kg gets into the lift in question 5.6. If the rope now exerts a force upwards on the lift of 20 000 N, find:

a the force the lift exerts on the man;

b the acceleration of lift and man.

This is a two-part problem: *two* force diagrams are therefore needed. Consider the forces acting first on the man and secondly on the lift (**Figs 5.14 and 5.15**). This is a good example of the use of Newton's third law. *F* is the force that the lift exerts on the man. It is equal and opposite to the force that the man exerts on the lift and is therefore also labelled *F*. Consider the upward direction as positive.

For the man (**Fig 5.14**) we have

$$\text{force on man} = \text{mass of man} \times \text{acceleration of man}$$
$$(F - w) = m \times a$$
$$F - 980\,\text{N} = 100\,\text{kg} \times a$$

For the lift (**Fig 5.15**) we have

$$\text{force on lift} = \text{mass of lift} \times \text{acceleration of lift}$$
$$(T - W - F) = M \times a$$
$$20\,000\,\text{N} - 17\,640\,\text{N} - F = 1800\,\text{kg} \times a$$
$$2360\,\text{N} - F = 1800\,\text{kg} \times a$$

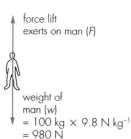

FIG 5.14 The forces acting on the man.

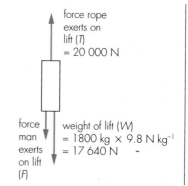

FIG 5.15 The forces acting on the lift.

This gives two equations with two unknowns, *F* and *a*, and they can be solved by adding them to give:

$$F - 980\,\text{N} + 2360\,\text{N} - F = 100\,\text{kg} \times a + 1800\,\text{kg} \times a$$
$$1380\,\text{N} = 1900\,\text{kg} \times a$$
$$a = (1380\,\text{N})/(1900\,\text{kg})$$
$$= 0.73\,\text{m s}^{-2}$$

Substituting this back gives

$$F - 980\,\text{N} = 100\,\text{kg} \times 0.73\,\text{m s}^{-2}$$
$$F = 73\,\text{N} + 980\,\text{N} = 1050\,\text{N} \text{ (to 3 sig. figs)}$$

So (a) the force the lift exerts on the man is 1050 N and (b) the acceleration of the lift and the man, which must be the same since they always have the same velocity, is 0.73 m s⁻².

The upward force the lift exerts on the man is greater than his weight, so he accelerates upwards. It is this larger than usual force acting upwards which is the force that we find strange when a lift starts off. Note that the weight of the man is still 980 N, although if he were to be in the unlikely position of standing on some scales while in the lift, the scales would read 1050 N.

Motive force

The motive force for a vehicle is the driving force and is provided by the engine of a car (or by a cyclist on a bicycle). Any self-propelled wheeled vehicle needs to have driving wheels that make good contact with the ground under all circumstances. The higher the frictional force, the better, as there is then less likelihood of the wheels skidding when a large driving force is required. The application of a torque to the driving wheels' axle causes the rubber of the tyre to push backwards on the road surface. Newton's third law applied to this situation states that the force that the tyre exerts backwards on the road is equal to the motive force that the road exerts forwards on the tyre. This is shown in **Fig 5.16a**.

It is common today for the driving wheels on a car to be the front wheels. Some very sophisticated engineering is needed for the control of steering, driving and braking all on the front wheels. Front wheel tyres normally wear down considerably more quickly than rear wheel tyres.

Government regulations

In order to reduce the number of traffic accidents, there are strict government-imposed regulations on:

1 *drivers* – concerning the amount of wear on a tyre tread and on brake pads;

2 *tyre manufacturers* – concerning the construction of tyres and the type of rubber that they can use in them;

3 *road builders* – concerning the material used as the road surface, and the slope on the road necessary to drain rain-water off the surface;

4 *brake manufacturers* – concerning the strength of the brake itself, and the material used for brake pads; brake pads have to be able to withstand high temperatures and used to be made of asbestos, but this is now banned in brake pads as it is carcinogenic.

Braking force

Most cars have braking systems in which a disc fixed to each wheel rotates freely until braking is required. Then each disc is squeezed between two fixed brake pads, and this causes a torque to be applied to the wheel, giving rise to a braking force on the wheel, as shown in **Fig 5.16b**. Since there is seldom any occasion when both the anticlockwise and clockwise torques shown in **Fig 5.16** need to be applied together, the driver slows the car by taking the right foot off the accelerator pedal and applying it to the foot brake. By use of both the accelerator and foot brake, a driver can vary the size and direction of the force the road exerts on the wheels.

The maximum possible size of both the motive force and the braking force on any one wheel are equal. They are both equal to the maximum possible frictional force between the road and the tyre. In practice, these maxima may not be reached, as they depend on the power output of the engine and the efficiency of the braking system. All cars have brakes on all four wheels, but motive power is only usually provided at two wheels.

Q *Why do brake discs sometimes glow red hot when used?*

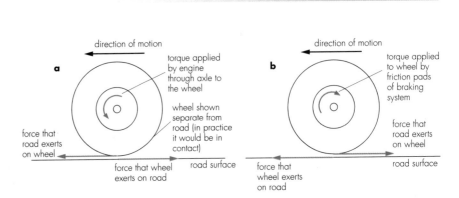

FIG 5.16a A car driving wheel showing the torque applied and the forces acting when the wheel is being driven.

FIG 5.16b The same arrangement but with the wheel being braked.

5.9　MOVEMENT OF OR THROUGH FLUIDS

a

b

FIG 5.17 **a** Laminar flow of a fluid through a pipe. **b** Turbulent flow of a fluid through a pipe

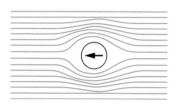

a

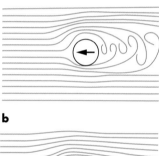

b

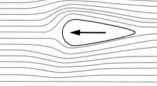

c

FIG 5.18 The speed and shape of an object moving through a fluid determine the type of flow

Turbulent and laminar flow

When a fluid, that is a liquid or a gas, flows through a pipe, the flow may be smooth or turbulent, depending on the internal diameter of the pipe, the speed of flow and the type of fluid. A smooth flow is called laminar flow and is shown in **Fig 5.17a**. Turbulent flow will always result in a lot of eddies in the fluid. It is shown in **Fig 5.17b**. Diagrams are difficult to draw for turbulent flow as individual fluid particles are moving in three dimensions. Laminar flow takes place normally for slow fluid speeds and changes to turbulent flow suddenly as the speed increases. The reverse process takes place if a solid object moves through a fluid. This is shown in **Figs 5.18a, b** and **c**. In **a** a sphere moves slowly through water and the flow of the water around the sphere is laminar. In **b** the speed of the sphere is increased and turbulent flow takes place. **Fig 5.18c** illustrates the fact that the shape of the object has a considerable influence on whether or not the flow is laminar. The speed of the object through the water is the same for **b** and **c**, but although water is turbulent in going past the sphere it is still laminar past the streamline shape.

Drag forces

The streamline shape of **Fig 5.18c** is the shape of many fish and boats. Fish have evolved into this shape because the drag force for laminar flow is considerably less than for turbulent flow. Submarines and many large ships with bulbous bows beneath the water-line copy this shape and hence reduce the drag force on them.

In order to reduce fuel consumption, car and aircraft manufacturers carry out many wind-tunnel tests and computer simulations to achieve a shape of vehicle with low drag. It is not possible, in practice, to design a car or an aircraft that moves through the air at all speeds with laminar flow of air around the vehicle. Vehicles are designed with a streamline shape and with the few parts that stick out into the air flow, wing mirrors on cars for example, also designed to be streamlined. This has the effect of giving increased laminar flow, particularly at low speed. The drag force F is proportional to the square of the speed of a vehicle and is given by the equation

$$F = \tfrac{1}{2}AC_D v^2$$

where A is the frontal area of the vehicle, v is its speed and C_D is called the drag coefficient. Car manufacturers and salesmen are very keen to emphasise a low drag coefficient when outlining the virtues of a particular mode of car.

EXAMPLE 5.4

A car of mass 850 kg has a frontal area of 2.3 m² and a drag coefficient of 0.34. These three horizontal forces are acting on the car when it is travelling at a speed of 36 m s⁻¹:

- a motive force of 950 N;
- a rolling drag force on the non-driving wheels of 86 N;
- a drag force due to wind resistance, as given by the drag force equation

Calculate the horizontal acceleration of the car.

The drag force F is a backwards force given by

$$\begin{aligned}F &= \tfrac{1}{2}AC_D v^2\\ &= \tfrac{1}{2}\times 2.3\ \mathrm{m^2}\times 0.34\ \mathrm{kg\ m^{-3}}\times (36\ \mathrm{m\ s^{-1}})^2\\ &= 507\ \mathrm{N}\end{aligned}$$

total backwards force = 507 N + 86 N = 593 N
total forward force = 950 N
resultant horizontal force on car = 950 N – 593 N
= 357 N

acceleration of car = force/mass
= (357 N)/(850 kg)
= 0.42 m s⁻² forwards

ANALYSIS

Passenger loading of a European Airbus

Use the data below concerning a European Airbus (which is shown in **Fig 5.19**) to answer the questions that follow.

Mass of aeroplane, including crew and all equipment	42 000 kg
Capacity of fuel tanks	18 000 kg
Maximum number of passengers	150
Average mass of a passenger and baggage	100 kg
Average use of fuel	5.0 kg km^{-1}
Safety reserve of fuel at end of journey	3000 kg
Take-off speed	75 m s^{-2}
Length of runway used	1500 m

a What is the safe range of the aeroplane?

b How much further could it travel, if, at the end of its scheduled flight, the airport it intended to land at was closed by poor weather conditions?

c What is the maximum total mass of the aeroplane, passengers and fuel at the start of a flight?

d When the plane is taking off, what is its acceleration (assumed constant) while on the runway?

e What force is necessary to cause this acceleration?

f Runways are always longer than is needed for take-off. What braking force would be needed if, just before leaving the ground, the pilot realised that something had gone wrong and that he needed to stop in a further 1200 m?

Sometimes aeroplanes find it necessary to take off from shorter than normal runways. They then have to reduce the number of passengers they carry or the amount of fuel carried. (They are not allowed to reduce their safety margin of fuel.)

g What total mass can be accelerated to take-off speed in a distance of only 1200 m? What is the disadvantage to such a take-off?

h Copy and complete Table 5.1 to show in the last row what range the aeroplane has when carrying different numbers of passengers.

i At the top of the final column find the number of passengers that may be carried for the full range of the aeroplane if the length of the runway available is 1400 m.

j What other physical factors will affect the range of the aeroplane, and what would you do as an airline operator to make operational efficiency as high as possible, without reducing safety standards?

FIG 5.19 The European Airbus A320.

TABLE 5.1 Data for Airbus with different operating conditions

LENGTH OF TAKE-OFF	1500 m	1200 m	1200 m	1200 m	1400 m
Number of passengers	150	130	110	90	
Mass of aeroplane/kg	42 000	42 000	42 000	42 000	42 000
Mass of passengers/kg	15 000	13 000	11 000	9 000	
Total mass of fuel/kg	18 000				18 000
Total mass/kg	75 000				
Mass of fuel in reserve/kg	3 000	3 000	3 000	3 000	3 000
Usable mass of fuel/kg	15 000				15 000
Range/km					

QUESTIONS

5.7 The take-off acceleration of a high-jumper (**Fig 5.20**) is 17.0 m s^{-2} upwards. His mass is 67 kg.

    a What force is the ground exerting on the man?

    b What is the ratio of this force to his weight?

    c Explain how it is possible for this force to be larger than the high-jumper's weight.

FIG 5.20 A high-jumper needs to exert a downward force on the ground much larger than his weight, so that the ground exerts an upward force on the high-jumper greater than his weight.

FIG 5.21 A builder's cradle, which enables a worker to move up and down the face of a building. Various safety devices are omitted from the diagram.

FIG 5.22 The forces on each part of the train must enable each part to have the same linear acceleration.

5.8 A model helicopter, of mass 5.0 kg, rises with constant acceleration from rest to a height of 60 m in 10 s. Find the thrust that is exerted by the rotor blades during the ascent.

5.9 A man in the builder's cradle (**Fig 5.21**) is lifting himself upwards (there are safety devices not shown to prevent him falling). He has a mass of 80 kg and the cradle has a mass of 30 kg. He is pulling on the rope with a force of 600 N. Draw free-body diagrams of

a the man,

b the cradle.

You can assume that the tension in the rope is the same at both ends of the rope, so that it exerts both a force of 600 N upwards on the man and a force of 600 N upwards on the cradle.

c Find the acceleration of the man and the force that the man exerts on the floor of the cradle.

5.10 A local train (like the one in **Fig 5.22**) consists of two tractor units of mass 10 000 kg (these are coaches with engines) and an unpowered coach of mass 8000 kg. The horizontal forces acting on the three coaches are shown in **Fig 5.23**, and consist of:

■ drive forces of 20 000 N on A and 18 000 N on C;

■ drag forces due to air resistance and friction of 9000 N on A and of 6000 N on both B and C;

■ forces that act by contact between adjacent coaches.

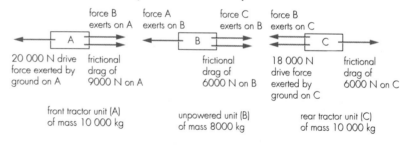

FIG 5.23 The forces acting on the three coaches of a train.

a Find the acceleration of the train.

b What is the magnitude and direction of the force that A exerts on B?

c What is the magnitude and direction of the force that C exerts on B?

5.11 Show that the unit for drag coefficient C_D is the same as that for density.

5.12 Repeat Example 5.4 for the car travelling at 16 m s^{-1}. Assume that the first two forces remain the same.

SUMMARY

Newton's laws

■ Every object continues in its state of rest or uniform motion in a straight line unless it is compelled to change that state by external forces acting upon it.

■ The rate of change of momentum of a body is proportional to the total force acting on it and occurs in the direction of the force.

■ If body A exerts a force on body B, then body B exerts an opposite force of the same size on body A.

Force

■ A force is necessary to accelerate an object. The direction of the acceleration is always in the direction of the force. Force is measured in newtons. One newton (1 N) is that force which, acting alone, will give a mass of one kilogram (1 kg) an acceleration of one metre per second squared (1 m s^{-2}).

■ Force (N) = mass (kg) $\times$ acceleration (m s^{-2}).

■ Forces outside the nucleus are either gravitational or electromagnetic. All contact forces are electromagnetic forces. The component of a contact force that is tangential to the surface is called the frictional force; the component of a force that is normal to the surface is called the normal contact force.

■ Near the surface of the Earth the gravitational pull on a mass is approximately 9.8 newtons per kilogram (N kg^{-1}). The gravitational pull on an object is called the object's weight.

ASSESSMENT QUESTIONS

5.13 A student makes the following statements of Newton's laws of motion.
'First law: every body continues in its state of uniform motion unless it is acted upon by a resultant force.'
'Second law: the acceleration of a body is inversely proportional to the resultant force acting on it and directly proportional to its mass. The direction of the acceleration is in the direction of the resultant force.'
'Third law: action and reaction are always equal and opposite.'
a The statement of the First law is incomplete in two respects. Re-write it with appropriate amendments.
b The statement of the Second law is wrong in two respects. Re-write it with appropriate corrections.
c The statement of the Third law is correct, but fails to emphasise an important aspect of action and reaction forces. Re-write it so as to make this emphasis.
CCEA 1998

5.14 a Write down a word equation which defines the magnitude of a force.
b Two forces have equal magnitudes. State three ways in which these two equal forces can differ.
London 1998

5.15 a State Newton's second law of motion.
b You are asked to test the relation between force and acceleration.
  i Draw and label a diagram of the apparatus you would use.
  ii State clearly how you would use the apparatus and what measurements you would make.
c Explain how you would use your measurements to test the relationship between force and acceleration.
London 1997

5.16 A train of mass 1.4×10^5 kg accelerates uniformly from rest along a level track. It travels 100 m in the first 26 s.
Calculate
a the acceleration of the train,
b the speed reached after 26 s,
c the resultant force required to produce this acceleration,
d the average power required.
NEAB 1997

5.17 Whilst braking, a car of mass 1200 kg decelerates along a level road at 8.0 m s^{-2}.

    a If air resistance can be assumed to be negligible, calculate the horizontal force of the road on the car.

    b Calculate the vertical force exerted by the road on the car.

    c Using a scale drawing or otherwise, calculate the magnitude and direction of the resultant force of the road on the car.

NEAB 1996

5.18 A spacecraft, of mass 4500 kg, which is landing on the Moon, uses its engines to keep its speed of descent constant at 5.0 m s^{-1} from the time when the craft is 14 m above the Moon's surface until it is 4.0 m above the surface. The engines are then switched off and the spacecraft falls freely to the Moon's surface.

The acceleration of free fall on the Moon is 1.6 m s^{-2}.

Calculate, for the spacecraft,

    a the speed of impact,

    b the time taken to travel the last 4.0 m,

    c the time taken to fall the full 14 m,

    d the power of the engines while the speed is constant,

    e the work done by the engines while the speed is kept constant.

NEAB 1998

5.19 This question is about the motion of a car.

A car of mass 800 kg travels along a straight level road. The figure shows a graph of speed against time.

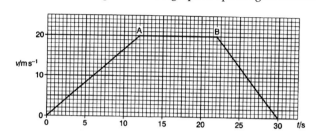

    a Use the figure to find

       i the initial acceleration,

      ii the resultant force accelerating the car during the first 12 s,

      iii the total distance travelled,

      iv the average speed for the journey.

    b The engine produces a driving force of 300 N in the time interval between A and B.

       i Suggest a reason why the car does not accelerate.

      ii Calculate the power required to drive the car during this time.

    c On another journey, the car reaches the same constant speed when climbing a slope with a gradient of 1 in 15.

Calculate the additional power required to maintain the same speed of 20 m s^{-1}.

O & C 1998

6

EQUILIBRIUM

LEARNING OBJECTIVES

At the end of this chapter you should be able to:

① calculate the turning effect of forces;

② establish whether or not a rigid body is in equilibrium;

③ understand the term 'centre of gravity'.

6.1 EQUILIBRIUM OF A POINT OBJECT

In discussing Newton's laws in Chapter 5, a simplification was made by assuming that the object under consideration was not rotating. Another way of making the same assumption is to assume that every part of the object concerned has the same velocity. In practical terms this is not too difficult to imagine, and usually does not involve very much error. A child travelling on a scooter at 4 m s^{-1} can be visualised easily, although if greater accuracy is required it might be necessary to take into account the fact that parts of the child and the wheels are moving with different velocities.

If we wish to make the simplification that all of an object is travelling with the same velocity, one way of doing it is to assume that the object is a **point object**. A point object is idealised. It has mass but no size. Point objects are not necessarily small. In gravitational problems, the Earth can be considered as a point object. If no information is given about the size of an object, then you *have* to assume that it is being regarded as a point object.

A point object is said to be in **equilibrium** if the resultant force acting on it is zero. If there is zero resultant force acting on it, then it cannot have any acceleration. It may however be moving with a constant velocity. The condition for zero resultant force on an object is that the vector sum of all the forces acting on it must be zero.

Sine rule

Calculations involving the triangle of forces often require the use of the sine rule for a triangle (see Appendix). This is

$$\frac{a}{\sin A} = \frac{b}{\sin B} = \frac{c}{\sin C}$$

where *a*, *b* and *c* are the lengths of the three sides of the triangle, and *A*, *B* and *C* are the three angles (with angle *A* opposite side *a*, etc.).

K **Equilibrium** **occurs for a point object when the resultant force on it is zero.**

FIG 6.1 A bridge must be designed so that either it is in equilibrium or it will return to an equilibrium position quickly after being distorted in some way, for example by a load or by adverse weather conditions.

EXAMPLE 6.1

In this example we consider a suspension bridge like that in **Fig 6.1**. Suppose that a vertical cable in the bridge supports a load of 8.4×10^5 N (**Fig 6.2**). The main cable, to which it is attached, is pulled so that the higher part is at an angle of 52° to the vertical and the lower part is at 61° to the vertical. **Fig 6.2b** shows an enlarged view of the point at which the vertical cable is attached to the main cable. Find the tensions *A* and *B* in the cable.

The point of suspension X must be in equilibrium, so the resultant force acting on it must be zero. The sum of the three forces shown in **Fig 6.2b** must be zero. A closed triangle will therefore be the result when we draw the vector diagram (**Fig 6.3**).

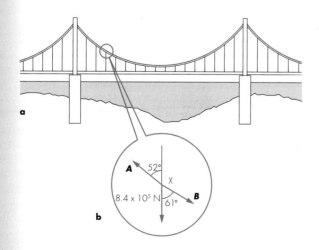

FIG 6.2 Each point of support on the main cables of a suspension bridge must be in equilibrium.

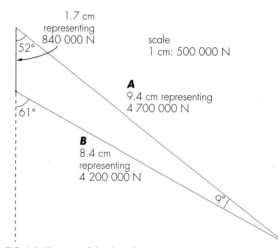

FIG 6.3 The sum of the three forces acting on each point of support must be zero. Since the angles are known or can be measured and the load supported is known, the triangle enables the tensions *A* and *B* to be found.

Finding the values of *A* and *B* is done here both by a scale drawing of the triangle of forces and by calculation. The calculation requires the use of the sine rule (see box and Appendix). The sine rule gives

$$\frac{8.4 \times 10^5 \,\text{N}}{\sin 9°} = \frac{A}{\sin 119°} = \frac{B}{\sin 52°}$$

$$A = \frac{\sin 119° \times 8.4 \times 10^5 \,\text{N}}{\sin 9°} = 4.7 \times 10^6 \,\text{N}$$

$$B = \frac{\sin 52° \times 8.4 \times 10^5 \,\text{N}}{\sin 9°} = 4.2 \times 10^6 \,\text{N}$$

Q **Why are the values found in Example 6.1 likely to have a high uncertainty?**

At the beginning of the chapter it was stated that a point object is in equilibrium if the resultant force acting upon it is zero. If the object is *not* a point object, then further consideration is necessary because such an object may rotate.

Moments

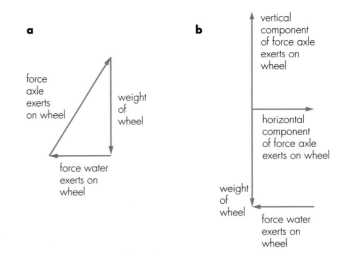

FIG 6.4 The forces acting on a water-wheel give it no linear acceleration but they do cause rotation.

Consider the forces acting on a water-wheel like that shown in **Fig 6.4**. The wheel makes contact with two objects: the water and the axle about which it rotates. Each of them exerts a force on the wheel, and in addition there is the gravitational force that the Earth exerts on the wheel, its weight. Since there is no linear acceleration of the wheel, the resultant force acting on it must be zero. A vector diagram, **Fig 6.5a**, can be used to find the force that the axle exerts on the wheel. This force is at an angle to the vertical, and so it can be resolved into a horizontal component and a vertical component as shown in **Fig 6.5b**. The weight of the wheel is equal and opposite to the vertical component of the force that the axle exerts on the wheel; and the force that the water exerts on the wheel is equal and opposite to the horizontal component that the axle exerts on the wheel.

Despite there being zero resultant force on the wheel, the wheel rotates. This is because of the turning effect of the water. The water exerts a horizontal force on the wheel at a distance from the axle of the wheel. The turning effect of a force

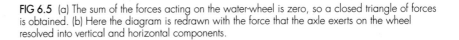

FIG 6.5 (a) The sum of the forces acting on the water-wheel is zero, so a closed triangle of forces is obtained. (b) Here the diagram is redrawn with the force that the axle exerts on the wheel resolved into vertical and horizontal components.

is called its **moment**. The size of the moment that a force exerts depends not only on the size of the force but also on the distance of the line of action of the force from the axis of rotation. Moment is defined by:

K moment = force × **perpendicular distance between axis of rotation and line of action of force**

The unit for moment is newton metre (N m), and by convention this is *not* written as joule (J).

We have seen that if an object is to be in equilibrium, the resultant force acting on it must be zero. But there is now a further condition that we need for equilibrium, because we have seen that rotation may occur despite there being zero resultant force. It is the following, which is known as the **principle of moments**:

K **When a system is in equilibrium, then the sum of the clockwise moments taken about any point equals the sum of the anticlockwise moments taken about the same point.**

FIG 6.6 A couple acting on the coil of an electric motor.

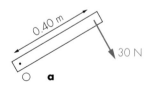

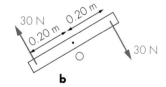

FIG 6.7 (a) A moment of 30 N × 0.40 m = 12 N m. (b) A couple of 2 × (30 N × 0.20 m) = 12 N m. A couple has a resultant linear force of zero.

Couple and torque

Forces are frequently applied to an object to turn it without there being any resultant force on the object. This is done by applying two equal and opposite forces to the object, often at equal distances from the axis of rotation. For example, a coil in an electric motor has driving forces acting on it as shown in **Fig 6.6**. The resultant force acting on the coil is zero, as there is an upward force of 200 N to cancel out the downward force of 200 N. The coil will therefore have zero linear acceleration. However, because the two forces are not in line with one another, there is a moment acting on the coil and so the coil rotates. A pair of equal forces such as this is called a **couple**. The moment provided by a couple is equal to the product of one of the forces and the perpendicular distance between them. The moment of a couple is called the **torque**. The terms 'torque' and 'moment of a force' are often regarded as having the same meaning, but a moment can be supplied by a single force and may result in there being a non-zero resultant force on the object. The term 'torque' always implies that only the turning effect is being considered.

The distinction between the terms is illustrated by **Fig 6.7**. The moment of the force in **Fig 6.7a** about O is

30 N × 0.40 m = 12 N m

This single force would cause a linear acceleration of the rod as well as a turning effect. In **Fig 6.7b** the equal forces acting in opposite directions give a resultant force of zero, so there is no linear acceleration. They are called a couple. The torque of the couple is a turning effect only, and as before has a value of

(30 N × 0.20 m) + (30 N × 0.20 m) = 12 N m

In some problems where a torque is applied to an object, the forces causing that torque are difficult to measure. A good example of such a situation is when a screwdriver is being held in the hand to turn a screw. The hand then exerts many thousands of tiny frictional forces on the screwdriver handle, wherever contact is made between the handle and the hand, but does not exert any resultant force to accelerate the screwdriver in a straight line. The sum of the moments provided by all of these forces is often referred to simply as the torque on the screw.

QUESTIONS

6.1 **Fig 6.8** shows various forces acting: **a–d** on different parts of the human body and **e** on the head and neck of a horse. Assuming that each object can be treated as a simple lever and is in equilibrium, find the value of the unknown forces.

[Hint: In **d** you will need to measure on the diagram the angle at which the force acts. In **e** the muscle supplying the force acting on the neck of a horse to hold it up is interesting. When the horse lowers its head to graze, the muscle behaves as a spring and stores potential energy. When the horse lifts its head this potential energy is used to raise the head.]

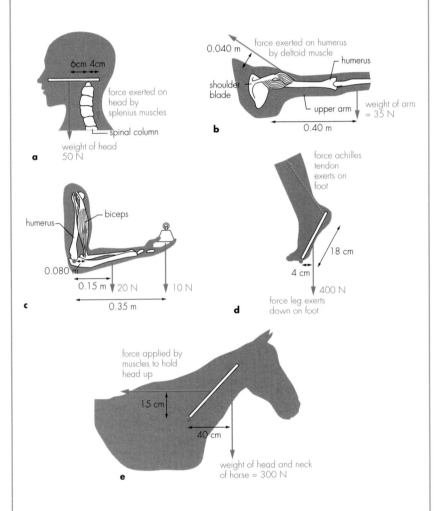

FIG 6.8 Forces in living things. These diagrams are simplified, but still show how the principle of moments can be applied to muscular forces.

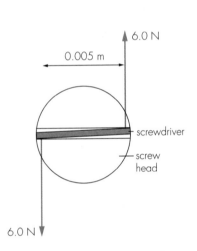

FIG 6.9 A screwdriver in the head of a screw.

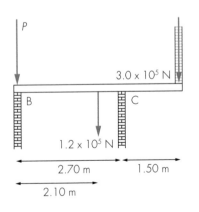

FIG 6.10 The cantilever is much used in building construction. Here the right-hand wall is supported by a cantilever beam.

6.2 Find the torque exerted by the screwdriver on the head of the screw shown in **Fig 6.9**.

6.3 A cantilever construction shown in **Fig 6.10** has a loading of 3.0×10^5 N on its outside edge. If the cantilever itself has a weight of 1.2×10^5 N, find the necessary loading P, on B, to prevent rotation of the whole structure. With this value of P, what support is given by the brickwork at B and at C?

6.4 Find the torque exerted by a rusty hinge on a pub sign, if the sign has a length of 1.4 m, a weight of 25 N and is in equilibrium at an angle of 12° to the vertical as shown in **Fig 6.11**

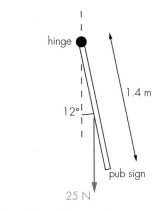

FIG 6.11 A hanging sign.

6.5 **Fig 6.12** shows a loaded bridge subjected to a series of loading forces. For a bridge loaded in this way, find the support forces on the bridge at A and B by applying the principle of moments about each of the supports in turn.

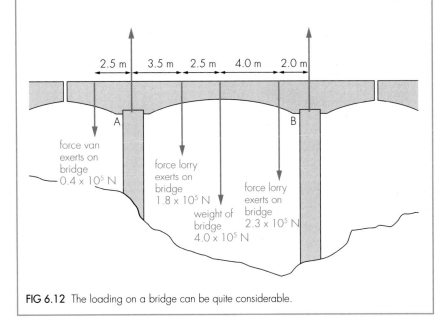

FIG 6.12 The loading on a bridge can be quite considerable.

6.3 EQUILIBRIUM OF A RIGID OBJECT

Any object other than our idealised point object has size. It therefore has its mass distributed over its volume and forces acting on parts of it. If it were necessary to consider every bit of the object's mass separately, the mathematical problems would be very complex. To simplify this problem we use the term **centre of mass**. When an object is acted on by external forces, the centre of mass moves as though all the object's mass were concentrated at that point and as if it were acted on by a resultant force equal to the sum of the external forces. The position of the centre of mass is usually in the same place as the centre of gravity of the object. The **centre of gravity** of an object is the point through which the entire weight of an object may be considered to act.

A **rigid object** is one that does not change its shape appreciably when undergoing acceleration. The conditions necessary for the equilibrium of a rigid object are:

- The resultant force on the object must be zero.
- The resultant torque on the object about every axis must be zero.

The first of these conditions means that the centre of mass of the object will have a constant velocity, which may often be zero. The body is then said to be in **translational equilibrium**.

The second condition is necessary for the object not to be changing its rate of rotation. An object rotating at a constant rate is said to be in **rotational equilibrium**, and zero torque is required for this to be the case. Note that rotational equilibrium is, strictly, not a true equilibrium as the individual particles within the object do have acceleration. Complete equilibrium exists when there is zero resultant force on an object and it does not rotate.

INVESTIGATION

Without recourse to major surgery, find the mass of one of your own hands. You can of course extend this investigation to finding the mass of an arm or even a leg if you so wish.

Also find the position of your own centre of gravity when you are in a normal standing or lying position.

Use the results of the two investigations to answer the following questions.

a How is it possible for an athlete to jump clear over the bar in a high-jump competition and yet for the centre of gravity of the athlete to go under the bar?

b An artificial leg is lighter than a real leg. How would this need to be taken into account by a physiotherapist in planning the rehabilitation of someone who has had to have a leg amputated?

6.4 CENTRE OF GRAVITY AND STABILITY

When a civil engineer designs a structure, such as a bridge, or a building, he or she must make certain that the structure is in equilibrium at all times. It is clear that the final structure must not accelerate, so it must always have zero resultant force and zero resultant torque acting on it. It is also necessary that the structure should be in equilibrium at all stages of its construction (**Fig 6.13**). This frequently causes structural engineers problems, as the strength of a building when completed is usually greater than its strength during construction.

FIG 6.13 Buildings must be stable *during* construction as well as when finished.

Making certain that the structure is in equilibrium is, however, only part of the problem. Any structure is subjected to variable external forces after it has been constructed. These forces can be very large. The most dramatic of these forces are earthquake forces. During a severe earthquake, forces can be sufficiently large to cause enormous damage. **Fig 6.14** shows a building that has been seriously damaged by an earthquake. In areas of the world where earthquakes are common, modern buildings are designed to withstand earthquakes up to a certain magnitude and to move in one piece thereafter. This reduces the amount of injury caused by falling floors or masonry.

Q **Why is it not possible to build a totally earthquake-proof structure?**

More usual variable forces acting on a building are those due to wind and weight of snow. Both of these forces can be surprisingly large at some time or other during a building's lifetime. It may be that a house of floor area 50 m^2 has to be able to support a depth of snow of 0.60 m (see **Fig 6.15**). For snow of density 100 kg m^{-3} this is a total mass to be supported of 3000 kg. Wind forces can be as large as this and can also act upwards! A stream of air flowing across a roof as in **Fig 6.16** causes a reduction in the pressure in region X. The normal pressure underneath the roof causes the roof to be lifted, and in a strong gale, roofs of houses can be lifted off completely (**Fig 6.17**). Wind forces can also result in vibrations being set up in a structure. The size of these vibrations can increase until damage is caused (see section 10.5).

FIG 6.14 An earthquake in California in 1989 caused this damage to 75 year old timber framed buildings.

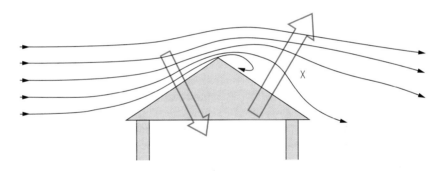

FIG 6.16 Wind blowing across a roof increases the loading on the side of the roof facing the wind and reduces the loading on the side away from the wind.

FIG 6.15 Snow loading on houses can be very large at times throughout the life of a house.

FIG 6.17 This house on a beach was destroyed by Hurricane Hugo in 1989. Wind speeds of 260 kmh^{-1} were recorded in this hurricane.

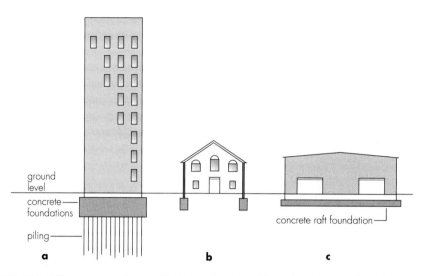

FIG 6.18 Different types and ages of buildings often have different foundations. Tall buildings almost invariably have piles pushed far down into the ground until they reach rock. Houses are usually built on strip foundations, whereas factories and schools are increasingly being built on concrete raft foundations.

FIG 6.19 Its wide wheelbase and low centre of gravity make a racing car very stable.

A structure therefore must not only be in equilibrium but also be able to return to its original position of equilibrium if some outside cause moves it. **Stability** is the term used to describe this.

> **K** A body is said to be in *stable equilibrium* if, when it is displaced a small amount, it will return to its original position.

The degree of stability of an object can be found by measuring the angle through which it must be rotated before the line of action of its weight lies outside its base and it topples over. This will depend on its shape and how it is supported. **Fig 6.18** shows three buildings in equilibrium: **a** is the least stable and **c** is the most stable.

Figs 6.19 and **6.20** show some more examples of stability and its relation to the position of the centre of gravity. The wide base of the racing car (**Fig 6.19**) means that it has to be tilted to a much larger angle than a saloon car before it will topple over. The low centre of gravity of the bus (**Fig 6.20**) means that it will not topple until at a much greater angle than would be the case for a uniform rectangular block of the same dimensions.

When a ship keels over, it is vital that the couple acting on it restores it to its original, upright position. The stability of ships is a subject on which a great

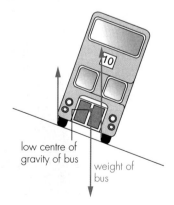

FIG 6.20 The stability of a bus is increased by having a low centre of gravity even when fully laden with passengers.

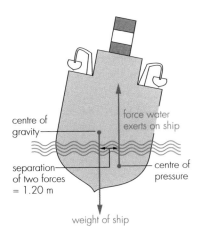

centre of gravity

force water exerts on ship

separation of two forces = 1.20 m

centre of pressure

weight of ship

FIG 6.21 The pair of forces shown must rotate the boat back to the vertical position.

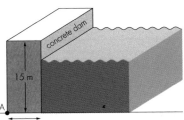

concrete dam

15 m

A

9 m

FIG 6.22 A simple dam.

Q **If you have answered question 6.8 correctly, you will find that the depth of water that can be retained is about 18 m. Does this answer make sense?**

deal of thought has to be given. In all cases the shape of the keel of the ship is constructed so that the force that the water exerts on the ship, which can be considered to act through a single point called the centre of pressure, lies outside of the weight of the ship, acting through the centre of gravity (see **Fig 6.21**).

QUESTIONS

6.6 Find the torque acting on the ship in **Fig 6.21**. The weight of the ship is 5.0×10^7 N.

6.7 Sketch a graph to show how the height of the centre of gravity of the bus in **Fig 6.20** above the ground varies with the angle of the bus to the vertical. Your graph should show both positive and negative values of the angle, and show clearly how the normal position of the bus is stable and has the least potential energy.

6.8 A dam is constructed out of a rectangular block of concrete whose dimensions are shown in **Fig 6.22**. What depth of water can be held back by the dam before it topples about point A? The density of concrete is 2200 kg m^{-3} and the density of water is 1000 kg m^{-3}.

[Hint: Start this question by drawing a force diagram for a one metre length of the dam when it is on the point of toppling over. The average pressure of the water may be considered to act at half the depth of water. This assumption is not strictly accurate but it over-estimates the moment the water exerts on the dam. If you can use integration, you should be able to find the correct moment.]

Dams using the principle shown in **Fig 6.22** are very common. They are often made of earth instead of concrete, and so must be made with sloping rather than vertical sides. The dykes in The Netherlands are earth dams (**Fig 6.23**). They have to be made so that some impermeable material is incorporated within them, so that they do not allow the passage of water through them or under them.

FIG 6.23 An earth dam in The Netherlands.

SUMMARY

- The moment of a force is a turning effect, and is the product of the force and the perpendicular distance between the axis of rotation and the line of action of the force.

- A couple is a pair of forces not in line with one another. They are equal in magnitude and opposite in direction. The moment of a couple is called a torque, and has the value of the product of one of the forces and the perpendicular distance between the forces. A torque produces a turning effect but no linear acceleration.

- A point object is in equilibrium if the vector sum of all the forces on it is zero.

- An extended object is in equilibrium if both the sum of the forces and the sum of the torques on it are zero.

- The centre of gravity of an object is the point through which the entire weight of a body may be considered to act.

ASSESSMENT QUESTIONS

6.9 This question is about forces and moments of forces.
  a **i** Define the *moment of a force* about a point.
     ii State the law of moments as a condition for equilibrium.
  b A crane has a jib (AC) of mass 2500 kg pivoted at A. The jib is supported by a horizontal cable BC. The centre of mass of the jib is at D. The dimensions are as shown on the diagram.

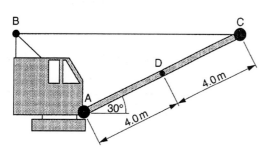

    i Make a sketch copy of the diagram and on it draw three arrows to represent the three forces acting on the jib.
    ii Calculate the tension in the cable BC.
  c The end of the jib C is raised by shortening BC. Explain whether the tension in BC will increase or decrease.

O & C 1998

6.10 a The diagram shows a force *F* applied to raise a loaded wheelbarrow of total weight 500 N.

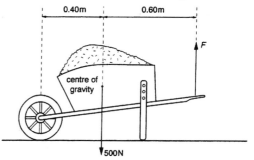

    i Calculate the minimum force needed to raise the wheelbarrow.
    ii Make a sketch of the diagram and on it draw a line to represent the position and the direction of the force exerted by the ground on the wheel.
    iii Calculate the magnitude of the force in **ii**.
  b The diagram shows two sets of forces acting on a disc. The forces are all of equal magnitude.

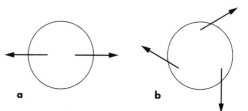

Explain why the forces in **a** are in equilibrium but those in **b** are not.

Cambridge 1997

6.11 The figure shows a theatre lighting bar of mass 15 kg which is suspended from the ceiling by two metal supports A and B. The centre of mass of the bar itself is at its centre. Two lamps are suspended from the bar. The masses and positions of these lamps are as shown.

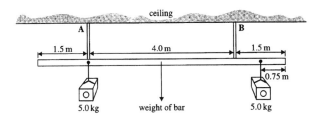

a State the principle of moments.
b Calculate the tension in each of the supports, A and B.

AEB 1998

6.12 The diagram shows a straight, horizontal, uniform swimming bath spring board of length 4.00 m and of weight 300 N. It is freely hinged at A and rests on a roller B, where AB is 1.60 m. A boy of weight 400 N stands at end C.

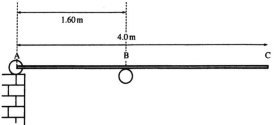

a **i** On a diagram, show the directions of the forces acting on the board at A and B.
  ii Calculate the magnitudes of the forces at A and B.
b With the boy still on the board, a girl of weight 630 N also stands on the board. How far is she from A if the force at B is doubled?

NEAB 1996

6.13 The figure illustrates a nut on a vertical bolt which is being tightened with a spanner.

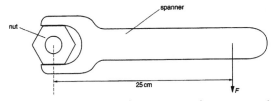

The distance between the centre of the nut and the point where the force *F* is applied to the spanner is 25 cm.

a On a copy of the figure, indicate the direction of the torque opposing the tightening of the nut.
b The torque applied on the spanner is 50 N m. Calculate the magnitude of *F*.

Cambridge 1997

6.14 This question is about the turning forces on a simple mechanical lever system.

a Write down the conditions required for a rigid body acted upon by two or more forces to be in equilibrium.
b The figure below shows a simple device to release the excess steam in a model steam engine boiler.

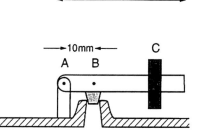

A light rod is hinged at one end, A. The safety valve is a plug at point B, 10 mm from A. C is a counterweight of 0.20 N which can be moved along the rod. When the pressure in the boiler increases above a critical value, the upward force on the plug at B tilts the rod about A and releases steam.

  i The force to lift the safety valve is 0.60 N. Calculate the position of the counterweight for this to happen.
  ii The area of the hole in the boiler is 4.0×10^{-6} m². Find the pressure difference between atmospheric pressure and boiler pressure to lift the safety valve.
c The situation is oversimplified. The uniform rod of length 40 mm has a weight of 0.10 N. Taking this into consideration, how far must the counterweight C be moved from the position calculated in part **b i** above for the valve to open at the required pressure?

O & C 1997

6.15 a **i** What is the name given to a *pair* of equal and opposite parallel forces?
  ii What effect will this system of forces have when applied to a body?

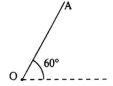

A uniform heavy rod OA of mass 12 kg is pivoted freely at O (i.e. there is no friction at the pivot) and held in the position shown by a force *F* acting at right angles to the rod and applied at A.
Calculate the magnitude of *F*.

WJEC 1998

6.16 This question is about a simple roof truss.

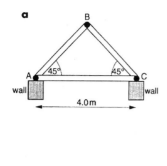

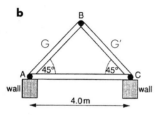

Fig a shows three beams forming a simple roof truss ABC of negligible weight. The truss supports two heavy roof panels whose centres of gravity are at G and G′ as shown in **Fig b**. G and G′ are half way along AB and CB respectively. The tie AC ensures that the reaction forces where the truss rests on the walls are vertical.

a State which members (AB, BC, AC) of the truss are in tension and which are in compression.

b Each of the roof panels has a total weight of 1800 N. By taking moments about B, calculate the force in tie AC.

c In a gale the wind produces a force equivalent to an extra horizontal force acting at G tending to blow the roof off the walls. Calculate the value of this equivalent force to make the roof truss pivot about C.

d In some buildings it is not possible to use a tie to counteract the horizontal force on the walls due to the weight of the roof. Explain another way in which such horizontal forces on walls can be balanced.

O & C 1998

6.17 a Explain what is meant by
   i the moment of a force,
   ii the torque of a couple.
b A desk lamp is illustrated in the diagram.

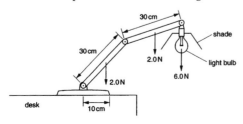

The lamp must be constructed so that it does not topple over when fully extended as shown in the second diagram. The base of the lamp is circular and has a radius of 10 cm. Other dimensions are shown on the figure. The total weight of the light bulb and shade is 6.0 N and each of the two uniform arms has weight 2.0 N.

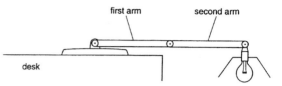

  i On a sketch of the second diagram, draw an arrow to represent the weight of the base.
  ii The lamp will rotate about a point if the base is not heavy enough. On your sketch, mark this point and label it P.
  iii Calculate the following moments about P.
    1 moment of first arm
    2 moment of second arm
    3 moment of light bulb and shade
  iv Use the principle of moments to calculate the minimum weight of base required to prevent toppling.

OCR 1999

6.18 a The simplified diagram represents the trunk of a person bent forwards, with the spine at an angle of 70° to the vertical. The extensor muscle, which joins the spine to the pelvis, makes an angle of 8.0° to the spinal column. The weight of the person's trunk and head, which is 400 N, is shown to act at the point along the spine where the extensor muscle is attached.

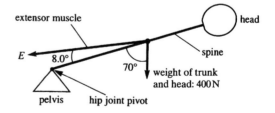

  i Draw and label an arrow R on a sketch to show the direction of the reaction force acting on the spinal column at the hip joint pivot.
  ii By resolving forces, calculate the tension E in the extensor muscle.
  iii Calculate the compression force acting along the line of the spine at the point where the extensor muscle is attached.
b Explain, in terms of the compression forces which act on the discs in the spinal column, why it is important to lift heavy objects with the spine almost vertical.

NEAB 1997

6.19 A wheel of radius 0.12 m is driven by a belt.

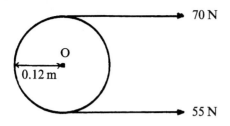

There is no tendency for the belt to slip on the wheel. When the wheel is rotating at a constant rate, the tensions in the upper and lower parts of the belt are 70 N and 55 N respectively. As a result of friction in the wheel's bearings, there is a constant opposing torque on the wheel.

a Calculate the moment of the tension in the upper part of the belt about the centre O of the wheel. State whether the sense of this moment is clockwise or anticlockwise.

b Calculate the constant opposing torque due to friction. State the sense of the opposing torque.

CCEA 1998

6.20 The diagram shows an old-style chemical balance. The total mass of the moving part (i.e. arm, pointer and two scale pans) is 128 g and when the pointer is vertical the centre of mass is 3.0 cm vertically below the pivot.

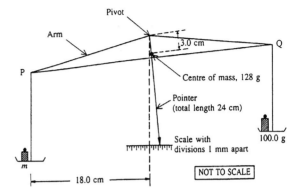

a The pointer, which is 24.0 cm long, moves over a scale whose divisions are 1.0 mm apart. The pointer is deflected five scale divisions to the right. Show that the horizontal displacement of the centre of mass is 0.625 mm.

b The mass in the right-hand scale pan is 100.0 g. Assume that the horizontal distance of P and Q from the pivot are constant. Calculate the mass *m* in the left-hand scale pan.

c Explain why the scale over which the pointer moves is not calibrated.

London 1999

7 MOMENTUM, WORK, ENERGY AND POWER

LEARNING OBJECTIVES

At the end of this chapter you should be able to:

① understand the term 'momentum' and relate it to force multiplied by time;

② understand the term 'work' and relate it to force multiplied by distance;

③ use the terms 'work' and 'energy' correctly;

④ use the principles of conservation of momentum and conservation of energy correctly;

⑤ understand the term 'power'.

7.1 CONSERVATION OF MOMENTUM

Momentum

omentum was defined (section 4.6) as the product of mass m and velocity v. It is a vector quantity in the same direction as the velocity. In equation form this is therefore

$p = m \times v,$ where p is the symbol used for momentum.

Momentum was used by Newton in his second law, which states that the rate of change of the momentum of a body is proportional to the total force acting on it (section 5.3). Using SI units, we therefore get for a constant force F the equation

$$\frac{mv - mu}{t} = F \qquad \text{or} \qquad \Delta p = F \times t$$

where Δp is the change in the momentum. This equation shows that the product of the constant force acting on an object and the time for which it acts gives the change in the momentum of the object. It also indicates that a unit that can be used for momentum is the newton second, N s.

EXAMPLE 7.1

A ship of mass 6000 000 kg has a constant force applied to it of 200 000 N. How long will it take to reach a speed of 5 m s^{-1} if it is starting from rest?

The initial momentum is zero, and since the product force × time gives the change in the momentum, it will therefore give the final momentum p of the ship. So

$$p = m \times v$$
$$= 6000\,000 \text{ kg} \times 5 \text{ m s}^{-1}$$
$$= 30\,000\,000 \text{ N s}$$
$$= F \times t$$
$$= 200\,000 \text{ N} \times t$$

The time taken t is therefore 150 s.

Units

Remember that

$$1 \text{ kg m s}^{-2} \equiv 1 \text{ N}$$

and so

$$1 \text{ kg m s}^{-1} \equiv 1 \text{ N s}$$

Questions such as Example 7.1 are rather artificial, as there is little chance of keeping such a force constant. The resistance to movement provided by the water will vary during the acceleration. The question does however give a guide to the time taken, and indicates the minimum time taken if the force applied by the propeller to the ship is 200 000 N. Drag through the water will, in practice, cause the time to be longer. Another point to note with this type of question is that it is not necessary to find the acceleration. As a general rule, if a question is a force–time question, then immediately consider the *momentum* of the object.

Impulse

In many practical applications of momentum, the force applied will not be constant. Consider a ball, of mass 0.20 kg, travelling with velocity 28 m s^{-1} directly towards a wall. It hits the wall and bounces off in the opposite direction with a velocity of 20 m s^{-1}. The change in the ball's momentum is

final momentum away from wall – initial momentum away from wall
$$= (0.20 \text{ kg} \times 20 \text{ m s}^{-1}) - (0.20 \text{ kg} \times -28 \text{ m s}^{-1})$$
$$= 4 \text{ kg m s}^{-1} - (-5.6 \text{ kg m s}^{-1}) = (4 + 5.6) \text{ kg m s}^{-1} = 9.6 \text{ N s}$$

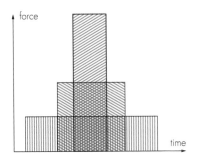

Therefore the wall has exerted a force on the ball in such a way as to provide the same effect as a constant force of 9.6 N would do if applied for 1 s. This could however have been achieved by a constant force of 96 N for 0.1 s, or a force of 48 N for 0.2 s, or any similar combination as shown in **Fig 7.1**. In graphical terms the area beneath the force–time graph has to be 9.6, and the shape of the graph, in practice, is much more likely to be as shown in **Fig 7.2**.

This area beneath a force–time graph is called the **impulse**, and its unit is the newton second (N s). The impulse applied to an object will be a measure of the change in the momentum of the object. For a constant force, it is the product force × time. For a varying force, it is (average force) × time. Using calculus notation it is $\int F \, dt$, so we get the equation:

FIG 7.1 Each of these force–time graphs will cause the same change in the momentum of the object on which the force acts. The area beneath each of the graphs is the same.

 K **impulse = force × time = change in momentum**

or

$$\text{impulse} = \int F \, dt = mv - mu$$

where v is the final velocity and u the starting velocity.

Often the word 'impulse' is used when the force is applied for a brief time, but there is no reason why it cannot be applied when longer times are involved.

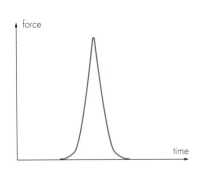

FIG 7.2 In practice, when one object collides with another, the force–time graph is more likely to be of this shape.

7.1 What driving force is necessary if a car of mass 1200 kg is to reach a speed of 30 m s^{-1} in 15 s?

7.2 What braking force is necessary if the car in question 7.1 is to stop in 5 s?

7.3 A ball of mass 0.047 kg, which is initially at rest, is struck with an average force of 2600 N for 1.2×10^{-3} s.

   a What is the final velocity of the ball?

   b What impulse is given to the ball?

7.4 A pilot ejected from an aeroplane experiences an acceleration of 12g for 0.25 s. If the mass of the pilot is 70 kg, find the average total force acting on him and the change in his velocity.

7.5 An alpha particle of mass 6.7×10^{-27} kg is ejected from a stationary nucleus at a speed of 3.2×10^6 m s^{-1}.

   a What impulse is given to it?

   b What average force is needed when it is ejected in approximately 10^{-8} s?

FIG 7.3 The collision between a golf club and a golf ball gives the ball a high acceleration for a short time. This is possible because of the large force exerted by the club. The force is sufficient to squash the ball appreciably.

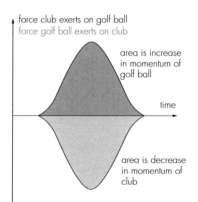

FIG 7.4 Force–time graphs showing how, if one object gains momentum, another object loses an equal amount of momentum.

Conservation of momentum

An object can only undergo a change in its momentum if a force is exerted on it. Using Newton's third law however, we know that, if an object exerts a force on something, then there must be an equal and opposite force on the object itself. This will change its own momentum. An example should help to make this clear. The head of a golf club strikes a golf ball horizontally (**Fig 7.3**). A graph showing the way in which the force that the club exerts on the golf ball varies with time may be as shown in **Fig 7.4**.

The impulse given to the golf ball will increase the momentum of the golf ball, and this will be the area beneath the graph. During the brief moment that this change in momentum is taking place, the force that the club exerts on the golf ball is, at all instants, equal and opposite to the force that the golf ball exerts on the club. The club therefore has its momentum changed by exactly the same amount but in the opposite direction. In other words, the club transfers some of its momentum to the golf ball. Any momentum it loses, the golf ball gains. The total momentum of the two objects remains constant. This is an example of a basic physical principle called the conservation of momentum, which is deduced from Newton's laws. The **conservation of momentum** states that:

K   **The total momentum of a system remains constant provided no external forces act on the system.**

The example with the golf ball and golf club can be extended to cover any objects. If anything loses momentum, then something else gains it. The total momentum of the Universe remains constant. Under all circumstances, no one has ever found any way of destroying momentum. Einstein's theory of relativity uses this fundamental law. It applies to sub-atomic particles as well as to vast galaxies.

EXAMPLE 7.2

A lorry of mass 10 000 kg collides with the back of a car of mass 1500 kg (see **Fig 7.5**). Immediately before the collision, the lorry was travelling at 30 m s^{-1}, and the car at 12 m s^{-1}. Find the speed of the vehicles immediately after the collision if they remain jammed together.

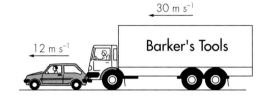

FIG 7.5 A lorry colliding with the back of a car.

All momenta are considered in the direction of travel. So before the collision we have

$$\text{momentum of car} = 1500 \text{ kg} \times 12 \text{ m s}^{-1} = 18\,000 \text{ N s}$$
$$\text{momentum of lorry} = 10\,000 \text{ kg} \times 30 \text{ m s}^{-1} = 300\,000 \text{ N s}$$
$$\text{total momentum before collision} = 318\,000 \text{ N s}$$

Momentum is conserved, so

$$\text{total momentum after collision} = 318\,000 \text{ N s}$$

Therefore, after the collision we have

$$(10\,000 \text{ kg} + 1500 \text{ kg}) \times v = 318\,000 \text{ N s}$$
$$v = (318\,000 / 11\,500) \text{ m s}^{-1}$$
$$= 28 \text{ m s}^{-1} \text{ (to 2 sig. figs)}$$

QUESTIONS

7.6 A 2000 kg van and a 1500 kg car both travelling at 40 m s^{-1} in opposite directions collide head-on and lock together. What are their speed and direction immediately after the collision?

7.7 A man of mass 60 kg, standing on a friction-free surface, throws a ball of mass 1.5 kg with a horizontal velocity of 25 m s^{-1}. What will be the recoil speed of the man?

7.8 An aeroplane of total mass 50 000 kg is travelling at a speed of 200 m s^{-1}. When a passenger of mass 100 kg walks towards the front of the aeroplane at a speed of 2 m s^{-1}, what change in the speed of the aeroplane does this cause?

7.2 WORK

Work is a word with a large number of meanings. It can be a noun or a verb. There is often a problem when a common English word is used as part of a strict scientific vocabulary. The difficulty arises because some of the common meanings conflict with the scientific meaning. Work in the scientific sense is precisely defined. The definition of **work** is given by the equation:

K   **work = force × distance moved in the direction of the force**

Distance moved in a stated direction is called the displacement, so the equation can be written:

work = force × component of the displacement in the direction of the force

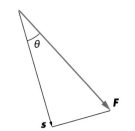

FIG 7.6 The work done by the force is *Fs* cos *θ*.

When the force acts in a direction at an angle *θ* to the direction of the displacement *s* (see **Fig 7.6**), the work done *W* is given by the equation

$$W = \boldsymbol{Fs} \cos \theta$$

Work is a scalar quantity, although obtained by multiplying two vectors. The unit of work must be the product of the force unit and the distance unit, namely, one newton × one metre. This unit, the newton metre (N m), is named the joule (J): **one joule** (1 J) is the work done when a force of one newton (1 N) moves its point of application through a distance of one metre (1 m).

EXAMPLE 7.3

Find the work done by a crane when it exerts a force of 3000 N on a load and lifts it 20 m.

We have

    work done = force × distance moved in the direction
                of the force

so

    work done = upward force crane exerts on load
    by crane   × distance load moved up
              = 3000 N × 20 m = 60 000 J

This was an extremely straightforward problem. Note how a difference occurs in the next example.

EXAMPLE 7.4

Find the work done by the Earth on a satellite going round it in a circular orbit (**Fig 7.7**).

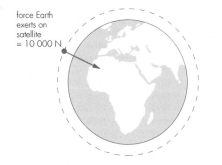

FIG 7.7 A satellite circling the Earth.

This time we have

    work done = force Earth exerts on satellite
    on satellite   × distance satellite moves in
                 direction of force
              = 10 000 N × 0
              = 0

No work is done because there is *no movement in the direction of the force.*

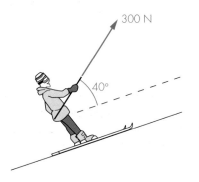

FIG 7.8 A ski tow.

QUESTIONS

7.9 Find the work done by gravity when a diver of mass 60 kg falls through a height of 11 m.

7.10 A winding engine operates a lift that raises coal out of a coal-mine that is 300 m deep.

    a How much work is done by the engine in raising 1000 kg (1 tonne) of coal?

    b Why does the winding engine not have to do any work to raise the lift itself?

7.11 A skier is dragged up a slope by a ski tow, which exerts a force on the skier of 300 N at an angle of 40° to the ground (**Fig 7.8**). What work is done by the tow when the skier moves 300 m?

7.12 What work is done on an electron in a cathode ray tube if the force acting on it is 3.2×10^{-14} N and the force causes the electron to move 5 cm in the direction of the force?

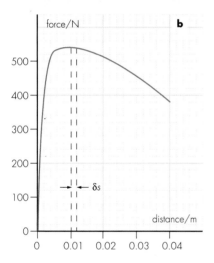

FIG 7.9 **a** The work done by muscles as a sprinter starts a race is partly changed into the kinetic energy of the sprinter. **b** This graph shows how the force exerted by a muscle varies with the distance the muscle contracts.

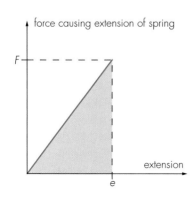

FIG 7.10 Graph of the force *F* against extension *e* for a spring not stretched beyond its elastic limit.

Work done by a varying force

In all of the previous examples and questions, the forces being considered have had a constant value. In many practical situations the force being exerted on an object will vary. In the scientific sense, 'work' is done so often in everyday life that normally it is not thought about, let alone calculated. All of the following actions involve doing work:

- Eating
- Breathing
- Walking
- Shutting a door
- Lifting a baby out of a pram
- Turning the page of a book
- Kicking a football
- Stretching a spring

The individual forces involved in these actions are many and varied, but all of them involve muscle action in some way – **Fig 7.9a** shows an extreme case. As an example of a varying force, consider a muscle in the sprinter's arm. The maximum tension in a muscle depends on its cross-sectional area and also on its length. The maximum tension is usually about 30 N for each square centimetre, and occurs when the muscle's length is only slightly changed from its resting length. A graph showing the way in which the force varies with the distance the muscle contracts is shown in **Fig 7.9b**.

To calculate the work done by this muscle in contracting by 0.040 m, it is necessary to find the sum of force × distance moved for all the small distances δs. The problem is therefore one of finding the area beneath the graph. Using calculus notation, the work done is ∫F ds. The area beneath the graph (up to 0.040 m in this example), found by counting squares (475), represents 19 J. An average force of 475 N moving the distance of 0.040 m would give the same work done. In order to simplify problems, average forces are often quoted when the actual force varies considerably.

Work done to stretch a spring

The extension caused to a spring in use is proportional to the force causing the extension. The force per unit extension is called the **spring constant**, *k*. A graph showing force plotted against extension is shown in **Fig 7.10**. The shaded area gives the work done *W* in stretching the spring. The work done is given by

$$W = \tfrac{1}{2}Fe$$

Since $k = F/e$, the work done can also be written as $W = \tfrac{1}{2}ke^2$.

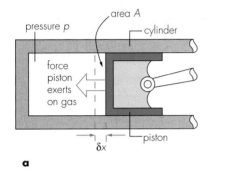

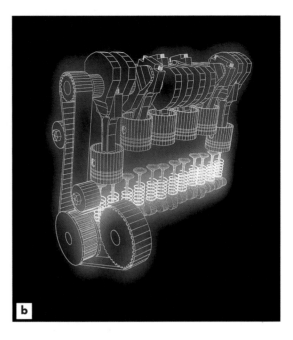

FIG 7.11 **a** The gas in the cylinder is being compressed by the piston moving to the left. **b** The work done by hot gases pushing against a piston provides the work output of a car engine.

Work done to compress a gas

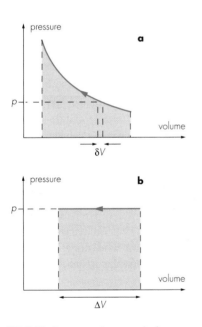

FIG 7.12 Pressure–volume graphs for compressing a gas **a** in a piston and **b** at constant pressure.

In engines, gases expand and contract at different parts of the cycle of movement. When a gas expands, it does work on its surroundings; when it is compressed work is done on it by its surroundings. Inside an engine it will normally be a piston that is used to compress a gas. The piston is part of the surroundings of the gas. As a gas is compressed, its pressure normally rises, and so an increasing force has to be used by the piston to reduce the volume of the gas further.

Fig 7.11 shows a piston, of cross-sectional area A, compressing a gas contained in a cylinder. Consider the piston to move a very small distance δx at a constant speed. If the pressure of the gas is p, then

> force exerted by the piston on the gas = pA
> work done by the piston on the gas = $pA\delta x$

However $A\delta x$ is the small change in volume δV of the gas in the cylinder. So the small amount of work done by the piston on the gas δW is given by

$$\delta W = p\delta V$$

If this is put in graphical terms, the total work done is the area beneath the graph of p plotted against V (**Fig 7.12a**). In calculus notation this is

$$W = \int p \, \mathrm{d}V$$

It is possible to compress a gas at constant pressure. In this case the area required for the work done is rectangular and is pressure $\times$ change in volume (**Fig 7.12b**).

7.13 Estimate how much work is done by the muscle referred to in **Fig 7.9** if it contracts by a distance of only 0.02 m.

7.14 How much work is done to stretch a spring, of spring constant 190 N m^{-1}, a distance of 1.7 cm?

7.15 Each of the suspension springs on a car requires a force of 100 N to depress it 1.0 cm. How much work needs to be done on one spring to compress it by 5.0 cm?

7.16 How much work is done by a gas if it expands from a volume of 1.0×10^{-3} m^3 to 3.7×10^{-3} m^3 at a constant pressure of 8.2×10^5 Pa?

7.17 A gas expands in such a way that pV^2 remains constant. It starts with a volume of 6.0×10^{-3} m^3 and a pressure of 3.0×10^4 Pa, and ends with a volume of 13.0×10^{-3} m^3. Find the work done by the gas, either by plotting a graph or by integrating.

7.3 POWER

Power is defined as follows:

K *Power* **is the rate at which work is done.**

Average power is therefore given by the equation:

$$\text{average power} = \frac{\text{work done}}{\text{time taken}}$$

The unit of power is the joule per second (J s^{-1}). This is given the name **watt** (W). Note that, although watts are used frequently in electrical work, the unit itself is not defined from electrical terms. Any power is measured in watts when using the SI system of units.

The useful power output from any device will only equal the power input to the device if it is 100% efficient. Usually this will not be the case because some of the input power will be wasted in heating the device. The **efficiency** of any machine is defined by the following equation:

$$\text{efficiency} = \frac{\text{power output}}{\text{power input}}$$

This can equally well be expressed in terms of work, in which case the equation becomes

$$\text{efficiency} = \frac{\text{work output}}{\text{work input}}$$

More detail is given about efficiency as applied to electrical circuits in section 14.6, and as applied to engines in section 26.5.

EXAMPLE **7.5**

Estimate the power of a suitable motor for lifting yachts of up to 2000 kg out of the water.

In answering questions that require estimates, it is often necessary to make some assumptions. In the answer, your assumptions must be clearly stated. Answers should generally be given to only one significant figure.

■ Assume that the distance the yacht has to be lifted is 5 m.
■ Assume that the time allowed should not be greater than 5 min.

The largest force that the hoist will need to be able to exert will occur when the yacht is out of the water. This will

equal the weight of the yacht, a force of 2000 kg × 9.8 N kg^{-1}. Then

total work done = force × distance moved in direction of force
= 2000 kg × 9.8 N kg^{-1} × 5 m
= 98 000 J

We assume that the time taken = 300 s. So

$$\text{power required} = \frac{\text{work done}}{\text{time taken}} = \frac{98\,000\,\text{J}}{300\,\text{s}} = 330\,\text{W}$$

The motor will have to be rated higher than this because it will not be 100% efficient. A suitable motor would be one rated at 500 W.

QUESTIONS

7.18 A car of mass 1200 kg has an output power of 58 kW when travelling at a speed of 30 m s^{-1} along a flat road. What power output is required if it is to travel at the same speed up a hill with a gradient of 10%? [This is a gradient of 1 in 10. The angle of the slope is tan^{-1} 0.10 = 5.7°.]

7.19 A motor used to pull on a cable is drawing in the cable at a speed of 4.8 m s^{-1}. The force being exerted on the cable is 8500 N.

  a What is the power output of the motor?

  b What is the general equation linking power P, force F and velocity v?

7.4   ENERGY

When work is done on an object, different things may happen to it. Work cannot be done without movement, so the object *must* move. The movement may result in there being an acceleration, or a deceleration, or no change in velocity. It may result in the object rising, or in it having its shape altered. It may result in the object's temperature rising or a chemical reaction or even the production of an electrical potential difference across it. In stretching a spring, work is done on the spring. The spring can then itself do work when it returns to its original shape. In other words, when stretched, the spring stores the ability to do work.

K   **The stored ability to do work is called *energy*.**

Energy is measured using the same unit as work, namely the joule, and can be calculated using the same equation as that used to define work, namely: force × distance moved in the direction of the force. Work is done when energy is

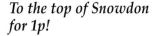

To the top of Snowdon for 1p!

Electrical energy is supplied to you at home for about 0.000 002 p per joule. Put another way, for only 1p you can buy 500 000 J, which is enough energy to lift you 1000 m, and that will put you on the top of Snowdon, all for 1p!

transferred; the two terms are closely connected. The energy stored in a compressed spring may be used to do work at any time. The electrical energy stored in a capacitor, an electrical component designed to store electrical energy, may be released in a fraction of a microsecond. The chemical energy in a lump of coal may have been stored for many millions of years before it is released.

There is much political discussion nowadays about how quickly we are, or should be, using energy. Much of the energy that is being used at a rapid rate in the world now is not renewable. One of the reasons why we are so profligate in our use of energy is that, at present, it is *very* cheap (see box). It is not really surprising therefore that, at this cost, we are willing to waste a lot of electrical energy. There will be much greater economic incentive to find forms of renewable energy when the cost of energy is higher. In discussions about energy use, it is important to use quantitative information to get some idea of the problems involved. Table 7.1 gives some possible energy values.

TABLE 7.1 Energy values of certain systems

DESCRIPTION	ORDER OF MAGNITUDE OF ENERGY/J
The Big Bang at the formation of the Universe	$\sim 10^{70}$
Rotational energy of the Galaxy	$\sim 10^{50}$
Energy released by the Sun up to the present	10^{45}
Binding energy of the Earth–Sun system	10^{33}
Kinetic energy of the Moon as it orbits the Earth	10^{28}
Energy radiated by the Sun in one second	10^{27}
Energy received at the Earth from the Sun in a year	10^{25}
Annual wind energy	10^{22}
Annual energy use by people	10^{21}
Annual energy dissipation by tides	10^{20}
Energy released by the largest nuclear bomb ever made	10^{18}
Annual electrical energy supplied by a large power-station	10^{16}
Energy released in burning a tonne of coal	10^{11}
Kinetic energy of Concorde at cruising speed	10^{10}
Output energy of a car in using one litre of petrol	10^{7}
Daily intake of food energy for an adult	10^{7}
Energy supplied by a loaf of bread	10^{6}
Maximum kinetic energy of a top sprinter	10^{4}
Maximum kinetic energy of a tennis ball in play	10^{2}
Two beats of a human heart	1
Kinetic energy of a fly in flight	10^{-3}
Electrical energy stored in a 0.02 μF capacitor at 10 V	10^{-6}
Energy released by a radioactive atom decaying	10^{-13}
Kinetic energy of a single electron passing through 1 V	10^{-19}
Kinetic energy of thermal vibration of a single atom	10^{-21}

 KINETIC ENERGY

Kinetic energy is referred to in Table 7.1. This term, which you are probably familiar with, is always associated with movement. Energy is defined in section 7.4 as the stored ability to do work. The **kinetic energy** (E_k) of an object is its stored ability to do work as a result of its motion. It can have *translational* kinetic energy as a result of its linear movement, and it may also have *rotational* kinetic energy as a result of rotation.

EXAMPLE 7.6

Find the kinetic energy of a train, of mass 300 000 kg, when travelling at 50 m s⁻¹.

The kinetic energy must be found from the amount of work the train can do against a force that stops it. We will assume, at first, that the force has a constant value of 120 000 N. With this force acting on the train in the *opposite* direction to its velocity, the acceleration a of the train is given by:

$$\text{acceleration} = \frac{\text{force}}{\text{mass}} = \frac{-120\,000\,\text{N}}{300\,000\,\text{kg}} = -0.40\text{ m s}^{-2}$$

Using the equation $v^2 = u^2 + 2as$ (section 4.4), where s is the distance travelled while stopping, we get:

$$0 = 50^2 + (2 \times -0.40 \times s)$$
$$0 = 2500 - 0.80s$$
$$0.80s = 2500$$
$$s = 3125\text{ m}$$

The work done by the train against the stopping force is

$$\text{work done} = \text{force} \times \text{distance}$$
$$= 120\,000\text{ N} \times 3125\text{ m}$$
$$= 375\,000\,000\text{ J}$$

This work is done by using its kinetic energy. The kinetic energy of the train at the start, i.e. its stored ability to do work as a result of its motion, is 375 000 000 J.

QUESTION

7.20 If the stopping force in Example 7.6 had been 60 000 N instead of 120 000 N, find:

a the distance the train takes to stop,

b the work done by the train against the stopping force in coming to rest,

c the kinetic energy of the train.

TABLE 7.2 Variation of force with stopping distance

FORCE /N	STOPPING DISTANCE /m
10	3600
100	360
1000	36
10 000	3.6
100 000	0.36

If you have worked through question 7.20 correctly, then you will have found that, with this different force to slow down the train, the work that it is capable of doing as a result of its motion is still 375 000 000 J. If the problem is repeated with any size of force, or a force of varying size, the work that can be done by the train is always 375 MJ. That is, the kinetic energy of the train is not dependent on the way its stored ability to do work is used, but only on the mass and speed of the train. This is important in the design of cars or other transport systems to reduce the injuries caused in accidents.

Accident analysis

Consider the kinetic energy of a person travelling in a car on a motorway to be 36 000 J. Note that this is about four times greater than the maximum possible kinetic energy of a sprinter. Table 7.2 gives values of the constant force necessary to stop the person in different distances.

In all cases the work done by the person against the stopping force is 36 000 J. In the first line of Table 7.2, the example is given of a very small force being exerted over a very long distance. The person pushes forwards against the seat and the floor of the car with a total force of 10 N. Using Newton's third law, the seat and the floor exert a backwards force of 10 N on the person. In the second line, the force of 100 N applied would correspond to a normal braking force. The third line would correspond to a fairly dramatic emergency stop; while stopping in 3.6 m, the fourth line, would be a serious accident. A person experiencing a force of 10 000 N applied would probably survive it if the force were applied by a seat-belt, but not if applied to the head by a steering wheel or windscreen. A force of 100 000 N would stop the person in a distance of 36 cm and would kill anybody.

Therefore, if the distance for stopping can be increased, then the force necessary to stop is reduced. Most manufacturers of cars now build into their cars collapsible sections that crumple and hence increase the distance a person can travel if involved in an accident – see **Figs 7.13** and **7.14**. The forces involved in a collision are far from being constant. **Fig 7.15** shows how, in a collision in which the average stopping force is 6000 N, the maximum stopping force is over 15 000 N. It is the maximum stopping force and how and where that force is applied that determine how serious an accident is, as it is this force which determines whether bones are broken or not.

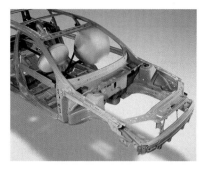

FIG 7.13 The box-like structure of the passenger compartment is made stronger than the engine compartment. In an accident this gives the passengers extra distance in which they can stop, so reducing the size of the force on them.

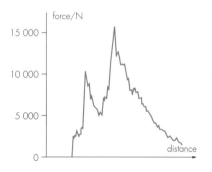

FIG 7.15 This graph shows how the force exerted on a person, when an accident occurs, can have a low average value but still cause injury as a result of there being large forces applied at certain places.

FIG 7.14 The effect of the structure described in Fig 7.13 is seen here. The passenger compartment is undamaged even after considerable damage to the engine compartment.

Equation for kinetic energy (E_k)

So far in this section, problems have been worked out from basic principles. It is often convenient to be able to find the kinetic energy of a body from its mass m and velocity v. If a constant force F is used to slow it down from an initial velocity v to a stop, then its acceleration a is given by

$$a = -\frac{F}{m}$$

Using

(final velocity)^2 = (initial velocity)^2 + $2as$

where s is the distance taken to stop, therefore gives

$$0 = v^2 - 2\frac{F}{m}s$$
$$v^2 = \frac{2Fs}{m}$$

That is,

$$Fs = \tfrac{1}{2}mv^2$$

Since Fs is the work done by the object against the stopping force as a result of its motion, it follows that its kinetic energy is given by:

K **kinetic energy, $E_k = \tfrac{1}{2}mv^2$**

Using calculus it is possible to show that this equation is valid whether or not the force is constant. In this case the work done is the change in the kinetic energy, so

$$Fs = \tfrac{1}{2}mv^2 - \tfrac{1}{2}mu^2$$

FIG 7.16 Concorde just after take-off. Maximum power is required on the runway to accelerate the aeroplane when it has maximum mass, and is sustained to enable the aeroplane to gain potential energy and further kinetic energy.

QUESTIONS

7.21 A lorry of mass 10 000 kg has a kinetic energy of 1.13×10^6 J.

   a What is the speed of the lorry?

   b What speed must a car of mass 1200 kg have in order for it to have the same kinetic energy as the lorry?

7.22 During take-off (**Fig 7.16**), an aeroplane's speed changes from zero to 40 m s^{-1} while travelling 1000 m along a runway. If its mass is 20 000 kg, find the driving force that needs to be applied to it.

7.23 Fishing line is usually sold by stating the breaking force of the line. What breaking force line is required if a 2.0 kg fish swimming at 1.8 m s^{-1} is to be stopped in 0.40 m?

7.6   POTENTIAL ENERGY AND POTENTIAL

Potential energy

The **potential energy** (E_p) of an object is its stored ability to do work as a result of its position or shape. When work is done on an object and the object accelerates, then the object stores an increasing amount of kinetic energy. An object does not necessarily accelerate, however, when work is done on it. When a lift, of total weight 8000 N, is being pulled up at a steady velocity of 2 m s^{-1}, there is no increase in the kinetic energy of the lift while the constant velocity is maintained. Work is being done on the lift and the lift in its raised position can do work on something else when it moves down, so the lift gains potential energy. The potential energy of the lift at a height of 12 m above the ground is the work the lift can do when it moves down. In this case the work done is $8000 \text{ N} \times 12 \text{ m} = 96\ 000$ J; so the potential energy of the lift 12 m above the ground is 96 kJ.

Usually a change in potential energy is required from some chosen zero. This might be sea-level, or the floor, or the lowest point of a swing, depending on the context. In the case of the lift, the 96 kJ is the increase in the potential energy of the lift as it goes upwards 12 m from the ground. We are in effect taking the potential energy on the ground to be zero. This is an arbitrary decision however; and it has the effect that, in this case, if the lift could descend into a basement, then its potential energy there would have a negative value.

In general, if a body of mass m is near the Earth, where the acceleration of free fall is g, then the weight of the body is mg. When the body is at a height h above the chosen zero of potential energy, then the work that the body can do as a result of its position, its potential energy, is given by:

   potential energy, E_p = force × vertical distance = mgh

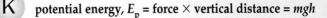

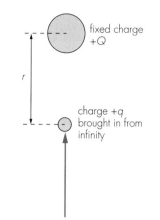

FIG 7.17 Work has to be done to bring a charge +q from a large distance to a point near a fixed charge +Q. The work done per unit charge is called the potential of the point.

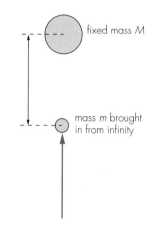

FIG 7.18 The principle illustrated in **Fig 7.17** can be applied to gravitational potential, but the gravitational potential is negative.

Other forms of potential energy

There are other situations in which a body can have work done on it but not increase its velocity.

If a spring is extended at a constant velocity by applying a suitably increasing force to it, then work is being done on the spring and the spring can release this energy at a later date, when it contracts. The energy stored by the spring, if it obeys Hooke's law (for more details, see section 23.1), is given by:

$$E_p = \tfrac{1}{2}ke^2$$

If an electrical charge +q is brought from an infinite distance to a distance r from a fixed charge +Q, then, because the two charges *repel* one another, work has to be done (see **Fig 7.17**). The electrical potential energy is a form of potential energy because the system stores the ability to do work as a result of the position of the charges (for more details, see section 14.2).

If a mass m is brought from an infinite distance to a distance r from a fixed mass M, then again there will be a change in the potential energy of the system. This will not be quite the same situation as above for the charges, because there is *attraction* between the two masses due to gravitation (see **Fig 7.18**). The potential energy stored in this system is therefore negative. It is called the gravitational potential energy (for more details, see section 9.3).

Chemical energy is a form of potential energy. When coal burns, carbon atoms combine with oxygen atoms to form carbon dioxide molecules. This reorganisation of the atoms reduces the stored electrical potential energy, and energy is emitted mostly in the form of infra-red radiation.

QUESTIONS

7.24 During her jump, a high-jumper whose mass is 48 kg raises her centre of gravity by 1.28 m.

  a What gain in potential energy takes place?

  b What must be the minimum speed of her centre of gravity at take-off?

7.25 A cricket ball is thrown upwards with a kinetic energy of 47 J. If its mass is 0.156 kg, what maximum height can it reach?

7.26 In a pinball machine, a force of 24 N is used to compress a spring a distance of 0.080 m.

  a What is the potential energy stored in the spring?

  b What is the maximum speed that it can produce in a ball of mass 0.15 kg?

7.27 In a tidal barrage scheme, the area of sea-water behind the barrage is 30 km², the tidal depth is 10 m and the density of the sea-water is 1030 kg m⁻³.

  a What energy is possibly available from this scheme in one cycle of ebb and flow?

  b If a cycle of ebb and flow takes 12 hours 40 minutes, what average power does this represent?

  c What practical problems are bound to occur if maximum energy output per cycle is to be achieved?

Potential

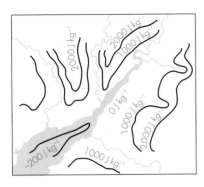

FIG 7.19 A potential map (of the Severn Estuary). Such a map is effectively a map showing contour lines. The potential is determined by the height above the reference zero.

Potential is a term applied to a place or position, rather than to a particular object. The **potential of a point** is defined as the potential energy per unit mass of an object at that point. A potential diagram is like an energy contour map. In fact, a geographical contour map can be turned into a gravitational potential energy diagram very simply. This is done in **Fig 7.19**, where the individual contour lines are renumbered to give the potential with reference to a zero of potential at mean sea-level. An object of mass 12 kg at a point 70 m above sea-level has a potential energy of

$$12 \text{ kg} \times 9.8 \text{ N kg}^{-1} \times 70 \text{ m} = 8232 \text{ J} = 8200 \text{ J (to 2 sig. figs)}$$

Unit mass at this point would have a gravitational potential energy of

$$1 \text{ kg} \times 9.8 \text{ N kg}^{-1} \times 70 \text{ m} = 686 \text{ J}$$

The potential of the 70 m contour line is therefore 686 J kg^{-1}. The potential near the Earth's surface is gh, if the zero of potential is taken at the Earth's surface.

7.7   CONSERVATION OF ENERGY

During the nineteenth century, Joule, among others, carried out a series of extremely careful experiments to measure many of the effects of doing work on a system. He took about forty years altogether, and during that time he increased the accuracy of his results as well as increasing the variety of experiments performed. Many of his experiments necessitated measuring small temperature rises, and he designed his own thermometers to measure rises in temperature to a hundredth of a degree. The type of experiment he did was to exert a known force for a known distance on a paddle wheel used to stir water. He measured the rise in the temperature of the water. He knew the work done to stir the water and he measured the heat produced in the water. He found that the same amount of work always produced the same heating effect, no matter how the work on the water was done.

In Example 7.6 we calculated how much work a train could do against the force that is being used to stop it. When this work is done, the effect is the heating of the train's braking system and the rails. The kinetic energy of the train is being reduced to zero while it stops. Joule's experiments lead us to accept that all of this loss of kinetic energy by the train is transferred into increased kinetic energy of molecules in the train and its surroundings. That is, the loss of kinetic energy of the train exactly equals the gain in thermal energy of other objects. The total energy thus remains constant. This is an example of the principle of conservation of energy, one of the most important principles in physics. The principle of **conservation of energy** states that:

K   **Energy may be transformed from one form to another but it cannot be created or destroyed, i.e. the total energy of an isolated system is constant.**

In this statement the system referred to is said to be 'isolated'. This means that, if an energy calculation is being carried out, then *all* the energy within the system must be taken into account. No energy must be allowed to escape. Another statement of the principle is: 'There is no change in the total energy of the Universe.' The principle is dealt with in greater detail in Chapter 26.

7.8 COLLISIONS

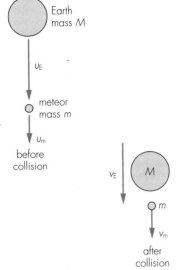

FIG 7.20 The collision of the Earth with a meteor.

FIG 7.21 The camera taking this photograph was pointed at the sky at night and the shutter opened. Whilst the shutter was open, a meteor has collided with the Earth's atmosphere. This provides the long track. Note the different colours of stars as a result of differing surface temperatures.

A collision, when spoken about in everyday speech, usually implies a car hitting something unintentionally. This is a very restricted meaning of the word and needs to be expanded here. On a microscopic scale, molecules are colliding with each other very frequently. On a cosmic scale, stars can collide. In between these two extremes there are many occasions when one object exerts a force on another object and affects the kinetic energy of both. When a collision occurs, the principles of conservation of momentum and conservation of energy can be applied.

Consider the collision shown in **Fig 7.20** where the Earth, of mass M, travelling with velocity u_E in the direction shown, collides with a small meteor, of mass m, travelling with velocity u_m in the same direction. For the collision to occur at all, u_E needs to be larger than u_m. Assume that after the collision the Earth moves with velocity v_E and the meteor moves with velocity v_m, both still in the same direction.

This problem has been simplified by having the objects always moving in the same direction, because when the principle of conservation of momentum is applied, it must be remembered that momentum is a vector. The two conservation principles give the following equations for the collision:

■ conservation of momentum

total momentum before collision = total momentum after collision
$$Mu_E + mu_m = Mv_E + mv_m$$

■ conservation of energy

total energy before collision = total energy after collision
kinetic energy (+ other energies) = kinetic energy (+ other energies)
before collision   after collision
$$\tfrac{1}{2}Mu_E{}^2 + \tfrac{1}{2}mu_m{}^2 = \tfrac{1}{2}Mv_E{}^2 + \tfrac{1}{2}mv_m{}^2 + \Delta E$$

where ΔE is the change in the forms of energy other than kinetic energy.

It should be clear from this that, whereas it is straightforward to apply the principle of conservation of momentum to a collision, provided that you remember to consider the directions of the momenta, it is less easy to apply the principle of conservation of energy. This is because the law of conservation of energy applies not just to the kinetic energy of the objects but to all forms of energy. In this example, kinetic energy must be lost. If you see a shooting star, you are seeing a collision between the Earth's atmosphere and a meteor (**Fig 7.21**). You see the light emitted from the hot meteor. Electromagnetic radiation energy is being produced by the meteor, and so some of the original kinetic energy is being used to produce this radiation.

Elastic collisions

Some collisions take place in which the total kinetic energy of the objects being considered remain constant. These collisions are called **elastic collisions**. To be doubly sure, they are sometimes called perfectly elastic collisions, or totally elastic collisions. The fact that all the kinetic energy remains as kinetic energy implies that no other energy is produced in the collision. They must be perfectly silent collisions, for example, as no sound could be produced. Elastic collisions are a theoretical perfection, although on an atomic or nuclear scale the approximation of a collision to an elastic collision is a very good one. In the laboratory, a nearly perfect elastic collision may be obtained by attaching magnets to the riders on a linear air-track, so that as the riders approach one another the two magnets repel one another (**Fig 7.22**).

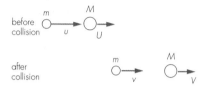

FIG 7.23 Velocity and mass symbols for use in the conservation principles for an elastic collision.

FIG 7.22 The trolleys on a linear air-track glide with very little friction. By placing disc magnets on each trolley with like poles facing, almost perfect collisions can be achieved. The trolleys bounce, totally silently, on the magnetic fields, without any direct contact.

Head-on elastic collision

The expression $u - U = V - v$ shows that, in the particular case of a head-on elastic collision, the velocity with which the bodies approach one another equals the velocity of their separation. It is worth while remembering that this applies to an *elastic* collision. Bouncing a ball on the ground illustrates this. If the ball approaches the ground at a particular speed and then it separates from the ground at the same speed, there is a perfect collision. The ground itself barely moves!

In *all* collisions, momentum is conserved; in an elastic collision, both momentum and kinetic energy are conserved. Consider a mass m colliding elastically with a mass M. Using the velocity symbols given in **Fig 7.23**, the two conservation principles give the following equations:

- conservation of momentum

 total momentum before collision = total momentum after collision
 $$mu + MU = mv + MV$$

- conservation of kinetic energy

 total kinetic energy before collision = total kinetic energy after collision
 $$\tfrac{1}{2}mu^2 + \tfrac{1}{2}MU^2 = \tfrac{1}{2}mv^2 + \tfrac{1}{2}MV^2$$

These equations are deceptively similar, and finding a relationship between the various velocities is not easy unless the equations are handled in the following way. First, cancel the $\tfrac{1}{2}$ on both sides of the kinetic energy equation, and then collect all the m terms on the left-hand side of each equation and all the M terms on the right-hand side of each:

$$m(u - v) = M(V - U)$$
$$m(u^2 - v^2) = M(V^2 - U^2)$$

Since $(u^2 - v^2) = (u - v)(u + v)$ and $(V^2 - U^2) = (V - U)(V + U)$, dividing the second of these equations by the first gives:

$$\frac{m(u^2 - v^2)}{m(u - v)} = \frac{M(V^2 - U^2)}{M(V - U)}$$
$$\frac{m(u - v)(u + v)}{m(u - v)} = \frac{M(V - U)(V + U)}{M(V - U)}$$

and then cancelling gives

$$(u + v) = (V + U)$$

or

$$u - U = V - v$$

QUESTION

7.28 A neutron, of mass 1.67×10^{-27} kg, moving with a velocity of 2.0×10^4 m s^{-1}, makes a head-on collision with a boron nucleus, of mass 17.0×10^{-27} kg, originally at rest. Find the velocity of the boron nucleus if the collision is perfectly elastic.

Inelastic collisions

Units

Remember that

$$1\ \text{kg m s}^{-2} \equiv 1\ \text{N}$$

and

$$1\ \text{N m} \equiv 1\ \text{J}$$

so that

$$1\ \text{kg m}^2\ \text{s}^{-2} \equiv 1\ \text{J}$$

If a collision takes place that is not elastic, then it is called an inelastic collision. An **inelastic collision** is one in which kinetic energy is changed into other forms of energy. The example given above of the Earth colliding with a meteor is an inelastic collision. So is a ball bouncing on the ground, because in reality balls do not bounce perfectly. Note that in this type of collision, although there is a net loss of kinetic energy, there is never any loss of momentum. The law of conservation of momentum is therefore used to solve problems of inelastic collisions. The fraction of kinetic energy lost in an inelastic collision may vary considerably. In some nearly elastic collisions, only a small fraction of the total kinetic energy is lost; whereas if a brick falls on a muddy building site, all its kinetic energy may be lost. Often, after a collision, the two colliding objects remain stuck together. These are sometimes called completely inelastic collisions. This does not imply that they have lost all their kinetic energy but that they both must have the same velocity after the collision.

EXAMPLE 7.7

In a motorway accident, a car of mass 1200 kg travelling at 40 m s^{-1} runs into the back of and gets stuck into an unloaded lorry of mass 3000 kg travelling at 25 m s^{-1}. How much kinetic energy does the car lose in the crash?

This is an inelastic collision, so we first consider momentum:

momentum before collision = momentum after collision

$$(1200\ \text{kg} \times 40\ \text{m s}^{-1}) + (3000\ \text{kg} \times 25\ \text{m s}^{-1})$$
$$= (1200\ \text{kg} + 3000\ \text{kg}) \times v$$

where v is the common speed of car and lorry after the crash. This gives

$$(48\,000 + 75\,000)\ \text{kg m s}^{-1} = 4200\ \text{kg} \times v$$
$$v = (123\,000\ \text{kg m s}^{-1})/(4200\ \text{kg}) = 29.3\ \text{m s}^{-1}$$

Now we can consider the kinetic energy of the car:

kinetic energy of car before crash
$$= \tfrac{1}{2} \times 1200\ \text{kg} \times (40\ \text{m s}^{-1})^2 = 960\,000\ \text{J}$$

kinetic energy of car after crash
$$= \tfrac{1}{2} \times 1200\ \text{kg} \times (29.3\ \text{m s}^{-1})^2 = 520\,000\ \text{J}$$

Therefore the car loses 440 000 J in the crash. The remaining 520 000 J of kinetic energy of the car will be lost immediately after the crash as the two vehicles skid to a stop. If the lorry had been more heavily loaded, the car would have lost more of its kinetic energy, up to a maximum of 585 000 J. Why cannot the car lose more than 585 000 J in the collision itself, whatever the mass of the lorry? It is interesting to note that if the driver of the car is 1/20th of the mass of the car, he has 1/20th, or 22 000 J, of the kinetic energy to lose. If the distance he moves during the collision is 2 m, then the force the seat-belt exerts on him will have an average value of 11 000 N. As this is about 20 times his weight, this force will certainly cause injury – but he will probably survive.

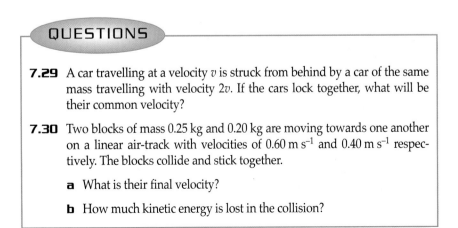

QUESTIONS

7.29 A car travelling at a velocity v is struck from behind by a car of the same mass travelling with velocity $2v$. If the cars lock together, what will be their common velocity?

7.30 Two blocks of mass 0.25 kg and 0.20 kg are moving towards one another on a linear air-track with velocities of 0.60 m s^{-1} and 0.40 m s^{-1} respectively. The blocks collide and stick together.

    a What is their final velocity?

    b How much kinetic energy is lost in the collision?

Explosions

QUESTIONS

7.31 A cannon of mass 3000 kg fires a cannonball of mass 50 kg with a horizontal velocity of 85 m s^{-1}. Find the recoil velocity of the cannon, and the ratio energy of cannonball/ energy of cannon.

7.32 An alpha particle of mass 4 u is emitted from a nucleus of mass 226 u. If energy E is released as kinetic energy, find the kinetic energy of each particle in terms of E.

An explosion can be considered in the same way as a collision, but with an increase in the kinetic energy of the system. When a rifle fires a bullet, there is a recoil in the rifle. The law of conservation of momentum can be applied to this situation. The momentum of the gun backwards equals the momentum of the bullet forwards. This is a good example of Newton's third law in action. The rifle exerts a force on the bullet, and the bullet exerts an equal and opposite force backwards on the rifle. Since these two forces are always equal and opposite and must be exerted for the same length of time, it follows that the momenta of the two parts are also equal and opposite.

The same argument cannot, however, be applied to the energies of the rifle and the bullet. In this case, although the forces are equal and opposite, the distance the bullet travels while the force is applied is much greater than the distance the rifle travels, and so the kinetic energy of the bullet is much greater than the kinetic energy of the rifle. They have the same momenta, but not the same kinetic energy. It is a pity it works out this way! There would be a good incentive not to use any guns if both the energies *and* the momenta were the same, because then the person firing the gun would get just as much kinetic energy as his or her target. Questions 7.31 and 7.32 illustrate this both on a large scale and on an atomic scale.

ANALYSIS

Energy and momentum of balls

Using data from Table 7.3 answer the questions that follow.

a Find the kinetic energy of each ball just after it is struck.

b Find the momentum of each ball just after it is struck.

c Find the loss of momentum of each striker as it strikes the ball.

d Find the impulse that each striker gives to each ball.

e Find the mass of
  i the cricket bat,
  ii the golf driver, and
  iii the tennis racket.

f Find the loss in kinetic energy of the golf driver as it hits the golf ball.

g Why are the masses found in part (e) not the true masses?

h Which of the collisions is most nearly elastic?

i What average force is exerted by each striker while it is in contact with the ball?

TABLE 7.3 Data showing typical ball speeds and striker speeds for different sports

BALL	BALL MASS /kg	BALL VELOCITY/m s^{-1} BEFORE	AFTER	STRIKER VELOCITY/m s^{-1} BEFORE	AFTER	IMPACT TIME /ms
cricket ball (hit from rest)	0.16	0	39	31	27	1.4
football (free kick)	0.42	0	28	18	12	8.0
golf ball (drive)	0.046	0	69	45	32	1.3
hand ball (serve)	0.061	0	23	19	14	1.4
squash ball (serve)	0.032	0	49	44	34	3.0
tennis ball (serve)	0.058	0	51	38	33	4.0

In this investigation you are asked to find how the velocity of an object A changes when it makes an elastic collision with another object B and how the change in velocity depends on the mass of B.

A linear air-track is required and it should be set up as follows: One slider (slider A) has fixed mass but the other slider (slider B) has its mass changed by the addition of Plasticine. The sliders on the track should have magnets fitted with opposing poles so that they repel one another. The velocity of A when it approaches B should always be the same, and the timing device for the track should be used to find the velocity of A after the collision.

A graph of velocity of A after the collision should be plotted against the mass of B.

SUMMARY

- Momentum is the product of a body's mass and velocity.
- Impulse is the product of force and time.
- Work is the product of a force and the distance moved in the direction of the force.
- Energy is the stored ability to do work.
- Power is the rate of doing work.
- Comparison of impulse and work:

	SHOWN ON A GRAPH BY	CAN CAUSE A CHANGE IN THE	UNIT
Impulse	area under force–time graph	momentum, $mv - mu$	N s
Work	area under force–distance graph	kinetic energy, $\frac{1}{2}mv^2 - \frac{1}{2}mu^2$	J

- In all collisions momentum is conserved.
- In elastic collisions kinetic energy is conserved; in inelastic collisions kinetic energy is not conserved.
- The increase in the potential energy of a body of mass m raised a height h near the Earth's surface is mgh.

ASSESSMENT QUESTIONS

7.33 a State *Newton's second law of motion*.

b During a shunting operation, a railway wagon of mass 1.2×10^4 kg moving at 3.0 m s^{-1} joins on impact with a stationary wagon of mass 2.0×10^4 kg.

  i Calculate the speed of the wagons after they have joined.

  ii The collision of the wagons is inelastic. State the energy changes that take place during this collision.

Cambridge 1997

7.34 a An athlete of mass 55 kg runs up a flight of stairs of vertical height 3.6 m in 1.8 s. Calculate the power that this athlete develops in raising his mass.

b One way of comparing athletes of different sizes is to compare their power-to-weight ratios. Find a unit for the power-to-weight ratio in terms of SI base units.

c Calculate the athlete's power-to-weight ratio.

London 1996

7.35 a State the *principle of conservation of momentum*.

b A bullet of mass 0.025 kg is travelling horizontally with a speed of 150 m s⁻¹ when it strikes the centre of a vertical face of a cubical block of mass 2.0 kg which is hanging at rest from vertical strings.
If the bullet embeds itself in the block, calculate the vertical height risen by the block and bullet.

WJEC 1998

7.36 A golf ball at rest is struck by a club which remains in contact with the ball for 200 ms. The mass of the ball is 0.045 kg and the mean force on the ball during contact is 9.0 N. The ball leaves the ground at 30° to the horizontal, and then travels above level ground. Air resistance may be ignored.

a Show that the initial speed of the ball as it leaves the ground is 40 m s⁻¹.

b Calculate the maximum height reached by the ball.

c Calculate the kinetic energy of the ball at the maximum height.

d What is the magnitude of the acceleration of the ball at the maximum height?

CCEA 1998

7.37 The figure illustrates a steel ball of mass 0.020 kg held vertically above a horizontal steel plate.

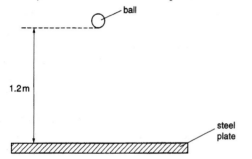

It is released from a height of 1.2 m and rebounds vertically to a height of 1.1 m. The ball is in contact with the surface for 0.90 ms.

a Calculate the magnitude of
  i the velocity of the ball as it arrives at the plate,
  ii the momentum of the ball as it arrives at the plate.

b i Calculate the change of momentum of the ball in its collision with the plate.
  ii Hence determine the average force which the surface exerts on the ball during the collision.

c State, with a reason, whether the collision between the ball and the plate is elastic.

Cambridge 1997

7.38 Two frictionless trolleys A and B, of mass *m* and 3*m* respectively, are on a horizontal track.

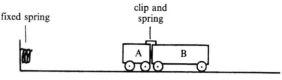

Initially they are clipped together by a device which incorporates a spring, compressed between the trolleys. At time *t* = 0 the clip is released and the trolleys move apart, the spring falling away. The time during which the spring expands is negligible. The velocity of trolley B is then *u* to the right.

a Show that the magnitude of the velocity of trolley A as the trolleys move apart is 3*u*.
State the direction of the velocity of trolley A.

b At time *t* = *t₁*, trolley A collides elastically with a fixed spring and rebounds. The compression and expansion of the fixed spring take place in a negligibly short time. Trolley A catches up with trolley B at time *t* = *t₂*.
  i What is the velocity of trolley A between *t* = *t₁* and *t* = *t₂*?
  ii Find an expression for *t₂* in terms of *t₁*.

c When trolley A catches up with trolley B at time *t* = *t₂* the clip operates so as to link them again, this time without the spring between them, so that they move together with velocity *v*. Calculate the common velocity *v* in terms of *u*.

d Initially, before the clip was opened, the trolleys were at rest and the total momentum of the system was zero. However, your answer to **c** should show that the total momentum after *t* = *t₂* is not zero. Discuss this result with reference to the principle of conservation of momentum.

CCEA 1998

7.39 A stationary ball of mass 6.0 × 10⁻² kg is hit horizontally with a tennis racquet. The ball is in contact with the racquet for 30 ms and leaves the racquet with a speed of 27 m s⁻¹.

a Calculate
  i the change in the momentum of the ball,
  ii the average force which the racquet exerts on the ball.

b Calculate the horizontal distance travelled by the ball before it hits the ground, if it leaves the racquet at a vertical height of 2.5 m.

c i Explain what is meant by an *inelastic collision*.
  ii Suggest a reason why the collision between the ball and the racquet is inelastic.

NEAB 1998

7.40 The water-wheel of a restored water-mill provides the torque needed to drive the machinery of the mill. The wheel is driven by the weight of water held in buckets at its rim, which are filled with water from a mill-pond above the wheel and empty into a channel beneath it. Before opening the mill to the public, measurements are made at the output shaft of the wheel to obtain operating data for an information booklet.

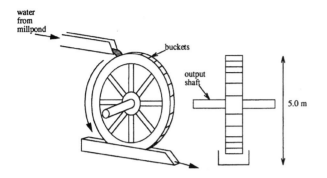

a The maximum flow rate possible without excessive spillage is 18 m³ of water per minute, which falls a vertical distance of 5.0 m from mill-pond to exit channel. Under these conditions the wheel turns at 7.0 rev min⁻¹ and produces a torque of 7.4 kN m at its output shaft.

  i Calculate the available power input to the wheel provided by the falling water, explaining your reasoning.
density of water = 1000 kg m⁻³

  ii Calculate the output power of the wheel.

  iii Hence show that the efficiency of this water-wheel as an energy converter is 0.37 (37%).

b The main reason for this poor use of available energy is thought to be due not to friction in the wheel bearings, but to the design of the buckets, which leak and spill water prematurely. To check that friction in the wheel-bearings really is a minor factor, the wheel is disconnected from the mill machinery and it is found that a flow rate of 1.5 m³ min⁻¹ is just sufficient to keep it turning at 7.0 rev min⁻¹. Assuming that the efficiency of energy conversion is 0.37, estimate

  i the power loss caused by the frictional torque acting on the wheel,

  ii the magnitude of the frictional torque.

NEAB 1997

7.41 This question is about the motion of a sheet of steel passing through a rolling mill.

a A sheet of hot steel of mass 400 kg is carried along a conveyor at 0.80 m s⁻¹. Find

  i the momentum of the sheet

  ii the kinetic energy of the sheet.

b At the end of the belt, the sheet is inserted at 0.80 m s⁻¹ between two rollers (see figure), which reduce the thickness of the sheet from 0.018 m to 0.012 m. The width of the sheet is unchanged.

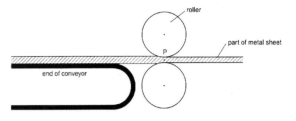

  i Show that the *increase* in speed of the sheet as it passes through the rollers is 0.40 m s⁻¹.

  ii In which direction is the resultant force on the sheet at P as it passes between the rollers.

  iii The original length of the sheet is 2.4 m. How long does it take for the sheet to pass between the rollers?

  iv Find the magnitude of the resultant force on the sheet during the squashing process.

c Find the change in kinetic energy of the sheet and hence the minimum extra power which must be supplied by the rollers to accelerate the sheet.

O & C 1997

7.42 This question is about the power output of a cyclist.

a *Power* can be calculated from the product of *force applied* and *velocity*; $P = Fv$. Justify this expression, starting from the definition of power.

b A cyclist pedalling along a horizontal road provides 200 W of useful power. The cyclist reaches a steady speed of 5.0 m s⁻¹. What is the value of the drag forces against which the cyclist is working?

c The drag forces are proportional to the speed of the bicycle.

  i Show that the useful power the cyclist must produce at speed v along the flat is proportional to v^2.

  ii Predict what power the cyclist must produce to reach a speed of 6.0 m s⁻¹ along the flat.

d What useful power would the cyclist have to develop to maintain a speed of 5.0 m s⁻¹ when climbing a hill of 1 in 30? Take the mass of cyclist plus bicycle to be 100 kg.

O & C 1997

7.43 This question is about collisions.

a **i** Define the *momentum* of a particle.

ii State Newton's second law of motion in terms of momentum.

A student investigates collisions between gliders on an air track. The gliders each have a mass of *m*.

b In a particular collision, a moving glider A collides elastically with a stationary glider B. It is observed that A stops and B moves off with the same velocity as A had before the collision. Show that this is consistent with both the conservation of kinetic energy and the conservation of momentum. Include a simple diagram in your answer.

c In another collision a glider, moving with speed *u*, collides inelastically with a stationary glider. The gliders stick together and move off as one.

i Find the speed of the combined gliders after the collision.

ii Calculate the kinetic energy dissipated in the collision in terms of *m* and *u*.

iii What happens to the kinetic energy dissipated?

O & C 1997

7.44 The graph shows the speed of a racing car during the first 2.6 seconds of a race as it accelerates from rest along a straight line.

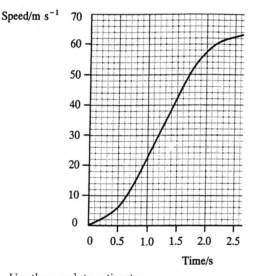

Use the graph to estimate

a the displacement 1.5 s after the start,

b the acceleration at 2.0 s,

c The kinetic energy after 2.5 s given that the mass of the racing car is 420 kg.

London 1999

7.45 The diagram shows a flat wooden board resting on an iron bar and with a heavy block resting on the left hand end of the board.

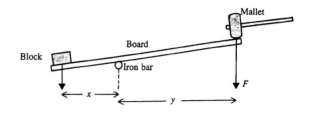

The block is thrown into the air when the right hand end of the board is struck by a mallet.

The diagram below is the free-body force diagram for the wooden board *just after the block has begun to move*.

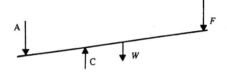

a Four forces are shown on the free-body force diagram. *W* is the weight of the board and *A* is the downward push on the board from the block. State whether the magnitude of the force A is greater than, equal to or less than the weight of the block. Explain your answer.

b Write down suitable labels for the forces *F* and *C*.

The graph below shows how the resultant force on the block varies with time.

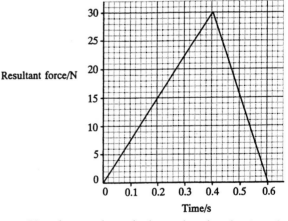

c Use the graph to find a value for the impulse applied to the block.

d The mass of the block is 7.1 kg. Calculate its maximum upward speed.

e The force of the board on the block is not equal to the resultant force on the block. Why not?

London 1999

7.46 A railway truck of mass 22 000 kg and moving at a speed of 3 m s⁻¹ catches up and collides with a truck of mass 66 000 kg moving at 1 m s⁻¹.

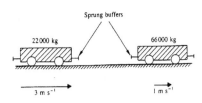

The graph shows the speeds of the trucks before, during and after the collision.

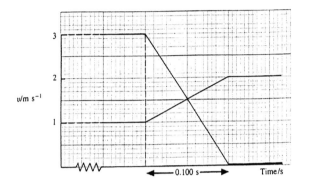

a Use the information in the graph to show that the collision is elastic.
b Show that the total kinetic energy halfway through the collision is less than the total kinetic energy after the collision.
Suggest a reason for this.
c How is it possible for the two trucks to be in contact but to be travelling at different speeds?
d Calculate the magnitude of the impulse exerted by the lighter truck on the heavier truck.
e Explain whether or not you would expect this impulse to change in value if the collision remains elastic but takes half the time.

London 1999

7.47 This question is about safety design of cars.
In a test rig, a car travelling at 25 m s⁻¹ collides with a barrier. The car is brought to rest in one second. *Without detailed calculations, use your knowledge of Physics* to answer the following questions.
a State and explain whether and in what ways conservation of **i** energy and **ii** linear momentum apply in this collision.
b Show how the average force *on the car* during the collision can be found by considering changes in momentum. Discuss the factors which determine the magnitude of the force.
c A dummy in the passenger seat is not wearing a seat belt during the collision. Explain why it will experience forces which, in a real collision, would produce severe head and chest injuries.

d The dummy in the driver's seat is wearing a seat-belt which restrains it. Explain why wearing a seat-belt may reduce injuries in a real collision.
e In a similar test, an airbag inflates within 0.05 s of the impact, filling the space between the dummy and the steering wheel. Discuss ways in which the inflation and subsequent slower deflation of an airbag reduce the risk of injury.

O & C 1998

7.48 A builder's hoist raises a total mass of 400 kg by means of an electric motor. While travelling at constant speed, the load is observed to rise through 12 m in 8.0 s.
a Calculate the constant speed at which the load rises.
b What is the gain in potential energy of the load when it rises 12 m?
c How much mechanical power is the motor developing when the load rises with this constant speed?
d If the electric motor draws 6.8 kW of electrical power when raising the load at this constant speed, what is the efficiency of the system?

CCEA 1998

7.49 The graph below represents the motion of a car of mass 1.2×10^3 kg.

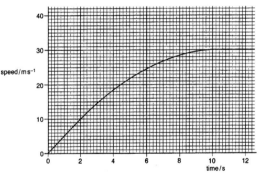

a Calculate
  i the initial acceleration of the car,
  ii the value of the driving force provided by the engine for the acceleration calculated in **i**,
  iii the maximum kinetic energy of the car.
b The car engine uses the energy available from the fuel with an efficiency of 20%. The fuel provides energy at a cost of 1.4 pence per megajoule. In order to maintain the constant maximum speed, a driving force of 1.2×10^3 N is required.
  i Calculate the power due to the driving force when the car is travelling at the maximum speed.
  ii Calculate the cost of the fuel consumed to provide the driving force during a twenty minute period in which the car travels at the maximum speed.
  iii Explain why your answer to **ii** is an under-estimate.

Cambridge 1997

7.50 In the calculations in this question, take the acceleration of free fall g as 10 m s^{-2}.

    a **i** State Newton's laws of motion.

       ii Two small trays, each of mass 80 g, are connected by a light, inextensible string which passes over a frictionless pulley, as shown in the left-hand diagram. The trays remain balanced.

         A mass of 40 g is then placed on one of the trays, which starts to move downwards (as in the right-hand diagram). Calculate

         1 the acceleration of the trays,

         2 the tension in the string,

         3 the force exerted by the tray on the 40 g mass when in motion.

    b **i** A car of weight 9000 N travels up a slope inclined at 10° to the horizontal with an acceleration of 0.60 m s^{-2}. At a particular speed the force of friction (opposing motion and acting down the slope) is 500 N. Calculate the driving force between the wheels and the ground at the instant when the car is moving with this speed.

       ii The maximum speed of the car up this slope is 20 m s^{-1}. When moving at this speed the total force of friction is 660 N. Calculate

         1 the maximum driving force between the wheels and the ground,

         2 the mechanical power produced at this speed.

    c **i** Explain what is meant by

         1 an *elastic* collision,

         2 an *inelastic* collision.

         For each type of collision, state an example of a situation in which it occurs.

       ii Two space probes of masses 2000 kg and 3000 kg, travelling in the same straight line in opposite directions at 5.0 m s^{-1} and 3.0 m s^{-1} respectively, collide, and after impact stick together. Calculate

         1 the speed after impact,

         2 the impulse given to the space probe of mass 3000 kg.

CCEA 1998

7.51 A car of mass 900 kg tows a caravan of mass 750 kg, as shown in the diagram.

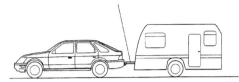

There is a total force of 1700 N opposing the motion of the car and a total force of 2000 N opposing the motion of the caravan. Initially, the car and caravan move at a constant velocity of 20 m s^{-1}.

    a **i** State, with a reason, the tension in the tow bar.

       ii Calculate

         1 the driving force provided by the engine,

         2 the power due to the driving force.

    b Part of the total force acting to oppose the motion of the car and caravan is due to friction and part is due to air resistance. The total constant frictional force which acts on the car and caravan system is 1500 N. The magnitude of the air resistance on the system is proportional to the square of the speed of the system. Calculate the total power developed by the new driving force when the car moves at a constant speed of 25 m s^{-1}.

Cambridge 1997

8 CIRCULAR MOTION

LEARNING OBJECTIVES

At the end of this chapter you should be able to:

① work with angles in radians;

② define angular velocity and period;

③ use the term 'centripetal acceleration' correctly;

④ see how Newton's laws apply equally well to circular motion as to linear motion.

8.1 ANGULAR VELOCITY

If something is spinning (**Fig 8.1**), it is not possible to state a unique velocity with which it is travelling, as different parts of it are travelling with different velocities. To overcome this problem, the term 'angular velocity' is used. Angular

FIG 8.1 Circular motion of a cyclonic storm.

velocity is determined by the rate of rotation of the body and could be measured in revolutions per second or revolutions per minute. Formally, the **angular velocity** of a body is defined as the rate of change of its angular displacement. The conditions stated in section 4.1 for linear velocity also apply to angular velocity, but with angular displacement replacing linear displacement. The angular velocity of a rigid rotating body is the same for all points of the body.

Measuring an angle

At first sight the measurement of an angle appears to be straightforward. A protractor marked in degrees is used and placed on the angle to be measured. However, using a calibrated instrument does no more to define an angle than does using a watch define time.

The way an angle is defined is shown in **Fig 8.2**. The angle θ is *defined* by the equation

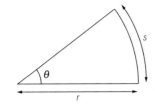

FIG 8.2 Definition of an angle.

$$\boxed{K} \quad \theta = \frac{\text{arc length}}{\text{radius of arc}} = \frac{s}{r}$$

where s is the curved distance along the arc of radius r. This equation is probably more frequently used in the form

$$s = r\theta$$

Since θ is a ratio of two lengths, it will be measured in metres divided by metres. To indicate that the angle has been measured in this way, the unit of angle is given the name 'radian'. One **radian** (1 rad) is the angle when the arc length is the same as the radius of the arc. A protractor measuring in radians looks like a protractor measuring in degrees, but with a different scale (**Fig 8.3**). Note that since

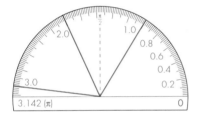

FIG 8.3 A protractor calibrated in radians.

$$\pi = \frac{\text{circumference of circle}}{\text{diameter of circle}}$$

it follows that the circumference of a circle equals $2\pi r$, where r is the radius of the circle. This gives the angle in radians, for one revolution of $2\pi r / r = 2\pi$ rad.

Table 8.1 gives conversion factors between the various angular measures. You should take particular care *always* to measure angles in radians in this chapter. The advantage of using angular measure in radians is that it avoids the need to keep using conversion factors, as the example shows.

If you are not already familiar with measuring angles in radians, it is recommended that, to gain familiarity with the unit, you answer all the parts of question 8.1 and that you work out Table 8.1 for yourself, with this book shut, and then check your values against those in the table.

EXAMPLE 8.1

The laser on a CD player is 5 cm from the central hole of the disc. What length of the disc is scanned by the laser when the disc turns through an angle of 0.45 radians?

Since

$$0.45 \text{ rad} = \frac{\text{arc length}}{\text{radius of arc}}$$
$$= \frac{\text{arc length}}{5\text{cm}}$$

we get

$$\text{arc length} = 0.45 \text{ rad} \times 5 \text{ cm}$$
$$= 2.25 \text{ cm}$$

TABLE 8.1 Conversion factors for angles

REVOLUTIONS	RIGHT-ANGLES	RADIANS	DEGREES
1	4	2π	360
3/4	3	$3\pi/2$	270
1/2	2	π	180
1/4	1	$\pi/2$	90
1/6	2/3	$\pi/3$	60
1/(2π)	2/π	1	180/π
1/8	1/2	$\pi/4$	45
1/12	1/3	$\pi/6$	30
1/360	1/90	$\pi/180$	1

EXAMPLE  8.2

A train is travelling on a track, which is part of a circle of radius 600 m, at a constant speed of 50 m s^{-1}. What is its angular velocity?

The angle the train turns through in unit time is the angular velocity. So its angular velocity ω is

$$\omega = v/r = (50 \text{ m s}^{-1})/(600 \text{ m})$$
$$= 0.083 \text{ rad s}^{-1}$$

Note that, although the unit (m) cancels out in the division, the unit of angle, rad, is still included in the answer to make it clear that it is an angle turned through in a second and that the angle itself is measured in radians.

EXAMPLE  8.3

A washing machine spins its tub at a rate of 1200 revolutions per minute (rpm). If the diameter of the tub is 35 cm, find (a) the angular velocity of the tub, and (b) the linear speed of the rim of the tub.

a The angular velocity ω is 1200 rpm, which we need to express in rad s^{-1}, so

$$\omega = 1200 \text{ rpm}$$
$$= (1200/60) \text{ revolutions per second}$$
$$= 20 \times 2\pi \text{ rad s}^{-1}$$
$$= 40\pi \text{ rad s}^{-1} = 126 \text{ rad s}^{-1}$$

b The linear speed of the rim v is

$$v = r\omega$$
$$= 0.35 \text{ m} \times 126 \text{ rad s}^{-1}$$
$$= 22.0 \text{ m s}^{-1}$$

Note that the unit, rad, is not written in the answer, as a linear speed is required.

8.1 Change the following angles into radians. *Do not* use a calculator. You can leave π in any of your answers.

   a two revolutions

   b 135°

   c five right-angles

   d 50 revolutions

   e 10°

   f the angle the minute hand of a watch rotates through in a day

Angular velocity

Angular velocity was defined formally at the beginning of the chapter as the rate of change of angular displacement. It is usually given the symbol ω, which is the small letter 'omega' from the Greek alphabet (Ω is the capital omega). The unit of angular velocity in SI is the radian per second (rad s^{-1}).

K The *angular velocity* is the angle turned through per unit time. The angle is normally measured in radians

For a constant angular velocity this gives

$$\omega = \frac{\theta}{t}$$

From the definition of angle θ in **Fig 8.2**, we had $s = r\theta$. If both sides of this equation are divided by time, then we get $s/t = r\theta/t$.

But $s/t = v$ and $\theta/t = \omega$, so we obtain

$$v = r\omega$$

Period

The **period** T of circular motion is the time taken for one revolution. Period is related to angular velocity by the equation

K   time for one revolution = $\dfrac{\text{angle turned through in one revolution}}{\text{angular velocity}}$

or

$$T = \frac{2\pi}{\omega}$$

(Compare this equation with time = distance/velocity. Here 2π is the angular distance and ω is the angular velocity.)

8.2 A clock has a second hand that is 5.0 cm long, a minute hand that is 5.0 cm long and an hour hand that is 4.0 cm long. Find the angular velocity of each of the hands and the linear speed of the tip of each hand. Assume that there is no jerkiness in their movement.

8.3 Give the angular velocity of each of the following:

    a the Earth in its orbit around the Sun,

    b the Earth about its axis,

    c the Moon in its orbit around the Earth,

    d the Moon about its axis.

8.2 CENTRIPETAL ACCELERATION

If you look at question 8.3 you should notice that it involves two types of motion in a circle. The Moon in its orbit around the Earth is an object travelling in a nearly circular path (**Fig 8.4a**). This is referred to as **circular motion**, and the assumption is usually made that the size of the object travelling in the circle is negligible compared with the radius of the orbit. The other type of motion is illustrated by the Earth rotating about its own axis (**Fig 8.4b**). This is referred to as **rotation**. Rotation is rather more difficult to deal with than circular motion because different parts of the object have different speeds. Detailed analysis of rotation is not dealt with in this book.

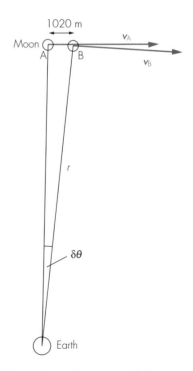

FIG 8.4 **a** The Moon as seen from a satellite in orbit about the Earth. The Moon travels in a nearly circular orbit around the Earth. **b** A camera pointing at the pole star clearly shows the rotation of the Earth.

FIG 8.5 A greatly exaggerated diagram showing the angle the Moon moves through in its orbit in one second ($\delta\theta$ should be much smaller).

A particular example

When an object moves along a circular path at a constant speed, its velocity is always changing because the direction in which the object travels is changing. It was indicated in Chapter 3 that some surprising results occur when changes in a vector quantity are calculated. It is now necessary to return to some of these results to find the acceleration of a body in circular motion. Consider the following question, done as an example.

EXAMPLE 8.4

Find the acceleration of the Moon. It travels at a constant speed of 1020 m s^{-1} (to 3 sig. figs) and takes 27.3 days for a complete revolution of the Earth. (This question is a rewording of question 3.5, which you may have already done.)

First we find the radius r of the orbit. Using **Fig 8.5** (and changing 27.3 days to seconds) we get

> circumference of orbit
> $= \text{speed} \times \text{time}$
> $= (1020 \text{ m s}^{-1}) \times (27.3 \times 24 \times 60 \times 60 \text{ s})$
> $= 2.41 \times 10^9 \text{ m}$

and so (using the property of a circle)

> radius of orbit $r = \text{circumference}/2\pi$
> $= (2.41 \times 10^9 \text{ m})/2\pi$
> $= 3.83 \times 10^8 \text{ m}$

The Moon moves distance $s = 1020$ m in one second. So the angle through which the Moon moves in one second ($\delta\theta$) is given in radians as

> $\delta\theta = s/r = (1020 \text{ m})/(3.83 \times 10^8 \text{ m}) = 2.66 \times 10^{-6} \text{ rad}$

A vector diagram can now be drawn to find the change in the velocity of the Moon in one second (**Fig 8.6**). The change in velocity (δv) is the velocity that has to be added to the velocity at the start to obtain the velocity after one second. It has a small value, and even in this narrow triangle the angle $\delta\theta$ has had to be exaggerated. The direction of the change in velocity is towards the centre of the circle, at right-angles to the individual velocities. Using the triangle in the vector diagram gives

> $\delta v = v\delta\theta = (1020 \text{ m s}^{-1}) \times (2.66 \times 10^{-6} \text{ rad})$
> $= 2.72 \times 10^{-3} \text{ m s}^{-1}$

This has made the very good assumption that the vector diagram is a sector of a circle and that therefore, by definition of an angle in radians, the angle at its centre is $\delta\theta = \delta v/v$.

Since the change in velocity of the Moon in one second is 2.72×10^{-3} m s^{-1}, in a direction towards the centre of the orbit, then this must be its acceleration, as acceleration is defined as the change in velocity per unit time, i.e.

> acceleration of the Moon
> $= 2.72 \times 10^{-3}$ m s^{-2} towards the Earth

FIG 8.6 The change in the Moon's velocity has the small value δv in a direction almost at right-angles to both v_A and v_B.

This example shows how the acceleration of an object may be found in a particular case when an object is travelling in a circle at a constant speed. The direction of the acceleration is always changing as the object moves, but is always directed towards the centre of the path in which it travels and always has the same magnitude. The acceleration towards the centre for an object travelling in a circle is called the **centripetal acceleration**. There is nothing basically different about a centripetal acceleration from any other acceleration. We can apply Newton's laws to a centripetal acceleration in exactly the same way as we can to other accelerations.

QUESTION

8.4 Assuming that the Earth goes round the Sun in a circle of radius 1.50×10^{11} m, find the centripetal acceleration of the Earth. You should know how long it takes for the Earth to go once round the Sun!

Centripetal acceleration

Note three important points concerning centripetal acceleration:

- The rate at which the object travels in its circular path is constant.
- The direction of the acceleration at any instant is in a straight line.
- The value of the acceleration can be found directly from the definition of acceleration.

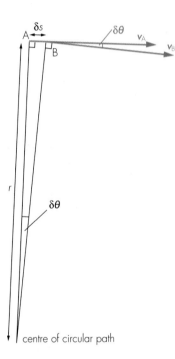

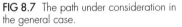

centre of circular path

FIG 8.7 The path under consideration in the general case.

FIG 8.8 The vector diagram in the general case.

The general case

Example 8.4 works from first principles to find the acceleration of the Moon in its orbit around the Earth. To establish a general formula that can be used to calculate centripetal accelerations, we need to do the same problem algebraically. The following terms will be used in the proof:

v = the constant speed of the body travelling in the circular path
v_A = the velocity of the body at A (v is its magnitude)
v_B = the velocity of the body at B (v is its magnitude too)
r = the radius of the circular path
δt = the small time it takes to get from A to B
$\delta \theta$ = the small angle turned through in this time
δs = the small distance travelled during this time = $v\delta t$
δv = the small change in velocity during this time.

Fig 8.7 shows the path under consideration, and **Fig 8.8** is the vector diagram to enable the change in velocity to be found. Note that the angle between v_A and v_B is the same in both diagrams. Using **Fig 8.7** and **Fig 8.8** gives

$$\delta s = v\delta t = r\delta\theta \text{ and } \delta v = v\delta\theta$$

Dividing these two equations gives

$$\frac{\delta v}{v\delta t} = \frac{v\delta\theta}{r\delta\theta}$$

Rearranging and cancelling gives

$$\text{acceleration} = \frac{\delta v}{\delta t} = \frac{v^2}{r}$$

which is in the direction of δv and so is towards the centre of the circular path. Since also $v = r\omega$ (see section 8.1) we can write the following:

> **K**   **Centripetal acceleration = $v^2/r = r\omega^2 = v\omega$ towards the centre of the circular path.**

QUESTIONS

8.5 The maximum speed of the blades on rotary lawn-mowers is restricted to avoid danger from flying stones. If the rate of rotation of the blade in one particular model is 3500 revolutions per minute and the blade has a radius of 0.23 m, find

   a the angular velocity of the blade;

   b the linear velocity of the tip of the blade;

   c the centripetal acceleration of the tip of the blade.

8.6 An electric motor has a rotor that rotates with an angular velocity of 50 revolutions per second. The rotor has a diameter of 12.0 cm.

   a What is the centripetal acceleration of the rim of the rotor?

   b How does the centripetal acceleration of the rotor vary through the body of the rotor from its centre to its rim?

8.7 A centrifuge is required to give an acceleration of 1000*g* to a particle at a distance of 8.5 cm from the axis of rotation. Find the necessary angular velocity of the centrifuge.

8.3 EXAMPLES OF CIRCULAR MOTION

Problem-solving

In solving problems concerned with Newton's laws, the following routine is strongly recommended:

1 State clearly which object is being considered.
2 Draw a free-body sketch of that object only.
3 Mark on the sketch the gravitational pull on the object, its weight.
4 Mark on the sketch all the points where the object touches anything else, and draw in the contact forces at these points. Label all forces clearly.
5 Decide which direction to call positive for the total force and acceleration.
6 Apply Newton's second law equation.

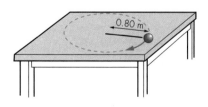

FIG 8.10 A ball of mass 0.300 kg, attached to a string, rotating on a friction-free surface. The ball has a constant speed of 3.45 m s⁻¹.

Having found the acceleration of an object travelling along a circular path at a constant speed, we are in a position to consider the application of Newton's laws. This needs to be done in exactly the same way as explained in Chapter 5, where the guidelines shown in the box were written. If this routine is followed, you will find that it can be used to solve all problems. You will not need to use a different approach when more complicated situations arise, and you will be far less likely to make unnecessary mistakes. In use, the routine is direct, reliable and quick. Supposed short-cuts in solving these problems lead to many mistakes being made. The essential need is for complete free-body force diagrams rather than composite pictures of more than one object.

Exactly this procedure needs to be followed in the case of circular motion. Circular motion really is no different from any other type of motion. In *all* motion, the resultant force on the object controls its acceleration, and the resultant acceleration is always in the same direction as the resultant force. When dealing with circular motion, the resultant force is sometimes called the 'centripetal force'. It is important to realise, however, that this is only the sum of all the forces acting on the body moving in a circle and is *not* an extra force on the body. It is *not* needed on your free-body diagram.

QUESTION

8.8 Fig 8.10 shows a mass of 0.300 kg rotating in a circular path of radius 0.80 m on a friction-free table. It is attached by a string to a peg at the centre of the circle.

a Draw a free-body force diagram for the mass.

b Find the force that the string exerts on the mass when the mass is moving at a constant speed of 3.45 m s⁻¹.

[Your free-body force diagram should have three forces shown on it. The mass is touching two objects, the table and the string, and each will exert a force on it. There is the weight of the mass as well. In this case, because the table is friction-free, the weight and the force the table exerts on the mass are equal and opposite, and so will cancel out. This leaves the force that the string exerts on the mass equal to the resultant force on the mass.]

EXAMPLE 8.5

A satellite of mass 800 kg is orbiting the Moon in a circular path of radius 1760 km (**Fig 8.9a**). The weight of the satellite is 1300 N. Find the speed of the satellite and its period of rotation.

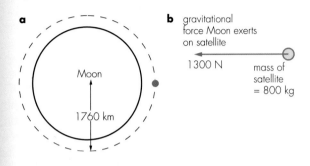

a

Moon

1760 km

b gravitational force Moon exerts on satellite

1300 N   mass of satellite = 800 kg

FIG 8.9 The force on a satellite in orbit around the Moon.

The force diagram is very simple in this example (**Fig 8.9b**). The satellite is touching nothing, so the only force exerted on it is its weight, i.e. the gravitational attraction that the Moon exerts on it:

force (N) = mass (kg) × acceleration (m s^{-2})

$$1300 \text{ N} = 800 \text{ kg} \times \frac{v^2}{r} = 800 \text{ kg} \times \frac{v^2}{1.76 \times 10^6 \text{ m}}$$

$$v^2 = \frac{1.76 \times 10^6 \text{ m} \times 1300 \text{ N}}{800 \text{ kg}} = 2.86 \times 10^6 \text{ m}^2 \text{ s}^{-2}$$

$$v = 1690 \text{ m s}^{-1}$$

Hence

$$\omega = \frac{v}{r} = \frac{1690 \text{ m s}^{-1}}{1.76 \times 10^6 \text{ m}} = 9.61 \times 10^{-4} \text{ rad s}^{-1}$$

$$T = \frac{2\pi}{\omega} = \frac{2\pi \text{ rad}}{9.61 \times 10^{-4} \text{ rad s}^{-1}} = 6540 \text{ s}$$

The satellite actually takes longer to travel around the Moon than it would to travel around the Earth, despite the Earth being larger. This is because the Earth would exert a larger force on it, and so there would be a much larger acceleration, hence a much greater speed.

EXAMPLE 8.6

A racing car of total mass 600 kg travels in a circle of radius 80 m at 30 m s^{-1} round a corner. Draw a diagram showing the forces acting on the car if **a** the track surface is horizontal and **b** the track is banked so that there is no tendency for the car to skid sideways. Calculate the required frictional force in **a**, and the required angle of banking in **b**.

a   The diagram is shown in **Fig 8.11a**. The resultant force on the car is the frictional force that the track exerts on the wheels, so

frictional force = mass × v^2/r
= (600 kg) × (30 m s^{-1})^2/(80 m)
= 6750 kg m s^{-2} = 6750 N

b   The diagram is shown in **Fig 8.11b**. The racing car is accelerating towards the centre of the circle of the curve and requires a force shown to be 6750 N in that direction. This force is provided by the horizontal component of C. The vertical component of C must be equal and opposite to the weight, which is $mg = 600 \text{ kg} \times 9.8 \text{ m s}^{-1} = 5880 \text{ N}$, since there is no acceleration in the vertical direction. This gives

$C \sin \theta = 6750$ N
$C \cos \theta = 5880$ N

Dividing gives $\tan \theta = 1.148$ and therefore

$\theta = 49°$

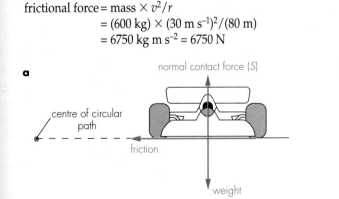

a

normal contact force (S)

centre of circular path

friction

weight

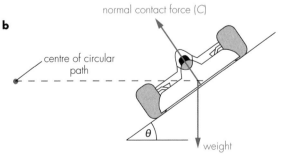

b

normal contact force (C)

centre of circular path

θ

weight

FIG 8.11 Diagrams for Example 8.6. **a** The normal contact force S must be equal and opposite to the weight since there is no acceleration in the vertical direction. **FIG 8.11b** The normal contact force C now has a different value.

EXAMPLE 8.7

Fig 8.12 is a photograph of an amusement park ride in which a person of mass 63 kg rotates in a vertical circle of radius 6.6 m. The time taken for one revolution is 3.2 s. Find the force that is exerted by the structure on the rider when the rider is (a) at the bottom of the circle and (b) at the top of the circle.

FIG 8.12 Some people enjoy being flung upside down at the top of a vertical circle!

The weight of the rider is 63 kg × 9.8 N kg^{-1} = 617 N. The acceleration of the rider is at all times towards the centre and has a value given by:

$$\text{acceleration} = r\omega^2 = 6.6 \times \left(\frac{2\pi}{3.2}\right)^2 = 25.4 \text{ m s}^{-2}$$

Fig 8.13 shows two free-body force diagrams for the person when in the two different positions under consideration.

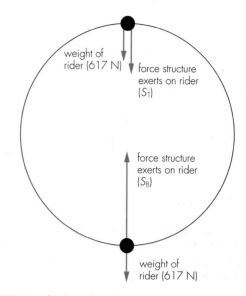

FIG 8.13 Two free-body force diagrams for a person sitting in the amusement park ride shown in Fig 8.12.

a At the bottom, the force the structure exerts on the rider is upwards and is larger than the weight of the rider. This gives a resultant upward force to cause the upward acceleration towards the centre of the circle:

$$\text{force (N)} = \text{mass (kg)} \times \text{acceleration (m s}^{-2})$$
$$(S_B - 617 \text{ N}) = 63 \text{ kg} \times 25.4 \text{ m s}^{-2}$$
$$S_B = (63 \times 25.4) \text{ N} + 617 \text{ N} = 2220 \text{ N}$$

b At the top, the force the structure exerts on the rider might be upwards, if there is a strap underneath holding him or her up, or it might be downwards, if it is exerted on his or her back. If the wrong direction is chosen for the diagram, then the result will come out as a negative value. It helps however to think out the direction carefully. In this case the rider will be pushing outwards on the structure, but that is a force on the structure *not* a force on the rider. The rider here must have the structure exerting a downward force on him or her because the gravitational force by itself is not large enough to give the rider a downward acceleration of 25.4 m s^{-2}:

$$\text{force (N)} = \text{mass (kg)} \times \text{acceleration (m s}^{-2})$$
$$(S_T + 617 \text{ N}) = 63 \text{ kg} \times 25.4 \text{ m s}^{-2}$$
$$S_T = (63 \times 25.4) \text{ N} - 617 \text{ N} = 986 \text{ N}$$

FIG 8.14 The seat belts are not essential in rides such as this one at a theme park. The downward acceleration is considerably larger than the acceleration of free fall, so that the seat of the cab has to push the rider downwards when at the top of the loop.

QUESTIONS

8.9 Work through the amusement park ride problem in Example 8.7 again, but with a time of rotation of 6.0 s. What would happen to the force that the structure exerts on the rider?

8.10 In an amusement park a passenger of mass 60 kg travels upside down in a carriage at the top of a circle of radius 6.0 m at a speed of 12.3 m s⁻¹ (**Fig 8.15**).

   a Draw a force diagram showing the forces acting on the passenger.

   b Find the magnitude and direction of the force that the carriage exerts on the passenger.

8.11 A pilot of mass 75 kg who has been diving vertically downwards with a velocity of 180 m s⁻¹ pulls out of his or her dive by changing course to a circular path of radius 1200 m.

   a If the pilot maintains a constant speed, what will be his or her maximum acceleration?

   b What is the maximum force that the seat exerts on the pilot?

8.12 A cyclist of mass 35 kg travelling at a constant speed of 11 m s⁻¹ rounds a corner of radius 17.0 m.

   a Draw a force diagram showing the forces acting on the cyclist.

   b At what angle to the vertical should the cyclist be?

8.13 A coin will rest on a disc rotating at 45 rpm (revolutions per minute) provided that it is not more than 10 cm from the centre of the disc. How far away from the centre may it be placed if it is to remain on the disc when rotated at 33.3 rpm?

INVESTIGATION

If an elastic band is placed on a rotating disc, the tension in it will provide sufficient force for it to rotate with the disc only up to a critical angular velocity. At angular velocities above this critical angular velocity, the elastic band will not stay on the disc. You are asked to show that the critical angular velocity, ω_c, varies with r, the radius of the disc, according to the relationship

$$\omega_c{}^2 = A - \frac{B}{r}$$

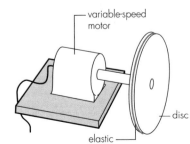

variable-speed motor

disc

elastic

FIG 8.15 Possible experimental arrangement for investigation on tension in a rotating elastic band.

Fig 8.16 illustrates a possible arrangement for the experiment. Varying sizes of wheels could be used for the disc, and some trial and error is necessary to obtain an elastic band of suitable length for the wheels in use and the range of angular speeds of the motor. The rate of rotation of the motor must be able to be varied and measured. Measurement is most conveniently done using a calibrated stroboscopic lamp, but it can also be done using a 'rev counter'.

Plot the straight-line graph of the equation to the left and use it to find A and B.

It can be shown that

$$A = 4\pi^2 k / M$$

where k is the spring constant of the elastic band (force per unit extension for a single strand of elastic), and M is the mass of the elastic band. Measure M and use your value of A to calculate k. Why is this method for finding k likely to be less accurate than finding k by a direct method?

ANALYSIS

The behaviour of aircraft tyres when rotating at high angular speeds

When a tyre is rotated at high speed, its diameter increases. (If the tyre has been retreaded, there is a danger that the retread can separate from the tyre body in the same way that the elastic band did in the investigation. Evidence for this is often to be seen as strips of rubber littering the hard shoulder of motorways.) The problem is particularly important on aircraft, as the undercarriage needs to be made as small and light as possible and yet to be safe and strong. This usually results in several wheels being used on each undercarriage, and the spacing of the wheels needs to be as close as practicable. In particular, when wheels are in tandem, as shown in **Fig 8.17**, any increase in the diameter of a wheel could make them rub against one another.

British Standard 2M45 Part 1 gives a graph showing how much clearance must be allowed for different tyre widths and speeds. It is reproduced in **Fig 8.18**, and shows that a clearance of at least 10 mm must always be used but that the wider the tyre or the faster the take-off speed the more clearance must be given.

Use data from the graph to answer the following questions. You might find it easier to imagine that the tyre is being tested on a stationary test rig so that the only factors you need to consider are the ones involved with circular motion. In fact, the horizontal velocity of the aircraft along the runway does not make any difference to the force necessary for circular motion, although it does alter the loading on the wheel.

a Consider first a tyre that has a diameter of 0.90 m and a width of 400 mm. What will be its angular velocity at each of the speeds shown?

b What will be the centripetal acceleration of the tread of the tyre at each of these speeds?

c What clearances are required for this tyre at each of these speeds?

d What force is necessary to hold each kilogram of tread on the tyre when the tyre is travelling at these speeds?

e If a particular aircraft has a clearance between two wheels in tandem of 60 mm, plot a graph to show how the maximum take-off speed possible varies with the width of tyre used.

f List some advantages and disadvantages of using a tyre of greater width.

g For a given take-off speed, how does the centripetal acceleration of the tyre tread vary with the diameter of the tyre?

h The graph has been used to provide data for a tyre of diameter 0.90 m. It can be used for tyres of any diameter. Suggest a reason why the graph need not be different for different diameter tyres?

i The very large forces that are necessary to hold the tyre in one piece are supplied by steel wires embedded in the rubber of the tyre. Sketch how this wire could be arranged within the tyre, and show the direction of the forces acting on the wire.

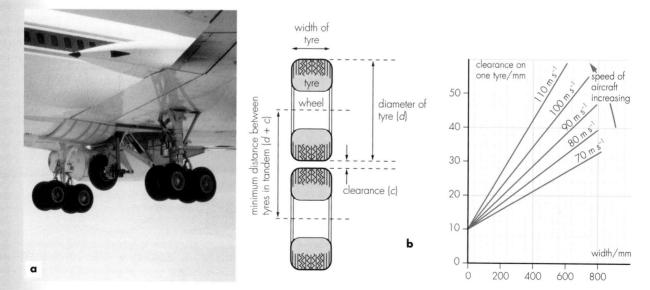

FIG 8.16 **a** The undercarriage of a modern jet. Note the hydraulic jacks for raising and lowering the undercarriage, and the arrangement of the wheels and tyres in tandem on the undercarriage of the aeroplane. **b** When rotating at high speed, the tyres stretch. The clearance between the wheels must be sufficient so that they do not hit one another.

FIG 8.17 Graph showing how the diameter of a tyre varies with the rate of rotation.

SUMMARY

- Angle θ in radians is defined by the equation

$$\theta = \frac{\text{arc length}}{\text{radius of arc}} = \frac{s}{r}$$

- Angular velocity ω is the number of radians turned through per unit time.

- The time for one revolution T, the period, is therefore given by $T = 2\pi/\omega$, but $\theta \times r = s$ by definition of θ and therefore $r\omega = v$ where v is speed.

- A body travelling at a constant speed v in a circular path has an acceleration towards the centre of its path. This acceleration is called a centripetal acceleration and its value is given by

centripetal acceleration $= v^2/r = r\omega^2 = v\omega$

- The total force on such a body must, as always, act in the direction of the acceleration, and the magnitude of the total force can be found by using $F = m \times a$.

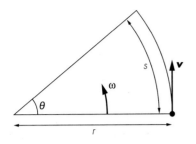

FIG 8.18 Diagram for definition of angle and angular velocity.

ASSESSMENT QUESTIONS

8.14 a **i** Define *acceleration*.
   ii State whether acceleration is a scalar or vector.
b State Newton's second law of motion.
c A body of mass m moves in a circle of radius r with constant speed v.
   i Give an expression for the force acting on the body.
   ii State the direction of this force.

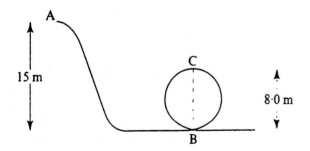

d The diagram shows part of a fairground attraction called 'loop the loop'. A car and its passengers is released from rest at A, travels down the slope and enters the circular loop at B.
   i If 20% of the initial energy at A is used up in overcoming frictional forces in reaching B, calculate the speed of the car at B.

ii As the car moves around the loop to the highest point C, 15% of the energy at B is wasted due to friction. What is the speed of the car at C?
iii Give a free-body diagram for the forces acting on a passenger when the car is at C.
iv Calculate the force exerted *on* a passenger of mass 70 kg *by* the car at C.

WJEC 1997

8.15 A grinding wheel of diameter 0.12 m spins horizontally about a vertical axis, as shown in the diagram. P is a typical grinding particle bonded to the edge of the wheel.

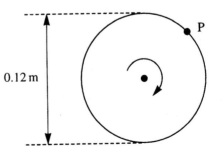

a If the rate of rotation is 1200 revolutions per minute, calculate
   i the angular velocity,
   ii the acceleration of P,
   iii the magnitude of the force acting on P if its mass is 1.0×10^{-4} kg.

b The maximum radial force at which P remains bonded to the wheel is 2.5 N.

    i Calculate the angular velocity at which P will leave the wheel if its rate of rotation is increased.

NEAB 1997

8.16 At one instant, the rotational speed of a disc in a CD player is 300 revolutions per minute.

  a Calculate the angular velocity of the disc.

  b Sketch a graph to show how the acceleration *a* of a point on the disc varies with its radial distance *r* from the axis of rotation when the disc is moving with a constant angular velocity.

  c P and Q are two points on the disc, 30 mm and 50 mm respectively from the axis of rotation.

    i Calculate the difference in the linear speeds of the two points.

    ii What is the difference in the angular velocities of the two points?

CCEA 1998

8.17 a Explain how a body moving at constant speed can be accelerating.

  b The Moon moves in a circular orbit around the Earth. The Earth provides the force which causes the Moon to accelerate. In what direction does this force act?

  c There is a force which forms a Newton's third law pair with this force on the Moon. On what body does this force act and in what direction?

London 1995

8.18 A child sits on the edge of a roundabout of diameter 3.2 m. The roundabout rotates at a constant rate of one revolution every 3.5 s.

  a Calculate, for the child,

    i the speed,

    ii the angular speed,

    iii the magnitude of the acceleration and state its direction.

  b The child is holding a football which she releases from a position above the edge of the roundabout. With reference to horizontal and vertical components, describe the subsequent motion of the ball as it falls to the ground.

NEAB 1998

8.19 This question is about circular motion.

  a Write down an expression for the force needed to maintain a particle of mass *m* moving in a circular path of radius *r* at an angular velocity ω. State the direction of the force.

  b A seat S is suspended from a fairground ride by an inextensible rope. When the ride rotates at an angular velocity of 1.2 rad s⁻¹, the rope makes an angle of θ with the vertical. The distance of S from the axis of rotation is 5.0 m.

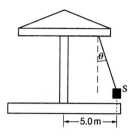

    i The tension in the rope is *T*. Explain how a component of *T* provides the force needed to keep S in a circular path and equate this component with the expression given in part **a**.

    ii S stays at a constant height during rotation at constant speed. Write a second equation expressing the vertical equilibrium of S.

    iii Hence, using the two equations from **i** and **ii**, calculate the value of θ.

  c The speed of the ride is suddenly increased. Describe the subsequent motion of S in as much detail as you can.

O & C 1997

8.20 This question is about the motion of a mass attached to a light spring.

  a A force of 0.50 N stretches a spring, of unstretched length 0.15 m, by 0.050 m. Find the stiffness (force constant) of the spring in N m⁻¹.

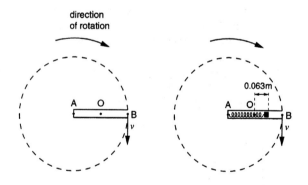

  b A smooth transparent tube of length 0.30 m rotates in a horizontal circle about one end A, making one revolution every 2.0 s. Calculate the speed *v* of the end B of the tube. See left-hand diagram above.

  c The spring of part **a** is inserted into the tube. One end is attached to A and the other to a small mass of 0.30 kg, which rests at point O, the mid point of the tube. The tube is now rotated steadily again at one revolution every 2.0 s. Show that the mass will move to a steady distance of 0.063 m from O. See right-hand diagram above.

  d The mass suddenly becomes detached from the spring. Copy the right-hand diagram and draw the path that you expect to observe the mass to follow as the tube continues to rotate.

O & C 1997

9 GRAVITATION

LEARNING OBJECTIVES

At the end of this chapter you should be able to:

① quote Newton's law of gravitation;

② understand the term 'gravitational field';

③ understand the term 'gravitational potential' and be able to relate it to 'gravitational field';

④ apply Newton's law of gravitation and principles of circular motion to satellites and other bodies;

⑤ calculate an escape speed;

⑥ understand why the measured value of the acceleration due to gravity varies from place to place on the Earth's surface.

9.1 NEWTON'S LAW OF GRAVITATION

As stated in section 5.6, gravitational attraction is a strange and remarkable force about which little is known (**Fig 9.1**). It is very difficult to do experiments on gravitational attraction because the force is so small between masses in the laboratory. The one case where the gravitational force *is* large is when it is due to the pull of the Earth. But, for any particular laboratory, it is then not possible to change the pull significantly on a given mass. We have electrical switches that enable us to switch on and off electrical fields so we can easily see their effect. There is no such thing as a gravity switch – which is a pity! The mystery of gravity deepens when you ask yourself, how can the Earth pull on the Moon even though the Moon is 400 000 km away from the Earth and there is a vacuum between them?

FIG 9.1 An optical photograph of the Orion Nebula, a bright cloud of gas and dust in the night sky. In such a nebula, stars are in the process of being born, as a result of more and more gas and dust particles being pulled together by the gravitational forces between them.

The law of gravitation and the gravitational constant

In section 8.2 (Example 8.4) the acceleration of the Moon was calculated to be 2.72×10^{-3} m s^{-2} in a direction towards the Earth. This calculation was first done by Newton, using different units. He compared the acceleration of the Moon with the acceleration of objects falling on the Earth – traditionally with the acceleration of an apple falling from a tree in his garden. In modern units, these are the figures he might have obtained:

	DISTANCE FROM CENTRE OF EARTH /m	ACCELERATION /m s^{-2}
Moon	3.84×10^8	2.72×10^{-3}
Apple	6.39×10^6	9.81

At first sight there does not seem to be anything special about the numbers. But looking for patterns and connections between figures was, and still is, one of the basic aims of scientists. Use a calculator to find the connection between these numbers *before reading on.*

In order to find connections such as this, the ratio of the two values should be found:

$$\frac{\text{distance of Moon from the centre of Earth}}{\text{distance of apple from the centre of Earth}} = \frac{3.84 \times 10^8 \, \text{m}}{6.39 \times 10^6 \, \text{m}} = 60.1$$

$$\frac{\text{acceleration of moon}}{\text{acceleration of apple}} = \frac{2.72 \times 10^{-3} \, \text{m s}^{-2}}{9.81 \, \text{m s}^{-2}} = 2.77 \times 10^{-4} = \frac{1}{3610} = \frac{1}{(60.1)^2}$$

To put this another way, the Moon is 60.1 times further away from the centre of the Earth, but its acceleration is only 1/3610 of the acceleration of the apple. Now note that $(60.1)^2 = 3610$ and you have the basis of Newton's inverse square law of gravitation.

Newton's **law of gravitation** states that:

> **K** Every particle of matter in the Universe attracts every other particle with gravitational force that is directly proportional to the product of the masses of the particles and inversely proportional to the square of the distance between them.

$$F \propto \frac{m_1 m_2}{r^2}$$

By putting in a constant of proportionality G, called the **gravitational constant**, we get the equation form of Newton's law of gravitation:

> **K** $$F = \frac{Gm_1 m_2}{r^2}$$

where F is the gravitational force, and has the same magnitude on each particle, m_1 and m_2 are the masses of the two particles and r is the distance between them. The value of G has been found experimentally to be 6.67×10^{-11} N m^2 kg^{-2}.

The force of gravity

The insight provided by Newton's law of gravitation is not so much that he found a mathematical connection between the acceleration of the apple and the acceleration of the Moon. Rather, it is because he realised that the force that he called 'gravity' controls the movement of both the apple and the Moon. Gravity is a truly universal force: gravity exists because the bodies have mass. In the same way that there is a gravitational pull between the Earth and an object, there will be a (smaller) gravitational pull between the Moon and the same object (**Fig 9.2**).

Gravitational pull is so familiar as 'weight' in everyday life that its peculiarities are simply accepted as being normal. In fact the human body does *not* have any gravity sensor; we do not feel gravity. What we feel is contact with whatever we are sitting or standing on. We are pulled onto the ground by gravity, and then our feet, through sense of touch, can feel the ground when distortion of our feet takes place on making contact.

That the force of gravity is usually a very tiny force can be readily shown. If one magnet moves past another, then there is an appreciable attraction or repulsion between them. If you walk past a house, you feel no attraction or repulsion at all. Yet there *is* an attraction; always an attraction, never a repulsion. The reason that you do not notice the attraction is that the force is so very small. For a house of mass 200 000 kg and a person of mass 60 kg, the force of attraction if the person is 10 m from the house is only 0.000 008 N. This is a force which it is quite impossible to notice, but it was the basis of the first attempt to measure G, in 1740. If a plumb-bob is suspended near one side of a mountain, then it is pulled very slightly out of vertical by the gravitational force that the mountain exerts on it, as illustrated in a very exaggerated way in **Fig 9.3**.

A large force of gravity will only exist if one of the masses is large. In practice this usually means the Earth. The Earth, on the other hand, can hardly be considered to be a particle. Newton was able to show that the force of gravity exerted outside of any uniform sphere is the same as if the entire mass of the sphere were concentrated at its centre. The force of gravity between two masses that are not spheres can be difficult to calculate. You may assume, in gravitational problems, that any masses may be considered as point masses unless you are specifically told not to.

FIG 9.2 The Apollo 8 mission to the Moon, 21 December 1968. A thrust of 3.3×10^7 N sent the vehicle carrying Borman, Lovell and Anders to orbit the Moon. They were the first astronauts to be more under the gravitational influence of the Moon than that of the Earth.

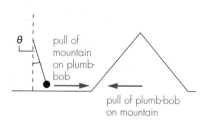

FIG 9.3 An exaggerated sketch of the way a mountain can pull a plumb-bob out of the vertical.

QUESTIONS

9.1 Find the following forces, to two significant figures:

   a The force that a molecule of mass 3.0×10^{-25} kg exerts on another molecule of the same mass when at a distance apart of 2.4×10^{-9} m.

   b The force that the Earth, of mass 6.0×10^{24} kg, exerts on a mass of 1 kg at its surface. The Earth has a radius of 6.4×10^6 m.

   c The force that the Sun, of mass 2.0×10^{30} kg, exerts on the Earth. The distance from the centre of the Earth to the centre of the Sun is 1.5×10^{11} m.

9.2 Consider the mountain referred to above and shown in **Fig 9.3**. Working to only one significant figure: the mountain may be considered as a point mass of 4×10^{12} kg at a distance of 3000 m from the plumb-bob of mass 2 kg; the Earth has a mass of 6×10^{24} kg and a radius of 6×10^6 m.

   a Find the force that the mountain exerts on the plumb-bob.

   b Find the force that the Earth exerts on the plumb-bob.

   c Draw a vector diagram showing these forces on the plumb-bob, and hence find the angle of deflection from the vertical θ.

   d If the angle of deflection is measured experimentally, then this question, worked through backwards from the end to the beginning, can be used to obtain the value of the mass of the Earth even though the value of G is not known. Write down the algebraic expression to find the mass of the Earth.

9.2 GRAVITATIONAL FIELD

When the magnetic field surrounding a magnet is drawn, the concentration of field lines gives an indication of the strength of the field. The direction of the field lines shows the direction in which the north pole of a compass would point at places around the magnet (**Fig 9.4**). The same type of diagram can be drawn to show the gravitational field around any body. **Fig 9.5** is a drawing of the gravitational field surrounding the Earth. It shows spherical symmetry because the Earth is very nearly a uniform sphere. Both of these field diagrams are really three-dimensional diagrams. The magnetic field diagram shows repulsion as well as attraction, but the gravitational diagram only shows attraction because gravitation is *always* an attractive force. The diagram of the Earth's gravitational field clearly shows that the gravitational pull on an object is always directly towards the Earth, and that it gets stronger the closer the body is to the Earth's surface. Fields are therefore used to show the strength of a force and the direction in which it acts. This needs to be expressed more formally.

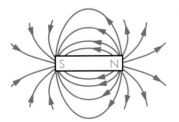

FIG 9.4 A typical drawing of a magnetic field.

 K The *gravitational field strength* at a point is defined as the gravitational force per unit mass acting at that point.

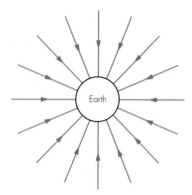

FIG 9.5 The shape of the gravitational field surrounding the Earth.

Gravitational field strength is a vector having direction as well as magnitude, and so addition of gravitational fields must be done by vector addition. The symbol used for gravitational field strength is g, so the equation defining g is

$$g = \frac{\text{gravitational force}}{\text{mass}} = \frac{F}{m}$$

and g has units of N kg^{-1}.

Since from Newton's second law we also know that $F = ma$, it follows that an acceleration g is produced by a gravitational field strength. Indeed, the units of gravitational field strength are equivalent to those of acceleration. The gravitational field strength at the Earth's surface has a numerical value of 9.8 N kg^{-1} and this causes an acceleration due to gravity of 9.8 m s^{-2} (see sections 5.6 and 9.5).

From the definition of gravitational field and the universal law of gravitation, we can work out the value of the gravitational field at any distance r from the centre of the Earth (or from any other body if needed). If a mass m is a distance r from the centre of the Earth, which has a mass M (**Fig 9.6**) then

$$\text{force of attraction} = \frac{GmM}{r^2}$$

$$\text{Gravitational field strength} = \frac{\text{force of attraction}}{\text{mass}}$$

$$= \frac{GmM}{r^2 m}$$

$$= \frac{GM}{r^2}$$

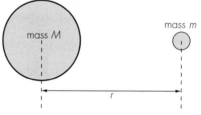

FIG 9.6 A mass m at a distance r from a larger mass M (which may or may not be the Earth).

EXAMPLE 9.1

Assuming the Earth to be a sphere, find its mean density from the following data:

gravitational field strength of Earth, g = 9.83 N kg^{-1}

radius of Earth, r = 6.37 × 10^6 m

G = 6.67 × 10^{-11} N m^2 kg^{-2}

Using the above equations we get

$$g = \text{force of attraction per unit mass} = \frac{GM}{r^2}$$

$$M = \frac{gr^2}{G} = \frac{9.83 \text{ N kg}^{-1} \times (6.37 \times 10^6 \text{ m})^2}{6.67 \times 10^{-11} \text{ N m}^2 \text{ kg}^{-2}}$$

$$= 5.98 \times 10^{24} \text{ kg}$$

$$\text{volume of Earth} = \frac{4}{3} \pi r^3 = \frac{4}{3} \pi (6.37 \times 10^6 \text{ m})^3$$

$$= 1.083 \times 10^{21} \text{ m}^3$$

$$\text{mean density} = \frac{5.98 \times 10^{24} \text{ kg}}{1.083 \times 10^{21} \text{ m}^3} = 5520 \text{ kg m}^{-3}$$

Variation of g with distance from the centre of the Earth

From the surface of the Earth upwards, the value of *g* decreases following an inverse square law. Within the Earth, the value of *g* follows a complex variation because the density of the Earth is not constant. The Earth has a core that is much denser than its crust. From measurements taken of the speeds with which shock waves from earthquakes travel, it is possible to calculate the values of *g* at different distances from the centre of the Earth. These values are used to obtain the graph in **Fig 9.7**.

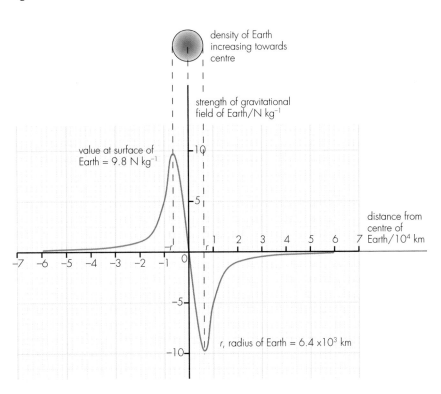

FIG 9.7 A graph showing how the Earth's gravitational field changes with distance from the centre of the Earth.

The reason for drawing this graph in this way is to illustrate two points. The first is to show that the gravitational field strength at the centre of the Earth is zero. This is because any matter at the centre would be pulled equally in all directions, giving a resultant force on it of zero. The second reason is that the graph shows that the direction of the field is in opposite directions on opposite sides of the Earth. On the diagram, the field on the right-hand side acts towards the left, and so is shown as having a negative value; the field on the left-hand side acts towards the right, and is shown as positive.

Gravitational field pattern of the Earth–Moon system

As stated earlier, the total gravitational field strength at a point is the vector sum of the individual fields caused by different bodies. Finding the field for even two bodies, such as the Earth and the Moon, can be a lengthy process, but the pattern

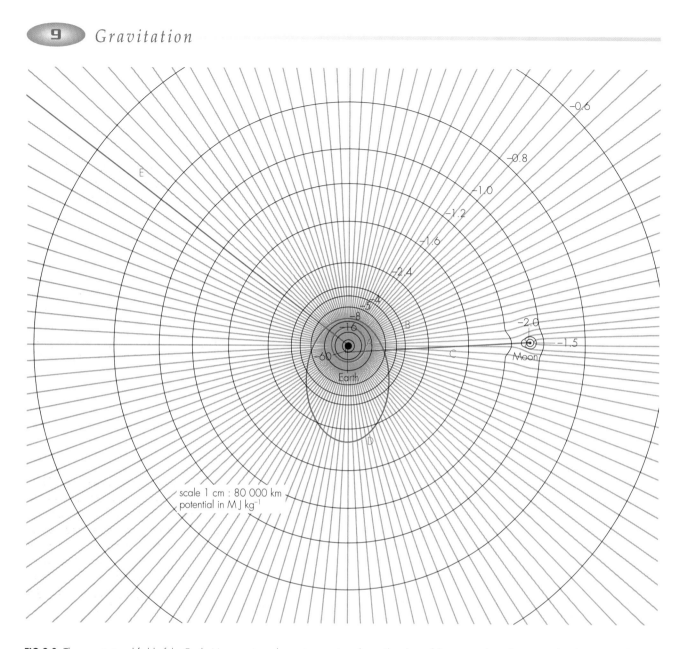

scale 1 cm : 80 000 km
potential in MJ kg⁻¹

FIG 9.8 The gravitational field of the Earth–Moon system, shown in grey, together with values of the potential at all points within the system, shown in black. Some satellite paths are shown in green.

produced by this process is much as would be expected. **Fig 9.8** is drawn to a scale of $1 : 8.0 \times 10^9$ and the grey lines show the gravitational field pattern of both the Earth and the Moon. The diagram shows how the Earth's field dominates this region of space, with the Moon's field having an influence only comparatively close to itself. The field of the Sun has not been included.

QUESTIONS

9.3 The mass of the Earth is 6.0×10^{24} kg and the mass of the Moon is 7.4×10^{22} kg. Find the position of the gravitational neutral point, X, between the Earth and the Moon if the distance between their centres is 3.8×10^8 m. Show that it divides the distance between the Earth and the Moon roughly in the ratio $9 : 1$.

9.3 GRAVITATIONAL POTENTIAL

Because the gravitational field strength is a vector quantity, it is difficult to make calculations of speed and energy of objects moving in the field. Whenever a field theory is used, it is therefore useful to define a scalar property of the field at each point, and this property is called the 'potential' of the field. In this case we define the **gravitational potential** of a point in the field as the work done on unit mass in moving it to that point from a point remote from all other masses. This is often stated in more abbreviated form as:

K   *The gravitational potential* **at a point is the work done in moving unit mass from infinity to that point.**

The definition itself implies that the zero of potential energy is taken to be when the mass is at infinity. Later, in section 16.3, the electrical potential of a point will be defined in a similar way as the work done in moving unit charge from infinity to the point. Gravitational potential, like gravitational field strength, is a property of a point in the field and does not depend on the mass at the point. The symbol used for gravitational potential is V, and its unit is $J\,kg^{-1}$.

EXAMPLE 9.2

Using the graph in **Fig 9.7**, find the work done on unit mass to move it from A, a distance of $10r$ from the Earth, to B, on the Earth's surface.

The mass would fall from A to B without any work being done on it by force other than gravity and would in the process gain kinetic energy. In order for it not to gain kinetic energy, work would have to be done *by* the mass on an external force. In other words, the work done *on* the mass is negative. Work is always the product of force times distance moved in the direction of the force. Here the force per unit mass is plotted against distance, and so the area beneath the curve gives the work done per unit mass, and it will have a negative value. The area can be found from the graph by counting squares. Each small square represents $2000\,km \times 0.5\,N\,kg^{-1} = 1000\,kJ\,kg^{-1}$. The area beneath the curve is approximately 60 small squares and therefore the gravitational potential difference between A and B is $-60\,000\,kJ\,kg^{-1}$.

In Example 9.2, note that if A had been much further away from the Earth than $10r$, there would not have been much change in the gravitational potential difference. This is because the gravitational force is so small at large distances from the Earth that very little work needs to be done in order to move the mass further away. We need to be able to find the potential at a point without having to plot a graph and count squares. This can be done satisfactorily only by the use of calculus.

Calculation of the gravitational potential due to a spherical mass

To calculate the gravitational potential at a point near a spherical mass requires calculus. Calculus provides a method for calculating the area beneath a curve, but you will not be required to memorise or prove this equation. The proof is given here so that those who have knowledge of calculus will be able to follow the deduction. **Fig 9.9** shows a spherical mass of radius R and mass M. If unit mass is to move at constant speed from A to B, a small distance δx, then it must have a

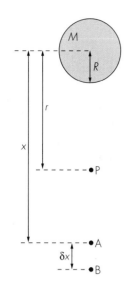

FIG 9.9 Diagram for finding the gravitational potential at a point P near a spherical mass M of radius R.

force applied to it equal and opposite to the gravitational attraction. We first find the total work done in moving unit mass from P to infinity as follows:

force applied to unit mass

$$= \frac{GM}{x^2}$$

work done in moving unit mass from A to B

$$= \frac{GM}{x^2} \times \delta x$$

total work done in moving unit mass from P to ∞

$$= \int_r^\infty \frac{GM}{x^2} \, \mathrm{d}x = \left[-\frac{GM}{x} \right]_r^\infty$$

$$= \left[-\frac{GM}{\infty} \right] - \left[-\frac{GM}{r} \right] = \frac{GM}{r}$$

The gravitational potential at infinity is therefore GM/r higher than at P. Since the definition of gravitational potential implies that the potential at infinity is zero, the gravitational potential at P, a point near a spherical mass, is given by

$$V = -\frac{GM}{r}$$

The potential at the surface of the spherical mass is therefore $-GM/R$. A mass m at a distance r from a mass M therefore has potential energy E_p given by

$$E_p = -\frac{GmM}{r}$$

QUESTIONS

9.4 Find the gravitational potential difference between a point on the Earth's surface and a point 1 m above the Earth's surface. Do this calculation by the two apparently different ways stated:

   a by use of the basic definition of potential;

   b by use of the formula just obtained, the data given in Example 9.1 and a calculator with enough significant figures displayed (or some algebra).

   You should get the same answer by both methods of course – always a good check on a solution.

9.5 **a** Find the gravitational potential at the Earth's surface.

   b At what distance from the centre of the Earth does the gravitational potential fall to one-half of this value?

   c What is the maximum potential that an object can gain as a result of moving a large distance from the Earth?

9.6 A piece of rock far out in interplanetary space is at rest relative to the Sun. Under the influence of the Sun's gravitational pull, it begins to fall towards the Sun along a straight radial line. Assuming that the Sun has a radius of 7.0×10^8 m and a mass of 2.0×10^{30} kg, with what speed will the rock hit the Sun?

The relationship between potential and field

The answer to question 9.4 is 9.83 J kg^{-1}. The reason this number keeps occurring is because it gives the gravitational force acting on a mass of 1 kg at the surface of the Earth. In question 9.4 you were asked to find the difference between the potential that a 1 kg mass has on the Earth's surface and that at a height of 1 m. To find this, you can simply take the gravitational force acting on the 1 kg mass, 9.83 N, and multiply it by the distance moved, 1 m. The work done is 9.83 J on the 1 kg mass.

Put in symbols, this gives the change in gravitational potential δV as the gravitational field strength g multiplied by the distance moved $-\delta x$, i.e.

$$\delta V = -g\,\delta x$$

The minus sign arises because the gravitational field is in the opposite direction to increasing potential. If the equation is written in the form

$$g = -\frac{\delta V}{\delta x}$$

we see that the gravitational field strength is numerically equal to the gravitational potential gradient. This is equally true for non-uniform fields.

  The gravitational field is minus the potential gradient:

$$g = -\frac{dV}{dx}$$

9.4 SATELLITES

FIG 9.10 A NASA photograph of a satellite over the Moon, which also shows the Earth, 400 000 km away.

Any satellite is kept in its orbit by the gravitational attraction of the body about which it is rotating (**Fig 9.10**). Many artificial (man-made) satellites are in circular orbits around the Earth, and once in orbit they do not need rocket motors to keep them in orbit. Provided they are sufficiently far away from the Earth's surface for there to be very little air resistance, they can remain rotating around the Earth for many years. Some of them are expected to remain in their orbit for millions of years. Besides artificial satellites there are also many naturally occurring satellites. All of the planets are natural satellites of the Sun. The Moon is a satellite of the Earth. Natural satellites may have circular or elliptical orbits. Circular orbits are easier to handle mathematically. As far as the Sun's planets are concerned, only Pluto has a markedly elliptical orbit. The Earth's orbit has a maximum radius of 1.52×10^{11} m and a minimum radius of 1.47×10^{11} m.

In order to be in a circular orbit around the Earth, any satellite has to have the correct velocity (in both magnitude and direction) for the distance it is from the Earth. Some satellites are placed in orbits directly above the equator and travel in an easterly direction so that they rotate about the Earth once every day. They are called geostationary (or geosynchronous) satellites (**Fig 9.11**) because to an observer on the ground, also rotating with the Earth once per day in the same direction, they seem to be stationary. The following example shows how to find the height at which a geostationary satellite must be placed.

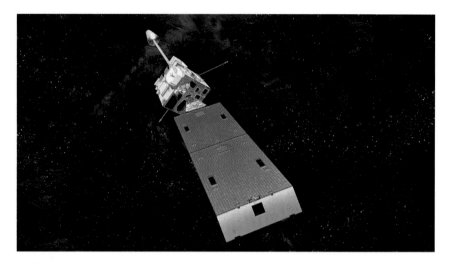

FIG 9.11 An artist's impression of a geostationary satellite. Such a satellite orbits the Earth once per day over the equator, so it appears to be stationary. They are used for international communications and were first suggested by the science fiction writer Arthur C. Clarke.

EXAMPLE 9.3

A communications satellite is to be placed in a circular geostationary orbit. Find its speed and height above the Earth's surface.

All such problems depend for their solution on using Newton's laws of motion. The satellite is acted on by a single force, the gravitational attraction of the Earth. Using the universal law of gravitation, we get

$$\text{gravitational force towards the Earth} = \frac{GmM}{r^2}$$

where m is the mass of the satellite, M is the mass of the Earth and r is the distance of the satellite from the centre of the Earth (see **Fig 9.12**).

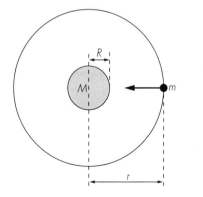

FIG 9.12 The communications satellite in geostationary orbit.

The centripetal acceleration of the satellite is also towards the Earth and has the value $r\omega^2$, where ω is the angular velocity of the satellite. Applying Newton's second law:

force = mass × acceleration

$$\frac{GmM}{r^2} = m \times r\omega^2$$

Since $\omega = 2\pi$ radians in one day:

$$\omega = \frac{2\pi \text{ rad}}{60 \times 60 \times 24 \text{ s}} = 7.27 \times 10^{-5} \text{rad s}^{-1}$$

$$\frac{GmM}{r^2} = m \times r \times (7.27 \times 10^{-5} \text{ rad s}^{-1})^2$$

$$r^3 = \frac{GM}{(7.27 \times 10^{-5} \text{rad s}^{-1})^2}$$

$$= \frac{6.67 \times 10^{-11} \text{N m}^2 \text{ kg}^{-2} \times 5.98 \times 10^{24} \text{kg}}{(7.27 \times 10^{-5} \text{rad s}^{-1})^2}$$

$$= 7.55 \times 10^{22} \text{m}^3$$

so

$$r = 4.23 \times 10^7 \text{ m}$$

Since R, the radius of the Earth, is 0.64×10^7 m, the height of the satellite above the surface is

$$(4.23 - 0.64) \times 10^7 \text{ m} = 3.6 \times 10^7 \text{ m}$$
$$= 36\,000 \text{ km (to 2 sig. figs)}$$

The speed of the satellite is therefore

$$v = r\omega = (4.23 \times 10^7 \text{ m}) \times (7.27 \times 10^{-5} \text{ rad s}^{-1})$$
$$= 3100 \text{ m s}^{-1}$$

Example 9.3 includes the equation showing how Newton's second law is applied to a satellite in a circular orbit, namely

$$\frac{GmM}{r^2} = mr\omega^2$$

To cancel out the mass of the satellite from this equation means that we are assuming that mass treated from the point of view of an object's reluctance to accelerate (its inertia) and mass treated from the point of view of gravitational attraction are the same thing. The fact that results obtained from such equations agree in practice to a very high order of accuracy confirms the belief that inertial mass and gravitational mass are identical.

Units

The units in the equation for r^3 are $N\, m^2\, s^2\, kg^{-1}$. But remembering that 1 N is equivalent to $1\, kg\, m\, s^{-2}$, this reduces to m^3.

QUESTIONS

9.7 The Earth is at a distance of 1.50×10^{11} m from the centre of the Sun. Find the distance from the Sun of Venus, Mars and Neptune given that:

1 year on Venus = 0.615 Earth years
1 year on Mars = 1.88 Earth years
1 year on Neptune = 165 Earth years

9.8 In 1978 a moon of Pluto was discovered. The radius of its orbit was estimated as 4×10^7 m and it rotated around Pluto in 4.8 days. Use this information and the value of G to find the mass of Pluto.

Energy of a satellite in orbit

Potential energy

A satellite of mass m in a circular orbit of radius r about a large mass M has a potential energy E_p given by

$$E_p = -\frac{GmM}{r}$$

as shown in section 9.3. **Fig 9.13** is a scale drawing for a satellite orbiting the Earth at a height of 400 km. At that height, although the satellite is effectively outside the Earth's atmosphere, the value of g can be found by direct use of the inverse square law of gravitation:

$$g \propto \frac{1}{r^2}$$

$$\frac{g_{\text{satellite}}}{g_{\text{surface}}} = \frac{(r_{\text{surface}})^2}{(r_{\text{satellite}})^2}$$

$$g_{\text{satellite}} = \frac{9.8\, N\, kg^{-1} \times (6.4 \times 10^6\, m)^2}{(6.8 \times 10^6\, m)^2} = 8.7\, N\, kg^{-1}$$

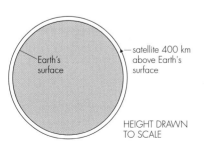

Earth's surface — satellite 400 km above Earth's surface

HEIGHT DRAWN TO SCALE

FIG 9.13 This diagram is drawn to scale. The satellite is in a virtual vacuum at its height of 400 km, yet the gravitational pull of the Earth is still large.

Kinetic energy

Using force = mass × acceleration for the satellite gives the kinetic energy E_k as follows:

$$\frac{GmM}{r^2} = m\frac{v^2}{r}$$

$$mv^2 = \frac{GmM}{r}$$

$$E_k = \frac{1}{2}mv^2 = \frac{GmM}{2r}$$

This shows that the numerical value of the kinetic energy E_k when in a circular orbit is half the value of the potential energy. They have opposite signs, however.

Total energy

The total energy is the sum of the potential and kinetic energies:

$$E = E_k + E_p$$

$$= \frac{GmM}{2r} + \left(-\frac{GmM}{r}\right)$$

$$= -\frac{GmM}{2r}$$

The total energy therefore has the same numerical value as the kinetic energy, but the *opposite* sign. This means that if, as a result of atmospheric friction, the satellite loses energy, it will move to a position nearer to the Earth, r is therefore smaller, and the kinetic energy increases. It is similar to a ball rolling down a hill. The ball loses potential energy: some of this potential energy is changed into kinetic energy, but not all of it. As a result of friction there is a small increase in the temperature of the ball; its internal energy increases. In the case of the satellite, half of its loss of potential energy is used against friction, and the other half is used to increase the kinetic energy of the satellite.

The velocity of a satellite in a circular orbit can be given in terms of the value of the gravitational field strength g it is in. We then get

$$mg = \frac{GmM}{r^2} = m\frac{v^2}{r}$$

where v is the velocity of the satellite. Therefore in this case

$$v^2 = gr$$

$$v = \sqrt{gr}$$

$$= \sqrt{(8.7 \text{ N kg}^{-1}) \times (6.8 \times 10^6 \text{m})}$$

$$= 7700 \text{ m s}^{-1}$$

A satellite orbiting near to the Earth would have a velocity of 7900 m s^{-1} and would take 85 min to orbit the Earth. It is impossible to have a satellite going around the Earth faster than this unless it can keep its rocket motors going all the time to increase the force on it *towards* the Earth (see **Fig 9.14**). In practice this is impossible, as there would be too great a need for fuel. With the existing pull of gravity it is never likely to be possible to reach Australia from the United Kingdom in less than 40 minutes!

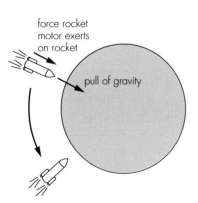

force rocket motor exerts on rocket

pull of gravity

FIG 9.14 This shows the only way a rocket could go round the Earth in a time shorter than 85 min. It would have to keep its rocket motor on in a direction to exert a force on the rocket towards the centre of the Earth.

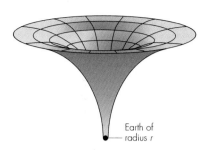

FIG 9.15 A sketch of the gravitational potential well of the Earth.

Speed of escape

This term is often called the **escape velocity** but, since only the magnitude of the velocity is needed and not its direction, it is more accurately called the **escape speed**. The speed of escape of a satellite is not what it appears to be. That is, it is not the speed of a satellite when it escapes from the Earth. Indeed, if it were possible to keep a rocket motor going for a long time, it would be possible to escape from the Earth at any speed. The escape speed of the Earth is 11.2 km s^{-1}. The time taken for astronauts going to the Moon, a distance of 380 000 km, was about three days, or 260 000 seconds. Their average speed was therefore only 1.5 km s^{-1}, well below the escape speed, and yet they escaped!

The term 'escape speed' is in many ways unfortunate. To begin with, it is never possible to escape from the pull of the Earth's gravitational field. If you go far enough, the pull of the Earth's gravity will certainly be very small, but it is not zero. The term can best be considered in conjunction with the term **gravitational well**. This is illustrated in **Fig 9.15**, where the gravitational potential is plotted on a three-dimensional sketch and shows how the potential falls as the Earth is approached.

The potential goes on rising for ever as the distance from the Earth increases, but the slope gets less and less. The depth of the gravitational potential well is GM/r. Now consider a piece of matter, stationary relative to the Earth and a long way from it. It is pulled very gently by the Earth and accelerates towards the Earth a little. As it approaches the Earth, it gets faster and faster, and will eventually collide with the Earth. Its speed on collision *is* the escape speed, because if it bounced perfectly back off the Earth it would retrace its path and just stop when it was a great distance from the Earth. If only the journey outward from the Earth is considered, it is possible to think of the escape speed as the speed needed at the Earth if something is to be thrown off the Earth. The speed of an object when it is thrown upwards decreases as it rises. A rocket at the Earth, moving upwards at 11.2 km s^{-1}, will gradually slow down as it rises and will just have enough kinetic energy to reach the top of the gravitational well. A similar rocket travelling at only 10 km s^{-1} does not have enough kinetic energy to reach the top of the gravitational well, and so will eventually fall back to Earth. One with an upward speed of 12 km s^{-1} has more than enough kinetic energy, and so will have some kinetic energy remaining when it has reached the top of the gravitational well.

The escape speed can be obtained by realising that the kinetic energy the mass must have at the Earth's surface needs to be equal to the gain in potential energy in escaping from the gravitational well. Working with unit mass we get:

kinetic energy = gain in potential in
per unit mass   leaving gravitational well

$$\tfrac{1}{2} \times 1\,\text{kg} \times v^2 = \frac{GM \times 1\,\text{kg}}{r}$$

$$v^2 = \frac{2GM}{r}$$

and since

$$\frac{GmM}{r^2} = mg \text{ or } \frac{GM}{r} = rg$$

then

$$v^2 = 2rg \text{ or } v = \sqrt{2rg}$$

The escape speed v is $\sqrt{(2rg)}$ and is therefore $\sqrt{2}$ times the speed needed for a circular orbit. The kinetic energy in orbit is therefore half the kinetic energy needed to escape.

QUESTIONS

9.9 What are the potential and kinetic energies of a satellite of mass 600 kg when the satellite is in a circular orbit of radius 8.00×10^6 m around the Earth?

9.10 What is the work done on a satellite of mass 600 kg in order to place it in a circular orbit of radius 8.00×10^6 m around the Earth?

9.11 As a result of air resistance, a satellite of mass 10^5 kg loses energy slowly over a period of time and falls from an orbit of radius 7.5×10^6 m to one of radius 7.0×10^6 m.

   a Find the changes in the potential and kinetic energies of the satellite during this time.

   b Hence find the total loss of energy.

   c What are the initial and final values of the speed of the satellite?

ANALYSIS

Satellite motion

Fig 9.8 needs to be used with this exercise.

The grey lines on the figure show the gravitational field near the Earth due to the Earth and the Moon. (The gravitational field due to the Sun is not considered.)

The black lines on the figure show the gravitational potential at all points. These can best be thought of as contour lines.

The green lines on the figure show the paths of various satellites. You are asked to ignore the movement of the Moon when answering the questions. It clearly does move and this adds an additional complication to any of these problems in practice.

Answer the following questions, assuming the mass of any satellite remains constant:

a A is rotating at a constant speed in a circle of radius 30 000 km and B is similarly rotating in a circle of radius 120 000 km; find the ratio of the period of B to the period of A.

b Find the kinetic energy per unit mass of A and hence its speed.

c Find the gain in potential of C as it goes to the Moon.

d C has a mass of 3000 kg. How much potential energy does it gain in reaching the Moon?

e If C started out with kinetic energy of 2.1×10^{11} J, how much kinetic energy has it left when it reaches the Moon? What is its speed at the Moon?

f D has a mass of 80 kg. Find the change in the potential energy of D between its point of closest approach to the Earth D_1 and its point of furthest distance from the Earth D_2.

g D is travelling with a speed of 7000 m s^{-1} when at point D_1. What is its speed at point D_2?

h E is off on a journey to Mars. It starts with a speed of 123 km s^{-1}. Plot graphs on the same horizontal axis to show how the following change over the first half a million kilometres:
   i its potential energy,
   ii its kinetic energy, and
   iii its velocity.

i Sketch the shape of the field and the lines of equipotential on a larger scale for the area near the Moon.

9.5 WEIGHT

Weight has been referred to on many occasions so far in this book, particularly in section 5.6. Unfortunately 'weight' is a word that is often used to mean different things, and some of these uses are extremely misleading, if not actually incorrect. Added to the problem of knowing exactly what is meant by the term 'weight' is the problem of knowing the meaning of such terms as 'weightless', 'weightlessness', 'apparent weight' and 'effective weight'. Because of the confusion, it is worth while re-stating some terms about which there should *not* be any confusion.

Mass

This is a measure of the inertia of a body, that is, its reluctance to undergo linear acceleration. It is measured in kilograms. If a force is applied to a body of unknown mass m, and causes it to have an acceleration a, and the same force when applied to the standard kilogram gives it an acceleration a_s, then

$$\frac{m}{1\,\text{kg}} = \frac{a_s}{a}$$

For example, if m is given an acceleration three times the acceleration of the standard kilogram by the same force, then its mass is one-third of a kilogram.

The mass of an object does not vary from place to place. It is just as difficult to stop a moving hammer on the Moon as it is on the Earth, so it will be as effective in hammering a nail into a piece of wood on the Moon as it is on the Earth. Mass is a scalar quantity.

Gravitational attraction

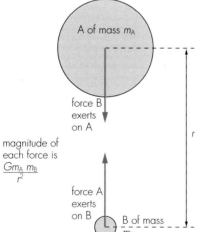

This is a force and so will be measured in newtons. It is a force of mutual attraction between two masses, equal and opposite forces acting on each of them as shown in **Fig 9.16**. The magnitude of the force is given by

$$F = \frac{Gm_A m_B}{r^2}$$

Gravitational attraction is a vector.

Because there is no confusion about these terms, it is preferable to use them whenever possible. Mass should always be the term used for a quantity measured in kilograms, despite the fact that in everyday speech a bag of sugar is said to have a weight of one kilogram. Similarly, when there is a force on an object due to the gravitational attraction of the Earth, then it is more precise to mark it 'gravitational attraction of the Earth'. This has the advantage that it is easy to realise that the gravitational attraction does not depend on how the object is moving or accelerating.

FIG 9.16 Equal and opposite gravitational forces.

Weight

Weight is defined as follows:

 The *weight* of an object is the gravitational force exerted on that object.

Normally the only appreciable gravitational force on an object is that due to the Earth, but there may be circumstances in which other bodies are important. If someone is on the Moon, then the gravitational force that the Moon exerts on him or her is the major contribution to his or her weight. The terms 'apparent weight' and 'effective weight' will not be used.

The gravitational pull the Earth exerts on a mass of 1 kg can be found using Newton's universal law of gravitation:

$$F = \frac{GmM}{r^2} = \frac{6.67 \times 10^{-11} \text{Nm}^2\text{kg}^{-2} \times 1 \text{ kg} \times 5.98 \times 10^{24} \text{kg}}{(6.37 \times 10^6 \text{ m})^2}$$

$$= 9.83 \text{ N}$$

This gives the gravitational field of the Earth, used earlier, as 9.83 N kg^{-1}.

The value varies a small amount from place to place on the Earth's surface because the Earth is a slightly flattened sphere and the distribution of matter within the Earth is not uniform. The Northern Hemisphere, for instance, has far more land area than the Southern Hemisphere. Accurate measurements of the gravitational pull are used to identify different densities of materials beneath the Earth's surface. These measurements are used in prospecting for oil, coal and other minerals.

The acceleration due to gravity (g)

If the force of gravitational attraction is the only force on a kilogram mass, then the mass is said to be in free fall. This would be true if it were falling in a vacuum. It would then have an acceleration g towards the centre of the Earth given by

force = mass × acceleration

9.83 N = 1 kg × g

g = 9.83 m s^{-2}

This value for the acceleration due to gravity is the acceleration of the mass towards the centre of the Earth. It is therefore not exactly the same as the acceleration of the mass towards the surface of the Earth. As the surface is going round in a circle once per day, the surface of the Earth has an acceleration at all points except at the North and South Poles. A vector diagram must be used to relate the various accelerations to one another. **Fig 9.17** is such a diagram for a mass at the equator. It can be seen that, if the acceleration of free fall towards the surface of the Earth is measured on the equator, a value less than 9.83 m s^{-2} should be expected.

Accurate values of g are now found by measurements taken directly on an object falling freely in a vacuum. They are needed because all force calibrations rely on using accurate balances, and the relation between mass and weight depends on knowing an accurate value for g.

9.83 m s^{-2} = acceleration of a freely falling mass relative to the centre of the Earth

acceleration of the surface of the Earth relative to the centre of the Earth

acceleration of free falling mass relative to the surface of the Earth

FIG 9.17 A vector diagram showing that the acceleration of free fall towards the surface of the Earth for a mass at the equator is not the same as the acceleration of free fall towards the centre of the Earth.

Weightlessness

This term should be more readily understood after working through Example 9.4. In that example it was shown that the reading on a spring balance holding a mass is not quite the same as the weight of the mass. The difference between the two forces causes the centripetal acceleration of the mass. There will always be a difference between the reading on the spring balance and the weight of the object if the spring balance is itself accelerating. In the unlikely event of someone standing on bathroom scales while going down in a lift, they would notice that the reading on the scales would be low while accelerating downwards, normal when travelling downwards with constant velocity, and high when stopping. The difference between the weight and the reading on the balance is greater when the acceleration is greater. The reading would become zero if the lift cables broke!

For a satellite in orbit around the Earth, the acceleration of the satellite is equal to the acceleration of free fall. An object placed on a spring balance therefore falls freely with the spring balance and the spring balance exerts zero force on it. It is the fact that a spring balance reads zero that gives rise to the idea of weightlessness. In fact, as was shown above, there is normally a difference between the balance reading and the weight; in a satellite it is much more dramatic.

EXAMPLE 9.4

(a) Find the measured acceleration of free fall at the equator, given that the gravitational field at the equator is 9.830 N kg^{-1}. (b) Find the weight of a 100 kg mass at the equator and find also the reading when it is placed on a sensitive spring balance. [Radius of the Earth = 6.37 × 10^6 m.]

a The acceleration a of a point on the surface of the Earth at the equator is

$$a = r\omega^2$$

$$= (6.37 \times 10^6 \, \text{m}) \times \left(\frac{2\pi}{24 \times 60 \times 60} \, \text{rad s}^{-1} \right)^2$$

$$= 0.034 \, \text{m s}^{-2}$$

since the Earth rotates 2π radians in one day. Using the vector diagram (**Fig 9.17**), we have

measured acceleration of mass relative to surface of Earth	+	acceleration of surface of Earth relative to centre of Earth	=	acceleration of mass relative to centre of Earth

so

measured acceleration of free fall + 0.034 = 9.830

measured acceleration of free fall = 9.830 − 0.034

$$= 9.796 \, \text{m s}^{-2}$$

b Weight of 100 kg mass at the equator is

$$100 \, \text{kg} \times 9.830 \, \text{N kg}^{-1} = 983.0 \, \text{N}$$

A force diagram for the mass when suspended by the spring balance is given in **Fig 9.18**. Applying Newton's law to the mass, which itself has an acceleration in rotating once per day, gives

$$\text{force (N)} = \text{mass (kg)} \times \text{acceleration (m s}^{-2})$$
$$(983.0 \, \text{N} - B) = 100 \, \text{kg} \times 0.034 \, \text{m s}^{-2}$$
$$B = (983.0 - 3.4) \, \text{kg m s}^{-2}$$
$$= 979.6 \, \text{N}$$

FIG 9.18 A mass of 100 kg supported on a spring balance at the equator.

The reading on the balance is therefore 979.6 N, as it supplies this force on the mass. It can be seen from this example therefore that the reading given by the spring balance is not quite equal to the weight of the mass. The difference between the two is small, however, and can, in most cases, be neglected. A similar problem could be solved for a point at any latitude on the Earth's surface, but the vector diagram for the accelerations is more difficult to solve if the latitude is not zero.

QUESTIONS

9.12 The asteroid Ceres has a radius of 550 km and a mass of 7.0 × 10^{20} kg.

   a What would be the Cerian weight of a person with a mass of 90 kg?

   b What would be the acceleration due to gravity on the surface of Ceres?

9.13 The mass of Jupiter is 1.90×10^{27} kg, it has a radius of 7.14×10^7 m and it can be assumed that the planet is spherical.

a What is the gravitational field at the surface of Jupiter?

b Hence find the weight of a 15 kg mass on its surface.

Jupiter rotates on its axis with a period of 9 h 50 min (35 400 s).

c What is the centripetal acceleration of an object at the equator of Jupiter?

d What force is necessary to cause this acceleration for a 15 kg mass?

e Draw a free-body diagram for the 15 kg mass when it is (i) at the pole, (ii) at the equator.

f Find the support force that the planet exerts on the mass at both the pole and the equator.

SUMMARY

- Newton's law of universal gravitation states that the gravitational force F between two point masses m_1 and m_2 separated by a distance r is given by

$$F = \frac{Gm_1m_2}{r^2}$$

where G, the gravitational constant, can be found experimentally to have the value 6.67×10^{-11} N m^2 kg^{-2}.

- The gravitational field strength at a point is the force acting on unit mass at that point. It is a vector.

- The gravitational field strength g at the surface of the Earth is 9.83 N kg^{-1}.

- Using the law of universal gravitation gives

$$g = \frac{GM}{r^2}$$

- Gravitational potential at a point in a gravitational field is the work done in moving unit mass from infinity to the point. It is a scalar and its value is given by

$$V = -\frac{GM}{r}$$

- The gravitational field is equal to minus the potential gradient. The minus sign indicates that the potential falls when moving in the direction of the field.

- The escape speed for any satellite is given by

$$v = \sqrt{2rg} = \sqrt{\frac{2GM}{r}}$$

where M is the mass of the planet, r is its radius and g is the gravitational field at the surface of the planet.

- The weight of a body on the Earth is the gravitational force exerted on it by the Earth. The weight of an object varies with its distance from the Earth but it is not altered by any acceleration the body may have.

ASSESSMENT QUESTIONS

9.14 a A satellite orbits the Earth once every 120 minutes. Calculate the satellite's angular speed.
b Draw a free-body force diagram for the satellite.
c The satellite is in a state of free fall. What is meant by the term *free fall*? How can the height of the satellite stay constant if the satellite is in free fall?

London 1996

9.15 All bodies moving in a circle experience centripetal acceleration.
a What does *centripetal* mean?
b The Moon's orbit about the Earth may be assumed to be a circle of radius 3.84×10^5 km. The Moon moves with constant speed in the orbit, and takes 27.3 days to complete one orbit. Find the magnitude of the centripetal acceleration of the Moon.

CCEA 1998

9.16 a State Newton's law of gravitation.
b Explain why it is extremely difficult to obtain a reliable value for the gravitational constant G in a laboratory.
c For the region outside the Earth's surface, the graph shows the variation of gravitational potential V with distance r from the centre of the Earth.

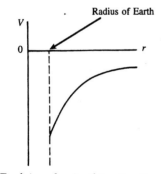

i Explain why in this situation V approaches zero as r approaches infinity.
ii Explain why in this situation V is always negative.

London 1998

9.17 The gravitational field strength g_0 at the Moon's surface can be expressed as

$$g_0 = 4\pi G r_\text{M} \rho / 3 \cdot$$

where ρ is the average density of the Moon and r_M its radius.
a What does G represent?
b Describe how, in principle, you could measure G. Why is it difficult to measure G very precisely?
c i Show that the above equation is homogeneous with respect to units.
Calculate g_0 given that $r_\text{M} = 1740$ km and that the mass of the Moon is 7.34×10^{22} kg.

ii It is suggested that a bullet fired from a high speed rifle on the surface of the Moon could orbit it like a tiny satellite.

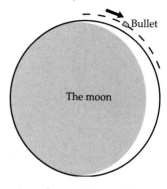

Ignoring the problem of lunar mountains, calculate a minimum value for the bullet's speed. Is the suggestion realistic?

London 1998

9.18 This question is about gravitational fields and equipotentials near a planet.
a i Copy the axes below. Sketch a graph to show the variation of the gravitational field g above the surface of a planet of radius R.

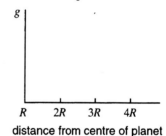

distance from centre of planet

ii Explain the shape of the graph you have drawn.
b i Draw a circle to represent a planet. On your diagram draw a set of *solid* lines to represent the gravitational field of the planet. Add a set of *dotted* lines to represent the gravitational equipotentials.
ii The radius of the planet is 4.8×10^6 m and the gravitational field strength at the surface of the planet is 16 N kg^{-1}. Calculate the mass of the planet.
c Two points, P and Q, lie above the surface of the planet. The gravitational potential at P is -5.2×10^7 J kg^{-1} and that at Q is -6.9×10^7 J kg^{-1}.
i Why do gravitational potentials have negative signs?
ii Explain which point is nearer the surface of the planet.
iii Calculate the potential energy change in moving a satellite of mass 1500 kg from P to Q.

O & C 1997

9.19 This question is about gravitational potential and the motion of a spacecraft near the Earth.

a State the value of the gravitational potential at an infinite distance from the Earth.

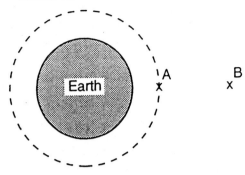

b The value of the gravitational potential at the Earth's surface, 6400 km from the centre, is −63 MJ kg^{-1}. Above its surface, the gravitational potential of the Earth is inversely proportional to the distance from its centre.

   i Use these facts to show that the gravitational potential at A, 3600 km above the surface, is −40 MJ kg^{-1}. See the diagram.

   ii Find the gravitational potential at B, 13 600 km above the surface.

   iii A spacecraft of mass 40 000 kg falls freely towards the Earth from B to A. Find the change in its potential energy, and hence the increase in its kinetic energy.

c The spacecraft uses its motors to correct its speed and trajectory so that it moves in a circular orbit around the Earth at a height of 3600 km above the surface. The motors are switched off. The speed of the craft is 6.3 km s^{-1}. Find its total energy in this orbit.

O & C 1998

9.20 This question is about planetary motion.
The table below contains data about 4 of the 12 moons of the planet Jupiter.

Moon	Period T/s	Radius of orbit r/m	T^2	r^3
Io	1.5×10^5	4.2×10^8		
Europa	3.1×10^5	6.7×10^8		
Ganymede	6.2×10^5	1.1×10^9		
Callisto	1.5×10^6	1.9×10^9		

a One of Kepler's laws states that for a moon orbiting a planet, $T^2 \propto r^3$.

   i Copy and complete the table above and then plot a graph of T^2 against r^3.

   ii Comment on how well the data support Kepler's law.

b **i** Consider the moons as moving in circular orbits. Write down expressions for the force between a moon and Jupiter and for the force necessary for an object to move in a circle. Hence derive the expression relating T and r.

   ii Hence use your graph to show that the mass of Jupiter is about 1.85×10^{27} kg.

c The gravitational field strength on the surface of Jupiter is 24.9 N kg^{-1}.

   i Calculate the radius of Jupiter.

   ii Suggest a problem which this value of gravitational field strength might cause for any future travellers to Jupiter.

d **i** What is a geostationary satellite?

   ii A geostationary satellite orbiting Jupiter would need to have a radius of orbit of 1.6×10^8 m. Calculate the period of rotation of Jupiter.

O & C 1998

9.21 **a** Define the *field strength* (*intensity*) of a gravitational field.
Gravitational field strength may be expressed in units of N kg^{-1} or m s^{-2}. Show that these units are equivalent.

b **i** Derive an expression for the acceleration of free fall at the Earth's surface. Give your answer in terms of the gravitational constant G, the Earth's mass M_E, and the Earth's radius R_E.

   ii The Earth's radius is about 6.4×10^6 m. Use your answer to **b i** and data from the back of this book to estimate the mass of the Earth. Hence show that the mean density of the Earth is greater than 5×10^3 kg m^{-3}.

   iii Approximate densities of some common substances making up the Earth are listed below.

Sea water	1×10^3 kg m^{-3}
Ice	9×10^2 kg m^{-3}
Earth, sand and rocks	3×10^3 kg m^{-3}

Comment on these data with reference to your answer to **b ii**.

c The gravitational potential V at a point a distance r from a point mass m is given by

$$V = -\frac{Gm}{r}$$

   i Define *gravitational potential*. Distinguish between *gravitational potential* and *gravitational potential energy*.

   ii Explain why gravitational potential is always negative.

d The escape velocity for a body on Earth is about 11 km s^{-1}.

   i State what is meant by *escape velocity*.

   ii Apply the principle of conservation of energy to derive an expression for the escape velocity v_e in terms of the gravitational constant G, the mass M_E of the Earth and its radius R_E. State any assumptions made.

   iii Calculate the escape velocity for a body on the Moon, which has a mass of $0.012M_E$ and a radius $0.27R_E$.

CCEA 1998

9.22 The Earth rotates about its axis once every 24 hours. A body of mass 2.0 kg is placed at a point on the equator so that it is at rest relative to, and on, the surface of the Earth.

Assuming the Earth to be a uniform sphere of radius 6400 km and mass 6.0×10^{24} kg, and given that $G = 6.7 \times 10^{-11}$ N m^2 kg^{-2}, calculate

a the linear velocity of the 2.0 kg mass,

b the centripetal force acting on the mass,

c the gravitational force of attraction exerted on the mass by the Earth.

WJEC 1998

9.23 a i State a formula for the magnitude of the gravitational force between a satellite and the Earth, given that:

R is the radius of the orbit;

G is the universal gravitational constant;

m is the mass of the satellite

and M is the mass of the earth.

ii By equating the gravitational force with the centripetal force ($mR\omega^2$), show that the period T of a satellite's orbit is given by:

$$T = 2\pi \sqrt{\frac{r^3}{GM}}$$

b Calculate the radius of the orbit of a satellite that has an orbital period of 2.0 hours.

$G = 6.7 \times 10^{-11}$ N m^2 kg^{-2}

$M = 6.0 \times 10^{24}$ kg

c Explain briefly why it requires more energy to lift a satellite into a geostationary orbit than it takes to lift a similar satellite into an orbit having a period of 2.0 hours.

AEB 1997

9.24 The gravitational field strength at the surface of a star of radius 1.0×10^{10} m and mass 5.6×10^{33} kg is 4.0×10^3 N kg^{-1}.

a i Sketch a graph showing the variation of the magnitude of gravitational field strength with distance d from the centre of the star, for values of d which are greater than the radius of the star.

ii Calculate the gravitational field strength of the star at a distance of 4.0×10^{17} m from its centre.

The universal gravitational constant, $G = 6.7 \times 10^{-11}$ N m^2 kg^{-2}.

b A second star, of mass 2.0×10^{30} kg is 4.0×10^{17} m from the first star. Calculate the force acting on the second star due to the gravitational field of the first.

AEB 1996

9.25 Use the following data in this question:

gravitational constant $= 6.7 \times 10^{-11}$ N m^2 kg^{-2}
mass of Earth $= 6.0 \times 10^{24}$ kg
radius of Earth $= 6.4 \times 10^6$ m.

A satellite of mass 1200 kg is to be placed in a circular orbit of radius 2.0×10^7 m around the Earth.
Calculate

a the speed of the satellite when in its orbit,

b the kinetic energy of the satellite when in its orbit,

c the change in gravitational potential of the satellite as it is placed into orbit from the surface of the Earth,

d the total energy which must be supplied to put the satellite into orbit (assuming the Earth's rotation provides negligible energy).

NEAB 1997

OSCILLATIONS

Oscillations are of considerably more importance than is generally recognised. The oscillation of a pendulum of a clock is visible; the oscillation of atoms within a solid is invisible. The oscillation associated with wave motion can be appreciated when a boat in a harbour gently rises and falls as the waves travel past it, but the oscillation associated with light waves cannot be perceived directly by our senses. There is an association with oscillation for all waves whether they are sound waves or electromagnetic waves such as light or radio waves. Maxwell's theory of electromagnetic radiation in 1864 was one of the outstanding achievements of the nineteenth century. It led immediately to the prediction of radio waves, and these were first produced by Hertz in 1887. The use of oscillations and waves in communication was not new. All communication by sight and by hearing makes use of wave oscillation. Radio waves greatly extended the range and speed of communication. ◼

10 FREE AND FORCED OSCILLATIONS

LEARNING OBJECTIVES

At the end of this chapter you should be able to:

① define and use the terms 'frequency', 'period', 'displacement' and 'amplitude';

② describe the conditions necessary for simple harmonic motion and calculate its period;

③ calculate the energy of a body in simple harmonic motion;

④ distinguish between free, damped and forced oscillations;

⑤ describe examples of resonance.

10.1 PERIODIC MOTION

I n many practical examples of motion, the moving object starts at one place and finishes somewhere else. Many such examples were given in Chapters 4 and 5. In this chapter the type of motion that will be considered is that in which there is no overall displacement of the object: the moving object finishes up where it started. One example of this type of motion is a child on a swing (**Fig 10.1**). Another would be the movement of the string of a guitar. The string is plucked, it moves backwards and forwards for a time, and then the motion dies out, leaving it where it was at the beginning. Of course, the guitar may have moved bodily during this time – the guitarist might be on a moving train – but here that will not be our concern. We need to examine the motion of the vibrating string.

Period and frequency

'Oscillation' is the word used to indicate that this type of movement is taking place. The motion is said to be **periodic motion**. There are many types of periodic motion. Some of them are easy to analyse, while others can be very complex. **Fig 10.2** shows certain familiar situations and, since on the printed page it is impossible

FIG 10.1 An example of to-and-fro motion.

Oscillating object   Waveform

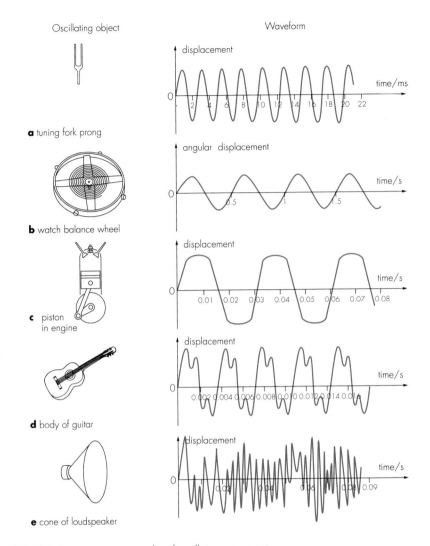

a tuning fork prong

b watch balance wheel

c piston in engine

d body of guitar

e cone of loudspeaker

FIG 10.2 Some common examples of oscillatory movement.

to show movement, graphs have been drawn to show how the displacement of the object might vary with time. The displacement can be a linear displacement (as for the prongs of a tuning fork) or an angular displacement (as for the balance wheel).

For the examples given in **Fig 10.2**, there is usually a clearly defined time interval, which is the length of time for one complete oscillation. This time interval T is called the **period** of the oscillation. Note that one period is a complete cycle back to the same starting point. For example, a clock pendulum that takes 1 s to swing from one side to the other has a period of 2 s since it will take another 1 s to swing back to its starting point. (Confusingly, such a pendulum is sometimes called a 'seconds pendulum'.) We can now relate period to frequency:

> **K** The *frequency f* **of any periodic motion is the number of oscillations per unit time.**

The time for one cycle is therefore $1/f$. This time is the period T of the oscillation, so we have in general:

> **K** $$\text{period} = \frac{1}{\text{frequency}} \qquad T = \frac{1}{f}$$

In SI units, the frequency is in hertz (Hz); 1 Hz is one cycle per second.

Displacement and amplitude

The term 'displacement' has already been used: **displacement** is the distance the object has moved from its rest position in a stated direction. It is a vector. The magnitude of the maximum value of the displacement is called the **amplitude**. It is a scalar. Be careful not to introduce an incorrect factor of 2 when dealing with amplitude. For example, if a guitar string moves a maximum of 4 mm from its rest position in one direction and then 4 mm from its rest position in the other direction, it will move 8 mm between stops. Its amplitude is nevertheless 4 mm and not 8 mm. These terms are illustrated in **Fig 10.3** for a pure oscillation. This oscillation is one which has the same shape as a sine wave and is called a **sinusoidal oscillation**. The vibration of the prongs of a tuning fork has this shape, when taken over a short time interval. If a longer time interval is taken, then it is seen that the amplitude of vibration of the tuning fork gradually decreases.

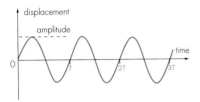

FIG 10.3 A pure sine wave.

Phase and angular frequency

A problem arises in obtaining an equation giving the displacement as a function of time. The equation must have a sine function in it to give the sinusoidal shape, but it is only possible to take the sine of an angle. The problem is, how can an angle be associated with a linear to-and-fro motion? The problem can be overcome by realising that a sine wave repeats itself every 2π radians. We need therefore to associate one cycle of an oscillation with an angle of 2π radians. This also gives rise to a way of relating the pattern of movement of two separate oscillations. If the two oscillations are in step with one another, they are said to be **in phase** with one another. Oscillations are said to be **in antiphase** if they are always moving in opposite directions, for instance if one moves up when the other moves down. This is a phase difference of half a cycle, so the oscillations are said to be π radians out of phase with one another. **Fig 10.4** shows an oscillation A, which is in phase with oscillation B, in antiphase with oscillation C, and out of phase with oscillation D by a quarter of a cycle, $\pi/2$ rad.

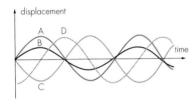

FIG 10.4 Four oscillations showing variations in phase.

 The equation for a sinusoidal oscillation is of the form:
$$x = x_0 \sin \omega t$$

where x is the displacement, x_0 is the amplitude, t is the time and ω is a constant, chosen so that, when t equals the period T of the oscillation,

$$\omega t = \omega T = 2\pi \quad \text{so} \quad \omega = 2\pi/T = 2\pi f$$

and hence the equation for a sinusoidal oscillation becomes

$$x = x_0 \sin(2\pi f t)$$

The units of ω are s^{-1}.

Angular frequency is a name sometimes given to ω, but in less formal discussion concerning a pure sinusoidal motion it is not given a name at all and is just referred to as 'omega'. The unit of ω is s^{-1}. The term $2\pi f t$ is an angle measured in radians, and so it is possible to take the sine of this expression and to plot the required graph. If you are not familiar with this particular piece of mathematics, you really have no alternative other than to work on your understanding of the measurement of angles in radians and their sines and cosines. The problem will not go away if you ignore it! It will recur in dealing with waves and also with

Notation

We use $\sin \omega t$ to mean $\sin(\omega t)$, i.e. the sine of (ωt). For longer terms, to avoid any possible confusion, we will use brackets, e.g. $\sin(2\pi f t)$, $\cos(\omega t + \theta)$, and $\sin(2\pi x/\lambda)$.

alternating currents in electricity. Radian measure is explained in section 8.1. You may find it necessary to study that section again before working through Example 10.1 to see how a sine wave graph can be plotted. Some questions on the topic are given after the example, and another question on the topic is question 20.1.

EXAMPLE 10.1

Plot a graph showing a sinusoidal oscillation with an amplitude of 0.0006 m and a frequency of 2500 Hz. This could be the oscillation pattern of a loudspeaker coil when used to produce a high-pitched note. What is the value of ω for this oscillation?

We start by giving the value of ω, which is

$$\omega = 2\pi f = 2\pi \times 2500 \text{ Hz} = 15\,700 \text{ s}^{-1}$$

The graph to be plotted is $x = x_0 \sin(2\pi ft)$ where $x_0 = 0.0006$ m and $f = 2500$ Hz. The numerical value of x is given by $x = 0.0006 \sin(2\pi \times 2500t)$, and care needs to be taken when plotting graphs with such awkward numbers.

Tabulated values are shown in Table 10.1. The first four lines of values are put in because this is how graphs are often started. You can see that the values of x for t of 1, 2, 3, 4, etc. are useless. They simply produce 0 (or error messages) on a calculator. This is because these times are long enough for thousands of cycles of the oscillation to have taken place, and all the readings that have been calculated are after a whole number of oscillations. The later lines are the sensible ones where short time intervals of under one period are used. It is convenient to start with the time for one cycle and then to work with convenient fractions of a cycle.

TABLE 10.1 Values of $x = x_0 \sin(2\pi ft)$ for Example 10.1, with $x_0 = 0.0006$ and $f = 2500$

t	$2\pi \times 2500t$	$\sin(2\pi \times 2500t)$	x
0	0	0	0
1	$5\,000\pi$	0	0
2	$10\,000\pi$	0	0
3	$15\,000\pi$	0	0
0.000 4	2π	0	0
0.000 1	$\pi/2$	1	0.000 6
0.000 05	$\pi/4$	0.707	0.000 424
0.000 15	$3\pi/4$	0.707	0.000 424
0.000 2	π	0	0
0.000 25	$5\pi/4$	−0.007	−0.000 424
0.000 3	$3\pi/2$	−1	−0.000 6
0.000 35	$7\pi/4$	−0.707	−0.000 424

Once one cycle of the oscillation has been drawn, it may be repeated as often as is required – see **Fig 10.5**.

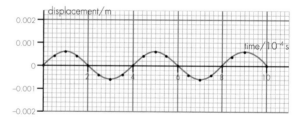

FIG 10.5 The graph for Example 10.1.

QUESTIONS

10.1 Plot graphs to show the following sinusoidal oscillations:

    a $x = x_0 \sin(2\pi ft)$, where $x_0 = 3.0$ m and $f = 2.0$ Hz;

    b $x = x_0 \sin(2\pi ft)$, where $x_0 = 0.3$ m and $f = 50$ Hz;

    c $x = x_0 \sin(2\pi ft)$, where $x_0 = 0.000\,04$ m and $f = 500$ Hz;

    d $x = x_0 \cos(2\pi ft)$, where $x_0 = 0.2$ m and $f = 50$ Hz.

10.2 Draw four more varied waveforms, similar to those drawn in **Fig 10.4**, each showing a possible pattern of oscillation.

10.2 SIMPLE HARMONIC MOTION

What is simple harmonic motion?

A person standing on a bench with a mass attached to the end of a long spring can make the mass move with a sinusoidal oscillation (**Fig 10.6**). If the amplitude of the oscillation is 1.20 m and the period of the oscillation is 3 s, then the frequency of the oscillation is $\frac{1}{3}$ Hz. The equation giving the value of the displacement x of the mass at any time t is

$$x = x_0 \sin(2\pi f t) = 1.20 \sin\left(\frac{2\pi t}{3}\right)$$

FIG 10.6 A mass oscillating on a long spring undergoes nearly simple harmonic motion.

This graph is plotted in **Fig 10.7a**. Since this is a displacement–time graph, the gradient of the graph at any time gives the velocity of the mass at that time. The tangent to the graph is plotted when $t = 6$ and the gradient is found to be $1.3/0.5 = 2.6$, so that the velocity is 2.6 m s⁻¹. This is the maximum value of the velocity, as the graph is at its steepest gradient where it crosses the time axis. Where the graph is at its maximum or minimum, the gradient is zero and so the velocity is zero. By making a few more calculations, and by realising that the velocity is frequently negative, it is possible to plot another graph, this time one of velocity against time. This is done in **Fig 10.7b**. This process can be repeated. The gradient of the velocity–time graph is the acceleration, and this can also be plotted against time. This is found to have a steepest gradient of 5.4, and hence the acceleration of the mass has a maximum value of 5.4 m s⁻². The acceleration–time graph is plotted in **Fig 10.7c**.

A fact crucial to pure sinusoidal oscillation can be seen from a comparison of these graphs. The acceleration–time graph is the same shape as the

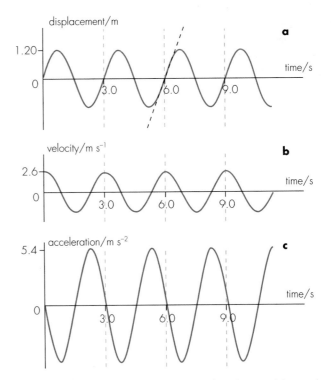

FIG 10.7 These three graphs show how the displacement, the velocity and the acceleration vary with time for the mass on the spring shown in **Fig 10.6**.

displacement–time graph but is of opposite sign. Put mathematically, this can be written as:

$$\text{acceleration} \propto -\text{displacement}$$
$$a \propto -x$$
$$a = -\omega^2 x$$

where ω^2 is a constant. The constant is written in this way because any number squared has a positive value, and it is essential that the acceleration and the displacement themselves have opposite signs. (As you may imagine, the ω factor is the one which was introduced earlier.)

Motion of this pattern is called simple harmonic motion, or SHM for short. **Simple harmonic motion** is defined as motion taking place in which the acceleration of an object is proportional to the displacement of the object from a fixed point and in the opposite direction to the displacement.

A mathematical look at SHM

Mathematics!

The mathematical treatment in the subsection that follows can be omitted by those who have not studied calculus. No A-level syllabus requires proof of the equations that are deduced, and all Examination Boards provide details of the equations on data sheets. If you are going to omit the section, you must, nevertheless, be able to use the equations once you have been given them.

With simple harmonic motion defined in this way, the equation

$$a = \frac{d^2x}{dt^2} = -\omega^2 x$$

is the starting point for a study of SHM. This equation is called a 'differential equation' because differentiating the factor x in the equation twice over gives another term a in the equation. A rigorous solution to this equation will not be given here but, for those able to use calculus, a check will be given on one of the solutions to the basic equation.

Assume that the value of x given by

$$x = x_0 \sin \omega t$$

is a solution to the equation, then

$$v = \frac{dx}{dt} = x_0 \omega \cos \omega t \text{ and } a = \frac{d^2x}{dt^2} = -x_0 \omega^2 \sin \omega t$$

But since $x = x_0 \sin \omega t$
it follows that

$$a = -\omega^2 x$$

showing that our solution is valid. Other solutions to this equation are

$$x = x_0 \cos \omega t$$
$$x = x_0 \cos(\omega t + \theta)$$
$$x = x_0 \cos \omega t + x_1 \sin \omega t$$

So far, the displacement and the velocity of the body undergoing simple harmonic motion have been found in terms of the time. Sometimes the velocity is required in terms of the displacement. This can be done by eliminating t from the equations

$$x = x_0 \sin \omega t$$
$$v = x_0 \omega \cos \omega t$$

by using $\sin^2 \omega t + \cos^2 \omega t = 1$. This gives as a final equation

$$v = \omega \sqrt{(x_0^2 - x^2)}$$

Note

Be particularly careful with the symbols used. Here v is the velocity and this can be written $\dot{x}$; and a is the acceleration and this can be written $\ddot{x}$. In some books a is used for the amplitude, and the solution to the basic equation becomes $x = a \sin \omega t$.

Once it is known that an oscillation is simple harmonic motion, it is not difficult to find the value of ω from the period or frequency and hence find anything required. In the case of many oscillations, however, it needs to be proved that simple harmonic motion is taking place and to find ω from the proof. In Example 10.2 this is done for a mass oscillating vertically on a spring.

EXAMPLE 10.2

A mass m is oscillating vertically on a spring whose spring constant is k. (The spring constant is the force required for unit extension, $F = kx$, and using SI units will have the unit $N\,m^{-1}$.) Show that the mass undergoes simple harmonic motion and find the period of the oscillation. The mass of the spring may be neglected in comparison with the mass m. **Fig 10.8** shows several different situations relevant to this problem:

a The mass is in free fall with only its weight acting on it. Under these circumstances

$$\text{force} = \text{mass} \times \text{acceleration}$$
$$W = mg$$

b This shows the unstretched spring.

c The mass is in equilibrium when the spring is stretched a distance d. Under these circumstances $P = kd$ and $P = W$, so

$$kd = mg$$

d The mass is in its displaced position, so that the total spring extension is $d + x$ and the mass is not in equilibrium. Under these circumstances

$$Q = k(d + x)$$

It is necessary in this type of problem to be careful to indicate which direction is being considered as positive. Here, consider downwards as being positive since x is measured downwards, and apply Newton's second law:

$$\begin{array}{cc}\text{resultant force} = \text{mass} \times \text{acceleration}\\ \text{downwards} \qquad\qquad \text{downwards}\end{array}$$
$$W - Q = m \times a$$
$$mg - k(d + x) = ma$$
$$mg - kd - kx = ma$$

but since $kd = mg$ we get

$$-kx = ma$$
$$a = -\frac{kx}{m}$$

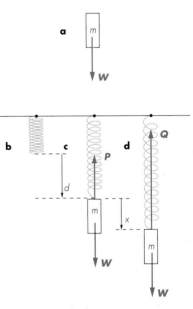

FIG 10.8 Different situations for the mass–spring system in Example 10.2.

This shows that the acceleration is proportional to the displacement and in the opposite direction to the displacement. The motion is therefore simple harmonic motion and

$$\omega^2 = k/m$$
$$\omega = \sqrt{\frac{k}{m}}$$

giving

$$T = \frac{2\pi}{\omega} = 2\pi\sqrt{\frac{m}{k}}$$

It also means that, knowing ω, the equation

$$x = x_0 \sin \omega t$$

can be used for the displacement, and the equation

$$v = x_0\omega \cos \omega t$$

can be used for the velocity.

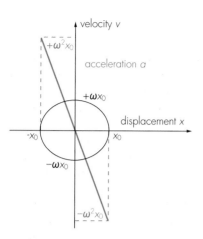

FIG 10.9 Graphs of velocity and acceleration against displacement for simple harmonic motion.

A graph of velocity against displacement is plotted in **Fig 10.9**. Note that the displacement has a maximum value of x_0, and that for any value of the displacement the velocity can be either positive or negative. On the same horizontal axis is also plotted a graph of acceleration against displacement, showing the basic fact about SHM that the acceleration is proportional to the displacement but in the opposite direction. A table of the equations that it is useful to be able to quote is given in the summary at the end of this chapter.

QUESTIONS

10.3 Find the maximum values of the velocity and the acceleration of the system undergoing sinusoidal motion as given in Example 10.1. These values seem large but they are typical values for the behaviour of a loudspeaker. The very rapid oscillations need large forces to be applied to the moving coil within the loudspeaker.

10.4 A rule of thumb used by sailors is called the 'rule of twelfths'. It says that in the first hour after low water (**Fig 10.10**) the sea-level rises by one-twelfth of the tide height, in the second hour it rises by two-twelfths, in the third hour by three-twelfths, in the fourth hour by three-twelfths, in the fifth hour by two-twelfths, and in the last hour before high tide by one-twelfth, i.e. 1-2-3, 3-2-1. Assuming that the rise and fall of a tide is simple harmonic over a period of 12 hours, calculate the actual values and hence the percentage error that the rule of thumb makes at each of the six, hourly intervals.

10.5 A light spring has a spring constant of $20 \, \text{N m}^{-1}$. A mass of 0.50 kg is placed on it and set oscillating vertically with an amplitude of 0.040 m. Calculate for this oscillation:

a the value of ω,

b the maximum speed of the mass,

c the maximum kinetic energy of the mass,

d the maximum acceleration of the mass.

For **b**, **c** and **d** state the position of the mass for these maxima.

10.6 A light spring of unstretched length 0.060 m is extended to a length of 0.075 m when a mass of 300 g is placed on it. It is now given a small displacement so that the mass starts to oscillate.

a Find the period of the oscillation.

b Which of the three pieces of information given is not needed?

c Explain in physical terms why this piece of information is not required.

FIG 10.10 The oscillation of the sea, causing tides, is something that controls the lives of sailors. In some coastal areas of the UK, there is a difference of 10 m between high and low water – an amplitude of 5 m.

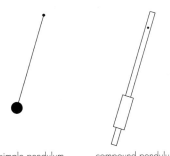

FIG 10.11 A simple pendulum is one in which all the mass is at a point some distance from the point of support. In a compound pendulum, the mass is distributed at different distances from the point of support.

The simple pendulum

The word 'simple' here is used to distinguish a pendulum in which a point mass oscillates through a small angle on the end of a piece of string, from a compound pendulum in which mass is distributed at different distances from the axis of rotation. **Fig 10.11** shows the difference, and you should appreciate that the pendulum in a clock (**Fig 10.12**) is, in practice, a compound pendulum.

Fig 10.13 shows the forces acting on the pendulum bob of a simple pendulum when the string is at a small angle θ to the vertical. One of these forces is the weight of the bob and the other is the tension in the string. Because they are not equal and opposite to one another, there will be an acceleration in the direction at right-angles to the string. The diagram shows mg split into two components, one of them at right-angles to the string and the other in line with it. Newton's second law can be applied to the motion at right-angles to the string, taking care to assign positive signs to directions to the right since x is measured from the centre to the right. We get

$$\text{force to right} = \text{mass} \times \text{acceleration to right}$$
$$-mg \sin \theta = ma$$

For small angles

$$\sin \theta \approx \theta = \frac{x}{l}$$

$$-mg\frac{x}{l} = ma$$

$$a = -\frac{g}{l}x$$

Since the acceleration is proportional to the displacement and in the opposite direction, the motion is simple harmonic motion and

$$\omega^2 = \frac{g}{l}$$

$$T = \frac{2\pi}{\omega} = 2\pi\sqrt{\frac{l}{g}}$$

FIG 10.12 The controlling pendulum in a clock.

The equation for the period of a simple pendulum is an approximate one because of the approximation of $\sin \theta$ to θ in radians. This is a good approximation only for small angles. For $\theta = 0.1$ rad (5.73°), $\sin \theta = 0.0998$. In other words the error in making the approximation is only 0.2%.

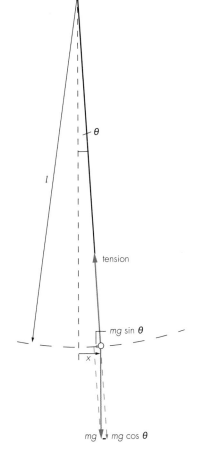

FIG 10.13 Forces acting on a simple pendulum.

```
QUESTION
```

10.7 A loaded test-tube floats in a liquid with a length l submerged (**Fig 10.14**). By calculating the resultant force on the test-tube when it is displaced downwards by a small distance x, show that the motion is simple harmonic motion and show that the period of the oscillation T is given by

$$T = 2\pi\sqrt{\frac{l}{g}}$$

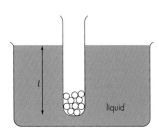

FIG 10.14 The floating loaded test-tube for question 10.7.

151

10.3 ENERGY IN SIMPLE HARMONIC MOTION

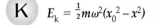

I n section 10.2 the velocity of a mass undergoing simple harmonic motion was found in terms of its displacement. The relevant equation was

$$v = \omega \sqrt{(x_0^2 - x^2)}$$

As you may have used in answering question 10.5, the kinetic energy E_k of the oscillating mass is:

$$E_k = \tfrac{1}{2}mv^2 = \tfrac{1}{2}m\omega^2(x_0^2 - x^2)$$

so

K $\quad E_k = \tfrac{1}{2}m\omega^2(x_0^2 - x^2)$

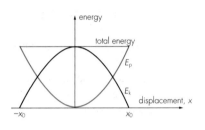

FIG 10.15 Energy alternates between kinetic energy and potential energy as any object oscillates.

The maximum kinetic energy occurs when $x = 0$, as it is at the centre of the oscillation that the velocity is greatest. So

$$\text{maximum kinetic energy} = \tfrac{1}{2}m\omega^2 x_0^2$$

If the potential energy is taken to be zero at the centre of the oscillation, then the maximum kinetic energy must also be the total energy of the oscillation. If no energy is lost during an oscillation, then the total energy must remain constant, and any loss of kinetic energy must be converted into potential energy. This is shown graphically in **Fig 10.15**. The value of the potential energy at any displacement is given by

$$\begin{aligned}
E_p &= \text{total energy} - E_k \\
&= \tfrac{1}{2}m\omega^2 x_0^2 - \tfrac{1}{2}m\omega^2(x_0^2 - x^2) \\
&= \tfrac{1}{2}m\omega^2 x_0^2 - \tfrac{1}{2}m\omega^2 x_0^2 + \tfrac{1}{2}m\omega^2 x^2
\end{aligned}$$

so

K $\quad E_p = \tfrac{1}{2}m\omega^2 x^2$

ANALYSIS

Energy in SHM

When a mass oscillates on the end of a spring, it interchanges its kinetic and potential energies as explained in section 10.3. The potential energy, however, is of two forms: as it rises and falls it gains and loses gravitational potential energy; as the spring stretches and contracts it gains and loses potential energy too. This elastic potential energy is greatest when the mass has least gravitational potential energy. It is quite difficult to show all of these different energies in graphical form, largely because the zero of potential energy is arbitrary. Zero potential energy is usually taken at the centre of the oscillation, but of course the spring is stretched at this point and so does have some elastic potential energy. This implies that the

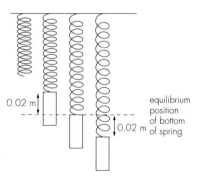

FIG 10.16 Oscillating mass–spring system: oscillating mass $m = 3.0$ kg; spring constant $k = 1200$ N m^{-1}; amplitude of oscillation $x_0 = 0.020$ m.

gravitational potential energy at the centre of the oscillation must be negative. In this exercise you are guided to use some data concerning a spring in order to plot graphs to show the interplay between the various types of energy. Assume that there is no overall loss of energy to friction. The oscillation and data are shown in **Fig 10.16**. By answering the questions in order, and copying and completing the following partial table, you should be able to sketch the final graphs. It will help to realise that the graphs in **Fig 10.15** apply in this case for total energy, kinetic energy and total potential energy.

	TOP	CENTRE	BOTTOM
Displacement/m	0.02	0	−0.02
Kinetic energy/J	①	②	③
Spring potential energy/J	⑩	⑪	⑫
Gravitational potential energy/J	⑭	⑬	⑮
Total potential energy/J	④	⑤	⑥
Total energy/J	⑦	⑧	⑨

a Find the extension that would be produced in the spring by placing the mass on the spring but not letting it oscillate. This gives the centre of the oscillation.

b Use the equation worked out in Example 10.2 to find ω.

c Find the maximum velocity.

d Find the maximum kinetic energy.

e Copy the table and fill in numerical values for ① to ⑨.

f Find the potential energy stored in the spring in each of the three positions – see section 7.2 if this causes problems.

g Fill in numerical values for ⑩ to ⑫ in the table.

h Using your values for ⑤ and ⑪, find ⑬.

i What increase in gravitational potential energy takes place between the centre and the top?

j Fill in numerical values for ⑭ and ⑮ in the table.

k Sketch, on the same displacement axis, graphs showing each of the five energies used.

10.4 DAMPED OSCILLATION

So far in this chapter friction has been largely ignored. The oscillations that have been referred to have been assumed to be of a fixed amplitude. In practice, this is not usually the case unless energy is being continually supplied to the oscillating system. The oscillator of a watch will cease to oscillate unless some energy can be supplied to it from the watch battery. The effect of frictional forces is to reduce the total mechanical energy of an oscillating system. This is a common everyday experience. A note played on a piano is loud immediately after it has been played, and then gradually fades away; a church bell, similarly, is loud as the clapper strikes it and it too fades away. Life would be very noisy if this were not the case! A knock against a table causes the table to reverberate; this reverberation also dies away, often after many hundreds of vibrations.

A damped oscillation is shown in **Fig 10.17**. An oscillating system in which friction has an effect is said to be a damped system. The damping may be light damping, in which the amplitude is reduced in a large number of oscillations, or heavy damping, in which oscillation may not occur at all. Not only does friction have an effect on the amplitude but it also has the effect of reducing the frequency slightly.

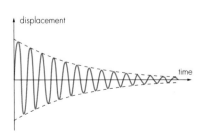

FIG 10.17 If a simple harmonic motion is subject to frictional forces, the amplitude of a freely oscillating object gradually decreases. This is called damped simple harmonic motion.

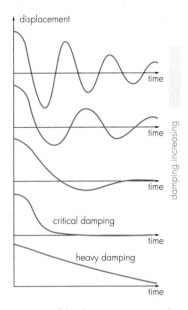

FIG 10.18 If the damping is increased, oscillation does not occur at all.

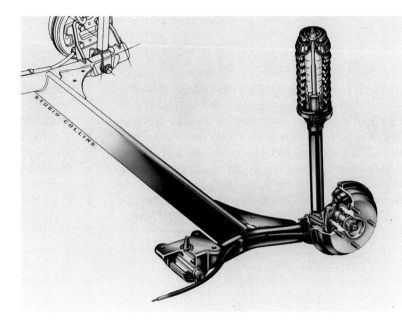

FIG 10.19 The rear suspension of a car. Damped oscillation is provided by the springs and shock-absorbers.

Fig 10.18 shows the way in which the degree of damping affects the oscillation. There is a condition known as critical damping: for less than critical damping, the system oscillates; above critical damping, oscillations do not take place. Also, for critical damping, the system returns to the mean position in the shortest time. The degree of damping of a mechanical system is important. Too little damping results in a large number of oscillations; too much damping leads to there being too long a time when the system cannot respond to further disturbances. This is illustrated well by the trouble that car manufacturers take with the suspension of cars (**Fig 10.19**). The suspension is the link between the wheels/axles of a car and the car body/passengers; the suspension consists of a spring that is damped by a shock-absorber. If a car's shock-absorbers are badly worn, then the car becomes too bouncy, and this is uncomfortable.

A good suspension is one in which the damping is slightly under critical damping, as this results in a comfortable ride and also leaves the car ready to respond to further bumps in the road quickly. **Fig 10.20** shows that, by the time the car has reached point A, the shock-absorbing system is ready for another bump. A very heavily damped shock-absorbing system would still have a compressed spring by the time point A is reached, and so would not be able to respond to further bumps. So long as there are bumps on a road, then these must have an effect on the passengers in a car. The shock-absorbing system can only *reduce* the forces applied. It cannot eliminate them because, clearly, in **Fig 10.20**, the passengers must rise eventually by the height of the bump.

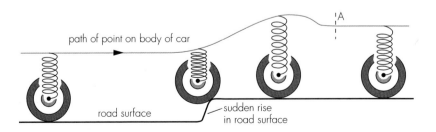

FIG 10.20 The springing of a car's suspension is damped so that when a car goes over a bump the passengers in the car quickly and smoothly regain equilibrium.

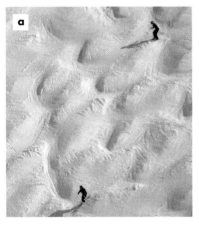

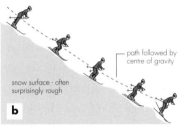

FIG 10.21 **a** A skier on bumpy snow uses his or her knees as shock-absorbers. **b** The shock-absorbing system of the human body is visible here. As the bumpy snow moves the skis up and down, the knees flex and straighten so that the upper body travels smoothly.

The suspension system of the human body is illustrated in **Fig 10.21**. The skier's body moves over the bumpy snow smoothly while his or her thighs and calves act like a damped spring. This is why skiing instructors are reputedly always saying 'Bend zee knees'.

A new development in suspension is that racing cars can now be fitted with active suspensions. This is a computer-operated hydraulic system in which bumps are sensed, and the suspension system is adjusted accordingly. Active suspension reduces the amount of driver movement dramatically. This would, if applied to the system shown in **Fig 10.20**, lift the spring as it reached the bump in the road. Similar computer-operated hydraulic suspension is also fitted to hydrofoils (**Fig 10.22**). At speed, a hydrofoil is lifted clear of the waves. Recent developments use computer control of the angle of the foils through the water to achieve remarkably stable movement of the craft.

FIG 10.22 A hydrofoil can be controlled so that little oscillation takes place. The angle of the foils can be computer-controlled to damp out unwanted oscillations.

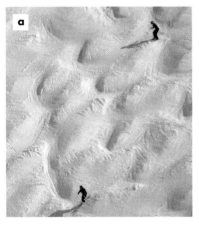

10.5 FORCED OSCILLATION AND RESONANCE

Resonance in mechanical systems

In many mechanical systems, vibration is a nuisance and potentially dangerous. Vibration can arise when an object is repeatedly displaced from its equilibrium position by some external force. If the repeated application of the force happens to be at the natural frequency of the object, then **resonance** is said to occur, and the vibration can build up to dangerous levels. The Tacoma Narrows Bridge disaster of 1939 was caused by the bridge being too slender for the wind conditions in the valley. A more mundane example of resonance is the way in which the bodywork of a bus can vibrate at a particular engine speed.

Resonance can occur when an object with a natural frequency f_0 has a force acting upon it with a frequency f (f is called the driver frequency). The amplitude of the oscillation of the object depends on how much it is damped and on the values of f and f_0, as is shown in **Fig 10.23**.

A tragic accident

A tragic accident took place in Angers, France, in 1850, when soldiers marching over a bridge broke the bridge and over 200 of them were killed. This is often quoted as an example of resonance, on the assumption that large oscillations built up at the same frequency as that of their marching. But it is rather unlikely that the period of oscillation of the bridge was as rapid as 2 Hz. It may well have been that the impact forces of a large number of feet simultaneously hitting a heavily laden bridge was enough to break it. The maximum force that a person exerts on the ground when stamping can be many times their own weight. Whatever the actual cause of the break-up of the bridge in Angers, it is still a requirement that troops break step when crossing a bridge.

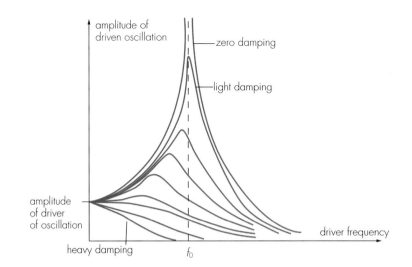

FIG 10.23 Resonance and the effect of damping on resonance.

Resonance in other systems

Resonance is not only a phenomenon for mechanical oscillations; it can equally well occur in electrical circuits, and at this stage the following examples are worth quoting.

The electrons in a radio receiving *aerial* are forced to vibrate by the radio wave passing the aerial. If the aerial is the correct length for the particular frequency being used, then the amplitude of the oscillation is larger. So a larger signal is passed by the aerial to the radio, where the circuitry again uses resonance to isolate and amplify the required frequency.

A *microwave oven* also makes use of resonance. The microwaves are produced with a frequency that is the same as a natural frequency of vibration of water molecules. When some food, which always contains some water molecules, is placed in the oven, the water molecules resonate, absorbing energy from the microwaves and consequently heating up. The microwaves do not heat up the containers in which the food is placed because plastic and glass do not contain any water molecules. Manufacturers of microwave ovens have to be very careful to ensure that nobody can be irradiated with microwaves, as water molecules in a person's body would also resonate. They make it impossible to open the door of an oven without automatically switching off the microwaves. Also they ensure that there is a conducting cage around any microwave region once the door is shut. This absorbs microwaves.

Increasing use is being made of a phenomenon called *magnetic resonance*. Here strong, varying radio-frequency electromagnetic fields are used to cause the nuclei of atoms to oscillate. In any given molecule there will be many resonant frequencies, and whenever resonance appears energy is absorbed. The pattern of energy absorption can be used to detect the presence of particular molecules within any specimen. Biochemists are using the technique to study complex molecules and the part they play in biological processes. Magnetic resonance is also being used instead of X-rays as an imaging system. **Fig 10.24** is a computer-enhanced photograph of a head taken using magnetic resonance. One major advantage of magnetic resonance used in this way is that no ionising radiation is involved.

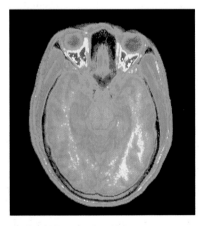

FIG 10.24 False-colour magnetic resonance image showing a head. Such 'slices' are built up using a computer to obtain a three-dimensional image.

INVESTIGATION

When clothes are being spin-dried in a drier or a washing machine, there is often a problem of excessive vibration if the clothes are all on one side of the rotating drum. In this investigation you are asked to find how the degree of vibration depends on the rate of rotation of the drum. In a school laboratory it is unsuitable to do this with an actual washing machine working on the mains, but a simple alternative with a small electric motor can be used. The motor needs to be fixed through a rubber mounting to a rigid support, and to have an eccentric (off-centre) mass mounted on its axle as shown in **Fig 10.25**.

It is convenient if the end casing of the motor has a definite mark or some writing on it, so that when it is viewed through a travelling microscope the writing seems to be blurred with the vibration that occurs. The extent of the blurring can be measured quite easily with a travelling microscope provided that the motor is well illuminated. It should be measured over a range of speeds of the motor. It is worth while carrying out a preliminary experiment to observe the vibration of the motor and to try to use a range of speeds within which there is a speed at which resonance occurs. If a stroboscopic lamp is used to find the speed of rotation of the motor, a graph of amplitude of vibration against rate of rotation can be plotted. If time permits, a further investigation could be made into how the position and mass of the eccentric mass affect the amplitude of the vibration.

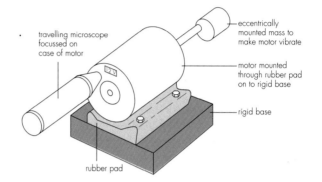

FIG 10.25 The experimental set-up for the investigation.

SUMMARY

- The time for one complete oscillation is called the period T.

- The number of oscillations per unit time is called the frequency f, which is related to period by $f = 1/T$.

- The angular frequency ω is given by $\omega = 2\pi/T = 2\pi f$.

- The maximum value of the displacement is called the amplitude.

- A body moves with simple harmonic motion if its acceleration is always directed towards a point and has a magnitude proportional to the displacement from that point.

- For a body moving with simple harmonic motion and with the displacement x given by

$$x = x_0 \sin \omega t$$

the velocity v and the acceleration a are given as follows:

	IN TERMS OF TIME	IN TERMS OF DISPLACEMENT
velocity	$v = x_0 \omega \cos \omega t$	$v = \omega \sqrt{(x_0^2 - x^2)}$
acceleration	$a = -x_0 \omega^2 \sin \omega t$	$a = -\omega^2 x$

- The kinetic energy of a body undergoing SHM is $\frac{1}{2} m \omega^2 (x_0^2 - x^2)$.

- The potential energy of a body undergoing SHM is $\frac{1}{2} m \omega^2 x^2$.

- A system with a natural frequency f_0, if forced to oscillate at frequency f, will oscillate with a large amplitude when $f = f_0$. This effect is known as resonance.

- An oscillating system is said to be damped if the amplitude of the oscillation decreases with time.

10.8 A fairground inspector measures the variation with time of the horizontal velocity of a swing. He discovers that the swing is undergoing simple harmonic motion. A graph of his data is shown below.

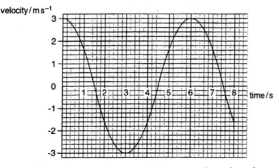

velocity/m s⁻¹

time/s

a Use the graph to obtain a value for the peak horizontal acceleration of the swing.

b Sketch a graph to show the variation with time of the horizontal acceleration of the swing.

Cambridge 1997

10.9 a The displacement of an object undergoing free oscillations varies with time, as shown.

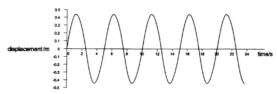

displacement/m   time/s

Take measurements from the graph to deduce
  i   the period,
  ii the amplitude,
  iii the frequency of the oscillations.

b Explain why, in practice, objects seldom undergo free oscillations.

c Sketch another graph which has the same frequency as the one already drawn but which shows lightly-damped oscillations.

d i Name a system in which damping of oscillations is important.
   ii Explain why the degree of damping is important in your named system.

Cambridge 1997

10.10 The diagram below shows a mass of 0.51 kg suspended at the lower end of a spring. The graph shows how the tension, F, in the spring varies with the extension, Δx, of the spring.

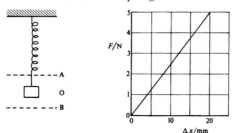

F/N   Δx/mm

a Use the graph to find a value for the spring constant k.

b The mass, originally at point O, is set into small vertical oscillations between the points A and B. Copy out and choose A, B or O to complete the following sentences.

   The speed of the mass is a maximum when the mass is at ....

   The velocity and acceleration are both in the same direction when the mass is moving from .... to ....

c Calculate the period of oscillation T of the mass.

d What energy transformations take place while the mass moves from B to O?

London 1996

10.11 a What is meant by *simple harmonic motion*?

b Calculate the length of a simple pendulum with a period of 2.0 s.

The graph shows the variation of displacement with time for a particle moving with simple harmonic motion.

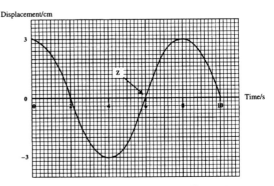

Displacement/cm   Time/s

c What is the amplitude of the oscillation?

d Estimate the speed of the particle at the point labelled Z.

e Draw a graph of the variation of velocity v with time for this particle over the same period of time. Add a scale to the velocity axis.

London 1997

10.12 A student was studying the motion of a simple pendulum the time period of which was given by $T = 2\pi(l/g)^{1/2}$.

a He measured T for values of l given by

$$l/\text{m} = 0.10, 0.40, 0.70, 1.00$$

and plotted a graph of T against $\sqrt{l}$ in order to deduce a value for g, the free-fall acceleration. Explain why these values for l are poorly chosen.

b How would the student obtain a value of g from the gradient of the graph?

c The graph below shows three cycles of oscillation for an undamped pendulum of length 1.00 m.

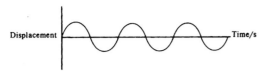

Copy the graph. Add magnitudes to the time axis and on the same axes show three cycles for the same pendulum when its motion is lightly air damped.

London 1996

10.13 a A simple pendulum executes simple harmonic motion with amplitude *A*.
Using axes like those in the figure, sketch:
  i a graph showing the variation of the potential energy of the pendulum bob with displacement. Label this graph P;
  ii a graph showing the variation of the kinetic energy of the pendulum bob with displacement. Label this graph K.

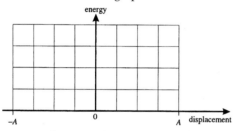

b A 1.5 m long simple pendulum oscillates with an amplitude of 0.042 m. The acceleration of free fall, *g*, is 9.8 m s^{-2}.
  i Calculate the frequency of oscillation of the pendulum.
  ii Calculate the maximum speed of the pendulum bob during an oscillation.
c You are supplied with a set of light springs, each of a different stiffness. You also have access to normal laboratory equipment. Explain how you would investigate the relationship between the period of oscillation of a mass–spring system and the stiffness of the spring.

AEB 1998

10.14 a i Explain what is meant by the *frequency* of vibration of an object.
  ii Distinguish between the *displacement* of a vibrating object and the *amplitude* of vibration.
b Some sand is placed on a flat horizontal plate and the plate is made to oscillate with simple harmonic motion in a vertical direction, as illustrated.

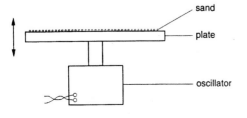

The plate oscillates with a frequency of 13 Hz.
  i Sketch a graph to show the variation with displacement *x* of the acceleration *a* of the plate.
  ii The acceleration *a* is given by the expression
    $$a = -\omega^2 x$$
    where ω is the angular frequency. Calculate
    1 the angular frequency ω,
    2 the amplitude of oscillation of the plate such that the maximum acceleration is numerically equal to the acceleration of free fall.
c Suggest, with a reason, what happens to the sand on the plate in **b** when the amplitude of oscillation of the plate exceeds the value calculated in **b ii 2**.
d One end of a horizontal string is now attached to the oscillating plate. The string passes over a pulley and the string is kept under tension by means of a weight, as illustrated.

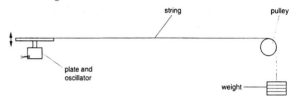

The frequency of oscillation of the plate is increased and at certain frequencies, stationary waves are produced on the string.
  i Copy the diagram and on your diagram show the stationary wave on the string when the frequency is such that the distance between the plate and the pulley corresponds to two wavelengths of the wave on the string.
  ii On your diagram, label the position of a node on the string.
  iii Briefly explain why a stationary wave is observed on the string only at particular frequencies of vibration of the plate.
e Some musical instruments rely on stationary waves on strings in order to produce sound.
Suggest why strings made of different materials or with different diameters are sometimes used.

OCR 1999

10.15 a Explain what is meant by *simple harmonic motion*.
  b

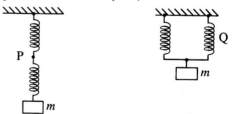

The diagram shows two mass/spring systems each constructed from identical springs, each with a spring constant *k*, of negligible mass supporting a mass, *m*.
  i Write down expressions for the time periods, T_P and T_Q, for the oscillators.
  ii Determine the ratio of the time periods $T_P : T_Q$.

c Four passengers whose combined mass is 320 kg are observed to compress the springs of a motor car by 50 mm when they enter it. The combined mass of the car and passengers supported by the springs is 1200 kg. Find the period of vertical vibration of the loaded car.

NEAB 1996

10.16 The figure represents an oscillation taking place at a particular point while a sound wave in a gas passes the point.

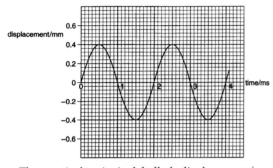

a The vertical axis is labelled displacement/mm. Explain what is meant by *displacement* in this context. Include in your answer details of what is being displaced.

b Using information in the figure, determine, for this sound wave,
  i the period,
  ii the frequency,
  iii the amplitude.

c A second sound wave of the same frequency but with one quarter of the amplitude passes through the same point together with the sound wave in the figure. This second wave is travelling in the opposite direction to the first wave and there is a 180° phase difference between the two waves.
  i Copy the figure, and draw a curve to represent this second wave. Label this curve X.
  ii Also on your diagram, sketch a curve to represent the resultant oscillation taking place at the point as a result of superposition of the two waves. Label this curve Y.

OCR 1999

10.17 Diagram A shows a mass suspended by an elastic cord. The mass is pulled downwards by a small amount and then released so that it performs simple harmonic oscillations of period T. Diagrams B–F show the positions of the mass of various times during a single oscillation.

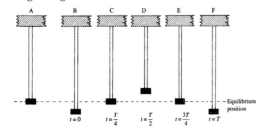

a Copy and complete the table below to describe the displacement, acceleration and velocity of the mass at the stages B–F, selecting appropriate symbols from the following list:

maximum and positive	$\rightarrow$	+
maximum and negative	$\rightarrow$	–
zero	$\rightarrow$	0

Use the convention that *downward* displacements, accelerations and velocities are positive.

	B	C	D	E	F
Displacement					
Acceleration					
Velocity					

In the sport of bungee jumping, one end of an elastic rope is attached to a bridge and the other end to a person. The person then jumps from the bridge and performs simple harmonic oscillations on the end of the rope.

People are bungee jumping from a bridge 50 m above a river. A jumper has a mass of 80 kg and is using an elastic rope of unstretched length 30 m. On the first fall the rope stretches so that at the bottom of the fall the jumper is just a few millimetres above the water.

b Calculate the decrease in gravitational potential energy of the bungee jumper on the first fall.

c What has happened to this energy?

d Calculate the force constant k, the force required to stretch the elastic rope by 1 m.

  e Hence calculate T, the period of oscillation of the bungee jumper.

London 1998

10.18 This question is about the motion of a mass on a spring.

a Explain what is meant by the term *simple harmonic motion*.

b A light spiral spring is suspended from its upper end. A mass of 7.5×10^{-2} kg, hung from the lower end, produces an extension of 0.15 m. The mass is pulled down a further 8.0×10^{-2} m and released so that it oscillates.
  i Calculate the spring constant, k.
  ii Calculate the time period of oscillation.
  iii **1** Calculate the maximum potential energy of the oscillating system.
      2 Sketch two graphs to show how the potential energy and kinetic energy of the system vary with displacement. Label your two graphs clearly.

c Suppose that the same mass and spring system is taken to the Moon. Will the time period be greater than, equal to or less than that on Earth? Explain your answer.

O & C 1998

10.19 The figure shows the displacement–time graph for the centre of a loudspeaker cone when it is displaced manually and then allowed to vibrate naturally.

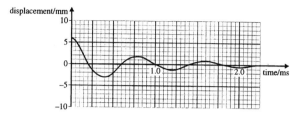

a i State the feature of the graph which shows that the oscillations are simple harmonic.

ii State the feature which shows that the oscillations are damped.

b i Calculate the natural frequency of oscillation of the cone.

ii Calculate the maximum acceleration of the cone during the time interval shown in the figure.

iii Draw a graph to show how the acceleration of the body changes with time.

c The graph below shows how the amplitude of the cone varies with frequency (the frequency response graph) when a signal generator provides alternating current of different frequencies but similar amplitude to the loudspeaker.

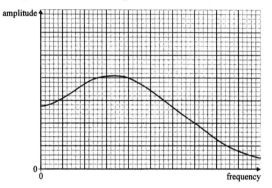

Copy the graph and draw graphs to show the effect of using:

i a coil of greater mass (label this graph, C);

ii a system of the same stiffness and mass, but less damping (label this graph, D).

AEB 1997

10.20 a What is meant by *resonance*?

b Sketch a graph which shows how the amplitude of a forced oscillation varies with frequency as the frequency varies from near zero to above the resonant frequency f_0.

c State a factor which determines the sharpness of the resonance.

d Name and outline one example where resonance is useful and one example where it is a nuisance or dangerous.

Cambridge 1997

10.21 A particle is undergoing simple harmonic motion. The motion has an angular frequency of 6.28×10^3 rad s^{-1}.

a Calculate the period of the motion.

b Sketch graphs to show the variation of the displacement x and of the velocity v of the particle during two cycles of oscillation.

OCR 1997

10.22 The figure shows the variation with time t of the displacement y of a lightly damped oscillating system.

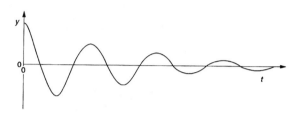

a Copy the table below. For each quantity listed, indicate with a tick how the quantity changes with time.

quantity	increases	remains constant	decreases
amplitude			
period			
frequency			

The diagram below is a corresponding graph for a heavily damped system.

b Outline one situation when heavy damping, such as that shown in the diagram, is useful in practice.

c Describe an experiment which would enable you to obtain a graph, such as that shown in the first figure, for a lightly damped oscillating system.

OCR 1999

11

WAVE MOTION

LEARNING OBJECTIVES

At the end of this chapter you should be able to:

① define and use the terms 'frequency', 'period', 'wavelength', 'displacement' and 'amplitude' when applied to progressive waves;

② relate the frequency and the wavelength to the velocity of the wave;

③ distinguish between transverse and longitudinal waves;

④ describe reflection and refraction as wave phenomena;

⑤ use the cathode ray oscilloscope for quantitative measurements;

⑥ list characteristics of the various parts of the electromagnetic spectrum.

11.1 PROGRESSIVE WAVES

Wave phenomena should be familiar to all of you, if only in the form of water waves (**Fig 11.1**). However, even if only water waves are considered, it is apparent that there is a great deal of variety of wave pattern. The ripples that raindrops make in a puddle are very different in speed, shape and size from a tidal wave or from the waves breaking on a beach. However, there are some similarities between these examples, which stem from their wave nature. It is the similarities that concern us in this chapter.

 What similarities do all waves show? What phenomena other than water waves show wave properties?

All waves involve a disturbance from an equilibrium position, and the disturbance travels from one region of space to another. A wave is said to be 'propagated' because the pattern of the disturbance occurring at one place at one time is

FIG 11.1 *Sea waves are rather unusual waves. Most waves do not have the possibility of becoming as unstable as these waves are.*

repeated at a later time at a different place. The complex movement of the electrons in the wire of a radio transmission aerial sets up electric and magnetic fields in the space surrounding the aerial. A fraction of a second later these changing fields cause the electrons in a receiving aerial to copy the pattern of movement of the electrons in the transmitting aerial. The link between the two aerials is an electromagnetic wave, the radio wave, which has travelled from the transmitter to the receiver. Microwaves, infra-red (IR), light, ultra-violet (UV), X-rays and gamma-rays (γ-rays) are other forms of electromagnetic waves.

Waves are said to be **mechanical waves** if the wave has a material substance, called the **medium**, through which to travel. Examples of mechanical waves include water waves and also such waves as sound waves, shock waves as in earthquakes, and waves in strings, springs and rods.

In many examples of wave travel, for instance when a ripple is made on the surface of a pond, energy from the source is being dispersed to the space surrounding the source. A wave that distributes energy in this way is said to be a **progressive wave**. As the wave is transmitted, this energy is spread over a larger and larger area, and so the energy associated with unit area drops. The following example illustrates the way in which the power per unit area may be calculated; it also shows how sensitive the eye is.

EXAMPLE 11.1

A 100 W light-bulb is 10% efficient (that is, 90% of its output is invisible infra-red radiation and only 10% is visible light). A person can see the light with the naked eye from a distance of 20 km on a dark night. If the area of the pupil of the person's eye is 0.5 cm², find the power of the light that the eye of the person is receiving.

First we need to work out the power of the *light* produced by the light-bulb. This is

power of light produced = 10% × 100 W = 10 W

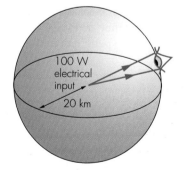

FIG 11.2 *The power from a light-bulb spreads out over a sphere.*

This power is assumed to be spreading out uniformly in all directions. So by the time it has reached a distance of 20 km from the light-bulb, the power is spread over the surface area of a sphere of radius 20 km. This is shown, in an exaggerated way, in **Fig 11.2**.

The surface area of a sphere is $4\pi r^2$. So in this case

$$\text{surface area} = 4\pi \times (20\,000\ \text{m})^2 = 5.0 \times 10^9\ \text{m}^2$$

$$= 5.0 \times 10^{13}\ \text{cm}^2$$

This means that the light power of 10 W is distributed uniformly over an area of $5.0 \times 10^{13}\ \text{cm}^2$. So for the eye of area 0.5 cm², we get

$$\text{power entering eye} = 0.5\ \text{cm}^2 \times \frac{10\ \text{W}}{5.0 \times 10^{13}\ \text{cm}^2}$$

$$= 1 \times 10^{-13}\ \text{W}$$

As indicated above, waves vary enormously in their complexity. The purest form of wave is one in which the oscillation of the source is a simple harmonic motion. Some of the terms introduced for simple harmonic motion can therefore also be applied to wave motion. These are:

period	T	the time for one complete oscillation
frequency	f	the number of oscillations per unit time
displacement	y	the vector distance a particle is moved from its rest position
amplitude	y_0 or a	the maximum displacement
angular frequency	ω	2π multiplied by the frequency

Another connection between wave motion and simple harmonic motion concerns energy. The **intensity** of a wave depends on the energy passing through unit area in unit time. For waves, as with simple harmonic motion, the energy associated with the oscillation is proportional to its amplitude squared (see section 10.3). Put in equation form this becomes

K   **intensity = $k \times$ (amplitude)^2 where k is a constant**

This readily becomes apparent at sea. A wave 1 m high can be produced by a light wind, and life on-board ship is not too uncomfortable. A wave 5 m high has 25 times the energy associated with it and will only be produced by a gale. It makes life on-board very uncomfortable.

As indicated in the light-bulb problem (Example 11.1), the intensity of a wave travelling through space decreases according to an inverse square law. That is

$$\text{intensity} \propto \frac{1}{\left(\text{distance from source}\right)^2} \qquad \text{or} \qquad I = \frac{\text{constant}}{d^2}$$

Astronomers make use of this equation in calculating the distance to stars, as it is much easier to measure the intensity of the light from a star than it is to measure its distance from the Earth.

Wave diagrams

There are some unavoidable problems with drawing diagrams of waves. A wave is essentially a moving pattern, and a piece of paper can only have fixed images on it. What is written and drawn to represent waves is therefore only a second best. Also a wave is usually three-dimensional, so representing it on a two-dimensional page must mean that the diagram is incomplete. Wave diagrams have to be carefully drawn, labelled and interpreted if they are to be useful. They also have to rely on the reader being able to visualise the pattern of movement from a variety of different diagrams.

Wavefronts and rays

Fig 11.3 is a photograph of the surface of a puddle with raindrops falling onto it. The ripples from the raindrops spread out in circles on the two-dimensional surface of the water. The raindrops fall randomly and the ripple produced by each raindrop moves completely independently of all the other ripples, so the pattern is one of overlapping circles. A diagram showing this looks like **Fig 11.4** and is rather confusing. Concentrating on one ripple, as in **Fig 11.5**, simplifies the diagram and enables the addition of circles and arrows to be made to indicate that, as time proceeds, the circles get larger.

A line showing the position of the crest of a wave is called a **wavefront**. A wave moves in a direction at right-angles to the wavefront, and lines drawn to show the movement are called **rays**. **Fig 11.5** therefore is showing the wavefronts at different times, and rays starting from the source of the wave disturbance and spreading outwards. It is nevertheless difficult to show that, the larger the circle, the smaller

Displacement: x or y?

Note that for wave motion we use the letter y for displacement. This is because a wave is often considered graphically to be moving in the x direction and its displacement is in the y direction.

FIG 11.3 Ripples from raindrops in a puddle behave independently of one another.

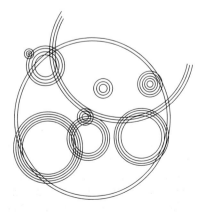

FIG 11.4 A diagram showing independent ripples produced by raindrops would very quickly become very confusing.

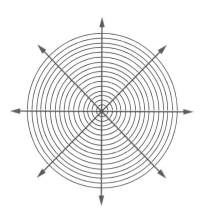

FIG 11.5 Waves spreading out. The blue lines showing the direction of travel are rays.

is the amplitude of the wave. This is done in **Fig 11.6**, where a series of graphs are drawn of the displacement, and the wave is shown in cross-section. Here it should be clear that the disturbance is of only short duration and that, as the wave spreads out from the centre, its amplitude is reduced. In order to show specific features of a wave, it is often necessary to draw more than one diagram or to draw composite diagrams like **Fig 11.6** in which several graphs are superimposed.

FIG 11.6 A composite diagram showing how the amplitude of a wave pulse reduces as it gets further from the source.

Displacement–distance and displacement–time graphs

A wave is a disturbance that changes both in space and in time. A graph representing a pure, continuous wave can therefore show either how the displacement changes with distance, or how the displacement varies with time. **Figs 11.7a** and **11.7b** are graphs for the *same* wave. **Fig 11.7a** is a graph of displacement against distance, and shows that one complete wave cycle occupies a distance of 10 m. The graph shows the wave at a particular moment in time, a 'graphical photograph' of the wave. **Fig 11.7b** shows how the displacement varies with time at a particular place, X. At the start, there is zero displacement at point X. As time passes, the displacement at X increases to a maximum as the crest of the wave passes, falls to zero, continues to a minimum as the trough of the wave goes past, and then returns to zero, to complete one cycle after 5 s. The period of this wave is 5 s, its frequency is 0.2 Hz and its angular frequency ω is $(2\pi/5)$ rad s^{-1} = 1.26 rad s^{-1}. **Fig 11.8** is a composite graph showing this information in a series of superimposed graphs.

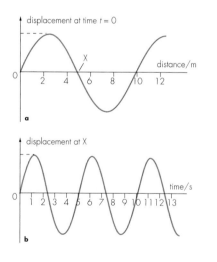

FIG 11.7 **a** A displacement–distance graph and **b** a displacement–time graph for the *same* wave.

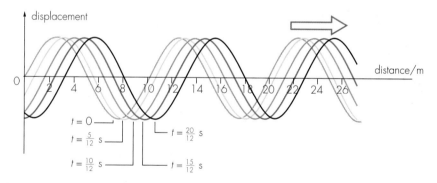

FIG 11.8 A composite graph showing movement of a wave.

Sketching wave diagrams

Graphs such as those shown here can be useful, but they are quite difficult to draw accurately. When sketching graphs of wave motion by hand, it is usually better to sketch them fairly quickly and to turn the page round so that you are not drawing them from underneath the wave but from the side. Try to get your hand holding the pen oscillating smoothly from side to side before you actually bring pen to paper. Then, as you start drawing the wave, move your whole arm down the paper while still maintaining the oscillation of your hand. It takes a bit of practice! The graphs in this book are drawn using computer software, so you cannot and will not be expected to draw them as well as this. Teachers and examiners are aware of the problem. They only expect you to draw the particular features of the wave that you wish to highlight.

Wave speed

Remember that $1\,\text{Hz} = 1\,\text{s}^{-1}$, and 'rad' appears in the answer for ω because it is the *angular* frequency.

Fig 11.7a shows that a complete cycle of the wave occupies a distance of 10 m. This distance is called the 'wavelength', and for a pure wave is the distance between adjacent crests. In general, **wavelength** λ is defined as the smallest distance between two points that are in phase with one another. It can be seen from **Fig 11.7b** that it will take 5 s for a complete cycle and so the speed is $2.0\,\text{m s}^{-1}$. The speed c of a wave is the speed with which the crests of the wave move. It is also the speed with which energy is transferred. It is *not* the speed with which the particles within the wave move. In general, the basic definition of speed is

$$\text{speed} = \frac{\text{distance}}{\text{time}}$$

For one cycle of a wave, the distance travelled is the wavelength and the time is the period, so

$$\text{speed} = \frac{\text{wavelength}}{\text{period}}$$

and since

$$\text{period} = \frac{1}{\text{frequency}}$$

we get

> **K** **speed = frequency × wavelength** or $c = f \times \lambda$

EXAMPLE 11.2

A radar wave is an electromagnetic wave whose speed is $3.00 \times 10^8\,\text{m s}^{-1}$. If its frequency is 1.64 GHz, find its wavelength and the value of ω for the wave.

Since $c = f \times \lambda$, substituting values gives

$$3.00 \times 10^8\,\text{m s}^{-1} = (1.64 \times 10^9\,\text{Hz}) \times \lambda$$

so

$$\lambda = \frac{3.00 \times 10^8\,\text{m s}^{-1}}{1.64 \times 10^9\,\text{Hz}} = 0.183\,\text{m}$$

$$\omega = 2\pi f = 2\pi \times 1.64 \times 10^9\,\text{Hz}$$
$$= 1.03 \times 10^{10}\,\text{rad s}^{-1}$$

Wave equations

The equation relating displacement y for a pure wave to the distance x along the wave is of a sinusoidal pattern. The equation is similar to those dealt with in Chapter 10 for simple harmonic motion. In this case the graph is one of displacement against time, and the angle whose sine is found is 2π multiplied by the fraction of wavelength x/λ. This gives the equation

$$y = y_0 \sin\left(2\pi\frac{x}{\lambda}\right)$$

Three variables

Mathematically the variable y depends on two other variables x and t. You will probably not be familiar with handling three variables, and certainly it is not possible on two-dimensional paper to draw graphs showing three variables. It is more than usually important with wave graphs, therefore, to be certain that you are explicit in what is being plotted on any horizontal axis.

the graph of which is drawn in **Fig 11.9a**. If the displacement at a particular point is required as a function of time, then the equation is the same as the equation for simple harmonic motion, namely

$$y = y_0 \sin\left(2\pi \frac{t}{T}\right)$$

This graph is drawn in **Fig 11.9b**.

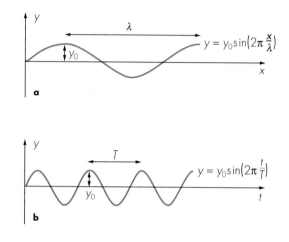

FIG 11.9 **a** Graph of the equation $y = y_0 \sin(2\pi x/\lambda)$.
b Graph of the equation $y = y_0 \sin(2\pi t/T)$.

QUESTIONS

11.1 At one end of the visible spectrum the wavelength of red light is 700 nm, and at the other end the wavelength of violet light is 400 nm. What is the range of frequencies of visible light? [Light travels with a speed of 3.0×10^8 m s^{-1} in a vacuum.]

11.2 Draw diagrams to illustrate two waves that are

  a $\pi/4$ radians out of phase,

  b $7\pi/4$ radians out of phase.

  c Is there any need to distinguish between these two diagrams?

11.3 What is the mathematical equation showing how the displacement y of a wave with an amplitude of 5 mm varies with time t if the wave has a wavelength of 600 mm and a speed of 340 m s^{-1}?

11.4 At a distance of 20 m from a point source of waves the amplitude of a wave is 1.6 mm and it has an intensity of 4.4×10^{-3} W m^{-2}. Find the intensity and amplitude of this wave at distances of 40 m and 100 m from the source.

11.5 Use the following data to answer the questions below:

wave speed = 40 cm s^{-1}
frequency = 50 Hz
speed of source of waves = 25 cm s^{-1}

a Draw a diagram, similar to **Fig 11.5**, to show the wave pattern produced when the source of waves is itself moving.

b Find the apparent wavelength and frequency of the waves in front of and behind the moving source.

c How does the diagram change if the speed of the source of the waves is changed to

   i 40 cm s^{-1}

   ii 60 cm s^{-1}?

TRANSVERSE AND LONGITUDINAL WAVES

he direction in which the displacement takes place within a wave motion affects the properties of the wave. In one type of wave, called a **transverse wave**, the particle movement is at *right-angles* to the direction of propagation of the wave. In the other type of wave, called a **longitudinal wave**, the particle movement is in the *same direction* as the direction of propagation. Both of these wave types can be illustrated using a slinky, a long flexible steel coil or spring, which in use rests on a smooth table. Wave energy can be transmitted by a slinky and, for illustration, each of the turns of the coil can represent a particle of the medium through which a wave is travelling.

a

b

FIG 11.10 **a** A transverse wave set up on a slinky spring. **b** A longitudinal wave set up on a slinky spring – instead of having the normal wave shape, as shown for the transverse wave in **a** the longitudinal wave has a series of compressions and rarefactions.

Transverse waves

The transverse wave is illustrated by the slinky in **Fig 11.10a**. At the left-hand side a turn of the coil is being oscillated sideways, perpendicular to the direction of propagation. Since the turns are connected to one another, the second coil copies the pattern of movement of the first, the third copies the second, and so on down the slinky. Each time the movement pattern is repeated, however, there is a small time delay, so the second coil is slightly out of phase with the first. There is a slight phase lag in the vibration of the second coil with respect to the first. It is this delay which gives rise to the wave propagation, and it is a characteristic of all progressive waves that adjacent particles have the same pattern of movement after a small delay.

Longitudinal waves

If, however, the first coil is pushed and pulled in a direction parallel to the length of the slinky, then a series of compressions and rarefactions is produced in the coil. A *compression* is a region where the turns of the coil are closer together than

FIG 11.11 Diagram to show longitudinal wave motion. A slot needs to be made in a piece of A4 paper or card, and the book moved down underneath the slot, as shown in **Fig 11.12**.

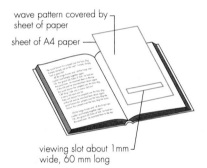

wave pattern covered by sheet of paper

sheet of A4 paper

viewing slot about 1mm wide, 60 mm long

FIG 11.12 Diagram showing how to use the longitudinal wave pattern in **Fig 11.11**.

average, and a *rarefaction* is where they are further apart than average. Again, as a result of each coil repeating the pattern of movement of the one next to it with a small time delay, it is found that both the compressions and the rarefactions propagate themselves forward in the direction of wave travel, as is shown in **Fig 11.10b**. This type of wave is called a longitudinal wave.

Fig 11.11 can help in the understanding of longitudinal wave motion. A narrow slot needs to be cut out of a piece of thick A4 paper or card according to the instructions given in **Fig 11.12**. (The slot will be needed again to view a different type of wave when you are reading Chapter 13.) Place the slot on top of the book so that a small portion of the wave pattern at the bottom of the diagram shows through the slot. Then slide the book towards you, keeping the sheet of paper still. The movement of the pattern shows the propagation of the wave to the right. If you concentrate your attention on a place where the lines are close together, a compression, it is easy to see the compression moving to the right, followed by a rarefaction, followed by a compression, . . . However, if you concentrate on just one line you will see that it is simply oscillating sideways. Each line represents one of the individual particles that make up the medium through which the wave travels. You can adjust the frequency and speed of this longitudinal progressive wave by pulling the book towards you at different speeds.

More about waves

All electromagnetic waves are transverse waves. The displacement in the case of electromagnetic waves is a variation in the electric and magnetic fields, and more will be written about these waves in section 11.7. Other transverse waves are water waves and waves on strings. Sound waves, on the other hand, are longitudinal waves.

There are several situations in which both types of wave can exist together, but they do not usually travel at the same speed. The shock wave from an explosion or from an earthquake consists of both longitudinal and transverse waves. Transverse shock waves (shear waves) reduce in amplitude quickly when they travel through a liquid. They are said to have rapid **attenuation**. Use is made of this as a way of prospecting for oil. If the shock wave from an explosion is found to contain much stronger longitudinal than transverse waves, then there is a good likelihood that the wave has travelled through a liquid. Test borings can then be made, and an oil company hopes it does not discover water! There is an analysis exercise related to this after section 11.6.

Wave speed

The speed with which different waves travel can be determined theoretically. It is usually found that there is a factor that has the effect of increasing the speed of the wave – this factor is associated with the strength of the elastic coupling between particles in the medium through which the wave travels. There is also a factor that reduces the speed – this factor is associated with the inertia of the moving particles. For instance, consider the speed of a transverse wave along a string. Increasing the tension in the string increases the strength of the coupling between the particles and hence also increases the speed of the wave. Increasing the mass per unit length of the string has the effect of increasing the inertia of the oscillating parts of the wave and hence of reducing the wave speed. Table 11.1 tabulates the equations for the speed of several different types of waves for reference, and question 11.6 asks you to check that they are correct for units.

Symbols used:
- c wave speed
- T tension
- m mass per unit length
- k force per unit extension
- l length of spring
- E Young modulus (see section 23.1)
- ρ density
- γ dimensionless constant
- p pressure
- σ surface tension (unit: J m^{-2})
- λ wavelength
- ε_0 permittivity of free space (see section 16.1)
- μ_0 permeability of free space (see section 18.6)

TABLE 11.1 Equations giving the speed of different types of wave

TYPE OF WAVE	EQUATION FOR SPEED OF WAVE
Transverse wave on a string	$c = \sqrt{\dfrac{T}{m}}$
Longitudinal wave on a spring	$c = \sqrt{\dfrac{kl}{m}}$
Longitudinal wave in a rod	$c = \sqrt{\dfrac{E}{\rho}}$
Longitudinal sound wave in a gas	$c = \sqrt{\dfrac{\gamma p}{\rho}}$
Transverse ripples on a pond	$c = \sqrt{\dfrac{2\pi\sigma}{\rho\lambda}}$
Transverse electromagnetic wave	$c = \sqrt{\dfrac{1}{\varepsilon_0\mu_0}}$

QUESTION

11.6   Use the data given in Table 11.1 to show that the right-hand sides of all the wave equations have units of metres per second. (If you have not yet studied electrical quantities, try looking up the relevant sections later in the book.)

11.3   REFLECTION

F ig 11.5 shows the way in which waves spread out from a source, with circular wavefronts and rays perpendicular to the wavefronts. The way in which successive wavefronts are built up from preceding ones is shown in both parts of **Fig 11.13**, in which each point on an existing wavefront is considered as a source of a new disturbance. The sum of all these disturbances becomes the new wavefront. **Fig 11.13a** shows this for a circular wavefront in which the new wavefront is also a circle but of larger radius. If the procedure is continued, the radius of curvature becomes greater and greater so that at a large distance from the source the wavefronts become virtually straight and the rays become parallel lines – **Fig 11.13b**. This method of constructing the new position of a travelling wave is making use of a technique called Huygens' construction, and it can be used in situations that are more involved than those with circular wavefronts.

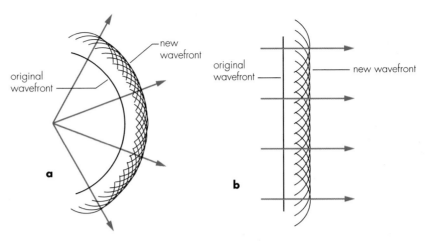

FIG 11.13 a The formation of a new circular wavefront on a circular wave. **b** The formation of a new wavefront on a plane wave.

Fig 11.14 shows a parallel wavefront approaching a plane surface and being reflected from it. After reflection, the intensity of the wavefront is appreciable where it has been drawn with thick lines but can be shown mathematically to be near to zero elsewhere. The construction shows that triangles ABC and DCB are exactly the same size and therefore the angle of incidence is equal to the angle of reflection. This is shown for rays in **Fig 11.15** and illustrates the law of reflection, i.e. that the angle of incidence is equal to the angle of reflection when a wave is reflected (see section 12.2).

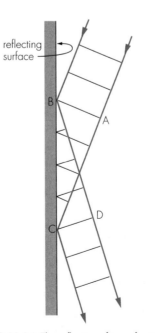

FIG 11.14 The reflection of wavefronts from a plane surface.

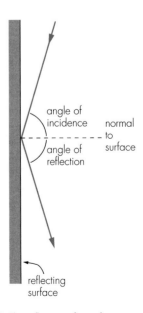

FIG 11.15 The reflection of rays from a plane surface.

FIG 11.16 A dish aerial. The parabolic shape of the aerial reflects radio waves to a sharp focus.

The law of reflection is made use of in aerial dishes. **Fig 11.16** is a photograph of one such dish, and **Fig 11.17** shows how a parallel beam of radio waves from a satellite positioned above the equator can be reflected to a single spot by the parabolic dish. The receiver is placed at this spot to obtain the largest intensity of radiation. At all points on the dish the angle of incidence is equal to the angle of reflection. For domestic use these dishes can be about a metre in diameter and the wavelength of the radio waves is typically of the order of half a metre.

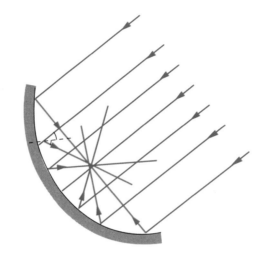

FIG 11.17 A ray diagram for reflection of parallel rays by a parabolic dish.

QUESTIONS

11.7 On a diagram copied from **Fig 11.17** show typical positions of the wavefronts before and after reflection.

11.8 A *radar kilometre* is a unit of time. It is the time taken for a radar pulse to travel a distance of 1 km to an object and to be reflected back. How long is a radar kilometre?

11.4   REFRACTION

The same procedure for finding the direction of wave travel can be performed if, instead of being reflected, a wave enters a region in which it has a slower speed. This might be when light passes from air into glass or where a water wave travels into shallower water. Assume that the change in speed takes place suddenly at a boundary. **Fig 11.18** shows this and illustrates the well known fact that light is refracted as it enters a dense material. The wavefronts are shown progressing towards the boundary and continue with a smaller wavelength after refraction. Here, not all of the wavefronts are shown so that the construction of one of them may be drawn.

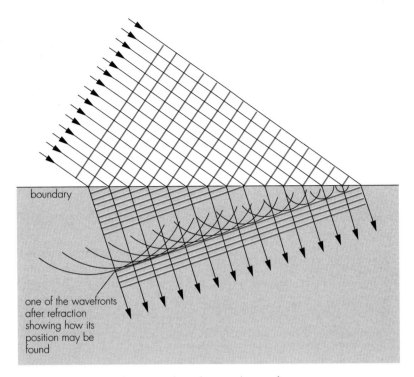

FIG 11.18 The refraction of waves, with emphasis on the wavefronts.

Using a simplified diagram (**Fig 11.19**) shows that in triangles ADC and ABC the angles at D and B are right-angles because rays are always at right-angles to wavefronts. If the wave takes time t to travel from B to C, then the other end of the wave also takes time t to travel from A to D. This gives

$$BC = c_1t \quad \text{and} \quad AD = c_2t$$

where c_1 is the speed of the wave in material 1 and c_2 is the speed of the wave in material 2. The ratio c_1/c_2 = constant, so for a ray travelling from one material into another the ratio

$$\frac{\sin i}{\sin r} = \text{constant}$$

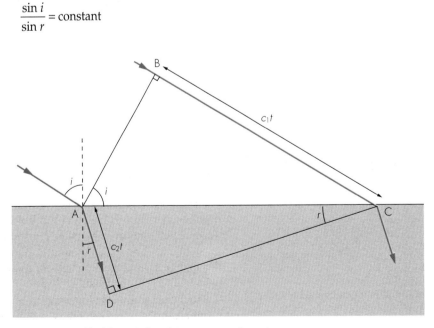

FIG 11.19 Simplified diagram for refraction, now with emphasis on just two rays.

For light, this is known as Snell's law of refraction. If light travels from a vacuum (material 1) into a transparent substance (material 2), then

$$\frac{\sin i}{\sin r} = \frac{c}{c_2}$$

where c is the speed of light in a vacuum.

> **K** The ratio
>
> $$\frac{\text{speed of light in a vacuum}}{\text{speed of light in a transparent material}}$$
>
> is known as the refractive index n of the transparent material.

More details about refraction, in terms of rays, is given in section 12.3.

QUESTION

11.9 What is the refractive index from material 1 to material 2 for the refraction shown in **Fig 11.19**? If the waves had been travelling in the opposite direction, the pattern would have been exactly the same. It is a basic property of waves that they are always reversible. What therefore would be the refractive index from material 2 to material 1 for the refraction shown in **Fig 11.19**?

11.5 POLARISATION

What is polarisation?

A s a result of the transverse nature of the vibration, transverse waves have an additional property that is not possessed by longitudinal waves. The movement of particles in transverse mechanical waves is at right-angles to the direction of propagation of the wave. In three dimensions this still leaves many possibilities for the direction of particle movement, as **Fig 11.20** shows. All of the double-headed arrows in this figure are drawn at right-angles to the direction of propagation and show possible directions for particle movement. The direction of particle movement for a string, for example, does not necessarily have to be up and down, it can equally well be from side to side or slanting. Frequently, oscillation takes place in a transverse wave in many different directions and the wave is then said to be **unpolarised**. If the oscillation does take place in only one direction, then the wave is said to be **polarised** in that direction. The term **a plane-polarised wave** is frequently used.

Polarisation of radio waves

Whether or not a transverse wave is polarised depends to a large extent on how it is produced. Electromagnetic waves are transverse waves and can therefore be polarised. Indeed, it is the fact that they can be polarised which enables us to be confident

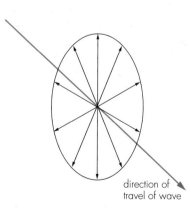

direction of
travel of wave

FIG 11.20 In an unpolarised transverse wave, oscillations may take place in any direction at right-angles to the direction in which the wave travels.

Liverpool
Community
College

that they are transverse and not longitudinal waves. If a radio wave is transmitted from a vertical aerial, then the wave will have vertical polarisation and any receiving aerial must also be positioned vertically. Use is made of the ability to polarise radio waves for television transmission. The main transmitters in the United Kingdom send out signals that are horizontally polarised. But because of the contours of the land, there are some areas which obtain less than the minimum field strength that is defined for a television service. To improve reception in these areas, there are about a thousand low-power transmitters each covering a limited area. Most of these low-power transmitters are vertically polarised, so that there is no danger of the signal that they transmit causing interference with the signals from the main transmitters.

The same procedure is used for radio transmission, though many areas of the country are now served by mixed polarisation transmitters. The following investigation may not work as intended in your area. It will nevertheless let you find out if you are in an area of mixed polarisation, and can also show how the strength of the received signal depends on the direction of the aerial.

INVESTIGATION

Use a portable radio to plot a graph that shows how the intensity of a received radio signal varies with the angle between the plane of polarisation of the radio wave and the direction of the aerial.

The investigation is easier if the portable radio has an FM band and a movable aerial. If the radio is rotated about a vertical axis, then the loudness of the radio changes not because of polarisation but because of the direction of the transmitter from the radio. If the aerial is rotated about a horizontal axis, however, the change in loudness is due to the polarisation of the wave.

Set up the radio so that maximum signal is being received, and arrange things so that you can measure the angle of rotation of the aerial as you rotate the aerial about a horizontal axis. Some detector must be used to measure the volume of noise that the radio is producing. This could be a decibel meter or a microphone connected to an oscilloscope. Relating the reading from the oscilloscope or decibel meter to intensity can be difficult, and therefore you should be careful how the vertical axis of your final graph is labelled.

A similar investigation may be carried out with microwaves, which, from a laboratory transmitter, are also polarised.

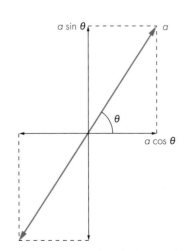

FIG 11.21 A wave of amplitude a can be resolved into two components at right-angles to one another. The amplitude of the horizontal component is $a \cos \theta$ and the amplitude of the vertical component is $a \sin \theta$.

The above investigation shows that, if a receiving aerial is not quite in line with the transmitting aerial, the intensity of the received signal does not suddenly drop to zero. Another illustration of the same thing is that you do not need to reposition the television aerial on the roof of your house if it is twisted a little in a gale. Nor does it become impossible to watch television when gusty wind conditions are causing the aerial to wobble. A wave oscillation can be resolved into two components at right-angles to one another, as is shown in **Fig 11.21**. Here a wave of amplitude a is oscillating so that its plane of polarisation is at an angle θ to the horizontal. This wave can be considered as two separate waves; one of amplitude $a \cos \theta$ that is horizontally polarised and one of amplitude $a \sin \theta$ that is vertically polarised.

Polarising filters and Polaroid

Light from a hot object, such as the Sun or a lamp filament, is unpolarised because it is emitted totally at random from the atoms of the hot object. If unpolarised light is passed through a perfect vertically polarising filter, often called a **polariser**, then the half of the light that is vertically polarised light will be allowed through and

FIG 11.22 The glare visible at the edges of this photograph has been eliminated in the centre by placing a Polaroid filter in front of the camera.

the other half, which is horizontally polarised, will be absorbed. Many crystals have the effect of polarising light. In Polaroid, crystals of quinine iodosulphate are used in which long molecules are aligned so that light polarised in one direction only can pass through. A sheet of Polaroid can be used to detect the presence of polarised light, in which case it is called an **analyser**. There are other natural means of producing polarised light. Light from blue sky is partly polarised and so is light reflected from a shiny surface.

This is why Polaroid is used in sunglasses. It reduces glare (see **Fig 11.22**) because the light reflected from a shiny horizontal surface, say the sea, is partially horizontally polarised. If the sunglasses are vertically polarised, then most of the reflected light is absorbed by the sunglasses. Anglers find these glasses useful, since it enables them to see into the water once the strong reflected light is reduced in intensity. The reduction in intensity of the reflected glare is shown in **Fig 11.23**.

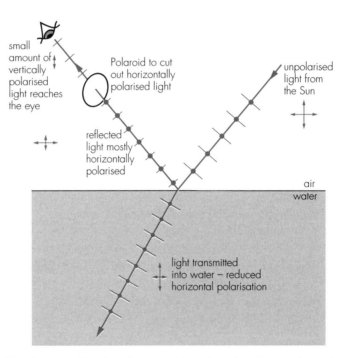

FIG 11.23 Polarisation of light by reflection enables polarised sunglasses to be used to reduce the glare.

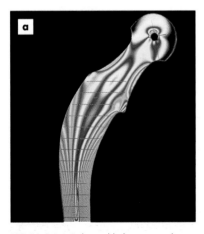

FIG 11.24 **a** Polarised light was used to take this photograph of a model of part of a hip joint – the stress pattern is clearly visible.

Optical activity

Use is often made in applications of polarisation of a property called optical activity. When polarised light is passed through some materials, the plane of polarisation is rotated. The plastic of which transparent rulers are made has this property; so do Sellotape and sugar solution. **Fig 11.24** shows photographs taken using this effect. The photographs are taken using an arrangement of crossed Polaroids shown in **Fig 11.25**. If the object is not present, then no light is able to pass through the analyser because it is set to transmit horizontally polarised light and the polariser is only letting through vertically polarised light. With the object in place between the crossed Polaroids, any optical activity of the object rotates the plane of polarisation, so that the analyser is able to pass some light. The colours in the photographs arise because different wavelengths of light are rotated by different amounts and also because different parts of the object under test are under different strains and this alters the amount of optical activity.

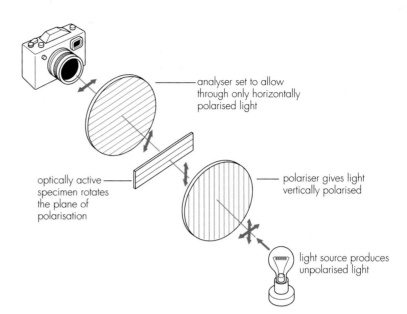

FIG 11.24 b The same technique using crossed Polaroids is used here to photograph geometrical instruments. You can see this effect very easily in the laboratory if you make a sandwich of a protractor between two crossed Polaroids.

analyser set to allow through only horizontally polarised light

optically active specimen rotates the plane of polarisation

polariser gives light vertically polarised

light source produces unpolarised light

FIG 11.25 The arrangement needed to take the photographs shown in **Fig 11.24**.

Effect of an electric field on polarisation

How much polarisation a material causes can be affected by the presence of an electric field. This is becoming an increasingly used effect. Liquid-crystal displays (LCDs) are usually polarised. Rotating a pair of polarised sunglasses over the display on a calculator will show the effect. The figures will appear and disappear twice per complete revolution of the sunglasses. The effect that an electric field has on the polarising ability of a material is known as the Kerr effect, and it can be used to interrupt a beam of light very rapidly. If a high-frequency potential difference is applied across a cell of liquid placed between crossed Polaroids, then light will not pass through the cell when the potential difference is zero but will when the field is set up. It is possible to obtain pulses of light with a duration as little as a picosecond (1 ps = 10^{-12} s) using this technique.

 EXPERIMENTAL METHODS

11.6

Some general principles concerning measurement of wave phenomena should be appreciated. For instance, if the speed c of a wave motion is required, one way of proceeding is to find both the wavelength λ and the frequency f of the wave motion in separate experiments, and then use the equation

$c = f \times \lambda$

If, on the other hand, a direct method of distance travelled divided by time taken is to be used, then it is essential in planning the experiment to be aware of the problems that are likely to be encountered. In a thunderstorm it is apparent that sound waves travel more slowly than light waves: the flash of lightning may be seen several seconds before the sound of the thunder is head. This may be used to find the speed of sound, but only if it is assumed that light travels instantaneously.

TABLE 11.2 Numerical values indicating the problem involved in measuring the speed of light

DISTANCE MEASURED /m	TIME ELAPSED /s	INVESTIGATOR/METHOD (DATE)
300 000 000 000	1000	Roemer (1676)
60 000 000	0.2	Satellite (modern)
300 000	0.001	Radar (modern)
30 000	0.000 1	Fizeau (1849)
30 000	0.000 1	Michelson (1878)
30	0.000 000 1	Laboratory rotating mirror (modern)

The speed of light

For many years it was assumed that light travelled instantaneously simply because no one could find any way in which a time delay could be observed. However, as early as 1676 Roemer, by observing the moons of Jupiter, was able to estimate the speed of light with quite remarkable accuracy. His biggest problem was in knowing the radius of the Earth's orbit. In modern units the problem is illustrated by Table 11.2, which shows how either very large distances or very small times must be used. By using astronomical distances, Roemer was able to give the first measured value of the speed of light. Modern direct methods can be performed in school laboratories by methods that effectively measure fractions of microseconds.

The speed of light has now, arbitrarily, been *defined* as 299 792 458 m s^{-1}. This means that all experiments which previously had been thought of as experiments to measure the speed of light are now regarded as experiments to measure distance! Nowadays this can be done very accurately, and the present definition of the metre is the distance light travels in a time of 1/299 792 458th of a second.

The oscilloscope

The structure and principles of the cathode ray oscilloscope will be dealt with in detail in sections 27.4 and 27.5, but some aspects of its use are essential when dealing with sound waves. The first thing to realise about an oscilloscope is that it is always measuring an electrical input. Therefore, if it is to be used to measure some feature of a sound wave, then it must be used in conjunction with a transducer of some sort. A **transducer** is a device that can convert one form of energy into a different form. Here, all that is necessary as a transducer is a microphone, which can convert a sound wave into an electrical signal of the same pattern.

The next important point is that, while the pattern on the screen is often stationary and horizontal distances can be measured on the screen, these distances do *not* correspond with any distances on a wave being measured, but are times. The trace on an oscilloscope is a graph of displacement against time. What might appear as a wavelength on the screen is the period of the wave. Finally, remember that the displacement in a longitudinal wave is in the direction of propagation of the wave, so that, although the wave may look on the screen as if it were a transverse wave, it may in reality be a longitudinal one.

Frequency and period

Frequency and period are readily measured with a calibrated oscilloscope, but the accuracy is not usually very high. The following example shows how this may be done.

EXAMPLE 11.3

The stationary pattern on the screen of an oscilloscope (**Fig 11.26a**), which is graduated in centimetre squares, might have appeared as shown in **Fig 11.26b**. The calibration for the time-base, which is applied to the X-plates, is also shown. Find the frequency and period of the oscillation.

The calibration of the time-base gives the information that a time of one millisecond (1 ms) is taken for the trace to move one centimetre (1 cm) sideways across the screen.

The pattern shows that five complete cycles occupy 8.2 cm. Five cycles take 8.2 ms, so

$$T = \frac{8.2\,\text{ms}}{5} = \frac{0.0082\,\text{s}}{5}$$

$$f = \frac{1}{T} = \frac{5}{0.0082\,\text{s}} = 610\,\text{Hz}$$

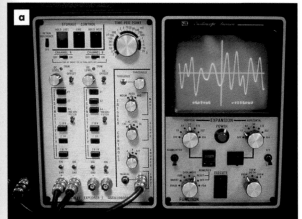

b

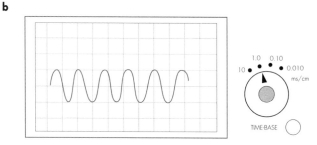

FIG 11.26 The trace on the screen of a calibrated oscilloscope enables the frequency of the oscillation to be found.

The speed of sound

The speed of sound can conveniently be determined using an oscilloscope as a clock to measure small time intervals. The experiment usefully illustrates several features of an oscilloscope, but is not particularly accurate because the accuracy of the calibration of the time-base on an oscilloscope is limited by the width of the screen and depends on whether the speed of the spot across the screen is constant.

An oscilloscope with a sweep output needs to be used. This output is a pulse given out at the start of each sweep of the spot across the screen, and leads connect this output to an amplifier and loudspeaker as shown in **Fig 11.27**. The sound output of the loudspeaker either can be a series of low-frequency clicks or, as the frequency is raised, can become more of a whistle. Since the distance the sound is to travel is something less than a metre, the time that needs to be measured is of the order of a few milliseconds. The spot on the oscilloscope screen needs to take a few milliseconds to cross the screen. If the time-base speed is set to something between 1 and

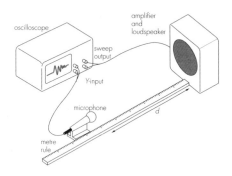

FIG 11.27 An arrangement for an experiment to find the speed of sound.

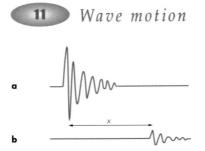

a

b

x

FIG 11.28 The screen patterns obtained in the experiment to measure the speed of sound.

0.1 ms cm^{-1}, then minor adjustments can be made as required. The sound output of the loudspeaker is picked up by a microphone, a distance d_0 from the loudspeaker, connected to the Y-input of the oscilloscope and the Y-sensitivity is adjusted to obtain a trace on the screen of suitable height. This should give a trace similar to **Fig 11.28a**.

The microphone is then moved different distances d away from the loudspeaker. As this is done, the pattern of the trace (**Fig 11.28b**) decreases in height because the microphone receives less volume of sound from the loudspeaker. The pattern also starts at a later time on the horizontal trace because, in the extra time that the sound takes to reach the microphone, the spot has travelled further horizontally. A graph can be plotted of x, the extra distance across the screen, against d and the gradient $\Delta x/\Delta d$ can be used to deduce the speed of sound. To do this it must be appreciated that the time it takes for the sound to travel any extra distance to the microphone is measured by using the calibration of the time-base velocity of the spot as it moves a distance x.

ANALYSIS

Seismic surveying

When the structure of the Earth near the surface is surveyed in prospecting for oil or minerals, one frequently used method is that of seismic reflection surveying. The process can be very complex because the strata in the Earth's crust are by no means regular, and also the quantity of data that is usually received is very large. The graphical results of one particular seismic survey are shown in **Fig 11.29**. Each of the vertical traces shows the output of a seismometer. It is interesting to note that if you look at **Fig 11.29**, you will see that one of the seismometers failed to work. It has been found easier to analyse the data produced if pulses are shaded in on one side. An output from a single seismometer, shown in **Fig 11.30a**, is printed as shown in **Fig 11.30b**.

Some of the principles behind the practice of seismic reflection surveying are explained and used in this exercise. The data have, however, been simplified.

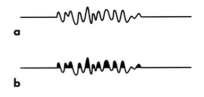

a

b

FIG 11.30 **a** The output from a single seismometer with a pulse of waves going past it. **b** The way the seismometer trace is printed.

In a place where there is a horizontal change in rock type of a certain depth, an explosion is set off (**Fig 11.32**). Eight detectors (D_1-D_8) are arranged as shown in **Fig 11.33**, and these detect vibrations from the explosion at source S, a short time after the explosion. The traces received from the eight detectors are printed alongside one another in **Fig 11.31**. You can see how the arrow shape of the actual seismic survey is developing even with only eight detectors.

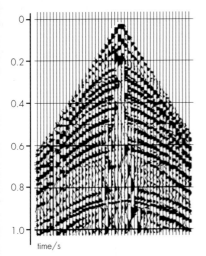

time/s

FIG 11.29 Seismic reflection survey result. Each vertical trace represents the output from a single seismometer.

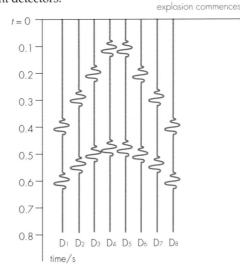

$t = 0$ is the time the explosion commences

$t = 0$

0.1

0.2

0.3

0.4

0.5

0.6

0.7

0.8

D₁ D₂ D₃ D₄ D₅ D₆ D₇ D₈

time/s

FIG 11.31 The output from eight seismometers printed alongside one other.

The rock through which the waves are travelling is known to have a density of 2700 kg m^{-3}, and in rock of this density the speed of P-waves is 3.1 km s^{-1}. P-waves are longitudinal waves, as shown in **Fig 11.34a**, and are the waves responsible for the pulses shown in **Fig 11.31**. S-waves are transverse waves, as shown in **Fig 11.34b**, and always arrive after the P-waves.

Answer the following questions, taking data from the diagrams where necessary.

a What other simple route can P-waves take to get from S to a detector apart from by the route shown in **Fig 11.33**? Identify each pulse on **Fig 11.31** with the route by which the P-wave arrives at the detector.

b What are the distances SD$_8$ and SXD$_8$? Use these distances to find the depth of the boundary between the two rock layers.

c Check your answer by repeating this with a different detector. (Why is it not sensible to use D$_1$?)

d The speed v_P of a P-wave is given by

$$v_P = \sqrt{\frac{A}{\rho}}$$

where A is a constant and ρ is the density of the rock. Find the value and the unit of A in this case.

e How will the traces change if, separately, the following additional factors have to be taken into account?
   i The S-waves, travelling at 1.8 km s^{-1}, are added.
   ii An extra layer of rock halfway down causes partial reflection.
   iii Double reflections are possible.
   iv The rock boundary is not horizontal.
   v Some refraction takes place at an intermediate layer as a result of density changes.

When, in practice, all of these factors can occur together, you perhaps can appreciate the problems of interpreting a seismic chart.

f How can you see from **Fig 11.31** that the sensitivity of all the seismometers is not the same? In what respect is the amplitude shown for each pulse received drawn incorrectly?

g When the wave reaches the boundary layer, some of the P-wave is transmitted downwards, with a speed of 5.7 km s^{-1}. Show that this is not possible at angles of incidence on the boundary greater than 33°. Draw a diagram to show another possible path for a P-wave from S to D$_8$. (This question involves the idea of critical angle, which is dealt with in section 12.3.)

h List other factors, besides those mentioned in part (e), that will affect the traces.

FIG 11.32 Drilling rigs doing a seismic survey on a glacier in Spitzbergen. Charges are detonated in drilled holes and shock waves bounce off rock strata.

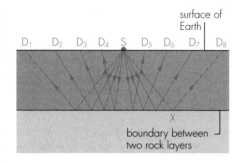

FIG 11.33 The arrangement of seismometers D$_1$–D$_8$ near S, the source of an explosion.

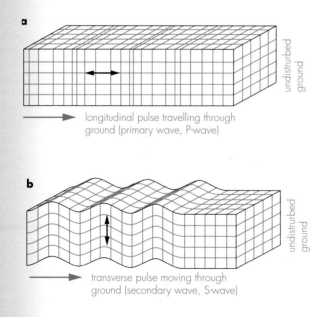

FIG 11.34 a Longitudinal waves (P-waves) and **b** transverse waves (S-waves) travelling through rock.

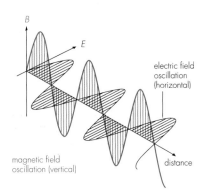

FIG 11.35 *An electromagnetic wave shown as interlocking electric and magnetic field oscillations at right-angles to one another.*

The electromagnetic spectrum covers a vast range of different waves, all of which travel with the same speed in a vacuum and have the characteristics of transverse electric and magnetic waves. The electric and magnetic waves are both essential for the transmission of energy; they interlock with one another and the electric wave is always at right-angles to the magnetic wave, as shown in **Fig 11.35**. This figure is a rather artificial diagram, so it may help to visualise an electromagnetic wave if **Fig 11.36** is examined. This figure shows how an electromagnetic wave originating from point P appears in cross-section. The electric field is shown in the plane of the paper and the magnetic field is always at right-angles to the paper. The oscillation of the magnetic field creates the electric field and vice versa, so that they sustain one another.

It was the mathematical analysis of this system that Maxwell achieved in 1864, and which led to the realisation that other electromagnetic waves besides light could exist and could perhaps be created artificially. The extent of the spectrum covers radio waves, infra-red (IR), visible light, ultra-violet (UV), X-rays and γ-rays (gamma-rays). But this division is in many respects artificial, as there is no sharp division between one electromagnetic wave and another of similar wavelength. While there are differences in methods of production and methods of detection, the waves themselves gradually change some of their characteristics as the wavelength and frequency change.

This can be illustrated by considering just that small part of the total spectrum which we call 'light'. Light does have a precise meaning. It is that part of the electromagnetic spectrum to which the retina of the human eye is sensitive, although even here there is a gradual start and finish since the sensitivity of the human eye varies with wavelength. Around about 400 nm wavelength, deep violet light is seen and gradually the colour changes until at a wavelength of beyond about 700 nm a deep red colour fades away as the sensitivity of the eye falls. It is clear from **Fig 11.37** that the visible spectrum contains the well known colours, but it is misleading to count the number of colours in the visible spectrum. Since one merges gradually into another, there must be an infinite number of colours in the visible spectrum. The same is true of the entire electromagnetic spectrum: the labels attached to any section must be used with caution. **Fig 11.38**

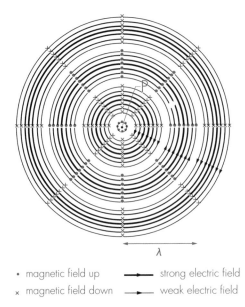

- magnetic field up   ⟶ strong electric field
× magnetic field down   ⟶ weak electric field

FIG 11.36 *An electromagnetic wave spreading out from a point source.*

FIG 11.37 The visible spectrum.

shows the complete spectrum, and the overlapping between the different regions results from this problem. Note that for all the waves the frequency multiplied by the wavelength has a constant value of 3.00×10^8 m s^{-1} and that the scales used are logarithmic. It is impossible to cover such a vast range of different values without using a logarithmic scale.

Wavelength / m	Frequency / Hz	Method of production	Type of radiation		Method of detection
10^{-16}					
	10^{24}				
10^{-15}	10^{23}	Energy released from the nucleus			
10^{-14}	10^{22}	of atoms in radioactivity			
10^{-13}			Gamma-rays		Ionisation
	10^{21}				Geiger–Muller tube
10^{-12}	10^{20}				
10^{-11}	10^{19}	Bombarding metal targets with high-			
10^{-10}	10^{18}	energy electrons	X- rays		photographic film
10^{-9}	10^{17}				
10^{-8}	10^{16}				
10^{-7}	10^{15}	High-temperature solids and gases	Ultra-violet		Skin – as sunburn Fluorescence
10^{-6}			VISIBLE LIGHT		EYE
	10^{14}	Sun			thermopile
10^{-5}	10^{13}	Hot solids	Infra-red		Skin – as heat
10^{-4}	10^{12}				
10^{-3}	10^{11}	Klystron oscillators			Oscillation of molecules
10^{-2}	10^{10}		Microwaves		
10^{-1}	10^{9}	Mobile phone			
1	10^{8}	TV FM radio		SHORT WAVE	
10	10^{7}	Electrons oscillated	Radio waves		Resonance in tuned electrical
10^{2}	10^{6}	by electric fields in aerials		MEDIUM WAVE	circuits
10^{3}	10^{5}			LONG WAVE	
10^{4}	10^{4}				
	10^{3}				

Increasing Wavelength → Increasing Frequency →

FIG 11.38 The electromagnetic spectrum.

11.8 INFORMATION TRANSMISSION

Radio waves

Hertz first produced radio waves in 1887 (section 29.1), and their importance as a means of communication was realised immediately. Initially, radio signals were sent short distances as a coded series of long and short bursts of waves (Morse code), but, as the technology improved, the distances covered and the complexity of the messages both increased. The radio waves that Hertz produced were not of one particular frequency. Waves of many frequencies are called *white* waves since white light is a mixture of many different frequencies of visible light. Noise in a radio signal is often white noise. If you listen to a radio when it is not tuned in properly, you can hear white sound as a hissing sound. Radio and television transmissions at present are always at specific frequencies and usually within the range of 10^5 to 10^{12} Hz.

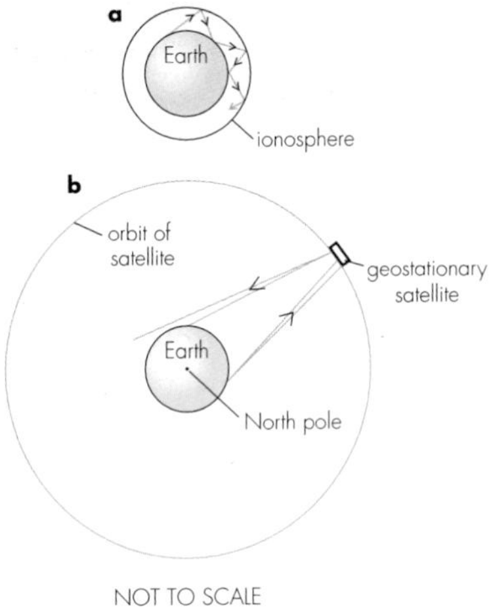

Transmission ranges

A radio wave can only travel a short distance through a solid or a liquid, so there is no possibility of communicating to other countries through the Earth itself. In general, radio waves can only travel to places within a line of sight. Television aerials are only effective if there is nearly direct vision of the television transmitting aerial. It is for this reason that these transmitters are set up on the tops of hills. In order to transmit radio waves over greater distances, two methods are used:

1 Surrounding the Earth is a layer of charged particles called the *ionosphere*. This acts as a reflector for certain low-frequency radio waves, as shown in **Fig 11.39a**. The Earth itself also acts as a reflector, so signals can go completely around the Earth, although the signal after this journey is very weak.

> **Q** Approximately how long does it take for a radio signal to circumnavigate the Earth in this way? Ignore extra distance due to the zig-zag route. You will need to know the speed of radio waves and that the metre was initially defined as one ten-millionth of the distance from the equator to a pole.

2 The use of satellites is now common for long-distance radio transmission. Geostationary satellites are positioned over the equator. These rotate once per day in a direction from west to east and so they seen to remain stationary over a point on the Earth. Radio signals can be beamed up to them using one frequency, amplified and then beamed back to a receiving station at a different place on the Earth at a different frequency. This is illustrated in **Fig 11.39b**. One particular UK transmitter at Goonhilly in Cornwall transmits frequencies in the range of 14.0–14.5 GHz, and receives in the range of 10.7–11.7 GHz (1 GHz = 10^9 Hz, so these frequencies have wavelengths in the range from 2 to 3 cm).

Analogue signals

A signal is any message that needs to be transmitted. It is an electrical potential difference variation which is a copy of some other oscillation. Often signals are transmitting speech or some other sound. A microphone will be used in this case to convert the initial sound wave into a p.d. variation called an audio signal as

NOT TO SCALE

FIG 11.39 a Long-wavelength radio waves can be reflected by the ionosphere and by the Earth itself, so that they can move around the Earth. The ionosphere is only a few hundred kilometres above the Earth's surface. **b** A geostationary satellite, placed 36 000 km above the Earth's surface, can be used to enable world-wide radio communication. (These diagrams are not drawn to scale.)

Effect of sunspots

At times of high sunspot activity, greater numbers of charged particles leave the Sun, and the ionosphere has a greater concentration of charge within it. This increases the ability of the ionosphere to reflect radio waves. At such times, amateur radio enthusiasts are often able to hear from other enthusiasts at much greater than normal distances.

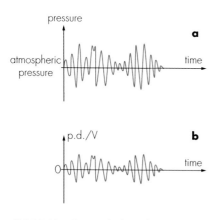

FIG 11.40 **a** The graph shows how the pressure varies above and below atmospheric pressure in a sound wave. The time scale on this graph might typically show the variation for 100 milliseconds. **b** This graph is a copy of (a) but is now an electrical p.d., which could be the output from a microphone.

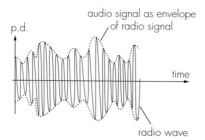

FIG 11.41 This graph shows the amplitude modulation of a radio wave. The time scale on this graph might typically show the variation for 10 milliseconds.

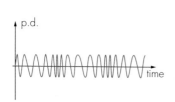

FIG 11.42 Frequency modulation of a radio wave. The information transmitted is in terms of how much the actual frequency of the radio wave differs from the norm.

shown in **Figs 11.40a** and **b**. A signal of this type is called an analogue signal. The p.d. can have any value. The frequency of this signal is far too low to be transmitted as a radio signal, so it is modified in one of several ways before transmission.

Amplitude modulation (AM)

Amplitude modulation was the first method that was used. The *amplitude* of a carrier radio wave is varied in step with the audio signal. **Fig 11.41** shows an amplitude-modulated radio wave which can be transmitted. In practice there are far more radio waves than the diagram shows, so the audio pattern can be transmitted quite accurately. Any radio programme on long or medium wave uses this process. The transmitted radio wave requires a range of frequencies for transmission, called a *bandwidth*. This limits the number of radio stations operating between any two frequencies.

A disadvantage of using amplitude modulation is that any noise which is picked up during transmission gets amplified along with the signal itself. If the so-called *signal-to-noise ratio* gets too low, then the quality of reception becomes very poor. The signal itself reduces in strength the further it travels, so it is likely that the signal-to-noise ratio increases with the distance from the transmitter, since noise can be picked up at any place in the signal's journey.

Frequency modulation (FM)

Frequency modulation became used as a method of reducing the amount of noise on analogue signals, and it is used for all FM and terrestrial television signals. **Fig 11.42** shows a frequency-modulated signal. The audio (and video) information to be transmitted is used to vary the *frequency* of the radio wave rather than its amplitude. A much higher radio frequency is required to do this, and so very high frequency (VHF) and ultra high frequency (UHF) are used. Each transmission still requires a bandwidth of frequencies, but more stations can be fitted in on VHF and UHF since there are more frequencies in these bands. (The space between 1 MHz and 10 MHz is only 9 MHz, but between 100 MHz and 1000 MHz there is a space of 900 MHz. This may seem obvious, but it does mean that more radio and television stations can transmit at higher frequencies.)

If noise is present in an FM signal, it will alter the amplitude of the signal, but the information required is in the frequency variation rather than the amplitude, so FM generally is much less noisy than AM.

Digital signals

A great deal of transmission of information at present is computer information. This information is coded as a series of 'bits' as shown in **Fig 11.43a**. Each bit is either OFF or ON, called logic 0 or logic 1. Transmission of digital information can be at a rate of billions of bits per second. Digital information can be manipulated so that any noise or distortion which gets added during transmission can be cleaned up to leave a perfect signal. This cleaning process can be performed several times in a long-distance transmission. The process is shown in **Figs 11.43b–d**.

Analogue-to-digital conversion

Analogue information, such as the output from a microphone or a video camera, is at present often converted into digital information before being transmitted. It is now proposed that all television transmission in the UK will be digitised in a few years time. An analogue-to-digital converter (ADC) is used before transmission and a digital-to-analogue converter (DAC) is used once the signal has reached its

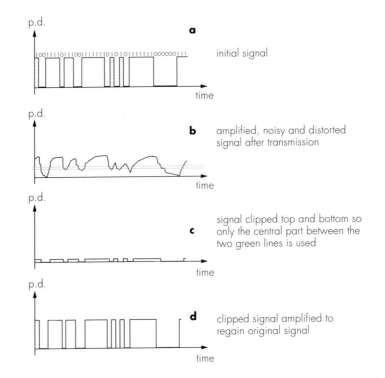

FIG 11.43 **a** Digital information expressed both as a graph and as a series of logic states. **b** A digital signal after transmission can become both distorted (not a rectangular shape) and noisy (not a smooth p.d.). The signal is electronically amplified and then clipped top and bottom so that only the central part between the two green lines is used. **c** The pattern after clipping. **d** The signal is then amplified back to regain its original shape.

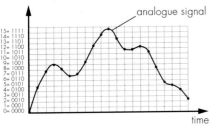

FIG 11.44 An example of analogue-to-digital conversion.

destination. The way in which analogue-to-digital conversion is done is illustrated in **Fig 11.44**, where the amplitude of the analogue signal is sampled at regular intervals and the digital output is shown.

In making a compact disc, the analogue sound signal is sampled electronically at a frequency of 44.1 kHz. In **Fig 11.44** only 4-bit numbers are used so each sample has a maximum value of 15, i.e. 1111. In practice eight or more bits are used. A digital-to-analogue converter repeats the process backwards and does leave a stepped output, but by using an 8-bit system and a high sampling frequency this distortion is minimal. The sampling frequency can be shown to be needed at twice the frequency of the highest analogue frequency that needs to be transmitted.

Consider what happens when you make a long-distance telephone call to a friend. When you speak, the information is in analogue form. This causes an analogue electrical signal from the microphone. This signal is converted into digital form before being transmitted by a combination of wires, optic fibres and microwave links to the person you are talking to. The last stage in the process is to convert the digital information of your speech back into an analogue electrical signal, which will produce sound at the earpiece of the telephone your friend is using. The following list gives some of the advantages of going through this seemingly complicated procedure.

■ Digital signals give better quality reception.

■ Digital signals can be transmitted easily by optic fibres.

■ More digital signals can be packed into a given transmission system than analogue signals because they use much higher frequencies.

■ As a result of the previous point, it is cheaper to send digital signals.

■ Modern electronic circuitry is very efficient with digital signals, so it is much more reliable.

■ Digital signals and digital transmission are more secure.

Time division multiplexing

It was stated in the list above that more digital signals can be packed into a transmission system than can be achieved with analogue signals. The Intelsat VI satellite can carry 44 000 two-way telephone calls simultaneously, and a single optic fibre can carry a few thousand two-way telephone calls simultaneously. This can be achieved by a process called *time division multiplexing*. The individual digital signals are chopped up into sections and transmitted section by section. Since transmission times for each section are far smaller that the time taken to produce the section, sections of other people's conversations can be fitted in as well. There obviously has to be an efficient sorting process after transmission to get all the sections of one conversation back together again. This is done by a coding system, which is transmitted with the signal itself.

There is little doubt that in a few years time almost all transmissions will be digital transmissions and that technology will continue to improve to cope with a continued huge expansion of international communication. Already business organisations, such as banks, send data to countries overseas, where often skilled labour is cheaper than in the UK. The computer programs may be devised overseas, data are analysed and evaluated there, and the results returned to the UK. Your bank account may be run from a computer office in Delhi!

QUESTIONS

11.10 What is the wavelength of a microwave of frequency 3.6 GHz?

11.11 **a** How long does it take for a message to be sent up to and return from a geostationary satellite that is 36 000 km from the Earth's surface?

    b Draw a diagram showing the path the message takes.

    c Are you justified in ignoring the radius of the Earth, which is 6400 km?

11.12 An analogue message is sampled 25 000 times per second and each sample contains eight bits. Using an optic fibre, 4.0×10^9 bits can be transmitted each second.

    a How many bits per second does one message generate?

    b What is the maximum theoretical number of messages of this sort which can be transmitted along the optic fibre?

SUMMARY

■ Waves may be classified as mechanical or electromagnetic. Any progressive wave transmits energy from one place to another by a cyclical movement of the medium in which the wave travels, but without there being any resultant transmission of the medium itself. In the case of electromagnetic waves, no medium is required, but the cyclical movement then takes place in electric and magnetic fields.

■ For longitudinal waves, the particles move in the direction of energy transmission; whereas for transverse waves, the particles travel at right-angles to the direction of energy transmission.

- Only transverse waves can be polarised.

- The amplitude of a wave is the maximum movement of a particle from its rest position. The intensity of a wave is proportional to the amplitude squared.

- Waves that are out of step with one another are said to be out of phase. One cycle is 2π radians, so two waves in antiphase are π radians out of phase.

- Wavelength for a progressive wave is the shortest distance between two particles that have no phase difference between them.

- Frequency is the number of oscillations per unit time.

- Speed = frequency × wavelength.

- The wave equation for displacement as a function of distance at one moment of time is

$$y = y_0 \sin\left(2\pi \frac{x}{\lambda}\right)$$

- The wave equation for displacement as a function of time at one place is

$$y = y_0 \sin\left(2\pi \frac{t}{T}\right)$$

- Reflection does not alter the speed of a wave, and for reflection the angle of incidence equals the angle of reflection.

- Refraction is caused when wave speed changes. At a boundary between two regions

$$\frac{\sin i}{\sin r} = \frac{\text{speed of wave before refraction}}{\text{speed of wave after refraction}}$$

For light

$$\frac{\sin i}{\sin r} = \frac{\text{speed of light in a vacuum}}{\text{speed of wave in a transparent material}} = \text{refractive index}$$

- The electromagnetic spectrum, commencing with the longest wavelength, consists of radio waves, microwaves, infra-red, visible light, ultra-violet, X-rays and gamma-rays.

- Signals can be analogue or digital. Digital signals can be transmitted directly if the frequency is high enough. Analogue signals can be transmitted using amplitude-modulated or frequency-modulated radio waves, or they may be digitised before being transmitted.

ASSESMENT QUESTIONS

11.13 a i The following is a list of terms used to describe waves:

longitudinal   electromagnetic
transverse   polarised

Which of the terms can be applied to sound waves.

ii Explain what is meant by a *transverse wave*.

b A sound wave from a certain source passes through a point. The figure shows the variation with time *t* of the displacement *d* of an air particle from its mean position at the point.

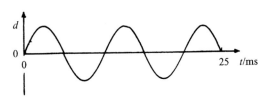

i Use information from the figure to calculate the frequency of the sound wave.

ii Suggest one possible source of sound for the wave shown.

iii On the figure it is possible to mark *either* the amplitude *A or* the wavelength λ of the sound wave. Copy the figure and on it mark and label the appropriate quantity. Explain why it is not possible to mark the other quantity.

CCEA 1998

11.14 a The figure represents the variation of the displacement *y* with distance *x* along a wave at a particular instant of time.

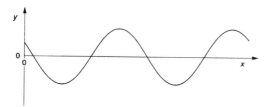

i By reference to the nature of the motion of particles in transverse and in longitudinal waves, explain why the figure may be used to represent both types of wave.

ii State how two items of quantitative information about the wave may be obtained from the figure.

b i State how an additional item of quantitative information may be obtained from a graph of the variation with time *t* of the displacement of one point in the wave shown in the figure above.

ii State a feature which is the same in the graphs of **a** and **b i**. Explain how information from the graphs may be used to determine the speed of the wave.

c There is an upper limit to the frequency of sound which can be heard and this limit is different for different people. Describe how, using a source of sound of variable but unknown frequency, a cathode-ray oscilloscope and a microphone, this upper limit is determined for an individual.

d The graph below shows the variation with time *t* of the displacements of two waves A and B.

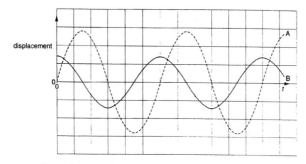

Use the graph to determine the phase difference between the waves. Suggest why, when two waves have different frequencies, the phase difference is not constant.

Cambridge 1999

11.15 a Energy can be transmitted through the Earth by mechanical longitudinal or transverse waves.

i Describe briefly the difference between the ways in which energy is transmitted by these waves.

ii Transverse waves such as light can be polarised. Describe briefly the difference between the light waves before and after being polarised.

b When exploring for minerals, geophysicists measure the time taken for a pulse of sound, made by a small explosion, to reach a detector placed a known distance away. The figure shows the situation for one such measurement. *The diagram is drawn to scale.*

The distances AB and CD are both 11.1 m. Other distances are as shown on the diagram.

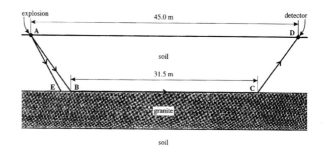

Sound can either take the direct route AD through soil only to the detector or it can travel by the path ABCD. The sound is travelling through granite from B to C. For certain values of AD, the sound can take less time by the longer path. Geophysicists use this information to determine the presence and depth of a lump of granite that might contain minerals.

i For the sound to follow the path ABCD, it has to be incident on the granite at the critical angle. Use the diagram to determine the critical angle.

ii On a sketch of the figure indicate the subsequent path of the sound travelling along AE after refraction at E.
You are not required to calculate the exact angles.

iii The speed of sound in soil is 3000 m s⁻¹. Show that the speed of sound in granite is about 5000 m s⁻¹.

iv Determine whether, for the situation shown in the figure, the time taken for the sound to travel from A to D is shorter by the path ABCD than by the direct path AD.

AEB 1998

11.16 a Earthquakes generate both *longitudinal* and *transverse* waves through the Earth. Outline the difference between the longitudinal and the transverse waves, making reference to the motion of the particles in their paths.

b For a certain sinusoidal progressive wave travelling with constant speed in a medium, the displacement of the particles of the medium at a particular instant is shown in the figure. The wave is travelling from left to right.

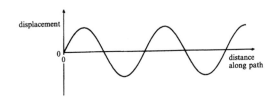

On a copy of the figure label the amplitude A and the wavelength λ of the wave.

c Instead of the graph of displacement against distance in the figure, the displacement d at a single point on the path of the wave in **b** may be plotted as a function of time t.

  i Sketch the graph of d against t for such a particle for the time interval $t = 0$ to $t = 2T$, where T is the periodic time.

  ii Sketch the corresponding graph for the same particle for the case where the new zero of time is $T/4$ later than in **c i**.

d Derive the relationship between the velocity v of a wave, its frequency f and its wavelength λ.

e A stationary radar transmitter sends out radio pulses which are reflected from a stationary object. The reflected pulses are received at the transmitter 0.40 ms after they leave. What is the distance of the object from the transmitter?

CCEA 1998

11.17 Many physical phenomena can be explained in terms of wave motion.

a Explain the meanings of six terms associated with waves, including displacement, phase difference and speed.

b Describe the nature of the motions in transverse and longitudinal waves, giving an example of each, and hence explain why polarisation is associated with only one of these types of wave.

c Describe experiments to measure

  i the critical angle for light at a glass–air boundary, and

  ii the frequency of sound produced by a plucked wire stretched between fixed supports.

Cambridge 1997

11.18 a Emission from stars and galaxies covers most of the electromagnetic spectrum.

  i Radio waves of certain frequencies emitted by galaxies cannot be observed on Earth. Explain why this is so.

  ii Ultraviolet radiation from stars cannot be detected on Earth. Explain why this is so and suggest how such radiation may be detected.

b **i** Explain why the wavelengths used in *radar astronomy* are confined to the range 1 cm to 20 cm.

  ii Describe how radar astronomy is used to measure the distance between a planet and Earth.

c **i** In the course of their orbits, the distance between the Earth and Venus changes. The radii of the orbits of the Earth and Venus are 1.5×10^8 km and 1.1×10^8 km respectively. If a radio pulse is transmitted from Earth towards Venus, calculate the greatest and least times that would be measured before the return pulse was detected.

  ii Explain why radar cannot be used to take measurements of a star such as α *Centauri* which is at a distance of 1.3 parsec from Earth.

NEAB 1998

11.19 a Explain what is meant by

  i unpolarised light,

  ii plane polarised light.

b

*   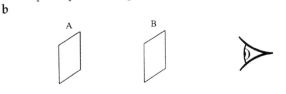

A light source appears bright when viewed through two pieces of polaroid, as shown. Describe what is seen when B is slowly rotated through 180° in its own plane.

c **i** Which of the following categories of waves can be polarised.

    radio   microwaves

    ultrasonic   ultra-violet

  ii State your criterion for deciding which.

NEAB 1997

11.20 a Explain the term *plane polarised wave*.

b Describe an experiment using light or microwaves which tests whether or not the waves are plane polarised.

c For each of the statements below, say whether the statement is true or false.

  i The speed of sound in air is less than the speed of sound in water.

  ii Since sound waves are longitudinal they cannot be diffracted.

  iii Sound waves transmit pressure but not energy.

  iv A sound wave of frequency 436 Hz travelling at 331 m s^{-1} has a wavelength of 75 cm ± 1 cm.

London 1998

11.21 a Graph A shows the displacement at time *t* = 0 at different points on a progressive wave. Graph B shows the situation at *t* = 0.10 s.

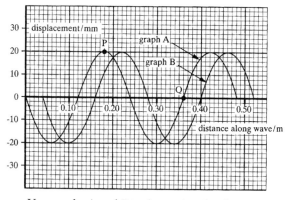

Use graphs A and B to determine, for the wave
  i the wavelength
  ii the amplitude
  iii the wave speed
  iv the frequency of the wave
  v the phase difference between points P and Q marked on graph A.

b Sketch a graph to show the variation with time of the displacement of point P on the graph in part **a**, showing numerical scales on both axes.

c When a stone is thrown into a still pond, circular ripples spread out. How could you tell, from observations of the ripples, whether or not they were travelling at constant speed? Explain your answer.

d Describe an experiment to demonstrate that light waves are transverse.

NEAB 1998

11.22 a Copy and complete the diagram below to show the different regions of the electromagnetic spectrum.

Radio waves	

b State *four* differences between radio waves and sound waves.

c Two radio stations broadcast at frequencies of 198 kHz and 95.8 MHz. Which station broadcasts at the longer wavelength?

d Why do obstacles such as buildings and hills present less of a problem for the reception of the signal from the station transmitting at the longer wavelength?

London 1998

11.23 Sound travels by means of longitudinal waves in air and solids. A progressive sound wave of wavelength λ and frequency *f* passes through a solid from left to right. The diagram X below represents the equilibrium positions of a line of atoms in the solid. Diagram Y represents the positions of the same atoms at a time *t* = t_0.

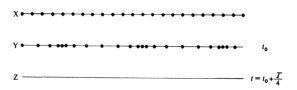

a Explain why the wave is longitudinal.
b Copy diagram Y and on it label
  i two compressions (C),
  ii two rarefactions (R),
  iii the wavelength λ of the wave.
c The period of the wave is *T*. Give a relationship between λ, *T* and the speed of the wave in the solid.
d Copy the line Z and along it mark in the positions of the two compressions and the two rarefactions at a time *t* given by *t* = t_0 + *T*/4.

London 1996

11.24 A water wave of amplitude 0.50 m is travelling in water which is 2.0 m deep, as illustrated in the diagram.

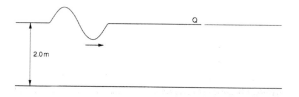

Water waves travel with a speed *v* which is dependent on the depth of water *h* and is given by the equation

$$v = \sqrt{gh}$$

where *g* is the acceleration of free fall. As there is a greater depth of water beneath the crest of a water wave than beneath the trough, wave crests will travel faster than wave troughs.

a Determine the depth of water beneath the crest of the wave.
b For the wave illustrated in the diagram, calculate the speed of travel of
  i the crest,
  ii the trough.
c Copy the diagram and on it draw a suggested shape of the wave a little later as it passes Q.

OCR 1999

12

OPTICS

LEARNING OBJECTIVES

At the end of this chapter you should be able to:

① quote and use the laws of reflection and refraction;

② describe the conditions under which total internal reflection occurs;

③ define and find experimentally the focal length and power of a lens;

④ use ray diagrams and the lens equation to find the position of images;

⑤ calculate magnification and angular magnification;

⑥ describe the eye, the camera and the telescope as optical instruments;

⑦ explain long and short sight and their correction.

12.1 INTRODUCTION

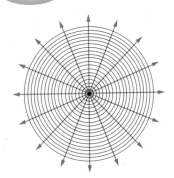

FIG 12.1 Waves spreading out from a point source. The green lines show the direction of travel, and for light are called light rays.

In Chapter 11 the propagation of energy in the form of waves was considered. If waves are created at a point source, then they spread out from the source in concentric circles, as shown in **Fig 12.1**. In three dimensions, the circles become spheres. When using waves of small wavelength, such as light, there is very little deviation from the wavefronts being perfect spheres. This implies that the directions in which the waves travel, as shown by the green lines on **Fig 12.1**, are always straight lines. For light, it is more usual to draw lines representing the direction in which the light is travelling rather than drawing the wavefronts. These lines are called light rays, and in a uniform transparent material **light ray**s travel in straight lines.

A diagram showing the passage of light rays is called a **ray diagram**. Ray diagrams often have to be exaggerated to make it clear what is taking place. Consider the most straightforward ray diagram: the one for an unaided eye viewing a green dot a distance of 15 cm from the eye. If drawn half scale, the

diagram will appear like **Fig 12.2a**. A more usual, easier-to-draw, diagram of the same situation is shown in **Fig 12.2b**, where the distance is decreased and the width of the cone of rays entering the eye is increased for clarity. However, the first diagram is the accurate one, and indicates two points well.

First, the amount of energy actually entering the eye is only a minute fraction of the energy emitted by the green dot, and that itself is minute in the first place. The eye can detect light when the power entering the eye is as little as 10^{-13} W.

The second point illustrated by **Fig 12.2a** is that the light entering the eye is nearly parallel; it diverges very slightly. Objects placed at any distance from the eye between infinity and about 15 cm can be seen in clear focus by a young person with normal vision. This means that the eye is capable of receiving and focussing light only if it is somewhere between being parallel and diverging by the small amount shown here. The unaided eye cannot focus light that converges onto it, nor can it focus light that diverges more rapidly than shown. Not only does the eye adjust to these slight differences of divergence, but the brain receives a signal that it can interpret as an estimated distance of the object from the eye.

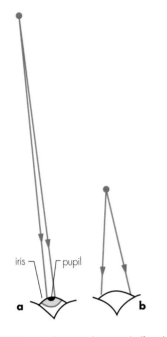

FIG 12.2 **a** A diagram drawn to half scale showing light rays from a green dot entering the pupil of an eye. Note how parallel the rays seem at the eye.
b The same diagram, not to scale and with the width of the eye exaggerated.

QUESTION

12.1 Assuming that the diameter of the pupil is 4 mm and the distance of the object is 150 mm, what is the angle of divergence in **Fig 12.2**? [This question can be done very easily if you use angular measure in radians.]

12.2 REFLECTION

When a light wave travelling in one material strikes a different material, it can be either absorbed or transmitted or reflected. If the surface interface is smooth, a reflection called a **specular reflection** occurs, in which the angle of incidence is equal to the angle of reflection, as shown in section 11.3 and also in **figure 12.3**. A reflection such as this can give the eye a false impression of where an object is. This is the principle of image formation, and is shown in **Fig 12.4**. Light from the green dot (the object) can spread out from the object and enter the eye directly, as shown by the green rays. But with the mirror present, each ray obeys the law of reflection, and the angle between the two green rays shown remains the same after reflection as it was before. As far as the eye is concerned, this reflected light is diverging and will be focussed in the same way as the direct light was focussed. The brain will interpret the received light as having come from a place on the far side of the mirror as far away from the mirror as the object was in front of it. Two dots can then be seen, and the second one is said to be an **image** of the first. There is no light at this image, and so it is called a **virtual image**. A screen placed at the position of a virtual image will show nothing.

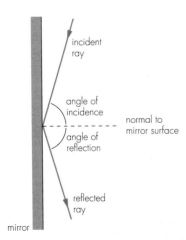

FIG 12.3 Reflection of a light ray by a mirror.

FIG 12.4 The formation of a virtual image by a plane mirror.

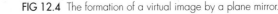

12.2 Draw diagrams to explain each of the following:

a Writing seen in a mirror is back to front.

b Parallel mirrors can produce an infinite number of images.

c Two mirrors at right-angles can produce three images, and the middle one of the three is not back to front.

d Two mirrors at right-angles will reflect light back parallel to its original direction. [In three dimensions this becomes a commonly used reflector for the back of a bicycle or for a radar reflector carried on the mast of many boats.]

e If you remain stationary and a mirror is moved towards you at a speed of 30 m s⁻¹ then your image moves at a speed of 60 m s⁻¹. [This principle is used in a radar speed detector. The mirror is the car and the radar source remains stationary. The image of the radar source is moving at a speed twice that of the car towards the detector.]

12.3 REFRACTION

Light can also be transmitted at a boundary between two transparent materials, and a change of direction often results. This is called **refraction** and is shown in **Fig 12.5**. It was shown in section 11.4 (using slightly different notation) that

$$\frac{\text{sine of angle of incidence}}{\text{sine of angle of refraction}} = \frac{\text{speed of light in less dense material}}{\text{speed of light in more dense material}}$$

$$\frac{\sin \theta_1}{\sin \theta_2} = \frac{c_1}{c_2}$$

where c_1 and c_2 are the speed of light in the less optically dense material 1 and more optically dense material 2, respectively. This can be written

K $$\frac{\sin h_1}{\sin h_2} = \frac{c_1}{c_2} = {}_1 n_2$$

where ${}_1 n_2$ is the refractive index for light travelling from material 1 into material 2. If the less optically dense material is a vacuum, this becomes

$$\frac{\sin \theta_1}{\sin \theta_2} = \frac{c}{c_2} = n$$

where n is called the (absolute) **refractive index** and c is the speed of light in a vacuum. For light, this is known as **Snell's law of refraction**. Using the principle of the reversibility of waves (see question 11.9), we can also write

K $$\frac{\sin \theta_2}{\sin \theta_1} = \frac{c_2}{c_1} = {}_2 n_1 = \frac{1}{{}_1 n_2}$$

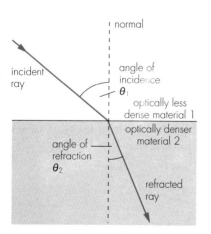

FIG 12.5 Refraction of a light ray at the surface between two transparent materials.

Optical density

Optical density has no connection with mass/volume. The optical density of a transparent material is related only to the speed with which light travels through it. The more optically dense the material, the slower light travels through it.

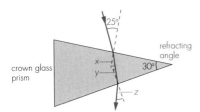

FIG 12.6 Refraction of a beam of light by a prism.

Significant figures

When working out an answer to a question like this, keep at least four significant figures in your working. This will enable you to have an answer to three significant figures.

Table 12.1 gives refractive indices for different transparent materials when the light being used is yellow sodium light. The colour of the light needs to be specified, as violet light slows down more than red light when it enters a transparent medium. This results in the dispersion of white light into the colours of the spectrum when refraction takes place (**Fig 12.6**). It can be seen from the table that very little refraction takes place when light travels from a vacuum into air. It consequently makes little difference if light enters some glass from a vacuum or from air.

TABLE 12.1 Refractive indices of a selection of different materials for yellow sodium light

MATERIAL	REFRACTIVE INDEX, n
Glasses (typical values)	
crown	1.52
light flint	1.58
dense flint	1.66
extra dense flint	1.80
Other solids	
diamond (C)	2.417
ice (H_2O)	1.309
quartz (SiO_2)	1.544
rock salt (NaCl)	1.544
rutile (TiO_2)	2.62
Liquids	
water	1.33
ethanol	1.36
turpentine	1.47
Gas	
air	1.0003

EXAMPLE 12.1

A ray of monochromatic light strikes a prism of refracting angle 30.0° that is made of glass of refractive index 1.52. The angle of incidence is 25.0° (see **Fig 12.7**). Find the deviation produced by the prism.

At the first surface, the angle of incidence is 25.0° and the angle of refraction is x. So

$$\frac{\sin 25°}{\sin x} = 1.52$$

$$\sin x = \frac{\sin 25°}{1.52} = 0.2780$$

$$x = 16.14°$$

In the triangle on the right of the prism, the sum of the angles of a triangle (180°) gives

$$(90° - x) + 30° + (90° - y) = 180°$$
$$30° = x + y$$
$$y = 13.86°$$

because we have already found that $x = 16.14°$. The light now emerges from the prism at the second surface, where the angle of incidence is $y = 13.86°$ and the angle of refraction is z. So

$$\frac{\sin y}{\sin z} = \frac{1}{1.52}$$

$$\sin z = 1.52 \sin y = 1.52 \sin 13.86° = 0.3461$$

$$z = 21.35°$$

The total deviation produced by the prism is the sum of the deviations produced at the two faces, so

$$\text{deviation} = (25.0° - x) + (z - y)$$
$$= (25.0° - 16.14°) + (21.35° - 13.86°)$$
$$= 16.4° \quad \text{(3 sig. figs)}$$

QUESTIONS

12.3 Find the deviation produced by a crown glass prism of refracting angle 60° on a beam of sodium light that strikes it at an angle of incidence of 35°.

12.4 Find the maximum refracting angle that a crown glass prism may have if sodium light is to pass through it with two refractions and no reflection. What deviation will be caused?

12.5 A beam of light strikes a water surface and some of it is reflected by the surface and some is refracted. Find the angle of incidence of the beam of light if it is found that the refracted and the reflected rays are at right-angles to one another.

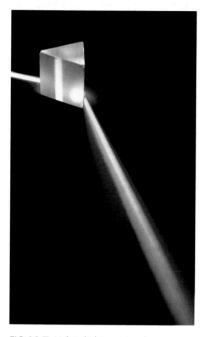

FIG 12.7 White light entering the prism is either reflected or refracted. The spectrum appears because blue light is slowed down more in the glass than red light. Blue light is therefore refracted more.

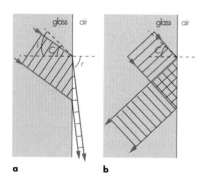

a   **b**

FIG 12.8 **a** Refraction of a wave emerging from a transparent material. The angle of incidence is less than the critical angle. **b** Total internal reflection for a similar situation, but in this case when the angle of incidence is just greater than the critical angle.

Total internal reflection

If a wave is leaving a medium in which it is travelling slowly, then Snell's law can be applied only within a certain range of angles, as was hinted at in question 12.4. **Fig 12.8** shows a beam of light leaving a glass block and being refracted away from the normal as expected. The relationship between the angle of incidence i and the angle of refraction r is given by the equation

$$\frac{\sin i}{\sin r} = \frac{1}{n}$$

From this equation, written as $n \sin i = \sin r$, it can be seen that if i is too large then $n \sin i$ will be greater than 1, and no value of r can be found, since it is not possible to have the sine of any angle greater than 1. The mathematics of this situation suggests therefore that refraction is not possible if i is greater than the value that makes $n \sin i = 1$. This value of i is called the **critical angle** C and we get

K $n \sin C = 1$ or $n = \dfrac{1}{\sin C}$

Not only does the mathematics of this situation make refraction impossible, but so does Huygens' construction. In **Fig 12.8a** a wave is shown approaching the boundary at an angle of incidence that is 5° less than the critical angle. The waves produced after leaving the boundary generate a new wavefront as expected. In **Fig 12.8b** the angle of incidence is increased to 5° above the critical angle, and now no new wavefront is set up. More detailed analysis of this situation shows that no wave energy can escape through the boundary, but that all the energy is reflected. This is called **total internal reflection**. This reflection is total reflection, with no energy lost at the reflecting surface.

Optical fibres

As a result of total internal reflection, it is possible to send a light wave along a glass fibre and for the light to be reflected millions of times without much loss of brightness (**Fig 12.9**). This is done in optical fibre systems, where the glass used is extremely pure to avoid absorption of the energy between reflections. **Fig 12.10** is a diagram of one type of optical fibre used to transmit messages. Because of the high frequency of light, it is possible to send vast amounts of digital information

FIG 12.9 Optical fibres trap light inside them. The light is reflected along the fibre and can only escape at the end.

down a single optical fibre. Fibres with diameter far less than that of a human hair can have enough digital information sent along them to transmit many thousands of telephone calls simultaneously.

Optical fibres of three different types are shown in **Fig 12.11**. The first, **Fig 12.11a**, is a step-index fibre of the type shown in **Fig 12.10**. Light that zig-zags along a fibre of this type takes an appreciably longer time to pass from one end of the fibre to the other than light that goes more or less straight down the centre with few reflections (see question 12.8). This has the effect of distorting the shape of the signal being transmitted. The second type of fibre, **Fig 12.11b**, which is now increasingly being used, overcomes this problem by being constructed out of glass only a few micrometres (1 μm = 10^{-6} m) in diameter. The amount of zig-zag that can take place is very much reduced. The third type of fibre, **Fig 12.11c**, is called a multimode fibre. It has a refractive index that varies between the centre and the circumference. The glass near the cladding is of lower refractive index than that along the axis of the fibre. This has the effect of slowing down any axial rays, so that rays travelling along any path all take the same time to travel along the fibre.

Really pure glass

Whereas normal glass, like sea-water, becomes difficult to see through if it is much more than a few metres thick, the glass used for optical fibre manufacture will allow light to pass through several kilometres without fading so much that it cannot be boosted by amplifiers. In fact, if sea-water were as pure as the glass used in optical fibres, it would be possible to see down to the bottom of all the oceans of the world!

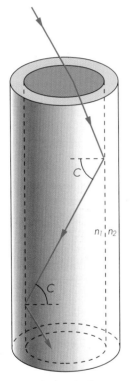

FIG 12.10 Total internal reflection enables light to be reflected many thousands of times in an optical fibre.

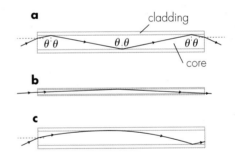

FIG 12.11 Optical fibres: **a** 'thick' step-index fibre; **b** 'thin' step-index fibre, only a few micrometres in diameter; and **c** multimode fibre. Light that enters the fibre can escape through the cladding, but some light will enter the fibre in such a way that the angles θ, shown in **a**, are equal to or greater than the critical angle for light travelling towards the cladding from the core. Once light is trapped in the fibre, it will be reflected again and again until it emerges at the end of the fibre.

EXAMPLE 12.2

A step-index optical fibre has a core made of a glass with a refractive index of 1.472 and a cladding made of glass of refractive index 1.455. What is the critical angle for the boundary between the two glasses?

The refractive index from the denser glass to the less dense glass is given by

$$n = 1.472/1.455 = 1.0117$$

and since

$$n = 1/\sin C$$

we have

$$\sin C = 1/n = 1/1.0117$$
$$\sin C = 0.988\ 45$$
$$C = 81.28°$$

12.6 The critical angle for light going from a certain glass to air is 41.3°. What is the refractive index of the glass?

12.7 Light is travelling along an optical fibre of refractive index 1.52. What must be the refractive index of the cladding if the critical angle is to be 85°?

12.8 The slowest speed with which light can travel along an optical fibre is when it zig-zags along the fibre making an angle of incidence equal to the critical angle whenever it is reflected (see **Fig 12.10**). The fastest speed is when it travels along the axis of the fibre. If the critical angle for one such fibre is 87.0° and the refractive index of the core is 1.520, find how long the fibre can be if the difference in the times taken by the two routes is not to be greater than 1 ns (1 ns = 10^{-9} s).

12.4 LENSES

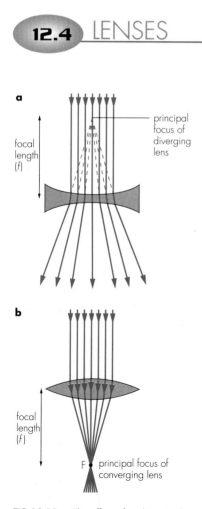

a

focal length (*f*)

principal focus of diverging lens

b

focal length (*f*)

F ● principal focus of converging lens

FIG 12.12 **a** The effect of a *diverging* lens on a parallel beam of light. **b** The effect of a *converging* lens on the same beam.

nell's law of refraction applies equally well at curved surfaces as at plane surfaces. Two curved surfaces close to one another form a **lens**. At one of the surfaces light travels from air into glass, and at the other surface the light travels out of the glass into the air. The effect on a parallel beam of light is shown in **Fig 12.12a** for a biconcave lens and in **Fig 12.12b** for a biconvex lens.

The analysis of the precise direction the rays of light have on emerging from a lens is difficult. Manufacturers take a great deal of care in designing and making lenses for cameras, for example. Their problems stem from the fact that light is not always parallel when it enters the lens, it does not always come from straight in front of the lens, it travels through different thicknesses of glass, and it may be of any colour. The detailed analysis necessary to design a camera lens is out of place here, but by making an assumption, the main features of lens behaviour can be understood. The assumption that can be made is that the lens has spherical surfaces close to one another. Such lenses are called thin spherical lenses. Practical problems in which real lenses, which may be very thick at their centre, are treated as if they are thin lenses, give surprisingly accurate results.

Thin spherical lenses

If light from a single distant point enters a thin lens, then after refraction all the light will pass through a single point if it is converging, or will seem to have come from a single point if it is diverging. In drawing **Figs 12.12a** and **b**, it has been assumed that the lenses are thin lenses. In these diagrams each point from which the light diverges or to which it converges is called the **principal focus**. The distance of the principal focus from the centre of the lens is called the **focal length** (*f*) of the lens. The **power** of a lens is defined by the equation

 power of a lens = 1/*f*

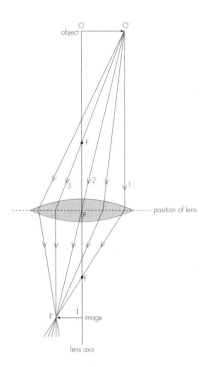

FIG 12.13 Image formation by a converging lens.

The unit used for lens power is the **dioptre**. A lens has a power of 1 dioptre if its focal length is 1 m. A 2 dioptre lens has a focal length of 0.5 m.

The ability of a thin lens to collect light from a point and to refract it to another point is the essential requirement for image formation, and is shown in **Fig 12.13**. Here green light starts from the tip of the object. The light entering the lens is not parallel and it is not arriving from directly in front of the lens, yet the lens has brought all the light to a point. The concentration of green light at this point will result in a green point on a screen or on a piece of film. Every point on the object results in a corresponding point on the screen, and so an image is built up. This type of image is called a **real image**. It should be noticed that the image is not at the principal focus. An image is only formed at a distance equal to the focal length from the lens when the object is at an infinite distance from the lens. **Fig 12.13** also illustrates another feature of the behaviour of a thin lens, namely that light passing through the centre of the lens is not deviated because the faces of the lens are parallel at the centre.

Magnification and the lens equation

The **linear magnification** produced by a lens is defined by

$$\text{linear magnification} = \frac{\text{height of image}}{\text{height of object}}$$

However, because the undeviated ray 2 makes two similar triangles O'OP and I'IP, this can also be written as

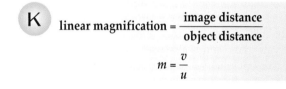

K   **linear magnification = $\dfrac{\text{image distance}}{\text{object distance}}$**

$$m = \frac{v}{u}$$

To calculate the solution of problems involving lenses, it is often necessary to use the **lens equation**, which is

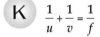

K $\dfrac{1}{u} + \dfrac{1}{v} = \dfrac{1}{f}$

where u is the object distance, v is the image distance and f is the focal length of the lens. A negative sign is always necessary for the term $1/v$ when a virtual image is formed. A negative sign is also needed for the term $1/f$ if the lens is a diverging lens. This is called the 'real is positive' sign convention.

Ray diagrams

The three rays labelled 1–3 in **Fig 12.13** are useful if the position of an image is to be found using ray diagrams:

1   A ray parallel to the axis of the lens is refracted so that it passes through the principal focus.
2   A ray through the centre of the lens is undeviated.

3   A ray first passing through the principal focus on the near side of the lens becomes parallel to the lens axis after refraction.

Also, instead of showing the lens and the refractions at both surfaces, we make the diagram simpler to draw by showing the position of the lens only; the refraction is assumed to occur at this line, as shown in **Fig 12.13**.

EXAMPLE 12.3

This example shows what happens if an object is placed closer to the lens than the principal focus. An object 4 cm high is placed 6 cm from a converging lens of focal length 9 cm. Find the position and magnification of the image.

The following steps are carried out, as shown on the diagram in **Fig 12.14**.

1 A scale of 1 : 2 is chosen.

2 The axis, object (O) and positions of lens and principal focus (F) are drawn in. (Note that there is a principal focus on *both* sides of the lens.)

3 The ray from the top of the object passing through the centre of the lens is drawn in undeviated.

4 The ray from the top of the object parallel to the axis (as it approaches the lens) goes through the principal focus (after passing through the lens). This ray will never cross the other ray drawn in step 3.

No real image can be formed by this lens arrangement, but if an eye is at X, as shown, the eye receives diverging light and it will be able to focus the light. The eye–brain system will deduce that the light appears to have come from I'. A virtual image of O' is formed at I' and a virtual image II' is formed of OO'. This is the optical arrangement for a magnifying glass, and is also used for the eyepiece of optical instruments such as a telescope or a microscope.

Fig 12.14 shows the image to be a distance of 18.6 cm from the lens and to have a height of 12.2 cm. The magnification is therefore

$$(12.2 \text{ cm})/(4 \text{ cm}) = 3.05$$

If the same problem is done by calculation, we get

$$\frac{1}{u} + \frac{1}{v} = \frac{1}{f}$$

$$\frac{1}{6 \text{ cm}} + \frac{1}{v} = \frac{1}{9 \text{ cm}}$$

$$\frac{1}{v} = \frac{1}{9 \text{ cm}} - \frac{1}{6 \text{ cm}}$$

$$= \frac{6-9}{54 \text{ cm}} = -\frac{1}{18 \text{ cm}}$$

$$v = -18 \text{ cm}$$

for the image distance; the magnification is

$$m = \frac{v}{u} = \frac{-18 \text{ cm}}{6 \text{ cm}} = -3.0$$

The discrepancy between the two methods is due to minor imperfections, which are unavoidable in drawing the diagram.

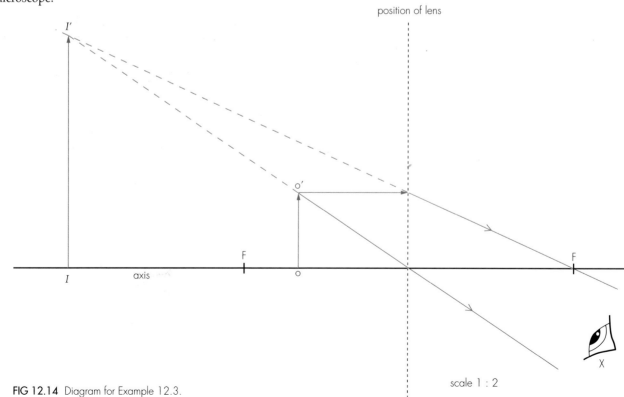

FIG 12.14 Diagram for Example 12.3.

EXAMPLE 12.4

Calculate the focal length required for an overhead projector lens for it to produce an image 1.20 m high from an object 20 cm high when the lens-to-screen distance is 1.80 m.

We need to find the object distance u so that we can substitute it in the lens equation to find the focal length f. First we use the equation for the magnification m to find u:

$$\text{magnification} = \frac{\text{height of image}}{\text{height of object}} = \frac{\text{image distance}}{\text{object distance}}$$

$$m = \frac{120 \text{ cm}}{20 \text{ cm}} = \frac{v}{u} = \frac{180 \text{ cm}}{u}$$

$$u = 30 \text{ cm}$$

and putting in values for u and v in the lens equation gives

$$\frac{1}{u} + \frac{1}{v} = \frac{1}{f}$$

$$\frac{1}{30 \text{ cm}} + \frac{1}{180 \text{ cm}} = \frac{1}{f}$$

$$\frac{6+1}{180 \text{ cm}} = \frac{1}{f}$$

$$f = \frac{180 \text{ cm}}{7} = 25.7 \text{ cm}$$

QUESTIONS

12.9 A camera has a focal length of 50 mm. It is used to take a photograph of an object 80 mm high placed 450 mm from the lens. Find

  a the power of the lens;

  b the distance the film needs to be placed from the lens;

  c the height of the image;

  d the linear magnification.

12.10 An enlarger lens has a focal length of 100 mm and is used to obtain a picture 200 mm long from a negative 35 mm long. How far must the negative be placed from the lens?

12.11 What happens to the image in Example 12.3 (**Fig 12.14**) if the object is moved back until it is 9 cm from the lens?

12.12 Find, by drawing, the position and magnification of the image formed by a diverging lens of focal length 10 cm when an object 3 cm high is placed 12 cm from the lens. Check your drawing by calculation.

12.5 THE EYE

Having studied the basics of reflection, refraction and lenses, it is now time to look at optics in action. In the next section we consider optical instruments, but first we must look into the human eye (**Fig 12.15**).

The parts of the eye

Fig 12.16 is a plan view of a person's right eye, with rays showing the optical situation when a person with normal sight is viewing a distant object. The function of the various components is as follows. The sclera is the outer casing of the eye. It is the white of the eye, and is approximately spherical with a diameter of about 2.5 cm. At the front of the eye there is a bulge where the sclera becomes transparent. This region is called the cornea, and is the area through which light passes into the fluids of the eye and the lens. The area of the hole through which light is allowed to pass is controlled by the iris. The hole itself appears black and is the pupil; the iris is the coloured part of the eye. The dark layer of the sclera at the back of the eye is the choroid. This prevents internal reflections. The inner layer, on the choroid, is the light-sensitive region of the eye, the retina. The centre of the retina is called the yellow spot, and is the part of the retina where the finest detail is seen. The part of the retina where the optic nerve leaves the eye is not sensitive to light, and is called the blind spot. One of the most remarkable features of the eye is the way the retina itself can adjust its sensitivity to light. The range of sensitivity of the eye is of the order of 10^9 (full sunlight to full moonlight is an intensity ratio of about 10^6). The ciliary muscles control the focussing of the eye, and there are other muscles, attached to the sclera but not shown, which enable the whole eye to be moved.

FIG 12.15 The human eye.

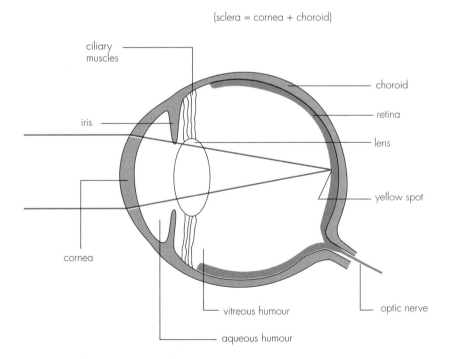

FIG 12.16 A plan view of a right eye.

Seeing under water

A simple demonstration of the importance of refraction at the cornea is given by opening your eyes when swimming under water. The presence of the water in contact with the cornea reduces the refraction at this surface to near zero and results in you being unable to focus on any object. Using goggles underwater enables you to see clearly because air is again present against the cornea surface.

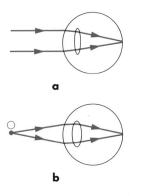

FIG 12.17 a When a person looks at a distant object, the eye lens is thin, as shown here. **b** The same person can focus on a close object by making the eye lens thicker, as here.

How the eye works

The basic optics of the eye is very similar to that of a single converging lens producing a small real image as in a camera. There are some important differences, however. Whereas in a camera light passes through the lens and emerges into the air, before hitting the film, light entering the eye through the cornea passes through the aqueous humour, the lens and finally the vitreous humour to fall on the retina. This series of transparent materials all play their part in focussing the light on the retina, and the initial refraction at the cornea is particularly important (see box).

The operation of the lens in the eye is different from that of the lens in a camera. The eye lens modifies the degree of convergence of the light passing through it by altering its shape rather than its position. When an object close to the eye is viewed, the ciliary muscles that surround the eye lens contract. This squeezes the lens into a thicker shape, so causing the lens to have greater power. If a distant object is viewed, the muscles relax and the lens has less power. This is shown in **Fig 12.17** in an exaggerated diagram. Sensing the degree of muscular effort necessary is another way the brain estimates the distance of an object.

The degree of flexibility of the lens decreases with age, and gradually the closest point that a person can see clearly in focus (the near point) recedes. A person whose eye lens becomes opaque is said to have a cataract. An operation can be done to remove a person's eye lens. This allows light to reach the retina, so vision is restored but the eye cannot then adjust for different distances of objects viewed. A person who has had such an operation therefore needs several pairs of spectacles for different object distances.

Another interesting point about the image on the retina is that, as in a camera, it is upside down. The brain interprets 'top' as being anything that causes an image on the bottom of the retina and vice versa.

Short sight

A person whose range of distinct vision is restricted to seeing only close objects clearly is said to be short sighted. Such a person has a longer eyeball than is necessary, and requires a diverging lens to correct the defect. The defect and its correction are shown in **Fig 12.18**.

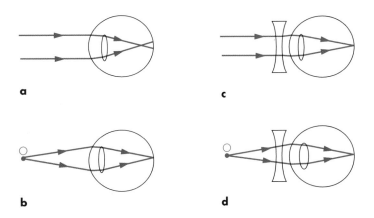

FIG 12.18 Short sight and its correction with a diverging lens. **a** Retina too far from lens – even the relaxed lens focusses light from a distant object too much. **b** Relaxed lens able to focus light from close object. **c** Sight corrected with diverging lens. **d** Presence of lens means lens must be squeezed in order to view a close object.

Long sight

This is the opposite defect to short sight. The eyeball is shorter than is required, so light from an object close to the person cannot be focussed, and even for viewing distant objects it is necessary to squeeze the eye lens. Additional focussing is provided by spectacles that have converging lenses. The defect and its correction are shown in **Fig 12.19**.

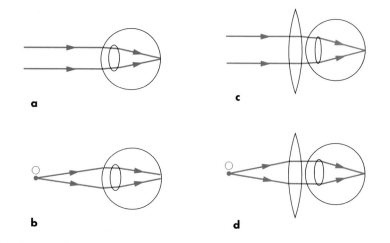

FIG 12.19 Long sight and its correction with a converging lens. **a** Retina too close to the lens – the lens has to be squeezed even when viewing a distant object. **b** The squeezed lens is not able to focus an image on the retina. **c** Lens is now relaxed when a correcting converging lense is used. **d** Lens makes possible a clear image from a close object.

12.6 OPTICAL INSTRUMENTS

The camera

A camera is basically a single-lens instrument that enables a real, diminished image to be formed on a piece of light-sensitive film for a short period of time (**Fig 12.20a**). The great variety of commercial instruments, from the very simple to the very sophisticated, cannot be described here.

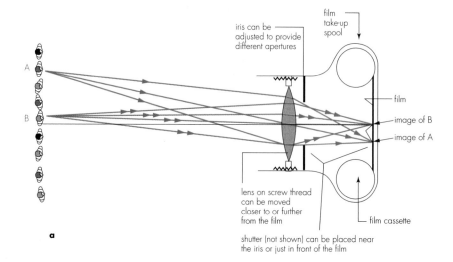

FIG 12.20 a A camera in use to take a team photograph.

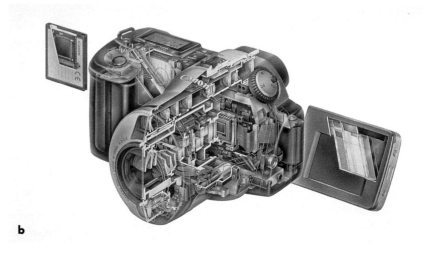

b

FIG 12.20 **b** An exploded diagram of a modern camera. The construction of the compound lens from many separate pieces of glass is clearly visible, as is the optical arrangement of mirrors and prisms for the viewfinder.

As you can see from **Fig 12.20b**, modern cameras use a compound lens made up of a number of pieces of glass rather than a lens made with just one piece of glass, but the basic optics is the same.

The following exercise should give you some insight into the operation of a camera and enable you to use the theory of the thin lens.

ANALYSIS

The camera

The following information is supplied with a single-lens reflex camera (the aperture of a camera is often written as $f/...$):

Shutter speed range	1/1000 s to 2 s
Focal length of lens	50 mm
Maximum usable diameter of lens (maximum aperture)	25 mm = $f/2$
Minimum aperture	$f/22$
Elements in the lens (no. of separate pieces of glass in its construction)	6
Angle of view	47°
Accompanying graph	**Fig 12.21**

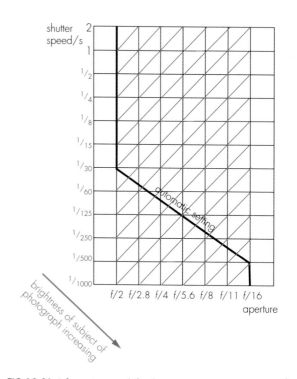

FIG 12.21 Information graph for the camera.

Answer the following questions:

a The lens moves in and out to focus on objects at distances between 0.45 m and infinity. How far must the lens be able to move?

b The scale for focussing around the screw thread of the lens is shown in **Fig 12.22**. Redraw the scale and put four other distances in their correct places, e.g. 20 m, 4 m, 1.3 m and 1.0 m. You can assume that the distance on the scale is proportional to the distance the lens moves towards the film.

c The illumination of the film depends directly on the area of the hole through which the light passes. The diameter of the hole is called the aperture, and it is usually measured as a fraction of the focal length. Show that $f/5.6$ gives twice the illumination of $f/8$. What would the value 5.6 be, if measured to, say, four significant figures?

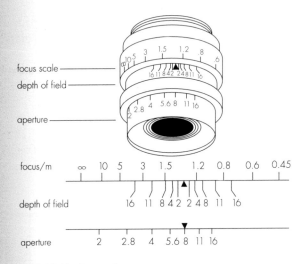

FIG 12.22 The lens scales.

d Show that each of the green lines on the graph corresponds to the same exposure. That is, a photograph taken of the same subject will cause the same total amount of light energy to fall on the film.

e A photograph is taken using the automatic exposure setting of $1/125$ s at $f/5.6$. The photographer then takes another photograph with the illumination received by the exposure metre 32 times brighter. What exposure and aperture will the automatic setting then use?

f **Fig 12.23** shows the camera receiving light from two objects at different distances. Object A is in focus on the film but object B is not. This is a normal situation for taking a photograph since not all of the object being photographed can be at the same distance from the camera. In practice, therefore, much of any photograph is slightly out of focus. If the distance XY on

Fig 12.23 is less than 0.1 mm, then object B also appears to be in focus. Explain, using a similar diagram to **Fig 12.23**, why it is that a photograph taken with a smaller aperture will have a greater range of objects apparently in focus. The distance within which objects seem to be in focus is called the **depth of field**.

g Using a diagram of the lens-to-image region and calculation, find the depth of field when the camera is focussed on an object at 0.8 m and is using an aperture of $f/11$.

h Explain how to use the depth of field scale shown in **Fig 12.22**.

i Why do you think that some of the information found in this exercise by using the thin lens formula is not entirely reliable?

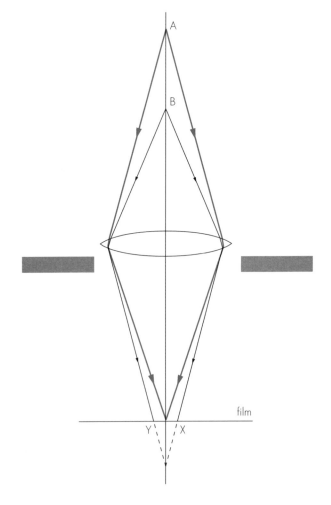

FIG 12.23 The image of A is sharply in focus. The image of B is not in focus, but if the distance XY is small enough, B may appear to be in focus.

The telescope

Refracting telescope

The basic refracting telescope is a two-lens instrument. The first lens, the objective, produces a small real image of a distant object. The second lens, the eyepiece, is used as a magnifying glass to view this first stage.

The linear magnification produced by a telescope is not of particular importance. As with a magnifying glass, the actual magnification can vary considerably without the user being very much aware of the fact that it is varying. Apart from the quality of the lenses in a telescope, which can be of overriding importance, two other factors are crucial. One of these is the size of the objective lens. An objective with a diameter of 12 cm has an area that is nine times greater than an objective of 4 cm diameter. This will make the image nine times brighter, so much more detail can be seen, provided the lenses are of comparable quality. The second factor is the **angular magnification**. This is the magnifying quantity that really matters and is defined by

K **angular magnification**

$$= \frac{\text{angle subtended by the image}}{\text{angle subtended by the object when the instrument is not used}}$$

Fig 12.24 is a ray diagram for a refracting telescope. It shows parallel light (in black), from a distant object A, arriving at the objective lens parallel to the axis of the telescope. The light is brought to a focus at the principal focus, and then spreads out until it reaches the eyepiece. The eyepiece is normally placed its focal length away from the principal focus of the objective, so parallel light emerges from the eyepiece, and the distance between the lenses is the sum of their focal lengths. These rays show clearly how the telescope concentrates light into an observer's eye.

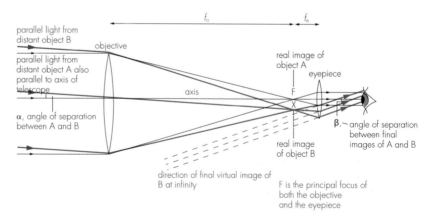

FIG 12.24 Ray diagram of a refracting astronomical telescope.

At the same time other parallel light (in green), from a different distant object B, enters the telescope at a small angle α to the axis. If these two objects are viewed without the telescope, then they appear to have an angle of separation of α. They may, for example, be two stars separated by an angle α. This is called the angle subtended by the two stars at the observer. The light from both A and B is parallel, but light from A is not parallel to light from B. On entering the telescope, the light from B is brought to a focus at X in the focal plane of the objective lens. When it passes on to the eyepiece, it too becomes parallel before entering the eye.

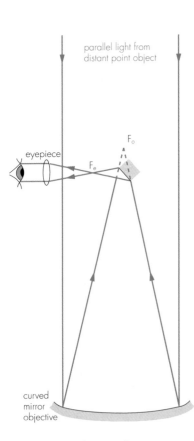

FIG 12.26 Ray diagram of an arrangement sometimes used in a reflecting astronomical telescope.

parallel light from distant point object

eyepiece

F_o

F_e

curved mirror objective

FIG 12.25 This reflecting telescope on Hawaii has a main mirror of diameter 3.6m.

If it is now at an angle β to the axis, the angular magnification of the telescope is given by

angular magnification = β / α

For a small angle θ, we can make the approximation that $\tan \theta = \theta$. So here we have

$$\tan \beta = \frac{FX}{f_e} = \beta \quad \text{and} \quad \tan \alpha = \frac{FX}{f_o} = \alpha$$

which gives the angular magnification as

$$\text{angular magnification} = \frac{\beta}{\alpha} = \frac{FX / f_e}{FX / f_o} = \frac{f_o}{f_e}$$

Therefore if the eyepiece has a focal length of 2 cm and the objective a focal length of 220 cm, the angular magnification will be

(220 cm)/(2 cm) = 110

If such a telescope were to be used to view the Moon, which subtends an angle of approximately 0.5° at an observer on the Earth, the Moon would appear to be 55° wide. It would also appear to be upside down.

By using prisms it is possible to reduce the length of a telescope and to obtain an image the right way up. This is done in binoculars, and the values 8 × 40 on a typical pair of binoculars indicate that the angular magnification is 8× and the diameter of the objective lens, which controls the light-gathering power of the instrument, is 40 mm.

Reflecting telescope

Large telescopes are all reflecting telescopes (**Fig 12.25**). They have objective curved mirrors rather than objective lenses, and because of this they do not suffer from chromatic aberration. Chromatic aberration occurs because, in lenses, different colours of light are refracted by different amounts. Using reflection at a mirror surface overcomes this problem, because all colours are reflected in the same way. An optical diagram for a reflector is similar to that for a refractor, but with the diagram folded back on itself (**Fig 12.26**).

QUESTIONS

12.13 a What power spectacles are required for a person who is short sighted and who can comfortably view objects between 80 cm and 15 cm?

  b What distances will he be able to view comfortably when wearing the spectacles?

12.14 A telescope with an objective lens of diameter 30 cm and a focal length 2.00 m can be used with eyepieces of focal length 4.0 cm, 2.0 cm or 1.0 cm.

  a What angular magnification can be achieved with each eyepiece and how much brighter will the image of a star appear when using the telescope rather than with the naked eye?

  b What advantages might there be in using the longer focal length eyepieces?

  c If such a telescope were to be used to take a photograph of a distant object, where would the photographic film be placed?

12.15 Explain why:

    a telescopes commonly have objective lenses with as large a diameter as possible;

    b telescopes are sited on the top of mountains; and

    c instruments for making astronomical observations in the ultra-violet and X-ray regions of the electromagnetic spectrum are usually placed in satellites.

SUMMARY

■   For reflection, the angle of incidence is equal to the angle of reflection.

■ For refraction,

$$\frac{\sin\theta_1}{\sin\theta_2} = {}_1n_2 = \frac{\text{speed of light in material 1}}{\text{speed of light in material 2}}$$

■ Total internal reflection occurs for light travelling towards a less dense medium if the angle of incidence is greater than the critical angle (C) and then

$$n = \frac{1}{\sin C}$$

■ The principal focus of a lens is the point to which parallel light, parallel to the axis, is refracted to, or from, after passing through the lens.

■ For a lens,

$$\text{power of a lens} = \frac{1}{f}$$

$$\text{magnification} = \frac{\text{height of image}}{\text{height of object}} = \frac{v}{u}$$

$$\frac{1}{u} + \frac{1}{v} = \frac{1}{f}$$

where v is negative for virtual images, and f is negative for diverging lenses.

■ A long sighted person has an eyeball that is too short and requires a converging lens for correction.

■ A short sighted person has an eyeball that is too long and requires a diverging lens for correction.

■ For a telescope

    angular magnification

$$= \frac{\text{angle subtended by the image}}{\text{angle subtended by the object if the instrument is not used}}$$

■ For a telescope, when the final image is at infinity

$$\text{angular magnification} = \frac{f_{\text{objective}}}{f_{\text{eyepiece}}}$$

ASSESSMENT QUESTIONS

12.16 The figure illustrates a small illuminated object in front of a plane mirror.

a **i** Copy the figure and on it mark the location of the image of the object. Label this point I.
   ii By drawing two rays from the object to enter the observer's eye, show how the image is formed.
b State the [two] laws of reflection of light.

Cambridge 1997

12.17 a State three properties of the image of an object as seen in a plane mirror.
   b When light is incident on an air/glass boundary, some of the light is refracted and some is reflected. The figure shows a ray of light from an object O incident on a thick glass mirror.

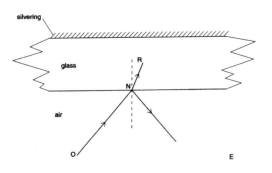

   i Copy the figure and on it continue the path of the ray NR until it has emerged into the air again.
   ii Suggest and explain what might be seen by an observer at E.
   c The speed of light in glass is 2.03×10^8 m s^{-1} and in water, the speed is 2.26×10^8 m s^{-1}. Calculate, for a ray of light passing from glass into water,
      i the refractive index,
      ii the maximum angle of incidence for the ray to be refracted.

OCR 1999

12.18 This question is about the reflection and refraction of light.
   a **i** State the laws of reflection of light.
      ii Draw a diagram to show how a plane mirror can be used to turn a ray of light through 90°.

b **i** State Snell's law for the refraction of light.
   ii A right-angled prism (see diagram) made of glass of refractive index 1.53, may also be used to turn a ray of light through 90° using *total internal reflection*.

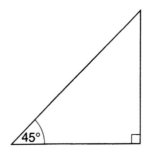

   1 Explain what is meant by total internal reflection.
   2 Calculate the critical angle for the glass of the prism.
   3 Copy the diagram and on it draw the path of a ray of light which is turned through 90° by the prism.
   c Discuss the relative merits of the plane mirror and the prism as devices to turn a ray of light through 90°.

O & C 1998

12.19 The diagram shows a ray of light incident on the surface of a cube made from glass of refractive index 1.50.

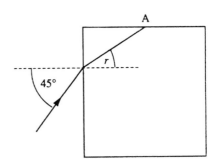

a **i** Calculate the angle, r.
   ii Calculate the critical angle for the glass–air interface.
   iii Copy the diagram. Use your answer from **ii** to determine the path of the ray of light through the glass cube and mark it on your diagram, showing the appropriate angles at the surfaces.
b When a drop of liquid of refractive index 1.40 is placed on the upper surface of the cube so as to cover point A, light emerges from the block at A. Explain, without calculation, how this occurs.

NEAB 1997

12.20 The diagram shows a ray of monochromatic light incident on a triangular glass prism (refractive index = 1.50) at an angle of incidence θ_1. The light just emerges from the face PR of the prism.

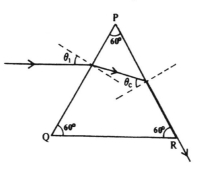

a Calculate
  i the speed of light in the prism
  ii the value of the angle θ_c
  iii the value of the angle θ_1.
b θ_1 is the minimum incident angle for light of this wavelength to just emerge from the prism. Explain why this is so.
c On a sketch of the diagram, draw the path that would be followed by a ray of light of a lower frequency at the same angle of incidence.

London 1995

12.21 A ray of yellow light travelling in air strikes side PQ of a glass prism and follows the path shown in the diagram.

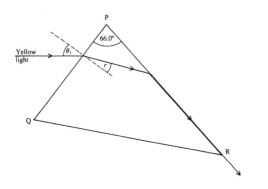

a The speed of yellow light in glass is 1.88×10^8 m s^{-1}. Calculate the refractive index of glass for yellow light.
b Calculate the critical angle for yellow light in this glass and indicate the critical angle on a copy of the diagram.
c Show that the angle r is 27.3°
d Calculate the angle of incidence θ_1.
e Add to your copy of the diagram the path that the yellow light would follow if the prism were replaced by a prism of similar shape but of greater refractive index.

London 1998

12.22 a Light travelling in a medium of refractive index n_1 is incident on the boundary with a medium of refractive index n_2. State the circumstances under which *total internal reflection* occurs.
  b **i** Calculate the critical angle for a water–air boundary if the refractive index of water is 1.33.
  ii A fish in a pond is swimming directly below a small boat which is 12 m from the shore. Calculate the minimum depth of the fish in the water for it to be able to see the ankles of a fisherman standing on the edge of the water.

NEAB 1998

12.23 The first figure shows a glass, part filled with water. A narrow beam of light is shone vertically down into the water and passes straight through. The second figure shows the glass tilted until the angle α is such that the light is refracted along the lower surface of the glass.

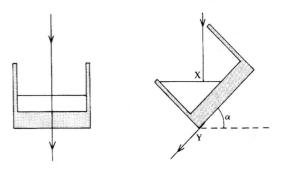

a Copy and complete the second figure to show the path of the ray of light from when it enters the water at X to when it leaves the glass at Y.
The refractive indices of air, glass and water are listed below:

Refractive index of air $n_a = 1.00$
Refractive index of glass $n_g = 1.50$
Refractive index of water $n_w = 1.33$

b Calculate the critical angle of the glass/air surface.
c Calculate the value of α.

London 1996

12.24 a A step-index optical fibre consists of a core surrounded by cladding. A ray strikes the plane end of the core of the fibre at an angle of incidence i and is refracted into the core. Show, with the aid of a ray diagram, the path of the ray into the core from air and its subsequent path along the core. Your diagram should show two reflections of the ray inside the fibre.
  b If the refractive index of the core is 1.50, suggest a value for the refractive index of the cladding, giving a reason for your choice.

NEAB 1996

12.25 The graph shows the variation of refractive index n with wavelength λ for light travelling in water.

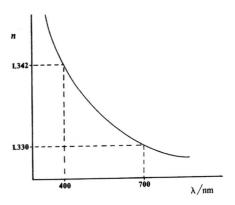

The diagram shows a mixture of red and violet light incident on an air/water interface.

a Calculate the angles of refraction for red and violet light. Copy the diagram above and on it draw the approximate paths of the refracted rays.
b If refractive index and wavelength were related as shown below, what changes would you need to make to your diagram?

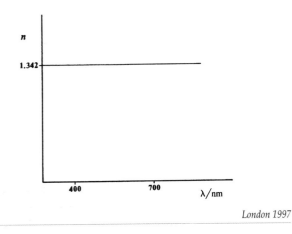

London 1997

12.26 a The figure shows the cross-section of a step-index optical fibre. Draw a sketch graph showing how the refractive index varies with the distance r from the centre of the core to a point well outside the cladding. Indicate clearly on your graph the interfaces between the core and cladding and between the cladding and air.

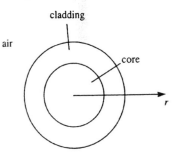

b Give *two* reasons, apart from cost, why it is advantageous to transmit signals optically through thin glass fibres rather than electrically through thin copper wires.

NEAB 1998

12.2

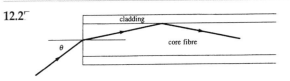

The diagram above shows how a ray of light may be transmitted along an optic fibre. If the core, cladding and external medium have refractive indices n_1, n_2 and n_3, respectively, show that the fibre will transmit light provided the angle θ is smaller than a value θ_m given by the expression

$$\sin \theta_m = \frac{(n_1^2 - n_2^2)^{1/2}}{n_3}$$

Hint: $\sin^2\theta + \cos^2\theta = 1$.

NEAB 1996

13

WAVE INTERACTION

● **LEARNING OBJECTIVES**

At the end of this chapter you should be able to:

① use the principle of superposition;

② explain what is meant by the term 'stationary wave' and explain how stationary waves are formed;

③ explain the meaning of the terms 'interference' and 'diffraction';

④ explain what is meant by the term 'resolution';

⑤ measure the wavelength of light using a diffraction grating.

13.1 SUPERPOSITION

When waves from two sources arrive at the same point, the resultant effect at that point is in general different from the effect that either of the waves would have made on its own. In this chapter we shall consider some situations where the resultant effect has important or interesting consequences. In order to do this, the first thing that is needed is some way of establishing how waves can be added to one another.

Adding waves

The photograph in **Fig 11.3** showed one fact about waves which we make use of all the time without realising it – namely that two waves can travel through one another without affecting each other at all. The ripples in the photograph are concentric circles. There might be some unusual disturbance where two ripples meet, but passing through another ripple does not alter the shape and height of a ripple. Sound waves criss-cross a room when a party is taking place in it – but it is

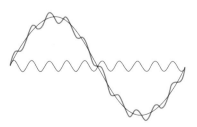

FIG 13.1 Two waves, one of frequency 10 times the other, are added together to give a total displacement shown by the green wave.

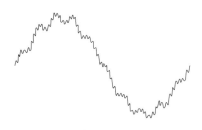

FIG 13.2 Here a third wave of higher frequency has been added to the two shown in **Fig 13.1**. One important property of waves is that they can be added to one another, with all sorts of surprising effects.

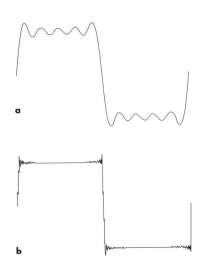

FIG 13.3 The results of adding together series of waves with different frequencies and amplitudes: **a** waves of frequencies *f*, 3*f*, 5*f*, 7*f*, 9*f* with amplitudes 1, 1/3, 1/5, 1/7, 1/9, respectively; **b** waves of frequencies *f*, 3*f*, 5*f*, 7*f*, ..., 97*f*, 99*f* with amplitudes 1, 1/3, 1/5, 1/7, ..., 1/97, 1/99, respectively.

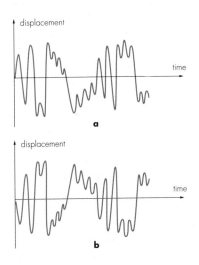

FIG 13.4 Sound and antisound can be added together to give zero amplitude.

still possible to hear a song and to know that the sound from it is the same as it would have been if all the other sound in the room were not present.

Light waves also can travel through one another. The complexity of the light wave pattern within any space can be huge. Light waves of different frequencies, different polarisations and travelling in different directions in three dimensions may be travelling through the space, but each one does not get distorted by the presence of all the other light waves. Because of this we are able to make sense of light waves when they enter our eyes.

A simple demonstration with an oscilloscope shows another feature of wave behaviour. If a microphone is connected to the Y-input of the oscilloscope, and the time-base and Y-sensitivity are suitably adjusted, then the pattern of any sound wave vibration can be seen. If two sounds are now made simultaneously, the pattern that appears on the screen is simply the sum of the two sounds separately.

This can be seen clearly if pure tones are used. **Fig 13.1** shows two separate wave patterns, one of a frequency 10 times the other and with smaller amplitude. It also shows the wave pattern obtained when both of the waves are present. This can be extended. **Fig 13.2** shows a wave in which three frequencies are present. This placing of waves one on top of another is known as **superposition**. The principle of superposition of waves is important both theoretically and practically.

Uses of superposition

One practical application of superposition is in the use of optical fibres, where thousands of messages, in the form of light waves, may be passed down a single fibre simultaneously. At the other end of the fibre, all of the messages can be unscrambled from the light waves.

Another interesting and important effect appears if waves of a series of increasing frequencies are added together. **Fig 13.3a** shows the result of adding together waves of frequency *f*, 3*f*, 5*f*, 7*f* and 9*f* and of amplitude 1, 1/3, 1/5, 1/7 and 1/9 respectively. If this is continued to 99*f*, then quite an accurate square wave is produced, as shown in **Fig 13.3b**. The high frequencies are always important in the transmission of square waves. It is the high frequencies that are responsible for the sharpness of the leading and trailing edges of the square wave.

The superposition of waves is increasingly being used to reduce noise pollution. It involves making 'antisound'. In a noisy environment, a microphone picks up the sound and an amplifier is used that reverses the sound, as shown in **Fig 13.4**. If the antisound is played through headphones, it can cancel out the original sound.

Beats

If two sound waves of similar amplitude but slightly different frequencies are listened to, a rhythmic pulsing of the intensity of the waves is heard. This is known as **beating**, and is a direct result of superposition of waves. **Figs 13.5a** and **b** show the variation of pressure with time for two waves of the same amplitude. At first sight the two graphs appear to be identical, but if you look carefully you will see that at X the waves are in phase with one another but that by Y they are out of phase. At Z they are back in phase. Between X and Z the wave in **a** has had 12 oscillations whereas the wave in **b** has had 13 oscillations. The sum of these two waves is shown in **Fig 13.5c**. Where **a** and **b** are in phase with one another, the resultant amplitude of **c** is large; where **a** and **b** are in antiphase, i.e. exactly out of phase, the resultant amplitude is zero.

 The *beat frequency* is the difference between the frequencies of the waves causing the beating.

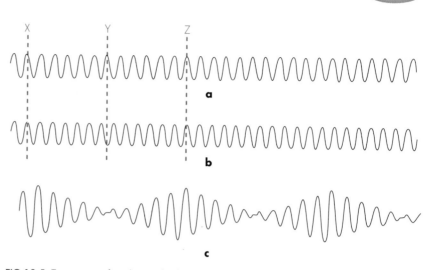

FIG 13.5 Two waves of similar amplitude and nearly equal frequency added to one another give a beating effect of increased and decreased amplitude.

Beats are used to detect small differences in frequency between two waves. They are also used for tuning musical instruments. On the higher frequencies a piano has two or three strings for each note. These strings must be accurately in tune with one another or a honky-tonk sound is produced. If the note is played, a person with a trained ear can listen for any beating that takes place between the sound from two strings and then adjust the tension in the strings to eliminate beating.

13.2 STATIONARY WAVES

Nodes and antinodes

The term 'stationary waves' sounds contradictory. In this context, *stationary* does not mean that nothing is moving, but simply that the positions of the crests and troughs of the wave are not moving. An alternative name for stationary waves is **standing waves**. When a string on a cello vibrates, parts of the string are moving to and fro but some parts are fixed. A finger on the left hand of the cellist is placed on a string to stop it moving at one point, and the bridge over which the string passes also keeps a point of the string fixed (**Fig 13.6**). In the form of its simplest vibration, the string appears as shown in **Fig 13.7**, and the maximum amplitude of the wave is in the centre of the string. This is one example of a stationary wave. A characteristic of a stationary wave is that there are some parts of the wave where the amplitude is always zero. These points are called **nodes**. Halfway between nodes are points where the amplitude is a maximum. These points are called **antinodes**. Between any two nodes, all the points on the wave are in phase with one another.

Stationary waves are set up as a result of the superposition of two waves of the same amplitude and frequency travelling at the same speed in opposite directions. Before considering how this comes about in practical situations, it is worth while examining **Fig 13.8**, which shows the formation of a standing wave. The green waves are travelling in opposite directions, and the bold black wave is the stationary wave that results from adding the other two waves together.

FIG 13.6 The frequency of vibration of the strings of the cello is controlled by the length of string used, the tension in the string and its mass per unit length.

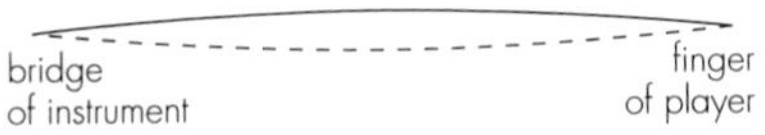

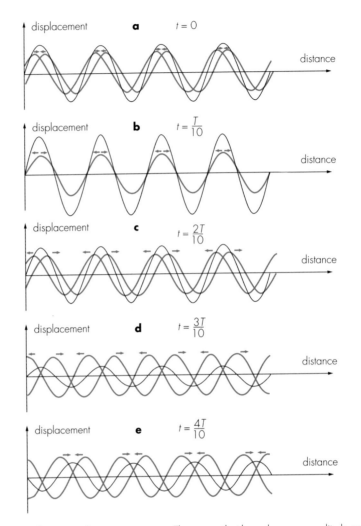

FIG 13.7 The movement of a string on a cello.

FIG 13.8 The formation of a stationary wave. The waves that have the same amplitude are travelling in opposite directions. The black wave is the stationary wave, and results from adding the other two waves together. Parts **a** to **e** show the waves at different times *t*, given as a fraction of the period *T* of the waves, which all have the same frequency.

Note that the distance between two nodes is a half a wavelength and not a full wavelength. If the patterns of the stationary wave are superimposed on one another, then the sequence of movement is as shown in **Fig 13.9** and in the photograph in **Fig 13.10**. Transverse stationary waves are easy to see when they are set up on a string.

The following investigation will enable you to see how the frequency of the vibration depends critically on the tension in the string and its length.

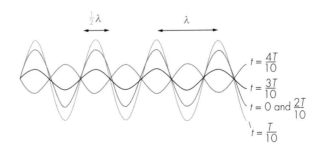

FIG 13.9 A transverse stationary wave. The notation is the same as in **Fig 13.8**.

Notation

Do not confuse the use of T for tension here with the use of T for period earlier in this chapter, especially in **Figs 13.8** and **13.9**.

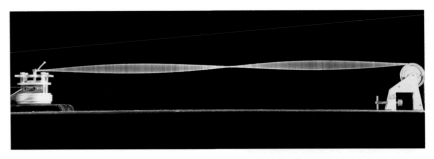

FIG 13.10 By causing this string to oscillate rapidly at a particular frequency, a stationary wave has been set up. The nodes and antinodes can be seen clearly.

INVESTIGATION

Attach a piece of light string or twine to a vibrator and pass the other end of the string over a pulley (which must have low friction bearings) to a tensioning weight, as shown in **Fig 13.11**. Some experimentation must be carried out to obtain a suitable weight, but the mass used as a tensioning weight will normally be around 100 g.

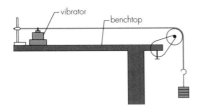

FIG 13.11 Experimental arrangement to investigate stationary waves on a string.

The aim of the investigation is to find the series of different frequencies that will give one, two, three, four, … antinodes. These frequencies can be used to find the speed of the transverse waves along the string. The following table headings should be used.

NUMBER OF ANTINODES	DISTANCE BETWEEN NODES/m	WAVELENGTH, λ /m	FREQUENCY, f /Hz	SPEED /m s^{-1}
1				
2				
3				
⋮				

The investigation can be done in a more dramatic way if a stroboscopic flashing unit is used to find the frequency. It will also enable you to see more clearly exactly how the string is moving when it has a stationary wave vibration.

The second part of the investigation is to find out how the frequency of vibration is determined by the tension T in the string. In this case the number of nodes and antinodes must be kept constant and the frequency adjusted for each different tension. A series of values of the frequency f and the corresponding tension T is obtained, and the problem is to find what relationship there is between f and T.

This can be done using a log graph in the following way. Assume that the relationship is

$$f = kT^n$$

where k is a constant and n is an unknown power. Taking logs to base 10 of both sides of the equation (remember that $\log_{10}$ is written as lg) gives

$$\lg f = \lg k + n \lg T$$

If this is compared with the equation of a straight-line graph

$$y = c + mx$$

it can be seen that if $\lg f$ is plotted against $\lg T$ then a straight-line graph will result in which the intercept on the y-axis is $\lg k$ and the slope of the graph is n, the power to which T has to be raised.

Harmonics

Stationary waves are set up in all stringed instruments and these instruments can be tuned by adjusting either the length or the tension of the string. Instruments with a large number of strings, such as a piano or a harp, have such a large total force between the 'sides' of the frame of the instrument that they need to have a very strong construction (**Fig 13.12**). The pitch of the note produced is also affected

FIG 13.12 With the large number of strings on a harp, the total force between the top and bottom is considerable.

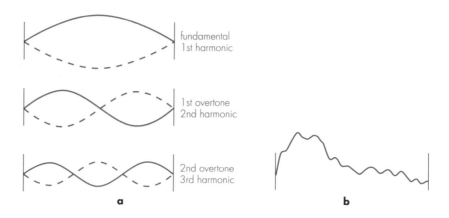

FIG 13.13 **a** The harmonic oscillations of a string on a harp. **b** The resulting oscillation of a sound wave from a harp string.

by the mass per unit length of the string. In practice, the oscillation of a string is not as simple as has been indicated here. A complex series of oscillations is possible within any string simultaneously and this gives instruments their characteristic sound. The complexity depends on such factors as the tension and mass per unit length of the string, but it also depends on the skill of the player, the shape and quality of the instrument, and the way in which the string is set in motion. A plucked string sounds different from a bowed string. This is partly because the vibrations start more suddenly and die away more quickly when a string is plucked than when it is bowed. Bowing a string is encouraging it to vibrate and it takes longer (more milliseconds) to set the vibration going.

The range of frequencies present in any complex vibration are called **harmonics**. The lowest-frequency harmonic is called the **fundamental** and the others are called **overtones**. It can be seen from **Fig 13.13a** how the different harmonics can be set up in a string of fixed length, tension and mass per unit length. The amplitude of the harmonics usually decreases as the frequency rises. All these vibrations taking place in a string cause the instrument structure to vibrate and emit sound waves. The sound waves are as complex as the string vibrations, and the sum of all the harmonic oscillations gives rise to a sound wave like that shown in **Fig 13.13b**.

The construction of musical instruments is as much an art as a science, but it is an art based on scientific principles. A Stradivarius violin is worth so much not only because it is antique but because the skill in shaping its sound box cannot quite be reproduced today. The balance of tone from all of the harmonics present in the oscillating string is dependent on the sound box. The shape of a wave from an instrument is called its **quality** and it depends on many factors. If electronic music is to be made artificially, it is necessary to use a harmonic synthesiser. This is a series of oscillators that can be adjusted in amplitude and then added together to get any wave shape required. The analysis exercise on the synthesiser at the end of this section enables you to see how differently shaped waves can be obtained by adding pure waves together, and also to see how a wave can be analysed into its harmonics.

Longitudinal stationary waves

Up to this point stationary waves have been considered only for *transverse* waves, but *longitudinal* stationary waves are also possible. The particles at a node in a longitudinal stationary wave do not move: the particles at an antinode have maximum amplitude. As with all longitudinal waves, the direction of the oscillation is parallel to the direction in which the wave is moving.

FIG 13.15 Particle movement within a pipe in which a stationary longitudinal wave has been set up.

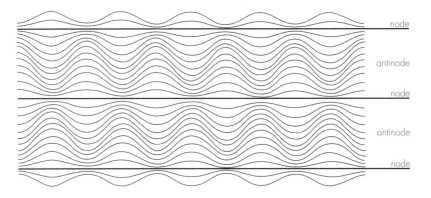

FIG 13.14 This diagram can be used to show a longitudinal stationary wave. To see the effect, move a piece of stiff card, with a slot of no more than 1 mm cut in it, across the diagram.

If a closed tube is to have the right conditions within it for a stationary wave to be set up, it must have two sound waves of equal frequency travelling in opposite directions. This can be achieved by reflecting the sound wave from the end of the pipe. In musical terms, an organ pipe with a closed end is said to be a stopped pipe. **Fig 13.14** can be used to see how the particles of air move in an organ pipe when it is sounding the third harmonic. The diagram is best looked at through a narrow slit that is moved slowly across the page. The slit needs to be about 5 cm long and preferably not more than 1 mm wide. **Fig 13.15** is a summary of **Fig 13.14**. It shows the extent of the displacement of air particles that takes place.

A surprising fact is that a sound wave can also be reflected from the *open* end of a pipe, so that almost any pipe can have a stationary wave set up within it. Whereas the closed end of a pipe is always a node, the open end of a pipe is always an antinode. Instruments such as the flute and the recorder are open pipes within which a stationary wave may be set up with an antinode at either end. The pitch of the note produced can be varied either by blowing in different ways to obtain different harmonics or by varying the length of the pipe being used by covering or uncovering holes placed at intervals along the length of the pipe. Ceremonial trumpets (see **Fig 13.16**) can only play the fundamental note and its overtones. This very restricted number of notes is increased in an orchestral trumpet by using different lengths of tube.

Blowing the trumpet

'Come to the cook-house door' is a tune on only three notes/harmonics:

|3 33|3 4|3 |2 |
|3 33|3 4|3 |

FIG 13.16 These ceremonial trumpeters play instruments of a fixed length. They can therefore play certain tunes with a very restricted number of notes. Each note corresponds to a different harmonic in the available length of tube in the trumpet.

QUESTIONS

13.1 The frequency *f* generated by a vibrating string is given by the equation

$$f = \frac{1}{2l} = \sqrt{\frac{T}{m}}$$

where *T* is the tension in the string, *l* is the length of the string and *m* its mass per unit length. On a particular piano, two adjacent low notes use the same tension in wire of mass per unit length 0.030 kg m^{-1}.

a One of the wires is 0.87 m long and it vibrates at a frequency of 55 Hz. What is the tension in the wire?

b The other wire is slightly longer at 0.91 m. What is the frequency of its vibration?

c What is the beat frequency produced when both of the notes are played together?

13.2 Instruments such as the clarinet use air columns that have one end open and the other end, by the reed, effectively closed. What is the effective length of a clarinet that has a fundamental frequency of 147 Hz?

13.3 a What are the possible frequencies of a note if it beats at a rate of 4 Hz when heard together with a tuning fork of frequency 440 Hz?

   b How could you tell which was the correct one?

ANALYSIS

Synthesisers

An electronic synthesiser can be a very complex electronic machine, but the principle it uses can be demonstrated quite simply. By combining together waves of different frequencies and amplitudes, it is possible to form any regularly repeating pattern. Some of these patterns are shown in **Figs 13.1** to **13.3**. In a synthesiser, many amplifiers are used; the outputs from them are combined in mixer circuits to give a signal that can be used as the input to a power amplifier which drives the loudspeaker.

When plotting waves in answering the following questions, do not be over-careful with the graphs. They are easier to draw if the amplitude is not too large, and you never need to draw more than one cycle of the lowest frequency because the pattern will repeat itself. In many cases you will only need half a cycle, as the other half of the cycle will be a mirror image of the first half. What you *do* need to be more careful with is the adding of the waves. The intention is to add or subtract displacements together for each value of θ. Look for places where one of the waves is zero; the sum will at that point be equal to the other wave.

1 Sketch three graphs superimposed on top of one another of
   a $y_1 = \sin \theta$,
   b $y_2 = \frac{1}{3} \sin 3\theta$, and
   c $y_3 = \frac{1}{2} \cos 2\theta$
2 Add graphs (a) and (c) together.
3 Subtract graph (b) from graph (a).
4 Add graph (c) to your answer to part 3.

These graphs show some of the basic shapes that can be achieved. They also show how different frequencies can be observed in the resulting wave pattern. Without drawing any more waves, what happens to the pattern in part 2 if $y_3 = \frac{1}{4} \cos 2\theta$ and to the pattern in part 3 if $y_2 = \sin 3\theta$?

Now look at the wave patterns shown in **Fig 13.17**. One of these patterns shows the variation in pressure when a recorder is being played and the other when a violin is being played. If you wanted to synthesise the sound for each of these instruments, which frequency or frequencies would you expect to have to use a large amplitude for? Give your answer in terms of the fundamental frequency.

In order to make synthesised sound acceptable to listen to, many more frequencies need to be used than have been suggested here. Besides adding to the frequencies, it is also necessary to control other features of the note being produced. List some of these additional features.

Many synthesisers have a whole range of special sound effects that they can produce. List some of these effects, and outline how you would need to modify the basic signal to produce the desired effect.

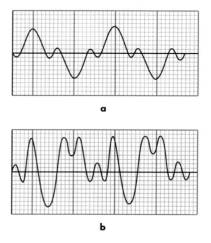

FIG 13.17 Wave patterns for the data analysis exercise.

Interference in sound

Constructive and destructive interference

If two loudspeakers are arranged so that they are about half a metre apart and are connected to the same signal generator producing an output frequency of around 1 kHz, then at different positions different intensities of sound are heard. At its simplest, the waves spread out in a circular pattern from each loudspeaker. **Fig 13.18** shows the pattern at one instant, but it must be appreciated that the pattern is continually being generated at the sources and spreading outwards and dissipating as it gets further from the sources.

It can be seen from this figure that there are places, e.g. at A, where a compression (a crest) caused by one loudspeaker meets a compression from the other loudspeaker. If the principle of superposition is applied at these places, then the amplitude of the resultant wave will be double that due to one wave by itself, and as these waves move past the hearer the intensity will be large. There are other places, e.g. at B, where a compression from one loudspeaker meets a rarefaction (a trough) from the other. At these places, application of the principle of superposition will give zero resultant amplitude, and when the hearer is at these places no sound will be heard. If this is done in a laboratory, then reflections of sound from the walls disturb the pattern. Having two ears also means that if one ear is at a place where the intensity should be zero, the other ear is probably not, so that in practice merely a reduction in sound intensity is heard rather than no sound at all. If you move around the room, the change from low intensity to high intensity is quite marked.

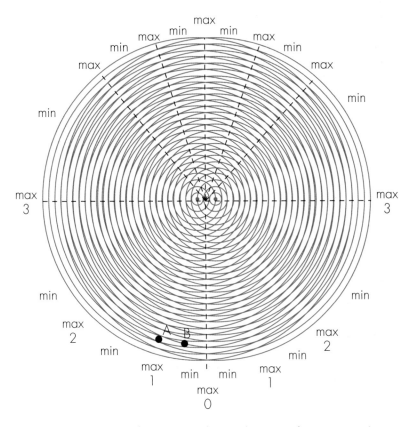

FIG 13.18 Two wave sources close to one another give this pattern of constructive and destructive interference.

This effect is called **interference** and the pattern is called an interference pattern. Interference is said to be **constructive** when two (or more) waves that are (all) in phase are added to give *maximum* amplitude; interference is said to be **destructive** when two waves that are in antiphase are added to give *zero* or *minimum* amplitude.

Path difference

The term **path difference** is often used when dealing with problems on interference, and is the difference in the distance one wave has to travel from its source to the observer compared with the distance the other wave has to travel. The path difference can be measured in metres, but it is often more convenient to measure it in wavelengths. Its use can be illustrated using **Fig 13.18**. From one source to A is 23 wavelengths; from the other source to A is 24 wavelengths; so the path difference is 1 wavelength. From one source to B is 23 wavelengths; from the other source to B is $23\frac{1}{2}$ wavelengths; and so the path difference is $\frac{1}{2}$ wavelength. We are not usually interested in the actual path length. The interesting features arise because of the path differences. In this example, for instance, a maximum intensity of sound will be heard whenever the path difference is a whole number of wavelengths. Cancellation of one wave by the other will occur whenever the path difference is an odd number of half wavelengths. This is shown on the diagram as a series of maximum and minimum amplitudes. The numbers indicate the number of wavelengths of path difference.

The interference pattern changes as the distance separating the sources is varied and also as the frequency generated by the source is changed.

EXAMPLE   13.1

Two loudspeakers on a radio are a distance of 40 cm apart. What frequencies will possibly be inaudible to a person whose ears are 14 cm apart and who is placed directly in front of the radio at a distance of 150 cm (see **Fig 13.19**)? The speed of sound is 334 m s^{-1}.

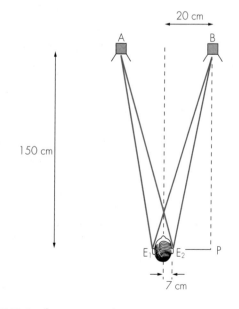

FIG 13.19 Interference in sound.

If AE_2 is half a wavelength further than AE_1, then because the diagram is symmetrical BE_1 is also half a wavelength longer than BE_2. This means that there is a possibility that the sound waves arriving at each ear are in antiphase and will destructively interfere. Using Pythagoras's theorem:

$$(E_1B)^2 = (BP)^2 + (E_1P)^2$$
$$= 150^2 + (20 + 7)^2$$
$$= 22\,500 + 729 = 23\,229$$
$$E_1B = \sqrt{23\,229} = 152.41 \text{ cm}$$

Using Pythagoras's theorem a second time gives

$$(E_2B)^2 = (BP)^2 + (E_2P)^2$$
$$= 150^2 + (20 - 7)^2$$
$$= 22\,500 + 169 = 22\,669$$
$$E_2B = \sqrt{22\,669} = 150.56 \text{ cm}$$

Since $E_2A = E_1B$, the path difference to ear E_2 is given by

$$E_2A - E_2B = E_1B - E_2B = 152.41 - 150.56 = 1.85 \text{ cm}$$
Destructive interference occurs if this 1.85 cm is half a wavelength. This gives the wavelength for destructive interference as $\lambda = 3.7$ cm $= 0.037$ m, so the inaudible frequency f will be

$$f = c/\lambda = (334 \text{ m s}^{-1})/(0.037 \text{ m})$$
$$= 9030 \text{ Hz (to 3 sig. figs)}$$

13.4 Draw a series of circles similar to **Fig 13.18** but with the sources six wavelengths apart. Use your diagram to find how the distance between positions of maximum intensity is affected by

a increasing the distance of the observer from the loudspeakers;

b increasing the distance between the loudspeakers;

c increasing the wavelength of the sound waves (decreasing the frequency produced by the signal generator).

Interference in light

If a similar demonstration to the one for sound is set up using lamps instead of loudspeakers, there is no possibility whatsoever of seeing any interference pattern. There are several reasons for this:

- The sound waves from the loudspeaker had a fixed wavelength, whereas light waves from a domestic lamp have many different wavelengths.
- The wavelength of a sound wave may be around 40 cm, whereas the wavelength of light is of the order of a million times smaller.

Coherence

If monochromatic light is used instead of white light, then only one wavelength is present. If the two sources are placed close to one another so that the interference pattern spreads out, then it should be possible, one would think, to observe an interference pattern that would be visible under a microscope. Early experiments to detect interference in light were carried out in this way in order to establish whether light was a wave motion or not. But because no interference was seen, it was assumed, by Newton among others, that light was not a wave motion. It was in 1800 that Thomas Young carried out an experiment that *did* show interference in light. The difference between his experiment and the others that preceded it was that he used light from a single source and split it into two. This proves critical, because two separate sources of light are said to lack coherence.

A polarised, monochromatic source of light has a fixed frequency and the oscillations take place in one plane, but the light still has a complicated nature. Although the average brightness of the lamp is constant, within short time-scales the amplitude varies because the pulses of light, which are generated as the electrons within atoms lose energy, do not occur in any set pattern. This results in light waves from any source having minor imperfections in them as shown in **Fig 13.20**. The chance of two waves from different sources having these imperfections all occurring at the same instant is virtually zero. The frequency is very high, around 10^{15} Hz, so even if an imperfection occurs only once in 10^8 waves there will still be 10^7 imperfections per second! If, on the other hand, light from a *single* source is split into two, the two parts will have all of their imperfections occurring simultaneously. The light in these two parts is said to be **coherent** light. In light that is coherent there is a constant phase relationship between the two parts.

FIG 13.20 Even a monochromatic light wave has frequent interruptions to its smooth oscillation. These interruptions make interference between two different light sources impossible in practice.

Young's slits experiment and interference pattern

Fig 13.21a shows how Young's experiment can be repeated in a modern laboratory. It consists of a monochromatic light source such as a sodium lamp or a laser with a narrow slit placed in front of it. The light from the narrow slit spreads out and illuminates a double slit held accurately parallel to and some distance in front of the single slit. If the light emerging from the double slit is viewed through a travelling eyepiece, an image similar to the one shown in **Fig 13.21b** can be seen. Suitable dimensions for the experiment are:

width of all slits	0.2 mm
separation of double slits, s	0.5 mm
distance of lamp to single slit	10 cm
distance of single slit to double slit	20 cm
distance of double slit to eyepiece, D	30 cm

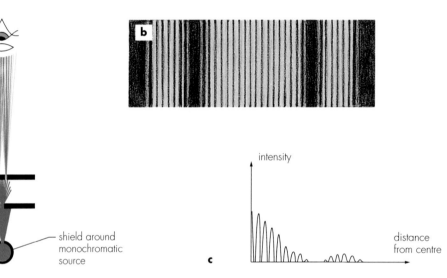

FIG 13.21 **a** The optical arrangement for the Young's slits experiment. **b** A typical interference pattern formed by the double-slit arrangement. **c** The variation in intensity with distance from the centre of the pattern in part **b**.

If you do this rather difficult experiment, you are advised to start with the overall distance from lamp to eyepiece rather shorter than is suggested here, find the interference pattern, which will be small and bright, and slowly turn the double slits for maximum clarity when the slits are all parallel to one another. Then you can increase all the distances to increase the **fringe width**, that is, the separation between one bright fringe and the next bright one. As the fringe width increases, the whole pattern gets dimmer and more blurred, so it is necessary to work in a dark room and to shield the lamp to cut out scattered light. Because the fringes are blurred, the fringe width should be calculated by measuring the distance across as many individual fringes as possible and then dividing by the number of fringes.

This experiment can be used to find the wavelength of the monochromatic light source. To see how this can be done, consider **Fig 13.22a**, which shows the path that light takes from each of the double slits to the first bright fringe away from the centre of the fringe pattern. A problem arises with drawing this diagram in that the horizontal distance from the slits to the eyepiece is many centimetres, whereas the slit separation s and the fringe width x are both about a millimetre. The diagram as drawn therefore exaggerates the fringe width and slit separation considerably. This means that the triangle APB is really even more elongated than is shown. Its

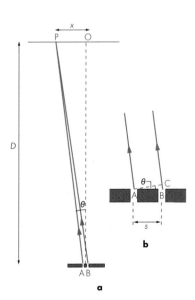

FIG 13.22 Using the Young's slit experiment to find the wavelength of the monochromatic light source.

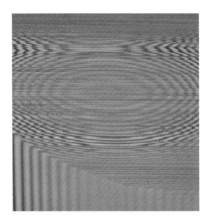

FIG 13.23 This interference pattern is created in a thin film of water between two pieces of glass which are not quite parallel to each other. The illumination is white light and the magnification is X60.

two long sides are very nearly parallel to one another, and this enables very good approximations to be made.

For the first bright line at P, the distance BP must be one wavelength λ longer than the distance AP. The extra distance is BC and is shown on a larger scale in **Fig 13.22b**. Because the slit separation s is so small compared with D, AOP can be regarded as a right-angled triangle. This makes the two triangles ABC and AOP to be nearly the same shape, so it follows that

$$\frac{AB}{BC} = \frac{AO}{OP}$$

giving

$$\frac{s}{\lambda} = \frac{D}{x}$$

so that the wavelength λ is given by

K $\quad \lambda = \dfrac{sx}{D}$

A fringe width of about 1 mm would be expected for $D = 1$ m.

A beautiful interference pattern obtained in a slightly different way is shown in **Fig 13.23**.

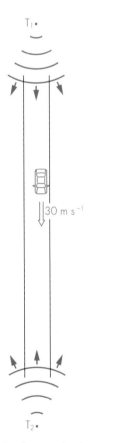

FIG 13.24 Interference of radio waves.

QUESTIONS

13.5 If laser light of wavelength 632.8 nm is shone through a pair of narrow slits a distance of 0.420 mm apart, what will be the separation of the spots on a screen placed at right-angles to the initial direction of the light at a distance of 3.00 m from the slits?

13.6 Two radio beacons are located on an east–west line 6.0 km apart. They are emitting radio waves of frequency 1.00×10^5 Hz in phase with one another. [The speed of radio waves is the same as the speed of light, 3.00×10^8 m s^{-1}.] The navigator of a ship finds that he is receiving signals from the two beacons that are in antiphase to one another (that is, out of phase by π radians).

a Draw a diagram to show several possible positions of the ship relative to the beacons.

b How can other beacons be used to establish the exact position of the ship?

13.7 **Fig 13.24** shows a car being driven at a speed of 30 m s^{-1} along a straight road between two radio transmitters. The transmitters are sending out the same programme, using a frequency of 1.50 MHz. The radio is heard to fade and strengthen regularly. What is the period of this regular fading? [This problem can be considered either as an interference problem or as a stationary wave problem. You should be able to see from a diagram why these two approaches are equivalent.]

13.4 DIFFRACTION

Spreading out of waves

If any wave meets an obstruction, then the effect on the wave depends on the relative sizes of the obstruction and the wavelength. An echo can be caused when a sound wave meets a large object. Bats use reflected ultrasonic waves to detect objects around them. Because the wavelength of these ultrasonic waves is small, the quality of the reflected signal is good enough for a bat to be able to detect a moth in flight. The same is true of light waves being reflected from the pages of this book. Because the wavelength of the light is very small compared with the size of the print, small details can be seen sharply in focus.

The formation of shadows takes place because light travels in straight lines, and when a large obstacle is placed in the path of light rays, the rays do not change direction. If the obstacle is very small, however, then some spreading of light takes place around the obstacle. **Fig 13.25a** shows a diagram, not drawn to scale, of a partial eclipse of the Moon with the Earth obstructing light from the Sun. **Fig 13.25b** is a photograph of such an eclipse. The blurring at the edge of the shadow of the Earth on the Moon results from the fact that the Sun is not a point source of light, rather than from any spreading of light around the Earth.

The spreading of waves near an obstacle is called **diffraction**. There is a corresponding effect when a wave passes through an opening. **Fig 13.26** shows this happening. In **Fig 13.26a** the width of the opening is about six wavelengths, and the amount of diffraction occurring is small. In **Fig 13.26b**, however, the width of the opening is comparable with a single wavelength, and now the amount of diffraction occurring is considerable. The actual sizes of the openings do not matter. It is a question of the relative size of the wavelength to the width of the opening that is all-important.

For instance, **Fig 13.26a** could be used to illustrate why you cannot see objects round the sides of a doorway when looking from the inside. **Fig 13.26b** illustrates why, through the same doorway, you can hear someone outside speaking although you cannot see them.

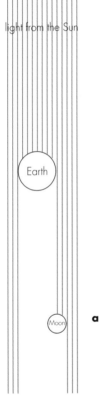

light from the Sun

Earth

Moon

a

b

FIG 13.25 A partial eclipse of the Moon. **a** The Earth obstructs light from the Sun reaching the Moon. The Moon is moving towards the reader (i.e. out of the plane of the paper). **b** The Earth's shadow is blurred because it is caused by an extended source – the Sun – not because of diffraction.

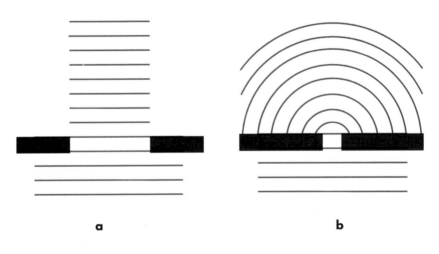

a

b

FIG 13.26 Waves approaching an opening. **a** The width of the opening is large compared with the wavelength, and little diffraction occurs. **b** The width of the opening is approximately the same size as the wavelength, and there is considerable diffraction.

Diffraction pattern

If a photograph is taken of a screen placed in front of a single slit through which monochromatic laser light is passing, an unexpected result is found. The experimental arrangement is shown diagrammatically in **Fig 13.27a**, and an example of the type of photograph obtained is shown in **Fig 13.27b**. A graph plotted to show how the intensity varies across the screen is shown in **Fig 13.27c**. The central maximum is much brighter than any of the subsidiary maxima, and it is also twice the width of the subsidiary maxima. The unexpected feature of this graph is the way that, although the intensity falls off as it might be expected to, there is a dark region followed by positions where light is again seen.

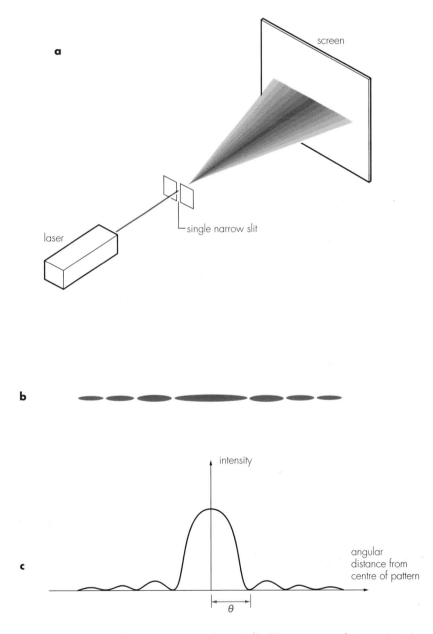

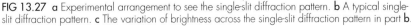

FIG 13.27 **a** Experimental arrangement to see the single-slit diffraction pattern. **b** A typical single-slit diffraction pattern. **c** The variation of brightness across the single-slit diffraction pattern in part **b**.

The angle θ (in radians), between the central maximum and the first minimum, and shown on **Fig 13.27c**, is given by

$$\theta = \lambda/d$$

where d is the slit width and λ is the wavelength of the light. It can be seen from this that θ gets larger and the pattern spreads out as d is reduced. The pattern also gets dimmer as d is reduced, because less light energy passes through a small slit.

Resolving power

The previous equation has important consequences. For example, the equation gives the minimum size possible for the image of a point object when using a pinhole camera. This by itself is of no real importance, but the fact that it also applies to a real camera is of much greater significance. Cameras are required to show fine detail, so the diffraction pattern produced by a camera must be as small as possible. The equation, with modification, can be used to show the theoretical limit for viewing small objects with a microscope, and this theoretical limit is larger than an atom, so optical microscopes cannot be used to observe atoms.

The ability of an instrument to observe small or distant objects is known as its **resolving power**. The resolving power of some large telescopes enables them to pick out craters as small as 100 m diameter on the Moon, but stars appear as if they are point objects. The image that a telescope produces of a star on a piece of film is the diffraction pattern. It is a small blurred centre with surrounding rings.

13.5 DIFFRACTION GRATING

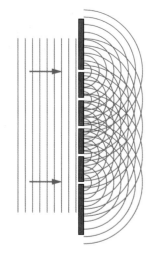

FIG 13.28 It is difficult to analyse the diffraction pattern from several sources.

If parallel, monochromatic light waves approach a series of narrow slits close to one another, the resulting pattern of waves is as shown in **Fig 13.28**. This diagram was drawn showing only five slits for the light to go through. But, even with as few as five slits, the resulting pattern is too confusing to interpret easily. The diagram is confusing because the light emerging from each slit is being spread over an angle of 180°, so too much information is being supplied for it to be able to be analysed.

In **Figs 13.29** and **13.30**, however, just the light emerging over a limited angle is being considered. By working with a limited angle in the first instance, the key features of the diagram can be found, and subsequently the whole pattern can be described. To draw **Fig 13.29**, the angle at which the light emerges has been chosen at random and is about 8°. There is no pattern to the position of the wavefronts, and, if all of these wavefronts were focussed by a lens to a point, the waves would arrive at the point with no pattern to their phase relationship. Some would be at a crest, some at a trough; some just starting to rise, some just starting to fall. If the number of slits through which the light starts is large enough, the total chaos of different phases produces almost *zero* intensity. This is why **Fig 13.28** was so difficult to decipher. For each trough, there will be somewhere among the waves a crest to cancel it out; and the same will be true for any wave. There will always be another wave in antiphase to it.

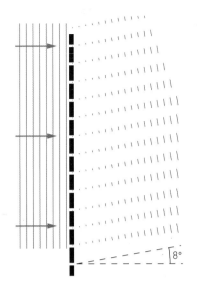

FIG 13.29 Diffraction from several sources. Here only the light emerging at an angle of approximately 8° is shown.

A large series of slits like this is called a **diffraction grating**. It is the ability of a diffraction grating to give a dark background, where the intensity is near zero, that makes it so useful for examining spectra. In contrast to **Fig 13.29**, now look at **Fig 13.30**. The angle here has not been chosen at random, but has been adjusted carefully so that the light from each slit, as you move down the pattern, travels exactly one wavelength further than the previous one. This means that all the waves arrive at the line XY in phase with one another. If these waves are focussed by a lens onto a single point, they will all arrive at that point in phase with one another and will produce a bright image, a maximum intensity region. Because the angle needs to be so carefully chosen to get an image, the image is very sharp; and since we are dealing with slits, at this angle of the emerging light the image on a screen or on the retina of the eye will be a narrow line if the source itself was narrow. **Fig 13.31** shows some typical examples of line spectra of some elements obtained in this way. These elements clearly do not provide monochromatic light. Each narrow line shown in the photograph corresponds to a different wavelength produced by the element.

Now, what will be visible at other angles? Over most of the field of view, there will be nothing. For zero angle of emergence, XY will be parallel to the grating, and again all the waves will be in phase at XY, and so again a maximum intensity will be found. The angle used in **Fig 13.30** was $22\frac{1}{2}°$. A similar diagram could be drawn for $-22\frac{1}{2}°$ to give another maximum. There is another possibility for a maximum. If a larger angle is used, the extra distance that light from one slit travels compared with light from the one next to it could be *two* wavelengths rather than one. This is said to give the second-order spectrum. Note that it does not occur at twice

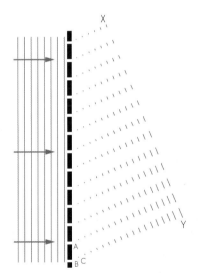

FIG 13.30 Diffraction from several sources. Here the angle of diffraction has been chosen so that all the waves are in phase with one another.

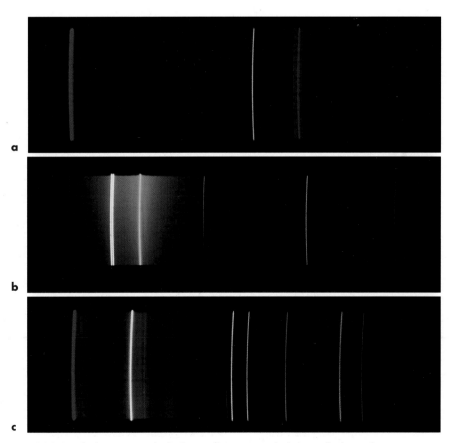

FIG 13.31 The line spectra of **a** hydrogen, **b** mercury and **c** helium, all taken in the same way. Photographs such as these are taken using an instrument called a spectrometer. The instrument uses a diffraction grating to produce the spectrum, and in these photographs the main lines from the first-order spectra are visible.

EXAMPLE 13.2

When a cadmium light was viewed through a diffraction grating having 500 lines per millimetre (**Fig 13.33**), a first-order red spectral line was observed at an angle of 18.78°. Find other angles at which red spectral lines will be observed, and draw a diagram to show the pattern.

First the value of d must be calculated. Since there are 500 lines per millimetre:

$$d = (0.001 \text{ m})/500 = 2.00 \times 10^{-6} \text{ m}$$

(This is a typical value for a diffraction grating.) Then we need to find the wavelength of the red spectral line. We use the above key equation for the first-order line (so $n = 1$), giving

$$n\lambda = d \sin \theta$$
$$1 \times \lambda = (2.00 \times 10^{-6} \text{ m}) \times \sin 18.78°$$
$$\lambda = 6.44 \times 10^{-7} \text{ m} = 644 \text{ nm}$$

Now λ is known, its value can be used to obtain other angles (i.e. other orders $n = 2$, $n = 3$) at which a red line may be observed. So using the equation $n\lambda = d \sin \theta$ again

$$2 \times (6.44 \times 10^{-7} \text{ m}) = (2.00 \times 10^{-6} \text{ m}) \times \sin \theta_2$$
$$\theta_2 = 40.1°$$

and

$$3 \times (6.44 \times 10^{-7} \text{ m}) = (2.00 \times 10^{-6} \text{ m}) \times \sin \theta_3$$
$$\theta_3 = 75.0°$$

Note that this is the limit to the number of orders obtainable because $\sin \theta$ can never be above 1.

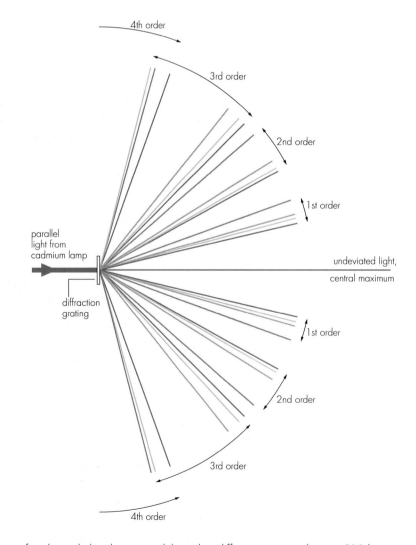

FIG 13.33 The diffraction pattern of cadmium light when viewed through a diffraction grating having 500 lines per millimetre.

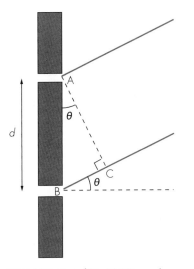

FIG 13.32 Part of Fig 13.30 on a larger scale.

the angle. If you try to find this angle by drawing, you will find that the second order occurs on **Fig 13.30** at an angle of 50°.

Once the principle of the diffraction grating is understood, the theory is easy. **Fig 13.32** is triangle ABC of **Fig 13.30** drawn on a larger scale. For a maximum, BC must be 1, 2, 3, … wavelengths, i.e. a whole number n of wavelengths. Since from the triangle,

$$\sin\theta = \frac{BC}{AB}$$

we get

$$\sin\theta = \frac{n\lambda}{d}$$

or

K $n\lambda = d\sin\theta$

where n is an integer and d is the grating spacing. The first-order spectrum is obtained when $n = 1$, the second-order spectrum when $n = 2$, etc.

QUESTIONS

13.8 When passing monochromatic sodium light through a diffraction grating having $d = 1.8 \times 10^{-6}$ m, the first-order spectrum was obtained at an angle of 19.1° and the second-order spectrum at 40.9°. What is the wavelength of sodium light?

13.9 Monochromatic light is observed through a diffraction grating at an angle of 12.2° for the first order.

    a How many other orders can be found?

    b At what angles will they be observed?

13.10 The visible spectrum extends from about 400 nm to 750 nm.

    a If observed with a diffraction grating having 480 lines per millimetre, find the angular width of the first-, second- and third-order spectra.

    b Give two reasons why such a spectrum will normally be viewed in the first order.

SUMMARY

■ Waves may be added to one another. This is called superposition and three particular cases should be understood:

  – Two waves of nearly equal frequency cause beats at a frequency equal to their difference in frequencies.

  – Two waves of equal frequency travelling in opposite directions set up a stationary wave in which the node-to-node distance is half a wavelength.

– An interference pattern is set up when two sources of the same frequency are close to one another. The fringe width is given by

$$x = D\lambda/s$$

■ The spread of a wave by diffraction is large if the opening is small compared with the wavelength; it is small if the opening is large compared with the wavelength.

■ Double-slit interference (**Fig 13.34**): the first maximum away from the central one occurs when the path difference for the two rays shown equals the wavelength.

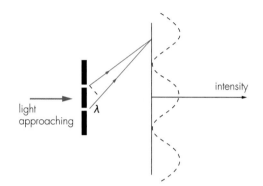

FIG 13.34 Double-slit interference.

■ Single-slit diffraction (**Fig 13.35**): the first minimum occurs when the path difference for the two rays shown is one wavelength.

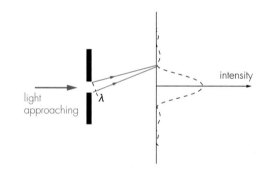

FIG 13.35 Single-slit diffraction.

■ Diffraction grating (**Fig 13.36**): maximum intensity occurs where $n\lambda = d \sin \theta$. Intensity everywhere else is virtually zero.

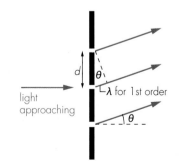

FIG 13.36 Diffraction grating.

ASSESSMENT QUESTIONS

13.11 A taut string of fixed length may be made to vibrate in a number of modes. Three of these modes have frequencies f_o, $2f_o$ and $3f_o$. The lowest possible frequency of vibration is f_o.

Copy the figure and on it sketch these three modes of vibration, labelling the nodes and antinodes with the letters N and A respectively.

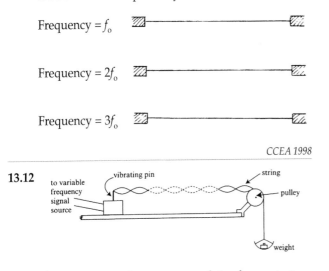

Frequency $= f_o$

Frequency $= 2f_o$

Frequency $= 3f_o$

CCEA 1998

13.12

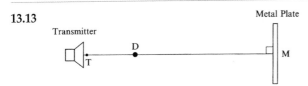

The apparatus above was used to demonstrate a transverse standing wave on a string. Both the weight, and the distance between the pin and the pulley, were kept constant.

At 480 Hz there was a standing wave pattern and each loop was 10 cm long. At a higher frequency there were two more loops than at 480 Hz and each loop was 8 cm long.

a Explain why standing waves occur at particular frequencies only.

b Calculate the speed of the waves along the string.

c Show that, at 480 Hz, eight loops would be created.

WJEC 1998

13.13

Metal Plate

Transmitter

T   D   M

A transmitter sends a beam of microwaves in the direction TM which is perpendicular to the stationary metal plate at M.

a A microwave detector D is moved from T to M and records a sequence of intensity maxima and minima. Explain this.

b If the distance between the first and eleventh intensity maxima detected is 30 cm, calculate the frequency of the microwave transmitter.

WJEC 1997

13.14 a With the aid of a diagram, explain what is meant by an electromagnetic wave.

b Hertz discovered how to produce and detect radio waves. He measured the wavelength of radio waves of frequency 4.5×10^8 Hz by reflecting the waves from a metal plate. When he moved the detector along the line between the transmitter and the metal plate, he found the detected signal fell to zero at intervals of 0.33 m.

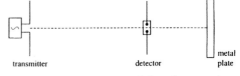

transmitter   detector   metal plate

i Explain how he deduced that the wavelength of the radio waves was 0.66 m.

ii Calculate the speed of the radio waves.

NEAB 1997

13.15 The apparatus below is used to set up a stationary wave on a stretched string. When the frequency of the vibrator is 60 Hz, resonance occurs and the stationary wave shown is produced.

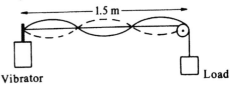

Vibrator   **Load**

For each of the statements below indicate whether the statement is true or false.

a When the frequency of the vibrator is 160 Hz there will be eight loops on the string.

b The speed of the wave is 30 m s^{-1}.

c Resonance will also occur when the frequency is 40 Hz.

d If the load is doubled and the frequency is kept constant, there will be twice as many loops on the string.

London 1996

13.16 a A radio source of frequency 95 MHz is set up in front of a metal plate. The distance from the plate is adjusted until a standing wave is produced in the space between them. The distance between any node and an adjacent antinode is found to be 0.8 m.

i Calculate the wavelength of the wave.

ii Calculate the speed of the radio wave.

iii What does this suggest about the nature of radiowaves?

b The minimum intensity that can be detected by a given radio receiver is 2.2×10^{-5} W m^{-2}.

Calculate the maximum distance that the receiver can be from a 10 kW transmitter so that it is *just* able to detect the signal.

London 1997

13.17 This question is about different types of waves and the motion of air in an open pipe.

  a Describe the similarities and differences between

  i *transverse* and *longitudinal* waves, and

  ii *progressive* (*travelling*) and *stationary* (*standing*) waves,

  illustrating your answers with diagrams where appropriate.

  b The air in the open pipe below is caused to vibrate at the lowest (fundamental) resonance frequency of

  •P •Q •R   •S

  Describe the similarities and differences in the displacement of the air molecules at

  i points P and S,

  ii points P and Q, and

  iii points P and R.

O & C 1997

13.18 a Waves can be added to one another to produce many different effects. The principle of superposition is used when adding waves.

  i Explain with the aid of wave diagrams what is meant by the *principle of superposition*.

  ii Apply the principle to explain how interference patterns and stationary waves can be produced.

  b Describe experiments which illustrate the following. In each case, give a diagram marked with suitable dimensions for key distances.

  i interference using light waves

  ii the setting up of stationary waves using microwaves

  iii the setting up of stationary waves using sound waves

  c The figure represents two sound waves which are heard simultaneously. Copy the figure and on it, draw the wave which is obtained by superposition of these two waves. Suggest what would be heard by a person listening to these waves.

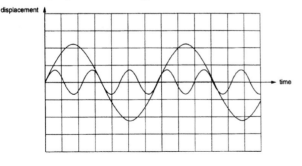

Cambridge 1997

13.19 The photograph shows the interference pattern produced when monochromatic light falls on a pair of slits 0.5 mm apart. The pattern was produced on a screen 1.5 m from the slits.

  a The photograph has been magnified by a factor of ×3. Use the photograph to obtain a value for the fringe spacing.

  b Calculate the wavelength of the light used.

  c **i** Make a sketch of the fringes. On your sketch, mark with an X the fringe or fringes where light from one slit has travelled a distance of two wavelengths further than the light from the other slit.

  ii Explain why the fringes near the centre of the photograph are clearer than those near the edges of the photograph.

  d Sketch the pattern which would be obtained on the screen if one of the slits were covered up. Label the bright and the dark regions.

  (An accurate scale diagram is *not* expected.)

  e What additional measurement would you need in order to draw an accurate diagram for this case?

London 1997

13.20 a A parallel beam of light of wavelength λ is incident normally on two parallel slits separated by distance a. Prove that the separation, y, of the bright fringes formed on a screen distance D from the slits is given by:

$$y = \frac{\lambda D}{a}$$

  b If the slits are 0.50 mm apart and the light has a wavelength 5.9×10^{-7} m calculate the angular separation of adjacent bright fringes as viewed from the slits.

WJEC 1997

13.21 This question is about interference.

  a State what is meant by the *superposition* of two waves.

  In an experiment, light from two sources P_1 and P_2 overlaps to form a stationary interference pattern.

  b Explain why it is necessary for the two sources to be coherent.

  c The monochromatic sources P_1 and P_2 are coherent. P_1 and P_2 are a distance a apart. The point O on a screen, distance d from the sources, is opposite the midpoint of P_1 and P_2 as shown in the figure.

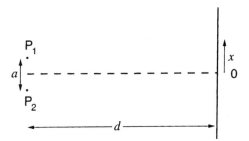

i Sketch the variation of intensity *I* of light on the screen with the distance *x* from O.

ii State the condition required to produce a minimum value of intensity on the screen.

d A wedge of glass is gradually placed over source P_1. It is observed that the interference pattern is shifted up the screen. Explain why.

O & C 1997

13.22 a State the conditions necessary for overlapping light from two sources to exhibit interference.

b The figure shows light hitting the face of a glass lens at approximately normal incidence. The lens is coated with a thin layer of transparent magnesium fluoride in order to reduce the amount of light reflected by the surface.

Light reflected by the air/magnesium fluoride surface is marked A. Light reflected by the magnesium fluoride/glass surface is marked B.

When rays A and B interfere destructively, no energy is lost by reflection and so the final image formed by the lens is brighter. The magnesium fluoride layer is made of an appropriate thickness for this destructive interference to occur.

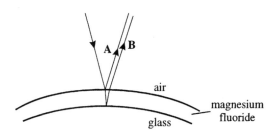

The refractive index for light passing from air to magnesium fluoride = 1.4. The speed of light in air, $c = 3.0 \times 10^8 \ \text{m s}^{-1}$.

i Calculate the speed of light in magnesium fluoride.

ii Calculate the wavelength, in magnesium fluoride, of light of frequency 5.7×10^{14} Hz.

iii Suggest a value for the thickness of the magnesium fluoride layer such that light of frequency 5.7×10^{14} Hz will not be reflected from the lens. Briefly explain your answer.

AEB 1998

13.23 a i Describe and explain, with the aid of a labelled diagram, the double slit arrangement which you would set up in order to produce and observe Young's interference fringes using monochromatic light.

ii Suggest approximate values for the slit separation and the distance from slits to screen.

b i Describe *two* features of the interference fringes.

ii Describe and explain *one* change in the appearance of the fringes if the slit separation is reduced.

c i State the measurements you would make in order to determine the wavelength of the light used.

ii Calculate the fringe separation which would result, using the approximate values which you gave in part **a ii**, assuming that the wavelength of light used is 600 nm.

NEAB 1998

13.24 a A student carried out an experiment to determine the grating spacing, *d*, of a diffraction grating by determining the diffraction angle in the second order for several spectral lines of known wavelength. These results are given in the table.

wavelength/nm	435	521	589	652
second order angle/degrees	20.4	24.6	28.1	31.4

Use these results to obtain the values needed to plot a straight line graph.

Draw the graph and use it to determine a value for *d*.

b Three diffraction gratings, illustrated below, are available for observing line spectra. The grating width and the *total* number of vertical rulings on the grating are given for each one. Determine which of the three gratings will give the largest diffraction angle in the first order with a given spectral line.

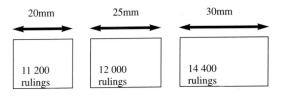

NEAB 1998

13.25 a i State the *principle of superposition of waves*.

  ii Describe briefly an experimental arrangement to demonstrate one example of the superposition of waves. State, with a reason, whether the waves involved in your demonstration are transverse or longitudinal.

  b i Explain the meaning of the term *diffraction*. Describe one situation in which the phenomenon of diffraction may be observed.

  ii The wavelength of the source of waves employed may have an important influence on the observation of diffraction effects. With the aid of an example, describe this influence.

  c i Describe an experiment to measure the wavelength of a monochromatic light source using a diffraction grating of known line spacing. Your description should use the headings: apparatus, procedure, calculation of results.

  ii A certain glass diffraction grating has a series of opaque lines of width 5.5×10^{-4} mm separated by spaces of width 7.0×10^{-4} mm. Calculate the number of lines per millimetre on the grating.
  A beam of monochromatic light incident normally on the grating produces a number of visible diffracted beams on the opposite side of the grating. One of these beams passes straight through the grating. Two of the other beams, symmetrically placed about the undeviated beam, are separated by an angle of 100.0°. Calculate the wavelength of the light.
  (The visible range of the electromagnetic spectrum is from about 400 nm to about 700 nm.)

  CCEA 1998

13.26 a Explain, using a diagram, the meaning of the term *diffraction*.

  b Describe an experiment which shows that the amount of diffraction which a wave system undergoes is dependent upon the relative sizes of the wavelength and the diffracting object. Any diagram which you draw to illustrate the experiment should concentrate on the apparatus used and not on the wave pattern.

  Cambridge 1997

13.27 This question is about a diffraction grating.

  a Both diffraction and interference occur in the operation of a diffraction grating. Explain the part played by each in the formation of a spectrum.

  b Light from a monochromatic source of wavelength 590 nm is incident normally on a diffraction grating. The first order line is observed at an angle of 17.2°.

  i Calculate the slit spacing.

  ii Find the highest order which can be observed.

  c The grating used in part **b** has the following changes made to it. In each case state and explain any similarities and/or differences observed in the first order line of the same source compared to that in **b**.

  i The outer areas of the grating are masked to halve the total width.

  ii Every alternate slit is blocked (made opaque).

  O & C 1998

13.28 Parallel light is incident normally on a diffraction grating.

  a Light of wavelength 5.9×10^{-7} m gives a first order image at 20.0° to the normal. Determine the number of lines per metre on the grating.

  b Light of another wavelength gives a second order image at 48.9° to the normal. Calculate this wavelength.

  c What is the highest order in which both these wavelengths will be visible? Justify your answer.

  NEAB 1998

13.29 A diffraction grating is used in transmission to examine a spectrum which contains only two wavelengths, 409 nm and 613 nm. Both wavelengths give a diffraction maximum at an angle of 50° with this grating which is known to have about 600 lines per millimetre.

  a What is the exact number of rulings per millimetre of the grating?

  b What is the angle of diffraction of the first-order spectrum of the shorter wavelength?

  c Explain whether or not there are other angles at which the two wavelengths coincide.

  NEAB 1997

ELECTROMAGNETISM

For a large fraction of the population, electricity is mystifying and simply something that is plugged into. An understanding of the principles of electricity is perhaps the particular attribute expected of all students of physics. A study of these principles leads to the realisation that the fundamental work done by such scientists as Galvani, Volta, Ampère, Ohm, Faraday, Kirchhoff and Oersted was all the more remarkable because it was experimentation with an invisible quantity, the electric current. ■

14 ELECTRIC CURRENT

LEARNING OBJECTIVES

At the end of this chapter you should be able to:

① define and use the terms 'charge', 'potential difference' and 'resistance' and the SI units for each of these quantities;

② relate current flow to the speed and number of charge carriers;

③ relate electromotive force with potential difference and internal resistance;

④ define and use the term 'resistivity';

⑤ calculate the energy and power supplied to any electrical component.

14.1 ELECTRIC CHARGE

An electric current always causes a magnetic field. The strength of this magnetic field is used in the SI system's definition of the value of electric current. The ampere was given as the unit of electric current in section 1.2. The definition of the ampere and more detail about the magnetic effect will be given in Chapter 18. In this chapter, use will be made of the fact that electric current is the base electrical quantity. This is not only the SI approach to the measurement of electrical quantities; it is also historically true that electric current was the first electrical quantity to be measured. The emphasis on electric current was necessary during the nineteenth century because at the time it was not known what was flowing in a wire carrying an electric current.

With the discovery of the electron in 1897, the nature of electric current in a wire as a flow of negative charge became apparent. It is possible that in the future some form of electron counter will be used as an ammeter, but at present it is the magnetic effect of the electric current that is used in ammeters to determine the value of the electric current. Experimentally, it can be shown that an electric current of one ampere (1 A) in a wire is a flow of 6.242×10^{18} electrons per second past any point in the wire.

EXAMPLE 14.1

A car battery supplies an electric current of 4.0 A for 2000 s; find the total charge that flows from the battery around the circuit and back to the battery.

The charge Q that flows past all points in the circuit is given by

$$Q = I \times t$$
$$= 4.0 \, \text{A} \times 2000 \, \text{s} = 8000 \, \text{C}$$

Since the choice of base electrical quantity is electric current, this quantity must be used in the definition of all other electrical quantities. Charge is the first of these additional quantities to be defined. The **charge** Q that flows past a point in time t if there is a constant current I is given by the equation

K   charge = current $\times$ time   or $Q = I \times t$

The unit of charge is the coulomb. One **coulomb** (1 C) is the charge that in one second crosses a section of a circuit in which there is a current of one ampere.

If the current in a circuit is not constant, then the charge that flows can be found by using the area beneath the current–time graph. In calculus terms, this can be written

$$Q = \int I \, dt$$

EXAMPLE 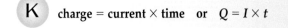 14.2

A more likely practical use of the car battery in Example 14.1 is for it to be used to supply varying currents. Find the total charge delivered from the battery if the variation in current supplied is as shown by the graph in **Fig 14.1**.

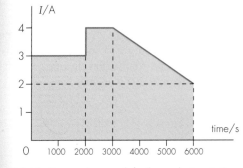

FIG 14.1 Current–time graph for the car battery in Example 14.2.

We need to work out the area under the graph. We do so in three steps:

■ In the first 2000 seconds,

   charge flowed = 3.0 A $\times$ 2000 s = 6000 C

■ In the next 1000 seconds,

   charge flowed = 4.0 A $\times$ 1000 s = 4000 C

■ In the 3000 seconds during which the battery is assumed to be supplying a current that is decreasing at a steady rate,

   charge flowed = average current $\times$ time
   $$= 3.0 \text{ A} \times 3000 \text{ s} = 9000 \text{ C}$$

So the total charge delivered is the sum of these three values:

   total charge flowed = (6000 + 4000 + 9000) C
   $$= 19\,000 \text{ C}$$

QUESTIONS

14.1 **a** What charge flows if a current of 6.5 A flows for a time of 800 s?

  b What is the steady current if a charge of 750 C flows in a time of 30 000 s?

  c How long will it take a current of 27 mA to supply a charge of 18.9 C?

14.2 Using the current–time graphs of **Fig 14.2**, find the charge that flows in each case.

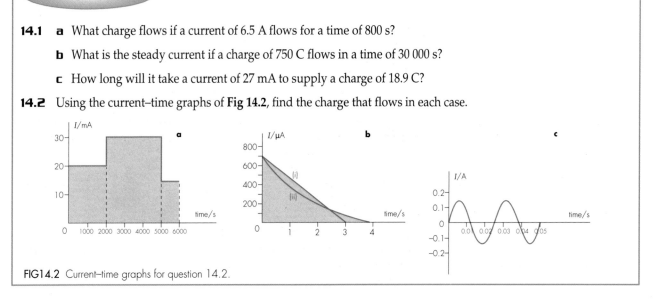

FIG14.2 Current–time graphs for question 14.2.

14.2 TRANSPORT OF CHARGE

Directions of flow

The fact that electrons have a negative charge and flow in a wire from negative to positive is an inconvenience, because it does not correspond with the historical convention of electric current flowing from positive to negative.

Electronics engineers frequently use the direction of electron flow rather than this conventional current, but electrical engineers generally still use conventional current, regarding current as a flow in the external circuit of positively charged particles from positive to negative. There are situations when positive charge is flowing from positive to negative. These situations include the movement of positive ions in a conducting liquid or gas, and the flow of positive charge carriers in semiconductor material.

In theory, a flow of negative charge from negative to positive is the same as a flow of positive charge from positive to negative, so there is not very much pressure to change from using conventional current to using electron flow.

Electricity is much associated with modern life, from the very small amounts of electrical power used by a calculator, to the vast quantities needed in modern cities (**Fig 14.3**). Whenever there is an electric current, electrons are moved by a force exerted on them by an electric field (see Chapter 16). When a current is flowing in a wire, the influence of the fixed atoms on the movement of the electrons is considerable. The electrons are continually bumping into atoms, so that the electrons gain kinetic energy and then lose it in inelastic collisions with the fixed atoms. This chaotic movement of the electrons in a wire results in a drift of negative charge through the wire. The drift speed is usually very low, typically only a fraction of a millimetre per second. Note that, although the electrons drift through a wire at a low speed, this does not imply that it will take a long time for the current to start at any point in a circuit. When a current is switched on, the slow drift of electrons starts virtually instantaneously at all points around the circuit.

The relationship between the current I in a wire and the speed v of the conduction electrons is determined by the cross-sectional area A of the wire and the number n of conduction electrons per unit volume (a particular numerical example is shown in **Fig 14.4**). Suppose that the number of conduction electrons passing any point P per unit time is N. Then, because each electron has charge e, the charge passing point P per unit time is Ne. But from the definition of charge in section 14.1, the charge passing a point per unit time is the current. So for the wire

$$I = Ne \quad \text{or} \quad N = I/e$$

The conduction electrons are moving with speed v, so in unit time the volume vA of conduction electrons will pass point P. Since there are n conduction electrons per unit volume, this volume vA will contain $n \times vA$ conduction electrons, which will pass point P per unit time. But this is just the number N we used above. So

$$N = nvA$$

Equating the two expressions for N gives $I/e = nvA$ or

K $\quad I = nAve$

FIG 14.3 New York skyline at dusk. A great deal of electrical energy is being used just for the lighting!

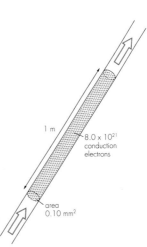

FIG 14.4 Illustrating individual electrons in a length of a copper wire. The number of conduction electrons in a 1 m long wire of area of cross-section 0.10 mm² is 8.0×10^{21}.

Units

In the following example we leave out the units in the intermediate working for simplicity. You should be able to see that the unit for v_c and v_t is

$$A/(m^{-3}\,m^2\,C)$$
$$= A/(m^{-1}\,A\,s)$$
$$= 1/(m^{-1}\,s) = m\,s^{-1}$$

It is worth while remembering the equation $I=nAve$, but at the same time do realise that n is the number of charge carriers per unit volume and that e, the electronic charge, would need to be changed to q, the charge on the charge carrier, if the current is a flow of particles other than electrons. In a liquid, for example, the flow might be of doubly charged ions.

Q **Explain, from the equation for current, why the drift speed of charge carriers increases in materials that have a low value of n and in wires of small cross-sectional area.**

EXAMPLE 14.3

A copper wire of diameter 1.40 mm connects to the tungsten filament of a light-bulb of diameter 0.020 mm. A current of 0.42 A flows through both of the wires. Copper has 8.0×10^{28} electrons per cubic metre and tungsten can be assumed to have 3.4×10^{28} electrons per cubic metre. Find the speed of an electron in each of the materials, and explain why it is that the copper wire stays cool although the tungsten filament reaches a high temperature.

We use the previous equation for each wire, using the subscript c for copper and t for tungsten. For copper we obtain

$$I = n_c A_c v_c e$$
$$0.42\,A = (8.0 \times 10^{28}\,m^{-3}) \times \pi \times (0.70 \times 10^{-3}\,m)^2$$
$$\times v_c \times (1.6 \times 10^{-19}\,C)$$
$$v_c = \frac{0.42}{8.0 \times 10^{28} \times \pi \times (0.70 \times 10^{-3})^2 \times 1.6 \times 10^{-19}}$$
$$= 0.021 \times 10^{-3}\,m\,s^{-1}$$

and for tungsten we get

$$I = n_t A_t v_t e$$
$$0.42\,A = (3.4 \times 10^{28}\,m^{-3}) \times \pi \times (0.010 \times 10^{-3}\,m)^2$$
$$\times v_t \times (1.6 \times 10^{-19}\,C)$$
$$v_t = \frac{0.42}{3.4 \times 10^{28} \times \pi \times (0.010 \times 10^{-3})^2 \times 1.6 \times 10^{-19}}$$
$$= 0.246\,m\,s^{-1}$$

The drift speed of an electron increases from 0.021 mm s⁻¹ in the copper to a value of 250 mm s⁻¹ as it enters the tungsten filament. The increased speed of the electrons in the tungsten means that they have a much greater kinetic energy in the tungsten and, as they collide with the fixed atoms, energy is lost to these atoms, resulting in their rise in temperature. It is interesting to note that, as electrons move into the tungsten, their speed increases. This is contrary to the popularly held belief that a resistance slows up electric current. Most of the speeding up effect is due to the constriction of the current into a smaller area of cross-section. There is a further effect, however, due to the fact that there are not as many charge carriers per unit volume in the tungsten as there are in copper.

QUESTIONS

14.3 The cable from a car battery can often be used to supply a current of 100 A. The diameter of the cable is 5.0 mm and it is made of copper with 8.0×10^{28} electrons per cubic metre. Find the speed of the electrons when the current is 100 A.

14.4 The density of copper is 8900 kg m^{-3}. One mole of copper atoms has a mass of 0.0635 kg.

  a Find the number of moles in a cubic metre of copper, and hence the number of atoms in a cubic metre of copper.

  b Compare this value with the number of free electrons per cubic metre of copper, given in question 14.3, and hence find the number of free electrons per atom of copper.

    [Avogadro constant $N_A = 6.02 \times 10^{23}$ mol^{-1}]

14.5 When a sufficiently high voltage is applied between two electrodes in a gas, the gas ionises. Electrons move towards the positive electrode, and positive ions move towards the negative electrode.

  a What is the current in a hydrogen discharge tube if, in each second, 4.4×10^{15} electrons and 1.5×10^{15} protons move in opposite directions past a cross-section of the tube?

  b What is the direction of the current?

14.3 POTENTIAL DIFFERENCE

Electrical circuits are basically used to transfer energy from one place to another. The energy may be very small in the case of a telecommunications signal or very large in the case of public energy supplies. A power-station built near a coal-mine can burn coal and transfer the chemical energy of the coal into electrical energy. The electrical energy can then be transferred very efficiently over long distances, where it can be used in a city by the industrial, commercial and domestic customers of the electricity companies. A similar situation exists within a simple torch circuit. In this case the chemical energy supplied by the torch battery is converted into electrical energy, and then supplied to the filament of the torch bulb, where it is converted into heat and light energy.

A term is needed to indicate how much energy is transferred with each unit of electrical charge, and this quantity is called the potential difference.

Circuit diagrams

In this book we shall use the recently introduced symbols for a cell, a capacitor, a filament lamp and a diode:

a

You may also meet the earlier versions shown below

b

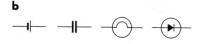

> **K** The *potential difference* between two points in a circuit is defined as the energy converted from electrical energy to other forms of energy when unit charge passes from one point to the other.

In equation form this becomes

$$\text{potential difference} = \frac{\text{energy converted}}{\text{charge}}$$

Unit of potential difference

There is no need to memorise the definition of the unit of potential difference. Units can always be obtained from the definition of the quantity itself. Here the crucial fact to remember is that potential difference is energy converted per unit charge. Its unit *must* therefore be the joule per coulomb, as that is the unit of energy divided by the unit of charge. You also need to know that *volt* is a short-hand form of *joule per coulomb*. Many problems connected with potential difference can be solved by rewriting volt as joule per coulomb.

Each component has its own p.d.

A common mistake in solving electrical problems arises when there is more than one component in a circuit, as is usually the case. When several components are present, it is essential to realise that each has its own particular potential difference. It is not possible to use a blanket term for the voltage, as if it were the same for every component – it is not!

The unit of potential difference is the joule per coulomb. This unit is given the name volt. One **volt** (1 V) is the potential difference between two points in a circuit in which one joule (1 J) of energy is converted when one coulomb (1 C) passes from one point to the other.

Because the unit of potential difference is the volt, potential difference is frequently called the **voltage**. The abbreviation **p.d.** is also frequently used. Since the definition refers to two points in a circuit, the potential difference can only be used if the two points are clearly stated. For a single component in a circuit, the two points are normally immediately before the component and immediately after the component. The expression 'the potential difference across the component' is therefore frequently used. It is correct to speak of the current through a component being caused by the potential difference across that component.

Sometimes the term 'the potential at a point' in a circuit is used. This is normally the potential difference between the point and an arbitrary zero of potential. In many cases the arbitrary zero of potential is the potential of the metallic casing of the apparatus. For safety, the casing will be connected to the ground and so the zero of potential is called the electrical **earth**.

Potential difference is a quantity relating the energy supplied to a component with the charge delivered to it. A slightly different viewpoint can be obtained, however, if, instead of considering total quantities of energy and charge, the rate of supply of energy and charge are considered. An example of the two approaches should make the similarities and differences clear (see Example 14.4).

This example shows that, by using unit time, the potential difference is given not only as energy per unit charge but also as power per unit current. The volt is not only the joule per coulomb but also the watt per ampere. The following two equations are entirely equivalent:

K
$$\text{potential difference} = \frac{\text{energy converted}}{\text{charge}} = \frac{E}{Q}$$

$$\text{potential difference} = \frac{\text{power converted}}{\text{current}} = \frac{P}{I}$$

EXAMPLE  14.4

An immersion heater is rated at 3000 W and is switched on for 2000 s. During this time a charge of 25 000 C is supplied to the heater. Find the potential difference across the heater.

The problem is solved first using the *total* quantities supplied:

energy supplied = 3000 W × 2000 s = 6000 000 J
charge supplied = 25 000 C

so

$$\text{potential difference} = \frac{\text{energy supplied}}{\text{charge supplied}}$$

$$= \frac{6000 000 \text{ J}}{25 000 \text{ C}} = 240\text{V}$$

The problem can also be solved using *rates* of supply:

energy supplied per unit time = 3000 J s^{-1} = 3000 W

charge supplied per unit time
$$= (25 000 \text{ C})/(2000 \text{ s}) = 12.5 \text{ C s}^{-1} = 12.5 \text{ A}$$

so

$$\text{potential difference} = \frac{\text{energy supplied per unit time}}{\text{charge supplied per unit time}}$$

$$= \frac{\text{power}}{\text{current}} = \frac{3000\text{W}}{12.5\text{A}}$$

$$= 240\text{V}$$

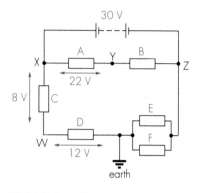

FIG 14.5 Circuit for question 14.7.

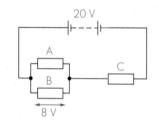

FIG 14.6 Circuit for question 14.8.

14.6 A light-bulb is marked 240 V, 60 W. It is switched on for 10 000 s. Assuming that the bulb is being used correctly, find:

a the total energy converted by the bulb from electrical energy;

b the total charge supplied to the bulb;

c the current.

Note that the answer to **c** can be obtained even if **a** and **b** had not been able to be determined as a result of the time not being known.

14.7 **Fig 14.5** shows a circuit containing a 30 V battery and six resistors. The potential differences across A, C and D are 22 V, 8 V and 12 V respectively. Find the potential difference across each of the components B, E and F, and explain why the potential at W is +12 V, at X is +20 V, at Y is −2 V and at Z is −10 V.

14.8 **Fig 14.6** shows a circuit containing a battery and three other components A, B and C. The circuit is switched on for 400 s. Table 14.1 gives some information about the charge supplied to, and the potential difference across, components in the circuit. Copy the table and complete it. If you find difficulty in answering this question, more detail is given about the procedure required in section 15.1.

TABLE 14.1 Data for question 14.8

	BATTERY	A	B	C
Charge/C	800	100		
Power/				
Energy/				
Current/				
Potential difference/V	20		8	

As indicated in the previous section, when a potential difference is applied across a material that conducts electricity, it causes an electric current in the material. Potential difference is the cause and electric current is the effect. Different currents are caused for a given potential difference depending on the nature of the material, on its shape and on its temperature. As is frequently the case with cause-and-effect problems, the ratio of the cause to the effect is a useful quantity. Resistance is the name given to the ratio here; **resistance** is defined by the following equation:

K $$\text{resistance} = \frac{\text{potential difference}}{\text{current}} \quad \text{or} \quad R = \frac{V}{I}$$

The unit of resistance must be the volt per ampere. This is given the name **ohm** (Ω). The equation defining resistance can be used to give the definition of the unit of

The same method of working is used with speeds; to say that you are travelling at 150 km h^{-1} does not mean that you necessarily have to travel for an hour. It is no defence against a prosecution for speeding to say that you could not possibly have been travelling at 150 km h^{-1} because the road was not 150 km long!

resistance, the ohm. A resistor has a resistance of one **ohm** (1 Ω) if there is a current of one ampere (1 A) through it when the potential difference across it is one volt (1 V).

Notice that if a resistor in the laboratory is marked 16 Ω it means that it requires 16 V across it, if it is to have a current of 1 A through it. Think of the resistance of any component as the voltage needed to cause a current of one ampere. A piece of thick copper wire has a low resistance because it only needs a low voltage across it to cause a current of one ampere to flow through it. A component with a high resistance is a poor conductor of electric current as it needs a high potential difference to cause the one ampere of current. A 1 MΩ (one megohm) resistor (1 MΩ = 1000 000 ohm) requires a million volts across it if one ampere is to flow through it. The fact that it would melt long before a voltage could actually be raised to this value is immaterial. The resistor is more likely to be used in a circuit in which 1 V causes a current of 1 μA (see box).

INVESTIGATION

Find how the resistances of (a) a light-bulb and (b) a coil of constantan wire vary with the potential difference across them.

A suitable light-bulb would be a car bulb, say 12 V, 24 W. Connect it in the circuit shown in **Fig 14.7a** and measure

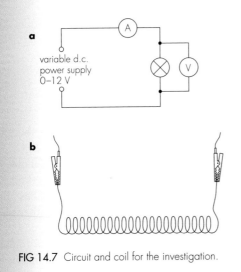

FIG 14.7 Circuit and coil for the investigation.

the current flowing through it for a series of values of potential difference across it from 0 V to 12 V. Your voltmeter must be a d.c. voltmeter and must be capable of reading up to 12 V. The ammeter must also be a d.c. meter, and you should be able to work out what the maximum current will be. Calculate the resistance for each reading taken and plot a graph of resistance against p.d. You may need to be reminded that 0 ÷ 0 does not equal 0. Here, this implies that the graph does not go through the origin.

The specification of the coil can be changed to a considerable extent. You may have suitable coils already available even though the wire used may be of different length and diameter. One possible coil would be made from about 1.5 m of 28 s.w.g. constantan wire. Wind the coil neatly and closely around a pencil, remove the pencil and slightly stretch the coil so that separate turns are not touching one another, and attach leads to each end using crocodile clips, as shown in **Fig 14.7b**. The coil will get hot in use, so it must either rest on a heat-resistant mat or be supported in the air. Repeat the experiment with the coil replacing the light-bulb.

Ohm's law

A very marked difference in the behaviour of the two wires is apparent from the investigation: whereas the resistance of the light-bulb varies, the resistance of the constantan coil remains nearly constant. The variation of resistance of the light-bulb occurs mainly because of the large temperature rise that takes place. The temperature of the constantan varies much less, but it was chosen because in any case the resistance of constantan does not change much with temperature. That is why it is called constantan and why it is used in standard resistors. It was experiments such as this that led Ohm to formulate his law. **Ohm's law** states that:

K **The current through a conductor is proportional to the potential difference across it provided its temperature remains constant.**

In Ohm's day, almost all of the conductors he used obeyed the law. Now there are many exceptions. A resistor that obeys Ohm's law has a variation of current with potential difference which gives a straight-line graph that passes through the origin. The resistance can be marked on such a resistor. If the graph of current against potential difference is not a straight line, then the resistance varies with potential difference.

In question 14.11(c) the gradient of the graph of V against I is required. This is sometimes called the **slope resistance** of a component. It is a quantity measured in ohms but it is *not* the resistance of the component. Resistance is not the gradient of a graph of V against I *unless* the graph is a straight line passing through the origin – that is, unless Ohm's law is obeyed. On a graph of I against V the gradient is $1/R$ if Ohm's law is obeyed. Be careful which way these graphs are plotted. Strictly, potential difference should be plotted on the x-axis, as it is the independent variable. That is, it is the cause, which can be determined independently. Current is the effect, and depends on the potential difference; so, being the dependent variable, it is plotted on the y-axis.

EXAMPLE 14.5

A graph of current against potential difference for a component is given as curve A in **Fig 14.8**. Find (a) the resistance when the p.d. across the component is (i) 2.0 V, (ii) 4.0 V and (iii) 6.0 V; (b) the value(s) of the voltage for which the resistance is 130 Ω; and (c) the minimum resistance of the component.

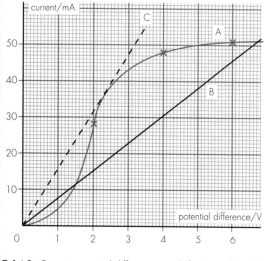

FIG 14.8 Current–potential difference graph for Example 14.5.

a   Reading off current from the graph at the values of p.d. required, we can work out the resistance, as follows:

	(i)	(ii)	(iii)
p.d.	2.0 V	4.0 V	6.0 V
current	28 mA	48 mA	51 mA
resistance	(2.0 V)/(0.028 A) = 71 Ω	(4.0 V)/(0.048 A) = 83 Ω	(6.0 V)/(0.051 A) = 118 Ω

b   A resistance of 130 Ω means that 130 V are required for a current of 1 A. This is the same as 0.13 V for 1 mA, or 1.3 V for 10 mA, or 6.5 V for 50 mA. A straight line for this resistance is plotted as B on **Fig 14.8**, and shows that the component has this resistance when the potential difference is either 1.5 V or 6.7 V.

c   The minimum resistance occurs when the highest value of current/p.d. is obtained. This is shown by the line marked C on **Fig 14.8**. It is a tangent to the curve. There is no other point on the curve from which a line can be drawn to the origin that gives a higher current per unit potential difference. The value of resistance given is (3.2 V)/(0.050 A) = 64 Ω.

QUESTIONS

14.9 A light-dependent resistor (LDR) has a p.d. of 6.0 V across it. When it is not illuminated a current of 0.87 mA flows through it, but when it is illuminated the current increases to 53 mA. Find the resistance of the LDR in both cases.

14.10 A graph of current against potential difference for a diode is given in **Fig 14.9**. Note that the diode does not behave in the same way to forward potential difference as it does to reverse potential difference. The scales for the two parts of the graph are also different.

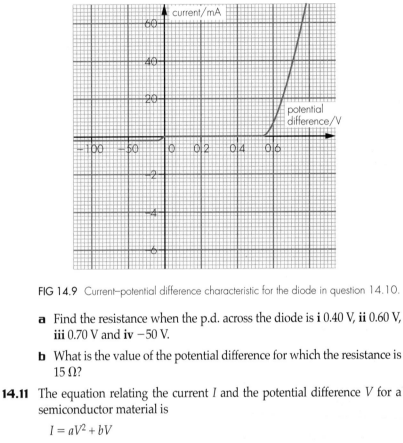

FIG 14.9 Current–potential difference characteristic for the diode in question 14.10.

a Find the resistance when the p.d. across the diode is **i** 0.40 V, **ii** 0.60 V, **iii** 0.70 V and **iv** −50 V.

b What is the value of the potential difference for which the resistance is 15 Ω?

14.11 The equation relating the current I and the potential difference V for a semiconductor material is

$$I = aV^2 + bV$$

where $a = 0.023$ mA V^{-2} and $b = 3.3$ mA V^{-1}.

a Find the resistance of the semiconductor when **i** $V = 1$ V, **ii** $V = 3$ V, **iii** $V = 5$ V and **iv** $V = 100$ V.

b Can you say that Ohm's law applies over any range of voltage for this semiconductor?

c Find the value of (change in potential difference) ÷ (change in current) when the potential difference changes from 100 V to 101 V.

14.5 RESISTIVITY

Before dealing with this term, a comment on word endings is appropriate (see box on this page). We can say that a resistor has the electrical property of resistance. It is possible to state the resistance of a component or of a particular piece of copper wire; but it is not possible to state the resistance of copper, because it might be of any shape or size. However, *resistivity* can be stated for copper. The resistivity of a material enables comparisons to be made between the conducting ability of different substances. We will now see how resistivity is defined.

The resistance R of a piece of material of uniform cross-section is found experimentally to be proportional to its length l:

$$R \propto l$$

It is also found that the resistance of the material is inversely proportional to the cross-sectional area A:

$$R \propto \frac{1}{A}$$

Combining these two expressions together gives

$$R = \frac{\rho l}{A}$$

where ρ is the constant of proportionality and is called the **resistivity** of the material. The resistivity ρ will be numerically equal to the resistance only in the unlikely event of having a wire of length one metre and of area of cross-section one square metre.

Rearranging the equation gives

$$\rho = \frac{RA}{l}$$

and therefore the unit of resistivity is the ohm metre (Ω m). Table 14.2 shows the resistivities of several different conductors and insulators. Good conductors have very low resistivities, whereas good insulators have very large resistivities. The vast difference in the conducting abilities of materials should be noticed.

Word endings are important!

- The suffix -or (on words such as 'resistor' and 'capacitor') indicates an electrical component.
- The suffix -ance (as in 'resistance' and 'capacitance') indicates an electrical property of the component.
- The suffix -ivity (as in 'resistivity') indicates an electrical property of a material.

FIG 14.10 Air at normal atmospheric pressure will conduct if a high enough potential difference is used, as on this artificial lightning test-rig.

TABLE 14.2 The resistivities for a selection of different materials

	SUBSTANCE	RESISTIVITY AT 25 °C / Ω m	USES
Conductors			
metals	copper	1.72×10^{-8}	connecting wires
	gold	2.42×10^{-8}	microchip contacts
	aluminium	2.82×10^{-8}	power cables
	tungsten	5.51×10^{-8}	light-bulb filaments
alloys	steel	20×10^{-8}	
	constantan	49×10^{-8}	standard resistors
	nichrome	100×10^{-8}	heating elements
Semiconductors			
	carbon	3.5×10^{-5}	resistors
	germanium	0.60	transistors
	silicon	2300	transistors, chips
Insulators			
	glass	$\sim 10^{13}$	power grid insulators
	polythene	$\sim 10^{14}$	wire insulation

FIG 14.11 A real lightning flash occurs when the electric field is sufficient to ionise the air. The potential differences can be many billions of volts.

Insulators

Measuring the resistivity of an insulator is a particularly difficult experiment to perform, because it is difficult to measure any current at all flowing through an insulator. When a current does flow, it often flows in the surface layer, where there is likely to be contamination with impurity atoms. The resistance can also be highly voltage-dependent. The current in many insulators will increase appreciably if the voltage rises above a particular value. This is of considerable practical importance in two situations.

Atmospheric air is a most important insulator. Overhead power cables on the National Grid are insulated by the air surrounding them. If too high a voltage is used, then the air ionises and becomes conducting, as has happened in the photograph in **Fig 14.10**. This also happens in a lightning flash (**Fig 14.11**). An interesting speculation is that if atmospheric air did always conduct electricity, then probably electricity itself would never have been discovered because, among other things, a battery would discharge itself through the surrounding air.

The second important consideration for the breakdown of insulators, so that they conduct, is in the space between the plates of a capacitor. There is an advantage in a capacitor in having the plates very close together (see section 17.1). If the insulator between the plates is only a few hundred molecules thick, then a relatively small voltage can cause the dielectric, as this insulator is named, to conduct and so damage the capacitor. For this reason, capacitors are marked with the maximum voltage that can be applied to them.

Strain gauges

We now consider how the resistance of a piece of copper wire changes when it is stretched, by working through Example 14.6.

EXAMPLE 14.6

Using data from Table 14.2, find the resistance of a piece of copper wire whose length is 8.5 cm and whose diameter is 0.068 mm. The wire is then stretched so that its length increases by 2% and its diameter reduces by 1%. Find the value of its resistance when under strain.

The resistance R of the unstretched wire is given by

$$R = \frac{\rho l}{A} = \frac{\rho l}{\pi r^2} = \frac{\rho l}{\pi (d/2)^2} = \frac{4\rho l}{\pi d^2}$$

$$= \frac{4 \times (1.72 \times 10^{-8}\,\Omega\,\text{m}) \times 0.085\,\text{m}}{\pi \times (0.068 \times 10^{-3}\,\text{m})^2} = 0.403\,\Omega$$

There is no need to recalculate the whole problem with the altered length and area of cross-section when under strain. If the equation for R is considered, then the new value for l will be $1.02l$, and the new value for d will be $0.99d$. This gives the new value for R to be

$$R' = 0.403\,\Omega \times \frac{1.02}{(0.99)^2} = 0.419\,\Omega$$

Q Work through the reasoning that leads to the above equation for R'. Such arguments can save you time and effort!

This principle is used in a device called a strain gauge (**Fig 14.12**). Strain gauges are used to measure distortion. **Fig 14.13** shows an aircraft wing undergoing tests in which considerable loading is applied to the wing. Strain gauges stuck firmly onto the wing become stretched when a load is applied and their resistances are altered.

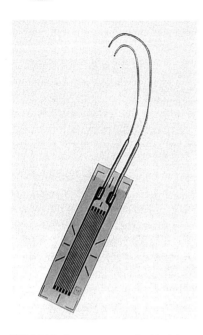

FIG 14.12 A strain gauge. The black lines are a thin conducting region (like a continuous zig-zag 'wire') and the light areas are insulating regions. When stretched, the 'wire' gets a small amount longer and thinner and so its electrical resistance increases.

FIG 14.13 Technicians are soldering strain gauges on to the interior of a section of large-bore sewerage pipe. The pipe will be tested for its mechanical strength to ensure that it will not break once in the ground.

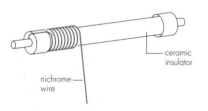

FIG 14.14 Winding the element of an electric fire.

QUESTIONS

14.12 A wire of length 4.6 m and diameter 0.83 mm is found to have a resistance of 0.24 Ω.

   a What is the resistivity of the material of the wire?

   b What substance might the wire be made of?

14.13 A manufacturer of electric fires makes elements for the fire out of nichrome wire. The wire is coiled on a spiral groove on a ceramic former, as shown in **Fig 14.14**. The centre-to-centre separation of adjacent turns is 0.12 cm and the wire diameter is 0.080 cm, so turns do not touch. The length of the coil is 22.8 cm and the diameter of each turn is 1.3 cm.

   a Find the length of wire used.

   b What is the resistance of the element when it is at a temperature of 25 °C?

ANALYSIS

Light-bulb construction

A light-bulb manufacturer makes 240 V, 60 W bulbs, like the one shown in **Fig 14.15**. The filament of the bulb is at a temperature of 2600 °C and the bulb lasts for 1000 hours before the filament breaks. There are two schools of thought concerning these bulbs. One group of people want the manufacturer to raise the temperature at which the filament operates so that, for the same power, more light is emitted. The other group of people think that the manufacturer makes too much profit on the bulbs by making them so that they break after 1000 hours. They

FIG 14.15 The coiled filament of an electric light-bulb. It is made of tungsten, so that it can operate at high temperatures without melting.

want the manufacturer to make bulbs that last 2000 hours. The manufacturer can happily satisfy both of these requirements – but only by manufacturing two other bulbs alongside the original bulb.

Using the data given below, design the electrical characteristics of the three bulbs A, B and C, and hence decide which bulbs you would advise people to buy. (Some of the data have been modified here because, in practice, the coiling of filaments can make considerable differences to filament temperature, life and radiated power.)

Material used for filament	tungsten
Resistivity of tungsten at 25 °C	$5.5 \times 10^{-8}\,\Omega\,m$
Resistivity of tungsten at 2600 °C	$7.9 \times 10^{-7}\,\Omega\,m$

Work to three significant figures in the exercise in order to keep two reliable significant figures at the end.

a Find the current through, and hence the resistance of, a 240 V, 60 W bulb. All the bulbs have this resistance.
b What diameter of tungsten wire is needed for the filament of bulb B if it is 0.14 m long? This is the standard production bulb operating at 2600 °C.
c The three filaments that could be used are illustrated in **Fig 14.16**. Explain how all can have the same resistance, and state which is the long-life filament and which is the high-temperature filament.

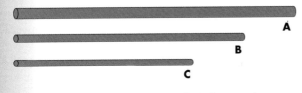

FIG 14.16 The three possible filaments for bulbs A–C (data analysis exercise).

d Copy and complete Table 14.3, assuming that the resistivity of all the filaments is $7.9 \times 10^{-7}\,\Omega\,m$.
e Using the graph of power lost per unit area (P) against temperature (**Fig 14.17**), find the temperature at which each filament operates.

TABLE 14.3 Results for the data analysis exercise

	A	B	C
Resistance/Ω			
Diameter/mm	0.0129		0.0113
Length/m		0.14	
Surface area/m^2			
Power lost per unit area/			

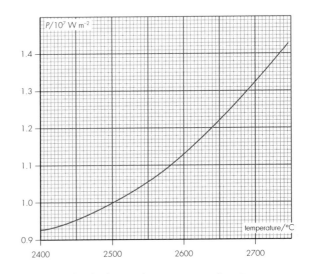

FIG 14.17 Graph of power lost per unit area P against temperature for the light-bulbs (data analysis exercise).

f The bulb at the highest temperature gives out 12% of its power input as light but it only lasts 500 hours, whereas the bulb at the lowest temperature gives out only 9% of its power as light. If a kilowatt-hour costs 8p and each bulb costs 50p, find the cost of running bulbs A and C for 2000 hours and hence work out the cost of equal quantities of light energy. [If you need a reminder about kilowatt-hour, see section 14.7.]
g Summarise your findings in the form of advice to a purchaser. Include not only financial considerations but also environmental considerations.

Filament *versus* fluorescent?

Greenpeace, in 1988, suggested that the wound filament type of bulb should be used less, and that fluorescent lamps should be used more, because the efficiency of light output from filament bulbs is so low.

14.6 ELECTROMOTIVE FORCE

s explained in section 14.3, electrical circuits are used to transfer energy from one place to another. 'Potential difference' (p.d.) was the term used to state how much electrical energy was being converted into other forms of energy per unit charge. The term 'electromotive force' (e.m.f.) is a complementary term to potential difference. It is *not* a force but, as for potential difference, it is a ratio of energy to charge, and is therefore measured in volts. Whereas potential difference is used when the electrical energy is being changed *into* other forms of energy, electromotive force is used when electrical energy is being produced *from* other forms of energy. Within a battery, for instance, energy is given to the electrons that flow from it, and the store of chemical energy within the battery is reduced.

> **K** The *electromotive force* of any source of electrical energy is the energy converted into electrical energy per unit charge supplied.

In equation form this is

$$E = \frac{\text{energy converted into electrical energy}}{\text{charge supplied}}$$

where E is the electromotive force. The units involved can, as always, be used as a check on any working done. Provided an e.m.f. in volts is thought of as joules per coulomb (e.g. $12\,\text{V} = 12\,\text{J}\,\text{C}^{-1}$), then there will be no problem in getting the equation correct.

For example, a torch battery usually has its electromotive force marked on its case. If 9 V is its stated e.m.f., this means that 9 J of energy are converted into electrical energy with every 1 C of charge passing through the battery. Since an electron has a charge on it of $1.60 \times 10^{-19}\,\text{C}$, then 6.25×10^{18} electrons are needed to supply 1 C. So 9 V actually means that the average energy converted is $9/(6.25 \times 10^{18}) = 1.44 \times 10^{-18}\,\text{J}$ per electron.

The law of conservation of energy can be applied to a situation such as this. An energy account for one coulomb of charge being supplied could be as follows:

ENERGY LOSS		ENERGY GAIN	
Chemical energy in battery	24 J	Internal energy in battery	15 J
		Electrical energy of charge	9 J

For a battery, the loss of chemical energy is always much larger than the electrical energy produced. The conversion of chemical energy to electrical energy is not very efficient. A battery's temperature will rise when it is supplying charge, as a result of the increase in its internal energy. (See section 26.1 for further details on internal energy.) From the above table we can see that the e.m.f. of the battery is 9 V.

A similar account can be made for what happens to the electrical energy on this 1 C of charge:

ENERGY LOSS		ENERGY GAIN	
Electrical energy of charge	9 J	Internal energy in battery	1 J
		Internal energy in light-bulb	8 J

You should note that there is a further loss of energy, as internal energy, within the battery itself. Some of the electrical energy produced within the battery never leaves the battery but is immediately lost for further electrical use in the resistance

The symbol E

Be careful with the symbol E. In this section it is used *only* to mean 'electromotive force', but it is often used for 'energy', as elsewhere in this book.

of the battery itself. These values show that the potential difference across the light-bulb is 8 V, and the potential difference across the internal resistance of the battery is 1 V.

You should be able to deduce from this that the potential difference across the components in a circuit will always be smaller than the e.m.f., because the electrical energy lost within a battery can never be reduced to zero. It can, however, be made very small if the current supplied by the battery is small. If the current being supplied approaches zero, then the e.m.f. and the potential difference across the circuit components approach being equal. This gives rise to the statement that:

K The electromotive force of a source is equal to the potential difference across its terminals as the current approaches zero.

From a practical point of view, if a good-quality voltmeter is the only component connected across the terminals of a battery, then the reading on the voltmeter will be equal to the e.m.f. of the battery. The quality of the voltmeter needs to be considered here, not only to ensure that it is giving an accurate reading, but also because good voltmeters have a very high resistance and therefore take very little current. A typical modern digital voltmeter might have a resistance of 20 MΩ. It therefore takes an extremely small current and can give accurate values of e.m.f.

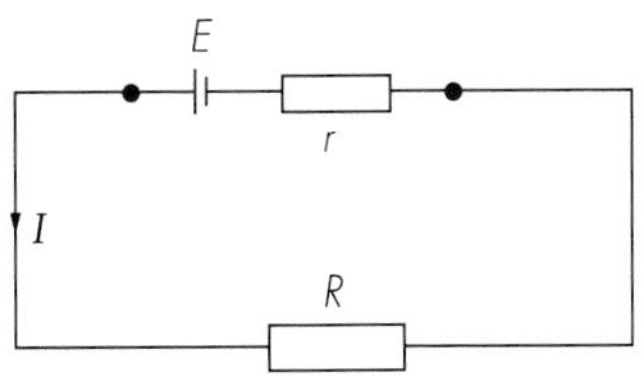

The simplest useful electrical circuit is shown in **Fig 14.18**. It consists of a battery, of electromotive force E and internal resistance r, connected to a resistor of resistance R. It is in many respects the most important electrical circuit of all, since it illustrates what all electrical circuits do in transferring energy from a source, which has internal resistance of its own, to an external consumer. The circuit could represent:

- a microphone delivering energy to an amplifier,
- an amplifier delivering energy to a loudspeaker,
- a power-station delivering energy to its customers, or
- an entire national generating company supplying all the electrical energy demand of a nation.

You are advised to carry out the investigation on the following page at this point.

Similar graphs to those plotted in the investigation may be obtained theoretically. Consider again **Fig 14.18**. If V is the p.d. across the resistance R, and V_b is the potential difference lost across the internal resistance r, then

$$E = V + V_b$$

But

$$V = IR \quad \text{and} \quad V_b = Ir$$

so

$$E = V + Ir = IR + Ir = I(R + r)$$

This gives by rearrangement and substitution

$$I = \frac{E}{R + r}$$

$$V = \frac{E R}{R + r}$$

$$V_b = \frac{E r}{R + r}$$

FIG 14.18 Arguably the most important electrical circuit.

Problem-solving

When answering electrical problems, the following sequence is recommended:

1. As you first read through a question, sketch out a large circuit diagram freehand. As extra facts are given, add the detail to the diagram.
2. If two sets of information are given, as in Example 14.7, use two diagrams. If you work as suggested, the diagrams take no time at all, since you are simply using time when other people are thinking of what to do.
3. Next put onto your diagram any further information that can immediately be calculated, such as the potential differences across the external resistors in Example 14.7.
4. With the information now on the circuit diagram, find links between known facts.
5. If you are stuck, put in an algebraic unknown and work out links between quantities using those unknowns.
6. Continue to use the diagram. Add extra facts as they become available.
7. At the end of the problem look at the diagram to see if what you have found is possible.

INVESTIGATION

Set up the circuit given in **Fig 14.19** using a torch battery as the source of electric current and with a variable external resistance R. Connect a suitable voltmeter across the terminals of the battery to measure the potential difference V. This gives the e.m.f. of the battery if there is zero current, and it is also the potential difference across R provided the ammeter has a low resistance, which is usually the case. Connect a suitable ammeter in the circuit to measure the current I. You may well need two ammeters (at different times) or an ammeter with more than one range to cope with the variation of current supplied.

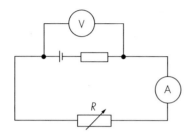

FIG 14.19 Circuit for the investigation.

A suitable range of resistance is shown in the partly completed Table 14.4. Copy the table and calculate the

other quantities listed. Plot graphs to show how V, I, V_b and P vary with the external resistance, where V_b is the potential difference across the internal resistance and P is the power delivered to the external resistance.

TABLE 14.4 Results for the investigation

RESISTANCE,	P.D. ACROSS R,	CURRENT,	P.D. LOST ACROSS INTERNAL RESISTANCE,	POWER TO R,
R/Ω	V/V	I/A	V_b/V	P/W
infinity	*	0	0	0
50				
30				
20				
10				
8				
5				
4				
3				
2				
1				
0.5				
0.2				
0				

* This represents the e.m.f. of the battery. Do not leave the battery connected to the small or zero resistances for more than a minimum of time, because shorting out batteries in this way runs them flat very quickly.

The power P_{ext} to the external resistance is given by

$$P_{ext} = V \times I = \frac{ER}{R+r} \times \frac{E}{R+r} = \frac{E^2 R}{(R+r)^2}$$

The total electrical power P_{tot} used is

$$P_{tot} = E \times I = \frac{E^2}{R+r}$$

Electrically therefore the circuit has an efficiency given by

$$\text{efficiency} = \frac{\text{power to external circuit}}{\text{total power used}} = \frac{P_{ext}}{P_{tot}}$$

$$= \frac{E^2 R/(R+r)^2}{E^2/(R+r)} = \frac{R}{R+r}$$

Therefore the efficiency of an electrical circuit cannot be equal to 1 unless the internal resistance of the source is zero.

EXAMPLE 14.7

A cell in a deaf aid supplies a current of 2.5 mA through a resistance of 400 Ω. When the wearer turns up the volume, the resistance is changed to 100 Ω and the current rises to 6.0 mA. What are the e.m.f. and the internal resistance of the cell?

To answer questions like this one, it is recommended that you follow the routine outlined below. The two circuits are shown in **Fig 14.20**. The currents and external resistances

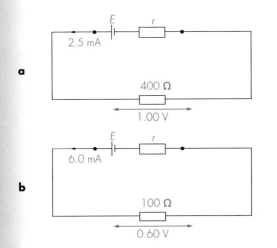

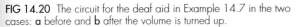

FIG 14.20 The circuit for the deaf aid in Example 14.7 in the two cases: **a** before and **b** after the volume is turned up.

are known in both cases, and so the potential differences across the external resistors can be found using $V = IR$ and inserted on the diagrams at once.

Assume the e.m.f. to be E and the internal resistance to be r. If I is the current, then Ir is the potential difference lost across the internal resistance. So we can rewrite the equation $E = V + Ir$ as

$$E - Ir = V$$

and use this in both cases. In circuits **a** and **b** we get

$$E - 0.0025r = 1.00$$
$$E - 0.0060r = 0.60$$

Subtracting the second of these from the first gives

$$0.0060r - 0.0025r = 0.40$$
$$0.0035r = 0.40$$
$$r = 0.40/0.0035 = 114\ \Omega$$

Substituting for r in the equation for circuit **b** then gives

$$E - (0.0060 \times 114) = 0.60$$
$$E = 0.60 + 0.69 = 1.29\ V$$

[*Check*: $1.29 - (0.0025 \times 114) = 1.29 - 0.285 = 1.00$, to 3 sig. figs, thus satisfying the equation for circuit **a**.]

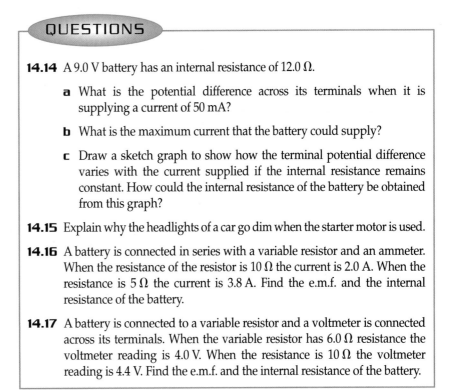

QUESTIONS

14.14 A 9.0 V battery has an internal resistance of 12.0 Ω.

   a What is the potential difference across its terminals when it is supplying a current of 50 mA?

   b What is the maximum current that the battery could supply?

   c Draw a sketch graph to show how the terminal potential difference varies with the current supplied if the internal resistance remains constant. How could the internal resistance of the battery be obtained from this graph?

14.15 Explain why the headlights of a car go dim when the starter motor is used.

14.16 A battery is connected in series with a variable resistor and an ammeter. When the resistance of the resistor is 10 Ω the current is 2.0 A. When the resistance is 5 Ω the current is 3.8 A. Find the e.m.f. and the internal resistance of the battery.

14.17 A battery is connected to a variable resistor and a voltmeter is connected across its terminals. When the variable resistor has 6.0 Ω resistance the voltmeter reading is 4.0 V. When the resistance is 10 Ω the voltmeter reading is 4.4 V. Find the e.m.f. and the internal resistance of the battery.

It was shown earlier that both energy and power can be expressed in terms of the potential difference. The expressions obtained, when combined with expressions for resistance, can be put into many different forms, and there is a real danger of confusion if too much memorising of formulae is done. Work as much as possible from first principles, that is, by using the defining equations:

K

$$\text{power}, P = \frac{\text{energy}}{t}$$

$$\text{charge}, Q = It$$

$$\text{potential difference}, V = \frac{P}{I}$$

$$\text{resistance}, R = \frac{V}{I}$$

where I is the current and t is the time.

The power P and the energy supplied to a resistor therefore can be given by the following expressions:

K

$$\text{power}, P = V \times I = IR \times I = I^2R$$

$$= V \times I = V \times \frac{V}{R} = \frac{V^2}{R}$$

$$\text{energy} = VIt = I^2Rt = \frac{V^2t}{R}$$

The kilowatt-hour (kW h)

This is a large unit of energy. The joule is a comparatively small unit of energy. So, for the commercial selling of energy by electricity (and gas) companies, the kilowatt-hour is used.

K **One kilowatt-hour (1 kW h) is the energy supplied by one kilowatt of power for one hour.**

So 1 kW h = 1000 W × 3600 s = 3 600 000 J.

QUESTIONS

14.18 An immersion heater has a resistance of 20.0 Ω and is used on 240 V mains.

    a What current does it use?

    b What is its power?

    c How long will it take to raise the temperature of 80 kg of water from 18 °C to 60 °C if it takes 4200 J of energy to raise the temperature of 1 kg of water by 1 °C?

EXAMPLE 14.8

An iron is marked 240 V, 800 W. If it is used when the supply voltage is only 220 V, what power will be supplied to it?

At 240 V we would have

$$\text{current to iron from 240 V supply} = \frac{800\,\text{W}}{240\,\text{V}}$$

$$= \frac{10}{3}\,\text{A} = 3.33\,\text{A}$$

$$\text{resistance of iron} = \frac{240\,\text{V}}{3.33\,\text{A}} = 72\,\Omega$$

Assume that the resistance is the same when used on the lower voltage. So at 220 V we have:

$$\text{current to iron from 220 V supply} = \frac{220\,\text{V}}{72\,\Omega}$$

$$\text{power supplied to iron} = V \times I = 220 \times \frac{220}{72}$$

$$= 670\,\text{W}$$

Note that this is *not* $800 \times (220/240)$, which is 730 W.

14.19 A storage battery, of the type used for emergency lighting in the event of a fire or a power cut, operates on 24 V. The battery has negligible internal resistance and is rated at 140 ampere-hours, which means that it can supply 1 A for 140 h or 2 A for 70 h, etc.

   a What is the total energy that is stored by the battery?

   b What external resistance would have to be connected to the battery if it were to be discharged in 14 h?

   c What power output is the battery giving at this rate of discharge?

14.20 A hot tap is accidentally left running for a long time. At first very hot water comes from the tap, but after a while the temperature drops to 15 K above the cold water temperature.

   a If the water is running at the rate of 0.044 kg s^{-1}, find the power of the immersion heater that is being used to heat the water.

   b What assumption is being made in obtaining this answer?

   [The specific heat capacity of water is 4200 J kg^{-1} K^{-1} (see section 26.3 if you need a reminder about this).]

14.21 In a particular household the average use per day of electrical appliances during a 100-day period was as shown in Table 14.5.

TABLE 14.5 Household usage of electrical appliances (for question 14.21)

APPLIANCE	POWER	AVERAGE TIME USED PER DAY
immersion heater	3 kW	2 h
cooker rings	2 kW	3 h
cooker oven	2 kW	1 h
lights	60 W	30 h
lights	100 W	20 h
television	200 W	5 h
refrigerator	120 W	10 h
deep freeze	400 W	14 h
central heating pump	50 W	16 h
vacuum cleaner	400 W	15 min
washing machine	500 W	30 min
washing machine heater	3 kW	20 min
hair drier	1200 W	15 min
iron	800 W	15 min
stereo	40 W	1 h 30 min
various other items	2 kW	1 h

   a If 1 kW h costs 6p, find how large a bill will be expected at the end of the 100 days.

   b What would you recommend as an economy measure?

   c How can 60 W lights be on for 30 h in one day?

   d A deep freeze and a refrigerator are normally left on all the time. Why is it therefore that the times given in Table 14.5 are less than 24 h?

SUMMARY

■ The ampere is the electrical base unit in the SI system of units.

■ Charge is defined as the product of current and time, $Q = I \times t$. One coulomb (1 C) is one ampere (1 A) for one second (1 s).

■ Potential difference is the energy changed from electrical energy to other forms of energy per unit charge. This is equivalent to the power per unit current:

$$V = \frac{\text{energy}}{\text{charge}} = \frac{\text{power}}{\text{current}}$$

One volt (1 V) is one joule per coulomb ($1 \, \text{J} \, \text{C}^{-1}$) or one watt per ampere ($1 \, \text{W} \, \text{A}^{-1}$).

■ Electromotive force (e.m.f.) is not a force. It is the energy changed from other forms of energy into electrical energy per unit charge. Its unit is the volt, and it is the potential difference across a battery when the battery is not supplying current.

■ Resistance is the potential difference per unit current, $R = V/I$. One ohm (1 Ω) is one volt per ampere ($1 \, \text{V} \, \text{A}^{-1}$). For some conductors the current through the conductor is proportional to the potential difference across it, if the temperature is constant. For these conductors, a graph of I against V is a straight line through the origin and the gradient is $1/R$:

$$V/I = R \qquad V \times I = P$$

■ Resistivity ρ of a material is defined by the equation

$$R = \frac{\rho l}{A}$$

■ Current I in terms of the charge q and the velocity v of charge carriers is $I = nAvq$, where n is the number of charge carriers per unit volume and A is the cross-sectional area of the conductor.

■ Equations relating the terminal potential difference V for a cell of e.m.f. E and internal resistance r are

$$V = E - Ir = \frac{ER}{R+r}$$

$$I = \frac{E}{R+r}$$

$$\text{useful power} = \frac{E^2 R}{(R+r)^2}$$

$$\text{efficiency} = \frac{R}{R+r}$$

■ Power P to a resistor of resistance R is

$$P = VI = I^2 R = \frac{V^2}{R}$$

■ Energy to a resistor of resistance R is

$$\text{energy} = VIt = I^2 Rt = \frac{V^2 t}{R}$$

■ One kilowatt-hour (1 kW h) = 1000 W × 3600 s = 3 600 000 J.

ASSESSMENT QUESTIONS

14.22 a A steady current of 1.25 A flows in a wire for 256 s.
  i How much charge passes a given point in the wire during this time?
  ii How many electrons pass this point during the same time?
 b The wire in **a** is made of copper and has a cross-sectional area of 1.50 mm².
  i The density of copper is 8.90×10^3 kg m⁻³, and 0.0635 kg of copper contains 6.02×10^{23} atoms. Each copper atom contributes one free electron. Show that the number density of free electrons in copper is 8.44×10^{28} m⁻³.
  ii Calculate the drift speed of free electrons in this wire when it carries the current of 1.25 A.

CCEA 1998

14.23 a Electric current in microchip circuits is carried by both metal and semiconductor materials. Discuss qualitatively the effects of increasing temperature on the resistance of samples of the same dimensions of a metal and a semiconductor. Illustrate your answer with sketch graphs.
 b The current I in a uniform metal wire is given by the equation

$I = nAve$

Identify the other terms in this equation. Show how the equation may be derived.
 c A copper wire of length 16 m has a resistance of 0.85 Ω. The wire is connected across the terminals of a battery of e.m.f. 1.5 V and internal resistance 0.40 Ω.
  i Calculate
   1 the potential difference across the wire,
   2 the power dissipated in it.
  ii In an experiment, the length of this wire connected across the terminals of the battery is gradually reduced from 16 m to 0.25 m.
   1 *Sketch* a graph to show how the power dissipated in the wire varies with the connected length. On the graph, label the point corresponding to the condition at which the power dissipated in the wire is a maximum.
   2 Calculate the length of the wire when the power dissipated in the wire is a maximum.
   3 Calculate the maximum power dissipated in the wire.
   4 The resistivity and density of copper are 1.7×10^{-8} Ω m and 8.9×10^3 kg m⁻³ respectively. A mass of 0.064 kg of copper contains 6.0×10^{23} atoms; each atom contributes one free electron to the conduction process. Calculate the drift velocity of the free electrons in the wire when the power dissipated is a maximum.

State, with a reason, whether or not this is the maximum drift velocity of the free electrons in the wire in this experiment.

CCEA 1998

14.24 a A semiconducting strip, 6 mm wide and 0.5 mm thick, carries a current of 8 mA. The carrier density is 7×10^{23} m⁻³. Calculate the carrier drift speed.
 b An approximate value for the drift speed in a copper wire of the same dimensions and carrying the same current would be about 10^{-7} m s⁻¹. Compare this figure with your calculated result and account for any difference in terms of the equation $I = nAqv$.
 c Explain why the resistance of a semiconducting strip decreases when its temperature rises.

London 1997

14.25 a A 100 W tungsten filament lamp operates from the 230 V mains. Calculate its resistance.
 b The drift speed of the electrons in the filament is much higher than the drift speed of electrons in the rest of the circuit. Suggest and explain a reason for this.

London 1999

14.26 a Draw a circuit diagram to show two heaters connected in series to a d.c. power supply. Use the resistor symbol to represent a heater.
 Each heater is rated at 120 V, 480 W. Mark on your diagram the voltage of the power supply which makes the heaters function at their normal rating.
 b Calculate
  i the current in the heaters in **a**,
  ii the charge passing through the heaters in **a** in 10 minutes.
 c Each heater has resistance 30 Ω and is made of 1.5 m of uniform wire. The material of the wire has resistivity 1.1×10^{-6} Ω m.
 Calculate the cross-sectional area of the wire.

Cambridge 1997

14.27 a Two domestic light bulbs are labelled 230 V, 25 W and 230 V, 150 W, respectively. Calculate the resistance of each of the light bulbs under its normal operating condition.
 b The two light bulbs referred to in part **a** are connected *in series* across a 230 V mains supply.
  i Which of the two lamps will dissipate the greater power? Explain the reasoning behind your answer.
  ii Without quoting specific values, state how the resistance of each of the light bulbs now compares with the value which you calculated in part **a**. Explain the reasoning behind your answer.

NEAB 1998

14.28 a State Ohm's law.

b The graph shows how current, I, varies with potential difference, V, for a manganin wire of length 1.0 m and cross-sectional area 0.049 mm², and for a tungsten filament lamp.

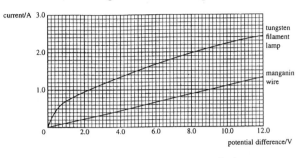

i Determine the resistance of the tungsten filament
    1 when $V = 1.0$ V
    2 when $V = 12$ V.
ii Explain why these two values are different.

c Use the graph in part **b** to show that the resistivity of manganin is approximately 4.5×10^{-7} Ω m.

d You are supplied with manganin wire of diameter 0.40 mm and asked to make a heating coil which will dissipate 1.0 kW when connected to a 200 V supply.
    i Calculate the resistance of the coil.
    ii Calculate the length of wire which is required.
    iii You are given *two* of these heating coils and instructed to arrange them to produce two other rates of heating. Show, by means of a separate diagram for each arrangement, how you could do this and calculate the power dissipated in each case, assuming the same 200 V supply.

NEAB 1997

14.29 a Sketch and label two graphs to show how the current varies with potential difference for **i** a metal wire, and **ii** a semiconductor diode, both at constant temperature.

b A semiconductor diode carries a current of 20 mA in normal operation. The potential difference across it should be 1.9 V. Copy and complete the diagram below to show how, with the addition of a single component, the semiconducting diode may be powered from a 5 V supply.

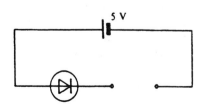

c Calculate the value of the additional component required.

London 1995

14.30 Thin films of carbon are sometimes used as resistors in electronic circuits. The diagram shows a thin film of carbon, of resistivity, ρ, which is being used as a resistor with current flowing perpendicular to the shaded face.

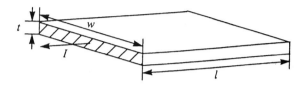

a i Show that the resistance, R, of the film to a current flowing perpendicular to the shaded area is given by

$$R = \frac{\rho l}{wt}$$

ii The film has a length, l, of 12 mm, a width, w, of 6.0 mm and a thickness, t, of 0.0010 mm. If the resistivity of carbon is 4.0×10^{-5} Ω m, calculate the resistance of the film.

b i If $l = w$ show that the resistance is independent of the length of the side of the square.

ii Explain what limits the minimum size of a square in a given application.

NEAB 1997

14.31 a Define the term *resistivity*.

b A student is asked to measure the resistivity of the alloy nichrome given a nichrome wire known to have a resistance of about two or three ohms. The wire is mounted between two copper clamps, X and Y, near the ends of the wire. The power supply is a variable power supply of output 0–5 V. The series resistor is 80 Ω.
Copy and complete the following circuit diagram.

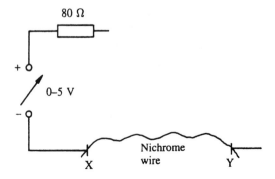

c The 80 Ω series resistor ensures that the current is kept small. Explain why this is important.

London 1999

14.32 a A wire 6.00 m long has a resistivity of 1.72×10^{-8} Ω m and a cross-sectional area of 0.25 mm^2. Calculate the resistance of the wire.

b The wire is made from copper. Copper has 1.10×10^{29} free electrons per metre cubed. Calculate the current through the wire when the drift speed of the electrons is 0.093 mm s^{-1}.

c The wire is cut in two and used to connect a lamp to a power supply. It takes 9 hours for an electron to travel from the power supply to the lamp. Explain why the lamp comes on almost as soon as the power supply is connected.

London 1999

14.33 A car battery has an e.m.f. of 12.0 V. When a car is started the battery supplies a current of 105 A to the starter motor. The terminal potential difference between the battery terminals drops at this time to 10.8 V due to the internal resistance of the battery.

a Explain briefly what is meant by:

i *an e.m.f. of 12.0 V;*

ii *the internal resistance of the battery.*

b Calculate the internal resistance of the battery.

c The manufacturer warns against short-circuiting the battery.

i Calculate the current which would flow if the terminals were to be short-circuited.

ii Explain briefly why the manufacturer provides this warning. Justify your explanation with an appropriate calculation.

d When completely discharged, the battery can be fully recharged by a current of 2.5 A supplied for 20 hours.

i How much charge is stored by the battery?

ii For how long could the motor be operated on a fully-charged battery? Assume that the motor could be operated continuously.

AEB 1997

14.34 a Define the *volt*.

b i State one similarity between potential difference and electromotive force.

ii Explain one difference between potential difference and electromotive force.

c The diagram shows a battery of e.m.f. E and internal resistance r connected to an external resistor of resistance R.

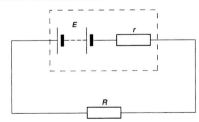

Explain, in terms of the energy transfer from the battery to the external resistor, how the internal resistance of the battery affects the potential difference across the external resistor.

Cambridge 1999

14.35 A battery of e.m.f. E and internal resistance r is connected to a variable external resistor of resistance R, as shown.

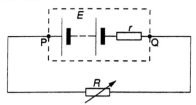

When $R = 16$ Ω, the current in the circuit is 0.45 A. The circuit is switched on for 1000 s and it is found that during this time, the battery supplies 4050 J of energy to the two resistors.

a Calculate

i the charge which the battery delivers in 1000 s,

ii the e.m.f. E of the battery,

iii the potential difference across PQ, the terminals of the battery,

iv the internal resistance r of the battery.

b Describe qualitatively how the resistance R should be changed in order to make the potential difference across PQ have a value closer to E, the e.m.f. of the battery.

Cambridge 1997

14.36 A town of 10,000 inhabitants is built near a waterfall of height 50 m which has an average water flow of 7.2×10^7 kg per hour. The energy of the waterfall is harnessed to provide an electrical supply for the town. The annual cost of maintaining the generating equipment is £260 000 and the efficiency of conversion of the water's energy to electrical energy is 12%.

On average, each family in the town contains four people and, at times of peak demand, the family requires 3.0 kW of electrical power.

a Use the data above to estimate

i the average electrical power available from the waterfall,

ii the total power required at times of peak demand,

iii the increase in the water flow required to meet a peak demand.

b Explain the unit kilowatt-hour (kWh) and hence calculate the price, in pence per kW h, which must be charged for electrical energy if the generating scheme is not to run at a loss.

c Make suggestions, two in each case, as to

i why the efficiency of energy conversion in the scheme is as low as 12%,

ii how surplus electrical energy, generated when demand is low, could be used to assist in meeting peak demands.

Cambridge 1997

15

ELECTRICAL CIRCUITS

LEARNING OBJECTIVES

At the end of this chapter you should be able to:

① apply Kirchhoff's laws to series and parallel circuits;

② adapt ammeters and voltmeters and use them in making practical measurements;

③ understand the principles and uses of the potential divider circuit.

15.1 CIRCUIT LAWS

In the previous chapter a series of electrical terms, such as 'electromotive force' and 'resistance', were introduced. These terms were used when dealing with circuits containing only the power source and a single component. In this chapter the same terms will be used but the circuits will be more complex. As is usually the case with increasing complexity, general rules are established that can be used repeatedly in analysing problems. Many practical electrical problems involve only a single circuit component, but where more than one component exists it is essential to take care to find the relevant electrical quantities for that component.

The general rules in the case of electrical circuits are called Kirchhoff's laws. Kirchhoff's laws are central to the understanding of all electrical circuits, even the simplest. In the previous chapter no specific reference was made to the laws, but use was nevertheless being made of them. Essentially the two laws are conservation laws: the first law is a statement of conservation of charge, and the second law is a statement, in electrical terms, of the law of conservation of energy.

Kirchhoff's first law

A formal statement of **Kirchhoff's first law** is that:

K **The algebraic sum of the currents at a junction is zero.**

Put another way, this law says that charge cannot be created or destroyed. If, within a complex circuit, there is part of the circuit that is as shown in **Fig 15.1**, then at each junction the total current flowing in to the junction equals the total current leaving the junction. At the left-hand junction of **Fig 15.1** there is 3.0 A + 2.6 A coming in to the junction and there must therefore be 5.6 A leaving the junction. Since a further 1.3 A joins with this current at the right-hand junction, the total current becomes 6.9 A.

Fig 15.2 shows this for four currents, *a*, *b*, *c* and *d*. The first law emphasises the algebraic nature of the sum of the currents. Account must be taken of whether there is current in to or out of the junction. Here *a*, *b* and *d* are currents in to the junction while *c* moves out. The algebraic sum of the currents, which must be zero, is shown by the equation:

$$a + b - c + d = 0$$

Example 15.1 illustrates that the use of Kirchhoff's first law is very straight-forward. It seems to be common sense that charge cannot escape from a wire but has to flow round the circuit. It is surprising therefore to find that many mistaken solutions to problems are given in which the law appears to have been broken. Almost always these mistakes arise because no check has been made on the solution to the problem to ensure that the currents stated are possible.

FIG 15.1 A numerical example illustrating Kirchhoff's first law.

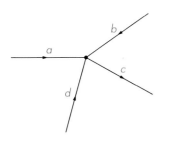

FIG 15.2 An algebraic example illustrating Kirchhoff's first law.

EXAMPLE 15.1

Find the currents *a*, *b*, *c*, *d*, *e* and *f* as marked on **Fig 15.3**.

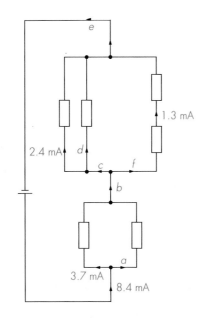

FIG 15.3 Circuit for Example 15.1.

At the bottom junction

$$8.4 \text{ mA} = 3.7 \text{ mA} + a$$
$$a = 4.7 \text{ mA}$$

Current *b* must be equal to

$$b = 3.7 \text{ mA} + 4.7 \text{ mA} = 8.4 \text{ mA}$$

This current splits up and, since the two series resistors have 1.3 mA through them, we must have *f* = 1.3 mA, and so

$$c = b - f = 8.4 \text{ mA} - 1.3 \text{ mA} = 7.1 \text{ mA}$$

Now we can find *d*; since

$$c = d + 2.4 \text{ mA}$$
$$7.1 \text{ mA} = d + 2.4 \text{ mA}$$
$$d = 4.7 \text{ mA}$$

A check that 1.3 mA + 4.7 mA + 2.4 mA = *e* = 8.4 mA confirms that Kirchhoff's first law has been obeyed.

Checking for mistakes

One quick check can always be made concerning a battery: a battery must *always* have the same current leaving the positive terminal as is entering the negative terminal. A battery does *not* produce electrical charge; it is a charge pump. A battery pumps the charge taken in at one terminal out through the other terminal. (In the same way a water pump is not a source of water; it can only pump out water that it takes in through its inlet.)

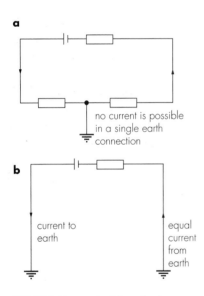

a

no current is possible in a single earth connection

b

current to earth

equal current from earth

FIG 15.4 How a circuit is earthed determines whether or not there can be a current in the earth connection.

Notation

We shall write the plurals of e.m.f. and p.d. as e.m.fs and p.ds, respectively. (They may be written in different ways in other books.)

Kirchhoff's first law can be applied at any instant to all electrical circuits, whether the current is a.c. or d.c., whatever the circuit components are, and whether or not there is an earth connection in the circuits.

Earthing

Contrary to popular opinion, charge cannot flow to earth and somehow get lost. If charge is flowing *to* earth in some part of a circuit, then an exactly equal charge must be flowing *from* earth in some other part of the circuit.

There can be no current in the earth connection shown in **Fig 15.4a**, because if some charge did flow to earth there would be more charge flowing from the battery's positive terminal than into its negative terminal. Current will exist in the circuit in Fig **15.4b** because, for every unit of charge that flows to one earth connection, a corresponding unit of charge can leave the earth through the other earth connection. This is a circuit used in cars. The metal bodywork of the car is used as the earth, and one terminal of the battery is connected to it. A single wire then connects the other terminal of the battery to the component and the return current path is through the car's bodywork. In a house circuit 'earth' does actually mean the ground, but even here, if there should be any current flowing into the ground, as a result of there being a fault in the circuit, then an equal current will be leaving the ground at some other point.

> **Q** There is a large current in the bodywork of a car when its lights are on. Why do you not get a shock when you touch the bodywork?

Kirchhoff's second law

A formal statement of **Kirchhoff's second law** is that:

> **K** Around any closed loop in a circuit the algebraic sum of the electromotive forces (e.m.fs) is equal to the algebraic sum of the potential differences (p.ds).

Another way of putting this is to say that each and every point in a stable electrical circuit has a particular value of potential. If therefore a charge moves around the circuit but comes back to the original point, any gains in electrical energy that it might have had in its journey must be balanced by corresponding losses of energy. Provided the charge returns to its original place, the gains and losses of electrical energy are always equal no matter what route is taken by the charge.

A mechanical system that involves similar principles concerns the movement of a body in a gravitational field. Each point on a mountain has a gravitational potential with respect to some arbitrary zero. Each kilogram of mass that moves around on the mountain will gain potential energy as it moves upwards and will lose potential energy as it moves down. If it finishes up at the point it started out from, then its potential energy gains must equal its potential energy losses no matter what route it takes. The independence of potential and potential energy on the route taken is an important feature of the law of conservation of energy as applied in both the mechanical and electrical cases.

To see how Kirchhoff's second law is applied to a circuit, work through Example 15.2.

Many electrical problems can be solved using this table method. The method has the advantage that it forces you to concentrate on components one at a time, so that there is less danger of using incorrect values. In this problem, for instance, there are three different numerical values of the potential difference to consider.

EXAMPLE 15.2

A battery with an e.m.f. of 20 V and an internal resistance of 2.0 Ω supplies a total current of 4.0 A to two resistors A and B, as shown in **Fig 15.5**. Resistor B has a resistance of 12 Ω. Find the resistance of A and the power supplied to each circuit component.

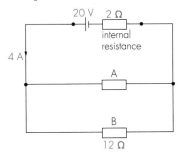

FIG 15.5 Circuit for Example 15.2.

Table 15.1 is drawn up first and the known information is inserted. Once two electrical quantities are known for an individual component, then other quantities can be found for that component. This gradual build-up of known information can be done quite simply on a single table. But here, to make the method clear, several tables have been drawn up to show the sequence used. You are asked to follow the circled numbers through from ① to ⑪.

① 4 A flowing through a 2 Ω resistance requires a potential difference of 8 V.

② A charge of 1 C flowing from the positive terminal of the battery is given 20 J of energy. A coulomb that flows through resistor A must have 8 J left to flow through

the internal resistance of the battery, and so will lose 12 J in heating up the resistor. This is applying Kirchhoff's second law. The e.m.f. of 20 V is the sum of the two potential differences ① and ②, i.e. 8 V + 12 V.

③ Using the same argument for a charge flowing through resistor B must also give 12 V across B. You may have known before that the potential difference across two resistors in parallel is always the same. This is a direct consequence of Kirchhoff's second law. Table 15.2 now shows the information that we have worked out, and the sequence of numbers continues.

④ If 12 V is the potential difference across a resistance of 12 Ω, then the current will be 1 A.

⑤ If the battery supplies 4 A, and 1 A is the current through resistor B, then 3 A must be the current through resistor A. This is a direct example of Kirchhoff's first law.

⑥ The current through the battery is the same as the current through its internal resistance, and the current from the battery is the same as the current to the battery.

⑦ If 12 V across resistor A are causing a current of 3 A, then the resistance of resistor A must be 4 Ω.

⑧–⑪ All of the powers can now be calculated using the fact that the potential difference across a component multiplied by the current through that component gives the power. The table can now be completed, as shown in Table 15.3.

TABLE 15.1 Known initial data for Example 15.2

	INTERNAL RESISTANCE	RESISTOR A	RESISTOR B		BATTERY	
P.d./V	①	②	③		E.m.f./V	20
Current/A	4				Current/A	
Resistance/Ω	2		12			
Power/W					Power/W	

TABLE 15.2 Updated version of Table 15.1

	INTERNAL RESISTANCE	RESISTOR A	RESISTOR B		BATTERY	
P.d./V	8	12	12		E.m.f./V	20
Current/A	4	⑤	④		Current/A	⑥
Resistance/Ω	2	⑦	12			
Power/W	⑧	⑨	⑩		Power/W	⑪

TABLE 15.3 Completed version of Table 15.1

	INTERNAL RESISTANCE	RESISTOR A	RESISTOR B		BATTERY	
P.d./V	8	12	12		E.m.f./V	20
Current/A	4	3	1		Current/A	4
Resistance/Ω	2	4	12			
Power/W	32	36	12		Power/W	80

Fundamental mistakes are made if 20 V is used whenever a potential difference is required. Another advantage of a table such as this is that it provides a good check at the end of the problem. Here the power used in heating resistors is 32 W + 36 W + 12 W = 80 W. Independently, the power supplied by the battery is 20 V × 4 A = 80 W. This confirms that the answer is correct.

There is one obvious gap in the final table. It is not correct to speak of the resistance of anything apart from the resistance of a resistor. It is correct, therefore, to put into the table the internal resistance of the battery, but you cannot speak of the resistance of the battery's e.m.f. However, dividing the e.m.f. of 20 V by the current supplied of 4 A does give a resistance of 5 Ω. This is the total circuit resistance. Note that you cannot get this figure by adding 2 Ω + 4 Ω + 12 Ω.

QUESTIONS

15.1 Find the currents a, b, c and d in the part of a circuit shown in **Fig 15.6**.

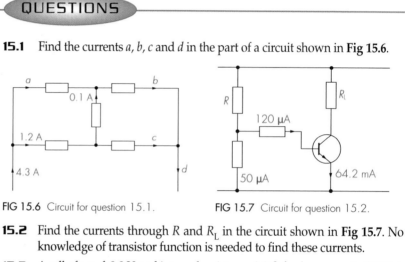

FIG 15.6 Circuit for question 15.1.　　FIG 15.7 Circuit for question 15.2.

15.2 Find the currents through R and R_L in the circuit shown in **Fig 15.7**. No knowledge of transistor function is needed to find these currents.

15.3 A cell of e.m.f. 9.0 V and internal resistance 2.0 Ω feeds a current of 0.50 A to three resistors A, B and C in series as shown in **Fig 15.8**. Resistor A has a resistance of 4.0 Ω and B has a potential difference of 1.5 V across it. Construct a table to find the value of the resistance of C and the power supplied to each resistor.

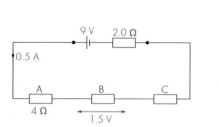

FIG 15.8 Circuit for question 15.3.

15.4 In the transistor circuit in **Fig 15.9**, you are asked to find various electrical quantities. Work from first principles to find, in this order:

　a the potential difference across the 87 kΩ resistor;

　b the potential difference between b (base) and e (emitter);

　c the potential difference across the 250 Ω resistor;

　d the power to the 87 kW and the 250 Ω resistors;

　e the power taken from the supply;

　f the power that must therefore be heating the transistor.

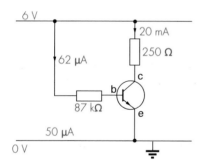

FIG 15.9 Circuit for question 15.4.

15.5 The circuit shown in **Fig 15.10** has two batteries that drive a current of 5.0 A through the resistor of resistance R. One of the batteries has an e.m.f. of 15 V and it supplies 3.5 A. Find:

　a the current being supplied by battery B;

　b the potential at the points C, D and E;

　c the e.m.f. of battery B;

　d the value of the resistance R;

　e the power supplied by the 15 V battery and by B;

　f the power supplied to each resistor.

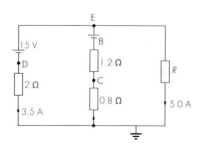

FIG 15.10 Circuit for question 15.5.

Kirchhoff's laws can be applied to all electrical circuits. When they are applied to circuits containing resistors in series and in parallel, they give results that are worth memorising, because they occur frequently.

Resistors in series

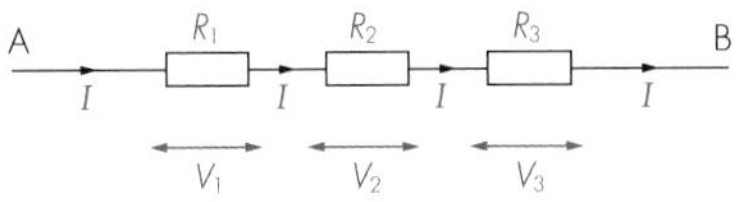

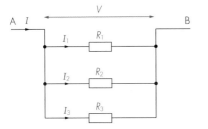

FIG 15.11 Resistors in series: the same current *I* exists in each resistor.

Fig 15.11 shows three resistors of resistances R_1, R_2 and R_3 connected to one another in series. Since there is only one possible path for the current through the three resistors, Kirchhoff's first law applied to this circuit shows that the current *I* through all of the resistors is the same. The potential differences across the resistors are therefore given by

$$V_1 = IR_1 \qquad V_2 = IR_2 \qquad V_3 = IR_3$$

Using Kirchhoff's second law gives the potential difference *V* across all three resistors to be the sum of the individual potential differences, i.e.

$$V = V_1 + V_2 + V_3 = IR_1 + IR_2 + IR_3 = I(R_1 + R_2 + R_3)$$

But $VI = R$, where *R* is the total resistance between A and B. So the total resistance of three resistors (of resistances R_1, R_2 and R_3) in series is

K $\quad R = R_1 + R_2 + R_3$

Resistors in series

The total resistance for resistors in *series* must always be *greater* than the largest of the individual resistances.

This equation can be extended to as many resistors as are present, provided there are no branches to the circuit and therefore each resistor has the same current passing through it. The result may seem to be just the application of common sense. It is, if you accept that the laws of conservation of energy and charge are also common sense.

Resistors in parallel

FIG 15.12 Resistors in parallel: the same potential difference *V* exists across each resistor.

Fig 15.12 shows three resistors of resistances R_1, R_2 and R_3 connected to one another in parallel. Kirchhoff's first law applied to this circuit shows that the total current *I* is the sum of the individual currents, i.e.

$$I = I_1 + I_2 + I_3$$

Using Kirchhoff's second law gives the potential difference *V* across all three resistances to be the same, so

$$I_1 = \frac{V}{R_1} \qquad I_2 = \frac{V}{R_2} \qquad I_3 = \frac{V}{R_3}$$

Since also V/I is the total resistance *R* between A and B, this gives

$$I = \frac{V}{R} = \frac{V}{R_1} + \frac{V}{R_2} + \frac{V}{R_3}$$

Cancelling *V* throughout, we get

K $\quad \dfrac{1}{R} = \dfrac{1}{R_1} + \dfrac{1}{R_2} + \dfrac{1}{R_3}$

This equation can be extended to as many resistors as are present, provided each resistor has the same potential difference across it. For two resistors only, this becomes

$$\frac{1}{R} = \frac{1}{R_1} + \frac{1}{R_2} = \frac{R_2 + R_1}{R_1 R_2}$$

and so

$$R = \frac{R_1 R_2}{R_1 + R_2}$$

Resistors in parallel

The total resistance for resistors in *parallel* must always be *smaller* than the smallest of the individual resistances.

$$\text{total resistance of two resistors in parallel} = \frac{\text{product of the two resistances}}{\text{sum of the two resistances}}$$

The total resistance is always smaller than the smallest resistance because each of the resistors has a current through it; each resistor therefore increases the total current and hence reduces the total resistance.

The circuits shown in **Figs 15.13a** and **b** need careful analysis. In **Fig 15.13a** only the resistors of resistances R_2 and R_3 are in parallel. These two are in series with R_1. This can be seen more clearly by redrawing the circuit, as in **Fig 15.13c**. (Imagine the circuit is in front of you on the bench or desk. Also imagine that the wires can be stretched as much as you wish. You are then free to move the components around, to see the electrical connections between them more clearly.) This circuit is basically the same as the circuit dealt with in Example 15.2. In **Fig 15.13b** the two resistors of resistances R_1 and R_2 cannot be considered to be in parallel because they do not have the same potential difference across them. The effect of E_1 in series with R_1 means that this is neither a series nor a parallel circuit. Should you encounter such a circuit, it is best dealt with from basic principles using the table method shown in Example 15.2.

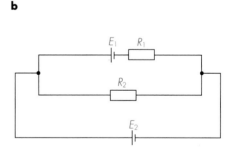

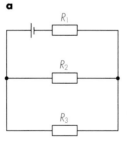

FIG 15.13 Care is needed to decide whether the resistors in the circuits in **a** and **b** are in parallel. Sometimes it can help to redraw circuits, as in **c**, being careful not to make mistakes.

a

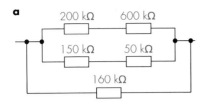

b

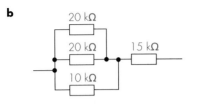

c

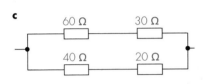

d

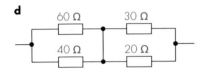

FIG 15.14 Circuits for question 15.6.

15.6 Find the total resistance of each of the resistance combinations shown in **Fig 15.14**. Why do **c** and **d** give the same answers?

15.7 Resistors are manufactured in preferred values. One series of values, used for resistors with tolerance of ±10%, is

1.0, 1.2, 1.5, 1.8, 2.2, 2.7, 3.3, 3.9, 4.7, 5.6, 6.8, 8.2 kΩ

then continuing

10, 12, 15, 18, …, 82 kΩ

and

100, 120, 150, 180, …, 820 kΩ

and so on. A particular resistor in a circuit is found to have a value of 2.10 kΩ and it is desired to reduce its value to 2.00 kΩ using resistors available from the preferred list. How can this be done?

15.8 Two variable resistors are often connected in series when it is desired to control closely the value of a current. Explain how the circuit in **Fig 15.15** functions, and give relative values for the two variable resistances.

FIG 15.15 Circuit for question 15.8.

THE POTENTIAL DIVIDER

This is an extremely useful, simple circuit. The odd thing about it is how much difficulty it causes! The circuit in its simplest form is shown in **Fig 15.16**, which shows an input potential difference V_{in} being applied across two resistors having resistances of R_1 and R_2 and an output potential difference V_{out} being taken across R_2. If the output current is zero, then the current I flowing through R_1 also flows through R_2 and the two resistors are in series. This gives

$$I = \frac{V_{in}}{R_1 + R_2}$$

$$V_{out} = IR_2 = \frac{V_{in}}{R_1 + R_2} \times R_2$$

$$V_{out} = \frac{R_2}{R_1 + R_2} \times V_{in}$$

This result is usefully thought of as being that the output p.d. is the same fraction of the input p.d. as R_2 is as a fraction of the total resistance $R_1 + R_2$. Example 15.3 shows how this can be used in practice. There is no need to calculate the current first.

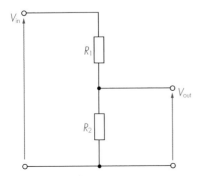

FIG 15.16 The potential divider circuit.

EXAMPLE 15.3

A resistance of 10 000 Ω in the form of a coil bent into a loop, as shown in **Fig 15.17**, is connected across a 12 V supply. A sliding arm can make contact with the coil, and an output can be taken from one end of the coil and the sliding arm. This gives the potential divider circuit shown in **Fig 15.18a**, where one part of the coil has a resistance of 3700 Ω, leaving 6300 Ω for the other end. Find the output voltage. Find also the reading a voltmeter gives when connected across the output, as shown in **Fig 15.18b**, when the resistance of the voltmeter is 9500 Ω.

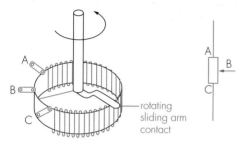

FIG 15.17 A potentiometer. This device, which enables a potential divider to be set up in practice, is frequently the electrical component behind the knobs on a radio or television that allows you to alter brightness, volume, etc. The circuit symbol for the device is also shown.

a We use the expression for the output p.d. V_{out} of a potential divider quoted in the text, and substitute the known values to obtain

$$V_{out} = \frac{3700}{3700 + 6300} \times 12\,V = \frac{3700}{10\,000} \times 12\,V = 4.4\,V$$

b When the voltmeter is connected in parallel with the 3700 Ω resistor, the resistance of the two is given by the formula quoted in the text for two resistors in parallel. Substituting known values, the total resistance of the two resistors in parallel is

$$R = \frac{3700 \times 9500}{3700 + 9500} = \frac{35\,150\,000}{13\,200} = 2660\,\Omega$$

The reading on the voltmeter is then found by using the equation for the output p.d. of a potential divider, now using the resistance value just found in (a). We get

$$V_{out} = \frac{2660}{2660 + 6300} \times 12\,V = \frac{2660}{8960} \times 12\,V$$

$$= 3.6\,V \text{ to 2 sig. figs}$$

Note in this case that the voltmeter is correctly reading the potential difference across itself and the 3700 Ω resistor. However, this is not the same value of potential difference that existed across the 3700 Ω resistor before the voltmeter was connected in place. This is a characteristic of many instruments. They frequently affect the very readings that they are trying to measure! A thermometer, for instance, will cool down a hot liquid a little bit when it is placed in the liquid. The thermometer reads correctly, but does not record the value that would have existed if the thermometer had not been used.

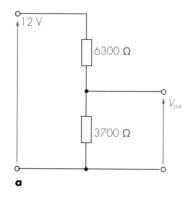

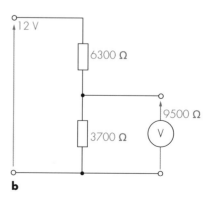

FIG 15.18 The potential divider circuits for Example 15.3.

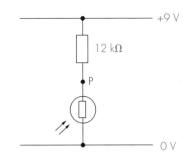

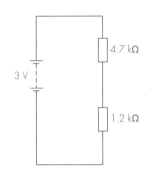

FIG 15.19 A light-dependent resistor in series with a resistor.

FIG 15.20 Circuit for question 15.10.

QUESTIONS

15.9 A light-dependent resistor (LDR) is connected in series with a 12 kΩ resistor and a 9 V d.c. source as shown in **Fig 15.19**. Find the potential at the point P when the LDR is

    a in the dark and has a resistance of 6.0 MΩ;

    b in the light and has a resistance of 1.0 kΩ.

    c What is the value of the resistance of the LDR if the potential at P is 0.60 V?

15.10 a Find the potential difference across the 1.2 kΩ resistor in **Fig 15.20**.

    b What is the smallest resistance a voltmeter can have if it is to measure the potential difference across the 1.2 kΩ resistor without introducing a systematic error of more than 1%?

15.4   ELECTRICAL INSTRUMENTS

Electrical and electronic instruments play a large part in our lives – even if we do not realise it! Without them, we would not have a mains electricity supply that we can use to listen to our favourite CDs nor would we be able to fly off to Florida for our holidays (**Fig 15.21**). These and many other situations require precise knowledge about currents and/or voltages.

FIG 15.21 The control room of a power station is a maze of electrical and electronic instrumentation and switching.

We have so far mentioned, and used, ammeters and voltmeters as instruments for measuring current and voltage (potential difference), respectively. We have not been concerned with the physics behind how they work – we have only needed to know how to use them correctly. All that is about to change!

In section 15.3 it was pointed out that a voltmeter can give misleading readings because of the current that it takes from the circuit into which it is placed. How much current it takes depends on the resistance of the voltmeter. An ideal voltmeter will not take any current at all.

An ammeter in a circuit causes a similar problem. When the ammeter is placed in the circuit to measure the current, it increases the circuit resistance and hence reduces the current. An accurate ammeter then correctly reads this lower current. An ideal ammeter would have zero resistance.

At the heart of a traditional ammeter or voltmeter is a coil of wire that can rotate through about 90° in a magnetic field when it carries an electric current. A pointer attached to the coil moves over a scale. The basic instrument is called a **galvanometer**. The principles behind the structure and calibration of a galvanometer concern magnetic field, which will be dealt with in Chapter 18. Here is a suitable place to deal with the use of a galvanometer as an ammeter or as a voltmeter by placing resistors in parallel or in series with it.

Ammeter

An ammeter contains a galvanometer and a resistor, called a **shunt**, placed in *parallel* with it. The range of the ammeter can be altered by the use of different shunts. Example 15.4 shows how a galvanometer can be used as an ammeter to measure currents over the range from 0 to 3 A.

It can be quite tricky to read from the scale of a meter because there are often several ranges that can be used and these may be printed on the scale of the instrument. Be careful that you read from a scale that rises to the known full-scale deflection, and also be careful to get the decimal point in the right place when measuring amps (A), milliamps (mA) or microamps (μA).

EXAMPLE 15.4

The galvanometer in an ammeter has a resistance of 80 Ω and the pointer is deflected fully across the scale, called full-scale deflection (f.s.d.), when a current of 2.5 mA is passing through the coil. Find the value of the shunt resistance necessary to make it an ammeter reading up to 3 A.

The maximum current that can flow through the coil is 0.0025 A; the maximum current that needs to be measured is 3 A. If 2.9975 A pass through the shunt resistor of resistance R, then full-scale deflection will occur when the current supplied is 3 A. This is shown in **Fig 15.22**.

The potential difference across the resistance of the coil is 0.0025 A × 80 Ω = 0.2 V. Since the coil and the shunt resistor are in parallel, the potential difference across the shunt resistor is also 0.2 V. The value of the shunt resistance needed is therefore

$$(0.2 \text{ V})/(2.9975 \text{ A}) = 0.0667 \text{ Ω}$$

If the current being measured is then reduced to 1 A, the shunt resistance and the coil resistance take one-third of their previous current, and the pointer moves only one-third of the way across the scale.

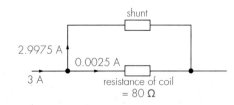

FIG 15.22 Ammeter circuit.

Commercial instruments are available in which the scale itself is changed as the shunt resistor is changed. Digital instruments do not have this difficulty, but they may still have different ranges. It is standard practice when using an ammeter to start by setting it for high currents and then to work downwards. A small current can do no damage to an ammeter set for a high current, but large currents through an ammeter set to measure low currents can damage the instruments irreparably.

An ammeter such as this does not behave ideally. As calculated earlier, the potential difference across the instrument when in use is 0.2 V. This indicates that when the ammeter is placed in series in a circuit and is measuring a full-scale deflection current of 3 A, it has a low total resistance of $(0.2\ \text{V})/(3\ \text{A}) = 0.67\ \Omega$. If the other resistances in the circuit are high, then this small resistance can be neglected.

Voltmeter

A voltmeter contains a galvanometer and a resistor, called a **multiplier**, placed in *series* with it. The range of the voltmeter can be altered by the use of different multipliers. Example 15.5 shows how a galvanometer may be used as a voltmeter over the range from 0 to 30 V.

A voltmeter is placed in parallel with the component across which the potential difference is required, and ideally has infinite resistance. If therefore the resistance of the voltmeter is too low, there will be an appreciable current through the voltmeter, and this will affect the rest of the circuit. In Example 15.5 the resistance is 12 kΩ and there is only a small current of 2.54 mA. Provided other resistances are appreciably lower than the 12 kΩ then the voltmeter is suitable.

For testing many electronic circuits, however, this type of instrument is not suitable because circuit currents are themselves small and circuit resistance values are high. It is now usual to use electronic voltmeters where the resistance of the meter can be of the order of 10 MΩ.

EXAMPLE 15.5

How can the instrument used in Example 15.4 be adapted to measure a potential difference up to 30 V?

If the galvanometer is to be used to measure potential difference, then Ohm's law has to be assumed to hold. If a p.d. of 30 V is used to cause a current of 0.0025 A to flow, then the total resistance of the circuit must be

$$(30\ \text{V})/(0.0025\ \text{A}) = 12\ 000\ \Omega$$

Since the meter resistance is 80 Ω, a series resistance (or multiplier) of 11 920 Ω needs to be placed in the circuit in series with the coil. The circuit is shown in **Fig 15.23**.

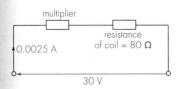

FIG 15.23 Voltmeter circuit.

In practice it is unlikely to be necessary to measure the series resistance to as many as four significant figures, as the basic instrument is probably not that accurate.

QUESTIONS

15.11 A galvanometer with full-scale deflection 3.00 mA and resistance 75 Ω is to be used as an ammeter to read up to 6 A.

   a What value of shunt resistance will be required?

   b What will be the potential difference across the meter when it is being used to measure a current of 6 A?

   c Describe qualitatively what will happen if the electrical connection from the galvanometer to the shunt gradually works loose.

15.12 Using the same data as in Example 15.5, find the necessary series resistances for measuring potential differences

   a up to 300 V,

   b up to 3 V, and

   c up to 0.3 V.

   d Draw a diagram to show how all of these four resistors (including that needed for measuring up to 30 V) can be connected into a single circuit to give a multi-range voltmeter with different input terminals for the different ranges.

SUMMARY

- Kirchhoff's first law: the current into any junction equals the current leaving the junction. This is a law that is in principle a law of conservation of charge.

- Kirchhoff's second law: around any closed loop in a circuit the sum of the potential differences equals the sum of the e.m.fs. This is a law that is in principle a law of conservation of energy.

- For resistors in series:

$$R = R_1 + R_2 + R_3 + \dots$$

- For resistors in parallel:

$$\frac{1}{R} = \frac{1}{R_1} + \frac{1}{R_2} + \frac{1}{R_3} + \dots$$

- For two resistors in parallel, this becomes

$$R = \frac{R_1 R_2}{R_1 + R_2}$$

- Ammeters have a low resistance (a shunt) in parallel with the galvanometer (**Fig 15.24**). They have a low resistance and are placed in series in the circuit.

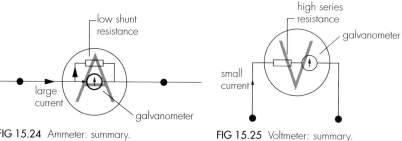

FIG 15.24 Ammeter: summary.

FIG 15.25 Voltmeter: summary.

FIG 15.26 Potential divider circuit.

- Voltmeters have a high resistance (a multiplier) in series with the galvanometer (**Fig 15.25**). They have a high resistance and are placed in parallel with the circuit.

- The output from the potential divider circuit (**Fig 15.26**) is

$$V_{\text{out}} = \frac{R_2}{R_1 + R_2} V_{\text{in}}$$

ASSESSMENT QUESTIONS

15.13 A lead acid car battery has an e.m.f. of 12 V and an internal resistance of 0.024 Ω. It has a capacity of 9.5 A h (ampere-hours), i.e. it can deliver a current of 0.95 A for 10 h or 1.9 A for 5 h etc.

    a **i** Show that the total energy the battery can transfer is 410 kJ.

        ii Will the fraction of this energy which is dissipated within the battery depend on how it is used? Explain your answer.

b A motorist turns on his sidelights and headlights before switching on the starter motor using switch S in the circuit shown.

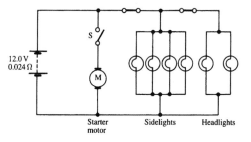

i   Each sidelight is rated at 6 W, 12 V and each headlight at 48 W, 12 V. Calculate the current delivered by the battery before the switch S is closed.

ii   The current in the starter motor, when first switched on, is 160 A. Explain why the head-lights become less bright when the switch S is closed.

c A scientist wishes to study the dimming of the headlights described in **b ii** above. She thinks that she can do so by analysing the wavelength of the light they give out. Outline an experiment which would enable her to investigate the spectrum of light produced by a headlight filament.
What changes may she notice in the spectrum as the headlight dims?

London 1998

15.14 a Three resistors R_1, R_2 and R_3 are connected in series. Derive a formula for their effective resistance.

b

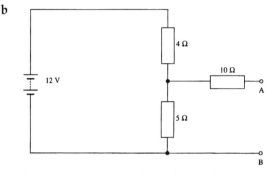

i   Calculate the potential difference between A and B in the above circuit if the 12 V battery has negligible internal resistance.

ii   What would the potential difference between A and B become if a 10 Ω resistor were connected between the two points?

iii   Calculate the current through the 10 Ω resistor connected to A and B.

iv   If the 10 Ω resistor connected between A and B is removed and a 10 μF capacitor is put in its place, what is the potential difference between A and B when the capacitor is fully charged?

NEAB 1996

15.15 A variable resistor, R, is to be used to control the brightness of a 12 V, 24 W light bulb, as shown. The maximum value of the variable resistor should be large enough to reduce the potential difference across the bulb to 2.0 V. The battery has negligible internal resistance.

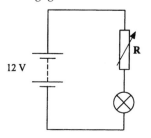

a At 2.0 V, the bulb has a resistance of 2.4 Ω.

i   Calculate the potential difference across the variable resistor when the potential difference across the bulb is 2.0 V.

ii   Suggest a value for the maximum resistance of the variable resistor and justify your answer.

iii   Calculate the power dissipated in the variable resistor when the potential difference across the bulb is 2.0 V.

b An alternative method of controlling the potential difference across the bulb is to use the 10 Ω potentiometer shown below. The section of the potentiometer which is in parallel with the bulb is R_2 and is set to 2.4 Ω.

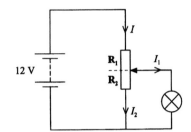

i   State the resistance of the parallel combination of R_2 and the bulb.

ii   Calculate the total resistance of the circuit.

iii   Calculate the values of I, I_1 and I_2 shown.

c Suggest a reason why it would be better to use the variable resistor rather than the potentiometer.

AEB 1996

15.16 a State what is meant by an electric current.

b A car battery has a capacity of 40 ampere-hours. That is, it can deliver a current of 1.0 A for 40 hours. When delivering the current, the potential difference across the terminals of the battery is 12 V. Calculate the electrical energy available from the battery.

c An electric heater is rated at 1.0 kW, 240 V but the supply voltage is only 230 V.

i   Calculate the resistance of the heater if it were operating at 240 V.

ii   Assuming that the resistance of the heater is constant, calculate the output power of the heater when it is being used with the 230 V supply.

d You have available a number of resistors, each of resistance 100 Ω. Draw an arrangement of resistors which would give a total resistance of

i   50 Ω,

ii   150 Ω,

iii   67 Ω.

OCR 1999

15.17 In the following circuits the battery has negligible internal resistance, and the bulbs are identical.

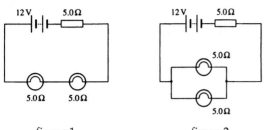

figure 1   figure 2

a For the circuit shown in figure 1 calculate
  i the current flowing through each bulb,
  ii the power dissipated in each bulb.
b In the circuit shown in figure 2 calculate the current flowing through each bulb.
c **i** Explain how the brightness of the bulbs in figure 1 compares with the brightness of the bulbs in figure 2.
  ii Explain why the battery would last longer in the circuit shown in figure 1.
d One of the bulbs in figure 2 develops a fault and no longer conducts. Describe and explain what happens to the brightness of the other bulb.

NEAB 1996

15.18 The diagram shows the way in which eight heating elements of the rear window heater in a car are connected.

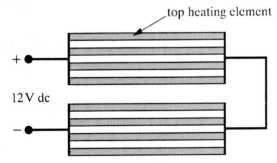

Each of the elements of the heater has a resistance of 8.0 Ω and the heater is connected to a 12 V d.c. supply.
a Calculate
  i the resistance of the heater,
  ii the potential difference across each of the elements,
  iii the current through each of the elements.
b The top heating element, marked with the arrow in the diagram, is damaged in use and stops conducting.
Calculate
  i the new resistance of the heater.
  ii the current flowing in each of the three top conducting elements.

NEAB 1998

15.19 a The terminal potential difference of a cell is always less than the e.m.f. when there are 'lost volts' across the internal resistance.

  i State a typical value for the e.m.f. of a *single* dry cell.
  ii State a typical value for the terminal potential difference for a dry cell when it is supplying a normal load.
  iii Sketch a diagram to show how many such cells you would use, and how you would connect them, to provide a power supply for a 9 V radio.
b Some dry cells can cause serious burns if short-circuited by a bunch of keys while being carried in a pocket. Explain why this problem occurs only with cells like nickel cadmium cells, which have a very low internal resistance.

London 1995

15.20 a Two resistors of resistance R_1 and R_2 respectively are connected in parallel. Derive a formula for the total resistance R of this combination of resistors.
b The circuit shown contains a component Y in parallel with a thermistor.

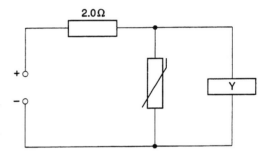

At a particular temperature, the potential difference across Y is 5.00 V. The resistances of component Y and of the thermistor are 125 Ω and 50.0 Ω respectively. Calculate
  i the total resistance of the circuit
  ii the p.d. across the power supply.
c The component Y will be damaged if the p.d. across it is greater than 7.5 V. The resistance of the thermistor reduces to 15 Ω when the p.d. across it is 7.5 V. This provides some protection for component Y.
  i Suggest why the resistance of the thermistor decreases with increasing p.d.
  ii Sketch a graph to show how the p.d. across component Y varies with the p.d. across the power supply.

Cambridge 1997

15.21 a The diagram shows three wires meeting at a junction in a circuit.

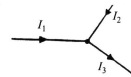

The wires carry currents I_1, I_2 and I_3 as shown. Write down the equation relating I_1, I_2 and I_3.

b In the circuit shown below the current leaving the battery is 1.5 A. The battery has internal resistance 1.0 Ω.

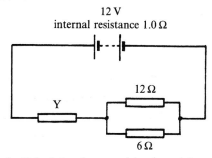

12 V
internal resistance 1.0 Ω

12 Ω

Y

6 Ω

  i Calculate the combined resistance of the resistors that are in parallel in this circuit.

  ii What is the *total* resistance of the circuit?

  iii Hence find the resistance of the resistor labelled Y.

  iv Find the current through the 6 Ω resistor.

c A 6.0 kV power supply of negligible internal resistance is connected in series with a 1.0 MΩ safety resistor. An external load resistor of resistance 4.0 MΩ is then connected to the supply as shown.

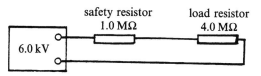

safety resistor
1.0 MΩ

load resistor
4.0 MΩ

6.0 kV

  i Calculate the potential difference across the 4.0 MΩ load resistor.

  ii A 0–10 kV voltmeter of resistance 20 MΩ is then connected across the 4.0 MΩ load. Calculate the reading on the voltmeter.

CCEA 1998

15.22 The circuit diagram shows a 12 V power supply connected across a potential divider R by the sliding contact P. The potential divider is linked to a resistance wire XY through an ammeter. A voltmeter is connected across the wire XY.

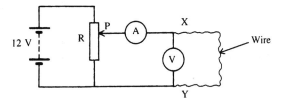

a Explain, with reference to this circuit, the term *potential divider*.

b The circuit has been set up to measure the resistance of the wire XY. A set of voltage and current measurements is recorded and used to draw the following graph.

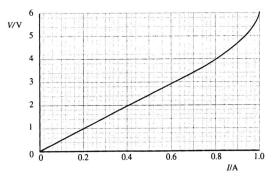

Explain why the curve deviates from a straight line at higher current values.

c Calculate the resistance of the wire for low current values.

d To determine the resistivity of the material of the wire, two more quantities would have to be measured. What are they?

e Explain which of these two measurements you would expect to have the greater influence on the error in a calculated value for the resistivity? How would you minimise this error?

London 1998

15.23 a State *Kirchhoff's laws* for d.c. circuits.

b In the diagram, the cells have zero internal resistance.

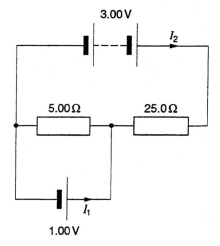

3.00 V

I_2

5.00 Ω

25.0 Ω

I_1

1.00 V

  i Use Kirchhoff's laws to write down two different equations involving the currents I_1 and I_2.

  ii Calculate the current I_1.

Cambridge 1997

16

ELECTRIC FIELD

LEARNING OBJECTIVES

At the end of this chapter you should be able to:

① calculate the force between point charges;

② define and use the term 'electric field strength' and state its unit;

③ relate electric field strength to potential gradient.

16.1 FORCE BETWEEN CHARGES

At the start of another chapter we come back again to the term 'force'. The influence that one object has on another is of universal importance. In section 5.5 it was stated that there are only two types of force outside the nucleus of an atom, the gravitational force and the electromagnetic force. In Chapter 9 details were given about the gravitational force. Here a further look will be taken at the electromagnetic force. An electromagnetic force can only exist between two objects if they have charge. If the charges are stationary, the force will have a different value from its value if the charges are moving relative to one another.

The development of ideas within science is often not straightforward, because discoveries are not necessarily made in a logical order. For example, the knowledge that light objects could be attracted to a piece of amber was known by the Greeks. Indeed, the very word 'electricity' comes from the Greek word for amber, which is 'ελεκτροσ' (*electros*). But a new property needed to be introduced to explain the attraction. This is the property we call 'charge'. **Fig. 16.1** shows that charge can also cause repulsion. Another example that requires the concept of charge is the existence of solids. We now know that solids are able to exist only because they have *charged* particles within them, and therefore have electromagnetic forces between their atoms (see section 22.3).

The same development of ideas is still proceeding. The nucleus of an atom cannot be held together by the gravitational force acting because of the masses of the nuclear particles. This force is not big enough. Neither can the nucleus be held

FIG 16.1 The effect of giving the little girl a charge from a Van de Graaff generator is particularly noticeable on her hair. Similar charges on each hair result in repulsion between adjacent hairs so they stand up on their ends.

*What Newton could **not** explain*

When Newton introduced the idea of gravitational attraction, and explained the movement of the planets and moons from his theories, he was still unable to explain the attraction of a light object by amber. Its mass is far too small for the force to be explained by gravitational attraction.

Another fact that Newton would not have been able to explain is why solids can exist. A calculation of the value of gravitational attraction between neighbouring atoms results in an extremely small force, far too small to have any noticeable effect on the fast-moving atoms in a solid.

together by the electromagnetic force acting. As a result of the charges on the nuclear particles, the force between them would scatter the nuclear particles because it is repulsive and not attractive. There must therefore be some other property of nuclear matter which results in strong nuclear binding forces.

Coulomb's law

This discussion of charge in effect states that there is a property of matter which can result in there being a force between two charged objects. When the results of accurate quantitative experiments were first published by Coulomb in 1784, they showed that the force F between two point charges Q_1 and Q_2 was proportional to each of the charges and inversely proportional to the square of the distance r between them. This is now known as **Coulomb's law** and can be written mathematically as

$$F \propto \frac{Q_1 Q_2}{r^2}$$

This expression should be compared with the corresponding one for the gravitational force of attraction between two masses M_1 and M_2:

$$F \propto \frac{M_1 M_2}{r^2}$$

The similarity of these expressions was first realised by Cavendish, who was working on both forces, and measured the value of the constant of proportionality in the gravitational case as part of his determination of the mass of the Earth. He was also probably the first person to state Coulomb's law for the electrostatic force.

There is an important difference between these two effects. In the gravitational case, the force is always an attractive one. Mass always attracts mass. In the case of electrostatic charge, however, the force can be one of attraction or of repulsion. This is because there are two known types of charge, which are called positive and negative. Positive charge repels positive charge; negative charge repels negative charge; but positive charge attracts negative charge.

In this chapter the particular concern is with the force between static charges. Experimental checking of Coulomb's law is difficult to do directly because it is difficult to keep the charges constant, but the constant of proportionality k for the Coulomb equation

$$F = \frac{kQ_1Q_2}{r^2}$$

can be found to be 8.99×10^9 N m^2 C^{-2} when the two charges are in a vacuum. (We shall often use only two significant figures in examples and questions, in which case $k = 9.0 \times 10^9$ N m^2 C^{-2}.) The very high value for this constant means that very large forces exist even between small charges. This again indicates why solids exist. The inter-atomic forces binding all the atoms in a solid into a rigid structure exist as a direct result of the charge present within the atoms. To get some idea of the size of these forces, look at the values in Example 16.1 and then work through questions 16.1 and 16.2.

EXAMPLE  16.1

Find the force between: **a** a proton and an electron in a hydrogen atom if their separation is 5.3×10^{-11} m; **b** a charge of 7 nC and a charge of 20 nC placed 2 cm apart in a vacuum; and **c** the Earth, with a charge of 2 MC, at a distance of 1.5×10^{11} m from the Sun, with a charge of 6000 MC.

a Substituting the numerical values into the Coulomb equation gives

$$F = \frac{(9.0 \times 10^9 \text{N m}^2 \text{ C}^{-2}) \times (1.6 \times 10^{-19} \text{C}) \times (1.60 \times 10^{-19} \text{C})}{(5.3 \times 10^{-11} \text{m})^2}$$

$$= 8.20 \times 10^{-8} \text{N}$$

This is a very large force when it is considered that it acts on an electron of mass 9.1×10^{-31} kg. If you calculate the acceleration this force produces on the electron, it comes to the enormous value of 9.0×10^{22} m s^{-2}.

b This would be typical for the size of charge on an insulating rod charged by rubbing. Substituting in the Coulomb equation gives

$$F = \frac{(9.0 \times 10^9 \text{N m}^2 \text{ C}^{-2}) \times (7 \times 10^{-9} \text{C}) \times (20 \times 10^{-9} \text{C})}{(0.02 \text{ m})^2}$$

$$= 0.0032 \text{ N}$$

This force is small for most objects, but would be capable of lifting tiny scraps of paper, for instance.

c These charges are estimates. The Earth is charged, and has a measurable electric field as a result. The value of the electric field does enable the charge to be calculated, as you will see later (section 16.2). In such problems it is only sensible to give answers to one significant figure. Substituting values gives

$$F = \frac{(9.0 \times 10^9 \text{N m}^2 \text{ C}^{-2}) \times (2 \times 10^6 \text{C}) \times (6 \times 10^9 \text{C})}{(1.5 \times 10^{11} \text{m})^2}$$

$$= 5000 \text{N}$$

This force is totally insignificant when compared with the size of the gravitational force acting on the Earth due to the Sun, which is of the order of 10^{22} N (section 9.1).

QUESTIONS

16.1 **a** Find the force of attraction between two isolated charges of 1 C each when placed 1 km apart.

b Suggest an object whose weight could be equal to this force.

c What does the answer to this question suggest about the practicality of obtaining an object with a charge on it of 1 C?

16.2 **a** Find the force between the charges on two protons, a distance of 10^{-15} m apart in a nucleus? Your answer should have a huge value when considered as a force acting on the mass of the proton.

b What is the value of the acceleration of the protons? [The mass of a proton is 1.67×10^{-27} kg.]

Permittivity of free space

The constant of proportionality in Coulomb's equation has the value of $8.99 \times 10^9 \, \text{N m}^2 \, \text{C}^{-2}$ when the space between the charges is a vacuum. The numerical value of this fundamental constant is linked with the way in which the units of length, mass, time and electric current are defined, and in the SI system has this fixed value. However, the constant is not normally written as k. By convention it is usual to write

$$k = \frac{1}{4\pi\varepsilon_0}$$

This makes

$$\varepsilon_0 = \frac{1}{4\pi k} = \frac{1}{4\pi \times 8.99 \times 10^9 \, \text{N m}^2 \, \text{C}^{-2}} = 8.85 \times 10^{-12} \, \text{C}^2 \, \text{N}^{-1} \, \text{m}^{-2}$$

The equation for the force between two charges then becomes

K $\quad F = \dfrac{Q_1 Q_2}{4\pi\varepsilon_0 r^2}$

It can be shown that farad per metre (F m^{-1}) (see section 17.1) is an equivalent unit to $\text{C}^2 \, \text{N}^{-1} \, \text{m}^{-2}$. The constant ε_0 is called the **permittivity of free space**. The reason for making this change in the constant is in part historical. The introduction of the 4π, though, is a modern development and has a specific advantage (see box).

Rationalisation of equations

Putting a 4π into the expression does not alter the value of the constant in any way, but it does *rationalise* certain equations, by having the following effects:

- A formula that applies to a system with *spherical* symmetry will contain a 4π term.
- A formula for a system with *cylindrical* symmetry will contain a 2π term.
- A formula for a *uniform* system will not contain a π term at all.

Examples of the use of these formulae will be pointed out when they occur.

QUESTION

16.3 In a sodium chloride crystal, a sodium ion has a charge of $+1.6 \times 10^{-19} \, \text{C}$ and a chloride ion has a charge of $-1.6 \times 10^{-19} \, \text{C}$. They are found by X-ray diffraction to be a distance apart of 0.2 nm. What force of attraction exists between them if they are in a vacuum?
[Assume that $\varepsilon_0 = 8.9 \times 10^{-12} \, \text{F m}^{-1}$ (2 sig. figs)]

 ## ELECTRIC FIELD STRENGTH

In section 9.2 comparison was made between the idea of a magnetic field around a magnet and the gravitational field around the Earth. The fields were treated as regions of influence within which a force was exerted on either a magnet or an object having mass. The term 'gravitational field strength' was introduced as the gravitational force per unit mass. **Figs 16.1** and **16.2** illustrate that there is an electric field surrounding charged bodies. We are now in a position to define 'electric field strength' in a similar way. Whereas a gravitational field is a region in which a force is exerted on a mass, an electrical field is a region in which a force is exerted on a charge.

K The *electric field strength* at a point is defined as the force per unit positive charge placed at the point.

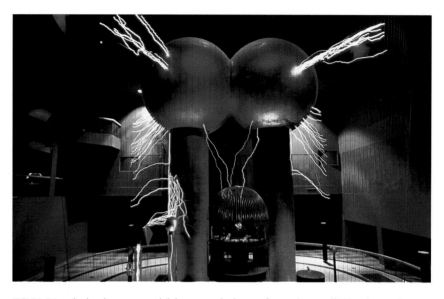

FIG 16.2 In a high-voltage research laboratory, discharges from spheres at high voltage take place to their surroundings.

Note that, since there are two types of charge, *positive* charge has to be stated in order to get the direction of the field correct. The experimental test for the existence of an electric field depends on placing a test charge in the field and measuring the force exerted on it. If this test charge has a large value, it may distort the field that it is being used to measure, so it is essential to use tiny charges. The unit of charge, the coulomb, is a very large unit in this context, so, in practice, test charges may well be measured in nanocoulombs (nC) or smaller units.

In equation form electric field strength *E* is given by

 $$E = \frac{F}{q}$$

where *F* is the force acting on a small charge *q*. Electric field strength is a vector and has the unit newton per coulomb (N C^{-1}).

Finding the electric field strength at any point can be done by either experimental or theoretical methods. Some of the mathematics in these calculations can be complicated, because of the vector nature of the quantity, but the principle of finding the electric field strength is straightforward.

 EXAMPLE 16.2

A charge of $+1.6 \times 10^{-19}$ C has a force of 8.7×10^{-15} N exerted on it when placed at a point in an electric field. Find the electric field strength at the point.

Substituting values in the previous equation gives

$$E = \frac{F}{q} = \frac{8.7 \times 10^{-15} \, \text{N}}{1.6 \times 10^{-19} \, \text{C}} = 5.4 \times 10^{4} \, \text{N C}^{-1}$$

The direction of the electric field is in the direction of the force. In this example, the force being exerted on a single fundamental charge is found. This would be the force due to this electric field exerted on a hydrogen ion.

EXAMPLE 16.3

The electric field strength between a pair of plates of length 4.0 cm in a cathode ray tube is 2.3×10^4 N C^{-1}. An electron enters the field at right-angles to it with a velocity of 3.7×10^7 m s^{-1}, as shown in **Fig 16.3**. Find the velocity of the electron when it leaves the electric field.

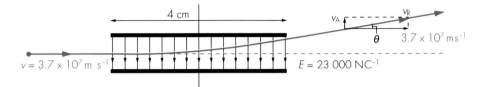

FIG 16.3 An electron passing through an electric field is deflected by the field. Its path is parabolic while in the field.

The force **F** on an electron in the electric field is

$$\mathbf{F} = q\mathbf{E} = (-1.6 \times 10^{-19} \text{ C}) \times (2.3 \times 10^4 \text{ N C}^{-1})$$
$$= -3.68 \times 10^{-15} \text{ N}$$

The minus sign indicates that the force is in the opposite direction to the field, i.e. it is vertically upwards on the diagram.

Since there is no horizontal force on the electron, its horizontal velocity v will be unchanged while it travels through the field. The time t taken to travel through the field is therefore

$$t = \frac{\text{horizontal distance}}{\text{horizontal velocity}} = \frac{0.04 \text{ m}}{3.7 \times 10^7 \text{ m s}^{-1}} = 1.08 \times 10^{-9} \text{ s}$$

The acceleration of the electron will be upwards in the direction of the force and will have the value given by

$$\text{acceleration} = \frac{\text{force}}{\text{mass}} = \frac{3.68 \times 10^{-15} \text{ N}}{9.1 \times 10^{-31} \text{ kg}} = 4.04 \times 10^{15} \text{ m s}^{\pm 2}$$

For any constant acceleration, in general we have

final velocity = initial velocity + (acceleration × time)

In this case the acceleration is vertically upwards. Since the initial velocity in the vertical direction is zero, the final vertical velocity v_A on leaving the field is given by

$$v_A = (4.04 \times 10^{15} \text{ m s}^{-2}) \times (1.08 \times 10^{-9} \text{ s})$$
$$= 4.37 \times 10^6 \text{ m s}^{-1}$$

The resultant velocity is in a direction θ given by

$$\tan \theta = \frac{v_A}{v} = \frac{4.37 \times 10^6}{3.7 \times 10^7} = 0.118$$
$$\theta = 6.7°$$

and has magnitude v_R given by

$$v_R = \frac{3.7 \times 10^7 \text{ m s}^{-1}}{\cos \theta} = \frac{3.7 \times 10^7 \text{ m s}^{-1}}{\cos 6.7°}$$
$$= 3.73 \times 10^7 \text{ m s}^{-1}$$

In this case the speed of the electron has increased by about 1% while undergoing a deflection of 6.7°.

Electric field strength near a point charge

A small test charge q is placed a distance r from a point charge Q (**Fig 16.4**). Coulomb's law is then used to find the force F acting on q:

$$F = \frac{Qq}{4\pi\varepsilon_0 r^2}$$

FIG 16.4 Finding the electric field strength near a point charge Q.

Using $E = F/q$, this gives the magnitude of the electric field strength as

$$\boxed{K} \quad E = \frac{Q}{4\pi\varepsilon_0 r^2}$$

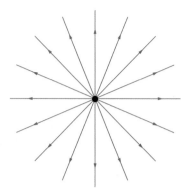

FIG 16.5 The electric field in the space surrounding a positive point charge.

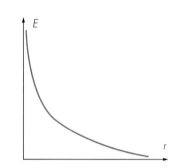

FIG 16.6 The electric field gets weaker further from the point charge.

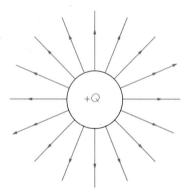

FIG 16.7 The field surrounding a charged, conducting sphere.

A diagram showing the field near a positive point charge is shown in **Fig 16.5**. The field does of course exist in three dimensions. The field has spherical symmetry, and because of rationalisation a factor of 4π occurs in the expression. The graph shown in **Fig 16.6** shows how the electric field strength falls off with distance according to an inverse square law. If the charge $+Q$ is not a point charge but is a charge $+Q$ uniformly distributed on the surface of a conducting sphere, then by symmetry the field is as shown in **Fig 16.7**. The value of the field at any point is identical to that which the point charge would cause (provided that the value of r is greater than the radius R of the sphere).

A biological example

The explanation of the transmission of electrical impulses along a nerve is explained in terms of electric field. The nerve cell itself has long, thread-like axons spreading out from its nucleus (**Fig 16.8a**). The membrane of the wall of a nerve cell is shown in **Fig 16.8b**. In its normal state, shown at point A, the outer surface of the cell carries a positive charge and the inner layer a negative charge. This gives an electric field across the cell wall from the outside to the inside. At another place, point B, the charges are reversed as a result of the movement of chloride ions, so in this part of the axon the field will be from the inside towards the outside. A nerve impulse is the movement of the B region along the axon. The speed of nerve impulse depends on the diameter of the axon. Typically it is 50 m s^{-1}.

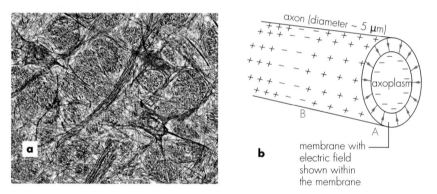

FIG 16.8 a The axons of the nerve cells are clearly shown as thread-like lines in this photograph.
b An axon is a part of a nerve cell, and is long and thread-like. The axon transmits nerve impulses by reversing the electric field across the membrane of the wall.

QUESTION

16.4 The Earth is a conducting sphere of radius 6.4×10^6 m and it is uniformly charged. The Earth therefore has an electric field besides its gravitational and magnetic fields.

   a Why cannot the Earth's electric field be used for navigation?

   b Why does the electric field vary from time to time?

   At a time when the intensity of the Earth's electric field is 300 N C^{-1}, find:

   c the total charge on the Earth;

   d the charge per unit area on the Earth's surface.

16.3 ELECTRIC POTENTIAL

When field theory was first introduced in Chapter 9, the field being considered was the gravitational field. There, it was stated that it was convenient to use a scalar property of the field at each point. The same is true with electric fields. Working with fields themselves is often difficult mathematically because of the vector nature of a field. If a scalar quantity is defined, then addition of that scalar property can be done by normal arithmetic instead of by vector addition. The property used with electric fields is called the 'electric potential' and it is defined in a similar way to the way gravitational potential was defined.

K **The *electric potential* of a point in an electric field is the work done in moving unit charge from infinity to the point.**

Infinity is a convenient way of saying a point far removed from all other electrical influence. If W is the work done in moving a small test charge q from infinity to the point, then the potential V of the point is given by

K $$V = \frac{W}{q}$$

The unit of electric potential is the joule per coulomb ($J\,C^{-1}$), which is the volt (V).

Whereas gravitational potential can only be negative, electric potential can be positive or negative, because the force between charges can be either one of attraction or one of repulsion. In practice, differences of potential are more frequently used than potential itself. In section 14.3, potential difference was introduced as the energy converted from electrical energy to other forms of energy per unit charge passing between the points. There is no conflict between that definition and the one just introduced for electric potential: there is an extension of the idea, however. When dealing with potential difference in an electric circuit, the potential of a point was regarded as being a point somewhere inside the wires of the circuit. Here the definition of potential and potential difference is extended to all points whether the point is in a conductor or in an insulator or in a vacuum.

To see how the idea of electric potential is used, work through Example 16.4.

EXAMPLE  16.4

A small charge of +6.2 nC is moved a distance of 2.0 mm from one conducting plate to another against a field of 20 000 N C⁻¹ (**Fig 16.9**). Find the work done on the charge and the potential difference between the plates.

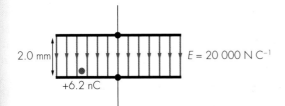

2.0 mm $E = 20\,000$ N C⁻¹

+6.2 nC

FIG 16.9 The small charge and conducting plates for Example 16.4.

Using the equations above we get

  force exerted by field on charge
    $= (6.2 \times 10^{-9}\,C) \times (20\,000\,N\,C^{-1})$
    $= 12.4 \times 10^{-5}\,N$

work done in moving the charge at constant velocity against the field	=	force exerted on charge	×	distance moved in direction of force

    $= (12.4 \times 10^{-5}\,N) \times (0.0020\,m)$
    $= 2.48 \times 10^{-7}\,J$

Then we use

  potential difference = work done per unit charge

$$= \frac{2.48 \times 10^{-7}\,J}{6.2 \times 10^{-9}\,C} = 40\,V$$

Relationship between potential and field

If the calculation done in Example 16.4 is repeated using the algebraic values E for the uniform electric field, q for the charge, x for the separation of the plates and V for the potential difference, it becomes

force exerted by field on charge = qE

work done in moving the charge at constant velocity against the field
= force exerted on charge × distance moved in direction of force
= qEx

and then

potential difference = work done per unit charge

$$V = \frac{qEx}{q} = Ex$$

It is not a coincidence that charge cancels out so neatly, because the potential difference between the plates is dependent on the field and the separation of the plates but not on the value of the small charge.

The previous equation can alternatively be written as

$$E = -\frac{V}{x}$$

Where the minus sign indicates that the direction of the field is the same as the direction of decreasing potential.

> **K** It is always true, for a non-uniform field as well as for a uniform field, that the potential difference per unit distance, the *electric potential gradient*, is the electric field.

This also gives an alternative unit for electric field as volt per metre ($V\,m^{-1}$), which is an equivalent unit to newton per coulomb ($N\,C^{-1}$).

Another way of looking at this is to regard lines of equal electric potential, called **equipotential lines**, as electrical contour lines. Force on a charge acts at right-angles to these contour lines. If the contour lines are close together, then there will be a large force acting on the charge; if the contour lines are well separated, then the force will be small. A charge will always have a force on it in the direction of the field, and therefore its acceleration will be in the direction of the field, but its velocity is not necessarily in the same direction as its acceleration.

Practical examples

One practical use of this topic is in producing focussing devices within an oscilloscope. If an electron beam is passed through two cylinders that are held at different potentials, then the electric field in the space inside the cylinders is as shown in **Fig 16.10**. The equipotential lines are also shown. An electron that goes straight through the centre of the two cylinders is accelerated but is not deviated. One that deviates to the left has forces acting on it to the right to bring it back towards the centre, whereas one that deviates to the right will have forces acting on it to the left. Remember that an electron, since it has a negative charge, will have a force on it in the opposite direction to the electric field. The same principle is used in electron microscopes and television sets, but in these cases the fields used are magnetic fields rather than electric fields.

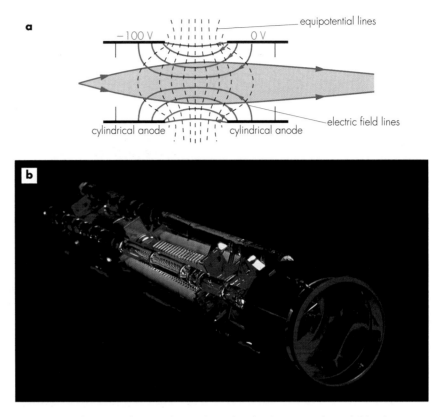

FIG 16.10 The focussing of a cathode ray tube is done by shaping an electric field within cylindrical anodes.

Another example of the use of electric fields is shown by how some fish (**Fig 16.11a**) can detect their prey. The fish acts as a dipole with a positively charged head and a negatively charged tail, as shown in **Fig 16.11b**. When another object is near the fish, the electric field is distorted, and the fish feels the change in the field (**Fig 16.11c**). In a similar way human beings can detect electric fields. If you hold the back of your hand near the screen of a television set, you are able to feel the movement of hairs in the electric field. Some people even claim to be able to forecast the weather by being able to detect changes in the atmospheric electric field.

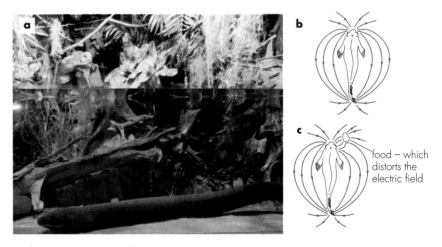

FIG 16.11 a An electric eel – which has the Latin biological name *Electrophorus electricus*. **b** Such a fish can use its electric field to detect its prey. **c** The distortion of the field caused by the presence of food can be sensed by the fish.

Potential in a vacuum at a distance from a point charge

The potential difference between two points in a uniform electric field was easy to calculate because the force acting on the charge was constant. In general, the force that acts on a charge in an electric field is *not* constant, and so calculation of the potential from the field usually requires calculus. Consider a small test charge q being brought from infinity to a point P a distance r from a point charge Q. The minimum work done to move the test charge will be represented by the area under the curve shown in **Fig 16.12**. If you can follow the following calculus through, well and good. If you cannot, do not worry. Just accept that the result for the potential near a point charge is as given at the end.

The small quantity of work δW done on q by some external system to move it a distance δx towards P is given by

$$\delta W = -Eq\delta x$$

The total work done by this external system in moving the charge q from infinity to the point P is therefore

$$W = -\int_{\infty}^{r} Eq \, dx$$

Since the field E at a distance x from a point charge Q is given by

$$E = \frac{Q}{4\pi\varepsilon_0 x^2}$$

then substituting this in the previous equation gives the total work done W as

$$W = -\int_{\infty}^{r} \frac{Qq}{4\pi\varepsilon_0 x^2} \, dx = -\frac{Qq}{4\pi\varepsilon_0} \int_{\infty}^{r} \frac{dx}{x^2}$$

$$= -\frac{Qq}{4\pi\varepsilon_0} \left[-\frac{1}{x} \right]_{\infty}^{r} = -\frac{Qq}{4\pi\varepsilon_0} \left[\left(-\frac{1}{r} \right) - \left(-\frac{1}{\infty} \right) \right]$$

$$= \frac{Qq}{4\pi\varepsilon_0 r}$$

and this gives the potential energy of charge q. To find the potential at P it is simply necessary to find the work done per unit charge. This gives

$$V = \frac{\text{work done}}{q} = \frac{Q}{4\pi\varepsilon_0 r}$$

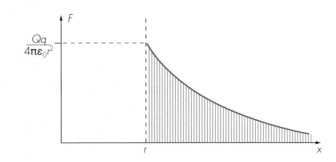

FIG 16.12 The work done to move charge q from infinity up to a distance r from point charge Q is represented by the area under the curve.

EXAMPLE 16.5

Fig 16.13 shows four charges placed at the corners of a rectangle. Find the potential of the point P.

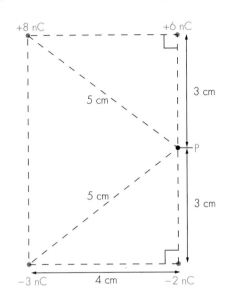

FIG 16.13 The charges for Example 16.5.

To find the potential at point P due to the four charges, we use the above equation for V to find the potential at P due to each charge separately, and then add them together to find the total potential:

■ Potential at P due to $+6\,\text{nC}$ charge (assuming that $\varepsilon_0 = 8.9 \times 10^{-12}\,\text{C}^2\,\text{N}^{-1}\,\text{m}^{-2}$)

$$= \frac{6 \times 10^{-9}}{4\pi \times 8.9 \times 10^{-12} \times 0.030} = 1790\,\text{V}$$

■ Potential at P due to $-2\,\text{nC}$ charge

$$= \frac{-2 \times 10^{-9}}{4\pi \times 8.9 \times 10^{-12} \times 0.030} = -600\,\text{V}$$

■ Potential at P due to $+8\,\text{nC}$ charge

$$= \frac{8 \times 10^{-9}}{4\pi \times 8.9 \times 10^{-12} \times 0.050} = 1430\,\text{V}$$

■ Potential at P due to $-3\,\text{nC}$ charge

$$= \frac{-3 \times 10^{-9}}{4\pi \times 8.9 \times 10^{-12} \times 0.050} = -540\,\text{V}$$

Then

$$\text{total potential at P} = (1790 - 600 + 1430 - 540)\,\text{V}$$
$$= 2100\,\text{V} \ \text{(to 2 sig. figs)}$$

If the potential is required for a series of charges then, because potential is a scalar quantity, all that is necessary is to find the potential due to each charge separately and then add the results together as is shown in Example 16.5.

The type of problem in Example 16.5 is becoming important in determining the shape and structure of complex molecules. The potential at different points in the space surrounding a group of atoms determines where additional atoms may be placed in order to have the lowest potential energy.

QUESTIONS

16.5 An electric dipole consists of charges $+e$ and $-e$ separated by a distance of $10^{-10}\,\text{m}$. If the dipole is placed in an electric field of intensity $10^6\,\text{N C}^{-1}$, find the torque on the dipole when it is:

a parallel to the field;

b at right-angles to the field.

[e is the fundamental charge and has the value $1.6 \times 10^{-19}\,\text{C}$]

16.6 An electron in the vacuum of a cathode ray tube moves from rest at a point where the potential is $-1400\,\text{V}$ to a point where the potential is zero.

a Find the gain in potential of the electron.

b Find the loss in potential energy of the electron.

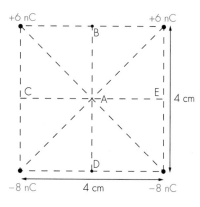

FIG 16.14 The charges for question 16.7.

c Find the gain in kinetic energy of the electron.

d Why are the answers in **a** and **b** of opposite sign?

e What is the final speed of the electron?

[Mass of electron = 9.1×10^{-31} kg]

16.7 Find the potential at each of the points A, B, C, D and E for the 4 cm square shown in **Fig 16.14**. Copy the diagram and sketch the field and equipotential lines.

SUMMARY

■ The force between two point charges in a vacuum is

$$F = \frac{Q_1 Q_2}{4\pi e_0 r^2}$$

■ Electric field strength (intensity) is the force acting on unit charge.

■ Electric field near a uniformly charged sphere or a point is $Q/(4\pi \varepsilon_0 r^2)$.

■ Electric field strength is a vector.

■ Electric potential of a point is the work done per unit positive charge brought from infinity to the point.

■ Electric potential is a scalar, so additions of potential can be done by direct addition.

■ Potential V near a point charge $+Q$ is given by

$$V = \frac{Q}{4\pi\varepsilon_0 r}$$

■ Electric field is the potential gradient

$$E = -\frac{\mathrm{d}V}{\mathrm{d}x}$$

ASSESSMENT QUESTIONS

16.8 A student uses the apparatus shown to investigate electrostatic induction.

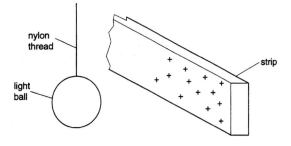

The light ball is made of expanded polystyrene covered with a thin layer of aluminium foil and is

suspended from a nylon thread. A positively charged strip is placed beside the ball.

a **i** On a copy of the diagram, draw an arrow to show the direction in which electrons in the foil move when the charged strip is put in place.

ii Explain why the electrons move in that direction.

iii Explain why the ball is attracted towards the rod.

b The student removes the aluminium foil from the ball and repeats the experiment. Once again, the ball is attracted towards the charged strip. Suggest a reason for this observation.

Cambridge 1997

16.9 a State Coulomb's law for the electric force between two charged particles in free space.
b What are the base units of ε_0 (the permittivity of free space)?

London 1996

16.10 a Explain what is meant by a *field* in physics.
b State two differences between electric and magnetic fields.

London 1996

16.11

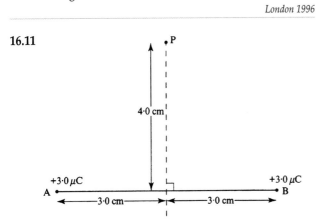

In the diagram above, equal charges of $+3.0 \ \mu C$ are placed at points A and B in vacuo. Find the electric field strength at P.

$$\left[\text{Take } \frac{1}{4\pi\varepsilon_0} = 9 \times 10^9 \ \text{F}^{-1} \ \text{m} \right]$$

WJEC 1998

16.12 Write an essay on the topic of electric fields. Your account should include
a how electric fields are described by diagrams, definitions and equations for both uniform fields and fields due to point charges,
b a description of the motion of beams of charged particles, both positive and negative, in uniform transverse and longitudinal electric fields,
c a definition of electric potential and its relation to electric field strength.

Cambridge 1997

16.13 a Explain what is meant by a *field of force*.
b There is an analogy between electric and gravitational fields. However, the effects of these fields are not similar in all respects.
  i State one way in which the electric field strength (intensity) near a point charge is *similar* to the gravitational field strength near a point mass.
  ii State one way in which the vector nature of the force produced by an electric field may *differ* from that produced by a gravitational field. Explain your answer.

CCEA 1998

16.14 a The figure shows a pair of parallel plates separated by a distance of 5.0 mm.

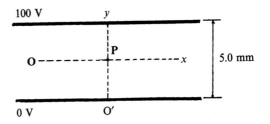

A potential difference of 100 V is applied between the plates.
  i The dashed lines Ox and O$'y$ define directions parallel and perpendicular to the plates respectively. Sketch graphs to show the variation of the electric field strength (intensity) E at different points along Ox and O$'y$ respectively.
  ii Point P is midway between the plates.
    1 Calculate the magnitude of the electric field strength (intensity) at P.
    2 Calculate the magnitude of the force on a point charge of $+6.0 \ \mu C$ placed at P. State the direction of this force.
b The diagram shows an isolated point charge $+Q$. The point A is unit distance (1 m) from the charge.

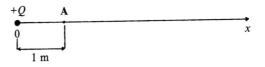

  i On a copy of the axes below, sketch a graph to show how the electric field strength (intensity) E varies with distance x from the charge $+Q$. Start the graph at the point marked B, corresponding to the point A at $x = 1$ m, and continue it in the direction of increasing x. Label this graph E.
  ii Also on your axes, sketch a graph to show how the electric potential V varies with x. Again start the graph at point B. Label this graph V.

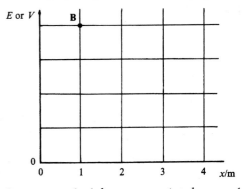

c The diagram overleaf shows two point charges of $+1.0 \ \mu C$ and $-2.0 \ \mu C$ respectively, separated by a distance of 9.0 mm.

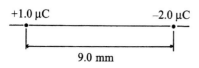

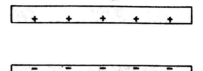

i D is the point, between the charges and on the line joining them, where the electric potential is zero. Find the position of D relative to the +1.0 μC charge.

ii Calculate the magnitude of the electric field strength (intensity) at D. State the direction of the electric field at D.

CCEA 1998

16.15 a State what is meant by the *potential at a point in an electric field*.

b A Van de Graaff generator is a machine which is used to produce very high potential differences. When in use, charge builds up on a hollow spherical conducting dome until the dome achieves the required potential difference between it and the Earth. The electric potential of the Earth is assumed to be 0 V.

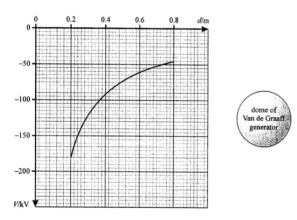

The figure above shows how the potential V varies with distance d from the centre of the charged spherical dome of a Van de Graaff generator, for values of d greater than 0.2 m. The spherical dome of radius 0.15 m is also shown.

i Use the figure to determine the potential at a distance of 0.30 m from the centre of the dome of the Van de Graaff generator.

ii Determine the potential at the surface of the dome of the Van de Graaff generator. Show your reasoning.

iii Draw lines to represent the electric field near the surface of the dome.

iv Determine the electric field strength at a distance of 0.30 m from the centre of the dome.

v Calculate the force exerted on an electron when it is 0.30 m from the centre of the dome. Charge on an electron, $e = -1.6 \times 10^{-19}$ C

AEB 1998

16.16 a The diagram below shows two charged, parallel, conducting plates.

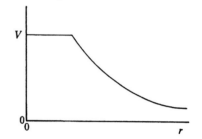

i Copy the diagram and add solid lines to show the electric field in the space between and *just beyond* the edges of the plates.

ii Add dotted lines to show three equipotentials in the same regions.

b Define *electric potential* at a point.

i Is electric potential a vector or a scalar quantity?

ii An isolated charged conducting sphere has a radius a. The graph below shows the variation of electric potential V with distance r from the centre of the sphere.

Copy the graph. Mark with an 'a' on the distance axis the point that represents the radius of the sphere.

iii Add to your graph a line showing how electric field strength E varies with distance for the same range of values of r.

London 1996

16.17 The figure shows graphs of the electric field strength and of the electric potential caused by a point charge. On each graph the vertical axis has a linear scale.

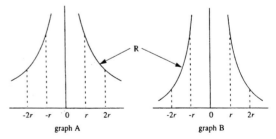

a Which of the graphs, A or B, shows the variation of potential against distance?

b State why, in the regions marked R, the shapes of the graphs are different from each other.

c At a point where the distance, r, from the point charge is 60 mm, the electric field strength is 2.5×10^4 V m^{-1}. Calculate the potential at this point.

NEAB 1997

17

CAPACITANCE

LEARNING OBJECTIVES

At the end of this chapter you should be able to:

① define and use the term 'capacitance' and state its unit;

② relate the capacitance of a parallel-plate capacitor to its dimensions;

③ calculate the capacitance of capacitors in parallel and in series;

④ calculate the energy stored in a capacitor;

⑤ describe how charge, current and potential difference vary as a capacitor is charged through a resistor.

17.1 CAPACITANCE

I f a conductor has no movement of charge within it, then it is all at the same potential. A copper wire at room temperature with a charge flowing through it must have a potential difference between its ends. If there is no potential difference, then no charge flows. This is not true for an insulator. Some parts of the same insulator may be at different potentials from other parts. Indeed, on an insulator the potential may have positive, zero or negative values at different places. This is why insulation is used around wires. The potential of the insulator near the wire will be different from the potential on the outside surface of the insulator.

The gravitational equivalent of this may help with understanding the idea. Mass is the property in gravitational terms that is equivalent to charge in electrical terms. If mass can flow, because it is in liquid form, then it will flow from higher to lower potential. Water flows down a river because of the potential gradient, so if the river surface becomes horizontal then flow ceases. In hot, dry countries, many rivers do not flow at all during certain seasons of the year. However, mass in solid form does not flow, and it will not usually have a horizontal surface and so different potentials are possible even when the mass is not moving.

The value of the potential of a conductor depends, among other things, on the charge on it. Since both the charge on the conductor and its potential have definite values, it is possible to define a quantity C, the **capacitance** of the object, by the equation

$$K \quad C = \frac{Q}{V}$$

where Q is the charge on the conductor and V is its potential.

The unit of capacitance is the coulomb per volt ($C\,V^{-1}$) and this unit is given the name **farad** (F). One **farad** (1 F) is the capacitance of a conductor that is at a potential of one volt (1 V) when it carries a charge of one coulomb (1 C). The farad is a very large unit, and therefore the microfarad, μF, and the picofarad, pF, are frequently used:

$$1\,\mu F = 10^{-6}\,F \quad \text{and} \quad 1\,pF = 10^{-12}\,F$$

Note that there is little meaning to the capacitance of an insulator since there is no single value of the potential for an insulator.

If more positive charge is added to the charge already on a conductor, its potential will rise. If the conductor has a small capacitance, its potential will rise by a large amount when a charge is added. In contrast, the potential of a large capacitor will rise by only a small amount for the same added charge. It is this ability to store charge that gave rise to the term 'capacity'. This is nothing to do with the volume of the conductor. It is simply the charge stored per unit potential. It is found experimentally that the charge on an isolated conductor is proportional to its potential. A graph of charge Q plotted against potential V is shown in **Fig 17.1**. Since it is a straight line through the origin, the capacitance C of the conductor equals the gradient and is given by

$$C = \frac{\text{total charge}}{\text{potential}} = \frac{\text{change in charge}}{\text{change in potential}}$$

$$C = \frac{Q}{V} = \frac{\Delta Q}{\Delta V}$$

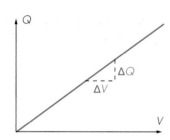

FIG 17.1 The charge stored by a capacitor is proportional to the potential difference across it.

Parallel-plate capacitors

So far, the term 'capacitance' has been applied to a single conductor, but in practice the term is more usually applied to a pair of conductors called a **capacitor**. This has the practical advantage that a greater quantity of charge may be stored per unit potential difference between a pair of conductors. Also, the difficulty of using potential with reference to an arbitrary zero of potential is overcome by using the potential difference between the two conductors. The two conductors are usually a pair of parallel plates with an insulator, called the dielectric, between them. The electrical symbols for different types of capacitor are shown in **Fig 17.2**. The reason why an electrolytic capacitor requires to be connected into a circuit with one of its plates held at a positive potential will be explained at the end of this section, after Example 17.1.

In order to reduce the physical size of a capacitor, the plates are usually made of aluminium foil and are then coiled up with insulators placed so that there is no contact between the plates. A huge variety of different values and types of capacitor are available commercially.

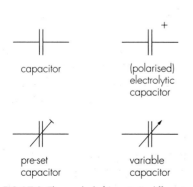

FIG 17.2 The symbols for various different types of capacitor.

One interesting point about the values of the capacitance of capacitors is that they are rather variable even when manufactured in the same way. For this reason, they are sold with a large tolerance, up to 50%. This means that a capacitor marked 10 μF may have a capacitance anywhere between 5 μF and 15 μF. Because of this, manufacturers make a series of nominal values only. One series of values used for standard capacitors is

0.010, 0.022, 0.047, 0.10, 0.22, 0.47, 1.0, 2.2, 4.7, 10, 22, 47, 100 μF, etc.

If values in between these values are specifically required, then either some capacitors are joined together in series or in parallel, or several are measured until one of the correct value is obtained. Usually the value of a capacitor is not critical, so the restriction on the values available is no great problem. If a greater variety of values is required, greater accuracy is needed in manufacture and then values in the series

1.0, 1.5, 2.2, 3.2, 4.7, 6.8, 10 μF

Geometrical series

Series such as those for values of capacitance are geometrical series of numbers. In the first series each number is 2.15 times the previous number, $(2.15)^3 = 10$; in the second series the multiplying factor is the square root of 2.16, namely 1.47.

are available – at greater cost.

If a pair of plates, each of area A, are oppositely charged and arranged so that they are parallel to one another and only a small distance d apart, then the electric field between them will be uniform except at their edges, as shown in **Fig 17.3**. If the edge effect is ignored, the value of the electric field strength E between the plates can be shown to be given by

$$E = \frac{Q}{\varepsilon_0 A}$$

where Q is the charge on each plate. Since also the electric field strength was shown in section 16.3 to be equal to the potential gradient, and since here the field is uniform, we have

$$E = \frac{V}{d}$$

Equating these two values of E gives

$$E = \frac{V}{d} = \frac{Q}{\varepsilon_0 A}$$

so

$$C = \frac{Q}{V} = \frac{\varepsilon_0 A}{d}$$

Note that a parallel-plate capacitor is said to have a charge Q when $+Q$ is the charge on one plate and $-Q$ is the charge on the other. You need not consider situations in which there are different charges on the two plates.

Dielectric and relative permittivity

The assumption has been made so far in this derivation that the space between the plates is a vacuum. This will not in practice be the case. The material between the plates is called the **dielectric**. For some parallel-plate capacitors air is the dielectric, but many capacitors have a dielectric that is a sheet of insulating plastic or waxed paper. The dielectric has three functions.

- It acts to keep the plates a small fixed distance apart.
- It increases the voltage that would cause sparking. Any insulator will experience breakdown if the electric field gets too high. For dry air at atmospheric pressure, a field of about 3000 V mm^{-1} will allow the air to become ionised and a current to pass through it. Modern dielectric materials can

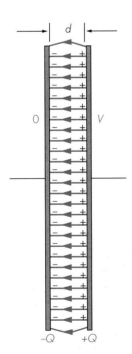

FIG 17.3 The electric field between the plates of a capacitor is very uniform.

withstand much higher fields before undergoing breakdown. However, it is this problem which necessitates that capacitors have to be marked with the highest safe voltage that can be used across them.

■ It increases the capacitance. It is found experimentally that a capacitor with a dielectric present may have a capacitance several times larger than one with a vacuum between the plates. This is because the effect of induced charge on the dielectric reduces the field in the dielectric, so a given charge can be stored with a smaller potential difference between the plates. The factor by which the capacitance is raised is called the **relative permittivity** ε_r or the **dielectric constant**. In equation form this becomes

$$\varepsilon_r = \frac{\text{capacitance of capacitor with dielectric between the plates}}{\text{capacitance of capacitor with a vacuum between the plates}}$$

ε_r is a ratio of two capacitances and therefore does not have a unit.

The value of ε_r for air at atmospheric pressure is 1.0006, so in performing an experiment to measure ε_r it is usually acceptable to use air between the plates instead of a vacuum. Some values for ε_r are given in Table 17.1. The high value for water results in there being a much smaller force between ions when they are separated by water rather than by air. This is why many crystals dissolve in water. Water is not suitable for use as the dielectric in capacitors because it is a liquid and because impurities in it allow it to conduct electricity. High values of relative permittivity in ceramic materials containing strontium titanate are used to obtain high-capacitance capacitors.

Once the effect of the dielectric is taken into account, the full equation for the capacitance of a parallel-plate capacitor is

K $\quad C = \dfrac{\varepsilon_0 \varepsilon_r A}{d}$

It can be seen from the equation that the capacitance of a parallel-plate capacitor is proportional to the area of the plates and inversely proportional to their separation. A larger area of plates will therefore increase the capacitance, but this also has the effect of making the capacitor larger in size. The miniaturisation of electronic circuits, however, requires capacitors of larger value and smaller size. This can be achieved by using high values of the relative permittivity or by having the plates closer together. The following example shows typical sizes for an electrolytic capacitor.

TABLE 17.1 Relative permittivity of some dielectrics

DIELECTRIC	RELATIVE PERMITTIVITY ε_r
vacuum	1 exactly
air	1.0006
teflon	2.1
polythene	2.3
waxed paper	2.7
polyvinyl chloride	3.18
neoprene rubber	6.7
glycerine	43
pure water	80
strontium titanate	310

Rationalisation of equations, again

Here again the effect of rationalisation of equations is apparent. There is a uniform electric field in a capacitor, and consequently π does not appear in this equation. If the constant of $1/(4\pi)$ had not been introduced in Coulomb's law, then a 4π would have appeared here.

EXAMPLE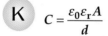

A capacitor is to be constructed to have a capacitance of 100 μF. The area of the plates is 6.0 m × 0.030 m and the relative permittivity of the dielectric is 7.0. Find the necessary separation of the plates and the electric field strength in the dielectric if a potential difference of 12 V is applied across the capacitor.

The capacitance C is given by

$$C = \frac{\varepsilon_0 \varepsilon_r A}{d}$$

$$= \frac{(8.9 \times 10^{-12}\,\mathrm{F\,m^{-1}}) \times 7 \times (6\,\mathrm{m} \times 0.030\,\mathrm{m})}{d}$$

and so putting in $C = 100 \times 10^{-6}$ F and rearranging gives

$$d = \frac{8.9 \times 10^{-12} \times 7 \times 6 \times 0.030\,\mathrm{m}}{100 \times 10^{-6}}$$

$$= 1.12 \times 10^{-7}\,\mathrm{m}$$

$$= 1.1 \times 10^{-4}\,\mathrm{mm}\ \text{(to 2 sig. figs)}$$

The electric field strength E is

$$E = \frac{12\,\mathrm{V}}{1.12 \times 10^{-7}\,\mathrm{m}} = 1.1 \times 10^{7}\,\mathrm{V\,m^{-1}}$$

This very small value of d is a layer less than 1000 atoms thick, so there is a real danger that the electric field will cause dielectric breakdown.

To obtain such thin layers of dielectric, an electrolytic method of production is used. A pair of aluminium plates is separated by a piece of paper soaked in aluminium borate solution, and a current is passed through the solution. Aluminium oxide is deposited on the positive plate, the anode, and this is an insulator, which reduces the current and acts as the dielectric for the capacitor. It is essential to maintain the anode at a positive potential to prevent the electrolytic process being reversed, so an electrolytic capacitor always needs to be connected with its anode positive and its cathode negative.

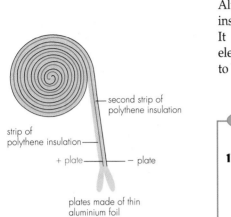

FIG 17.4 Many capacitors are constructed by rolling up two sheets of aluminium foil and two sheets of insulating material.

second strip of polythene insulation
strip of polythene insulation
+ plate — plate
plates made of thin aluminium foil

QUESTIONS

17.1 **a** Find the capacitance of a parallel-plate capacitor in which the plates are 2.0 m long and 1.0 cm wide and are separated by a dielectric that is 0.017 mm thick and of relative permittivity 2.3. [Be careful to put all lengths into metres.]

b Capacitors of this construction are always coiled up (2 m long capacitors are too clumsy!). When coiled, an extra layer of dielectric is added as shown in **Fig 17.4**. What is the effect of coiling on the capacitance?

17.2 CAPACITORS IN PARALLEL AND IN SERIES

The circuit laws, Kirchhoff's laws, which were applied to circuits containing resistors and in which there are direct currents, are laws that can be applied to *any* electrical circuit. Here we need to see how these laws can be applied to circuits containing capacitors.

Capacitors in parallel

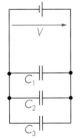

FIG 17.5 Capacitors in parallel: each capacitor has the same potential difference across it.

In **Fig 17.5**, three capacitors of capacitance C_1, C_2 and C_3 are shown connected in parallel and with a potential difference V across them. Because they are in parallel, the potential difference is the same for each capacitor and for the supply. When this circuit is switched on, a charge will flow from the cell for a short time. Some of the total charge Q that flows will flow to charge up capacitor C_1 with a charge Q_1. Similarly C_2 will gain a charge Q_2, and C_3 will gain a charge Q_3.

Using the principle of conservation of charge, Kirchhoff's first law, gives

$$Q = Q_1 + Q_2 + Q_3$$

But from the way that we defined capacitance

$$Q = CV \qquad Q_1 = C_1 V \qquad Q_2 = C_2 V \qquad Q_3 = C_3 V$$

we get

$$CV = C_1 V + C_2 V + C_3 V$$

and dividing through by V gives

K $\quad C = C_1 + C_2 + C_3$

Capacitors in parallel

The total capacitance for capacitors in *parallel* must always be *greater* than the largest of the individual capacitances.

where C is the total circuit capacitance. This result shows that, when capacitors are in parallel, the total capacitance is the sum of the capacitances of the individual capacitors.

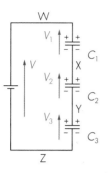

FIG 17.6 Capacitors in series: each capacitor stores the same charge.

Capacitors in series

The total capacitance for capacitors in *series* must always be *less* than the smallest of the individual capacitances.

Capacitors in series

In **Fig 17.6**, three capacitors of capacitance C_1, C_2 and C_3 are shown connected in series and with a potential difference V across them. When these capacitors are initially connected, a charge will flow in the circuit. Kirchhoff's first law states that, if ammeters are placed at W, X, Y and Z, they will at all times show the same current values, since the circuit is a series circuit. This being the case, the charges on each of the three capacitors will all be the same as the charge Q leaving the cell.

From Kirchhoff's second law we have

$$V = V_1 + V_2 + V_3$$

but since $V = Q/C$, etc., we get

$$\frac{Q}{C} = \frac{Q}{C_1} + \frac{Q}{C_2} + \frac{Q}{C_3}$$

giving

K $\quad \dfrac{1}{C} = \dfrac{1}{C_1} + \dfrac{1}{C_2} + \dfrac{1}{C_3}$

Note that it is when capacitors are in series that their capacitances are added by this reciprocal addition, which gives the total capacitance to be smaller than any of the individual capacitances. When they are in parallel, they are added by normal addition. This is the *reverse* situation of that for resistors. This arises because, when resistors are in parallel, more charge flows and therefore the resistance is reduced; with capacitors in parallel, more charge flows to be stored and therefore the capacitance is increased.

Tabulation of electrical quantities to solve numerical problems can be done with capacitors in the same way as with resistors, as is shown in Example 17.2.

A frequent difficulty for capacitors in series

If 6 mC leave the cell in **Fig 17.6**, people imagine that, if the charges on each capacitor are equal, 2 mC will be on each of the capacitors. This is *not* the case. If 6 mC leave the cell, 6 mC will be on each of the capacitors – even if there are 100 capacitors all in series, and even if they all have different values.

You should be able to see why this is so if you look at the piece of wire at X in **Fig 17.6** and the two plates to which it is connected. The two plates and the wire are isolated electrically from the rest of the circuit. If they start with zero charge on them, they will end with a total of zero charge on them. Any positive charge that moves to the upper plate leaves an equal and opposite charge on the lower plate. The total charge is zero.

EXAMPLE 17.2

An 80 μF and a 20 μF capacitor are connected in series to a 10 V supply, as shown in **Fig 17.7**. Find the charge on each capacitor and the potential difference across each.

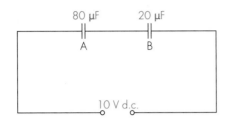

FIG 17.7 The circuit for Example 17.2.

Start by constructing a table (Table 17.2) and inserting the known information.

TABLE 17.2 Known initial data for Example 17.2

	SUPPLY	CAPACITOR A	CAPACITOR B
Charge/μC	Q	Q	Q
P.d./V	10		
Capacitance/μF		80	20

This example illustrates one other aspect of tabulating answers to electrical problems. When the initial information is inserted, it is found that for none of the components are two quantities known, so at first sight no progress seems to be possible. When this situation arises, use an algebraic term for something that needs to be known. In Table 17.2 the charge delivered from the 10 V supply has been inserted as Q, assumed to be in microcoulombs. Since the capacitors are in series, both capacitors receive the same charge Q. Using Q, further quantities may now be found (Table 17.3).

TABLE 17.3 Updated version of Table 17.2

	SUPPLY	CAPACITOR A	CAPACITOR B
Charge/μC	Q	Q	Q
P.d./V	10	Q/80	Q/20
Capacitance/μF		80	20

Now that all the potential differences have been found in terms of Q, some subsidiary working can be done to find Q. The potential difference across the supply must equal the sum of the potential differences across the capacitors, so:

$$10 = \frac{Q}{80} + \frac{Q}{20}$$
$$= \frac{Q + 4Q}{80} = \frac{5Q}{80}$$

which gives

$$5Q = 800$$
$$Q = 160 \ \mu C$$

This value can then be inserted into the table and all other details found (Table 17.4).

As explained in section 15.1, the gradual build-up of information can be done on a single table, although three tables have been used here to show the sequence. The tabulation method has the advantage of clarifying the need for electrical quantities for each component separately.

TABLE 17.4 Completed version of Table 17.2

	SUPPLY	CAPACITOR A	CAPACITOR B
Charge/μC	Q = 160	Q = 160	Q = 160
P.d./V	10	Q/80 = 2	Q/20 = 8
Capacitance/μF	(160 μC)/(10 V) = 16	80	20

Two points should be noted about the answers to Example 17.2. First, the value entered in the space labelled for the capacitance of the supply is the total effective capacitance of the whole circuit. This value is lower than the capacitance of either of the capacitors separately. Two capacitors in series like this do indeed store less charge than either of them would separately. The second point to note is that the smaller capacitor has the higher potential difference across it. Because the charge on both capacitors is the same, the potential difference across each capacitor is inversely proportional to its capacitance.

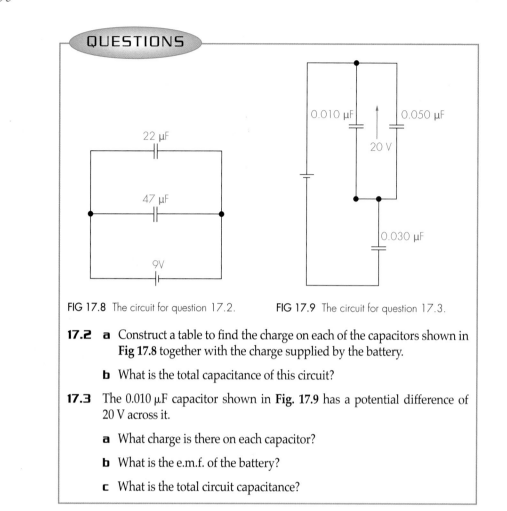

FIG 17.8 The circuit for question 17.2. FIG 17.9 The circuit for question 17.3.

17.2 **a** Construct a table to find the charge on each of the capacitors shown in **Fig 17.8** together with the charge supplied by the battery.

b What is the total capacitance of this circuit?

17.3 The 0.010 μF capacitor shown in **Fig. 17.9** has a potential difference of 20 V across it.

a What charge is there on each capacitor?

b What is the e.m.f. of the battery?

c What is the total circuit capacitance?

17.3 ENERGY STORED IN A CHARGED CAPACITOR

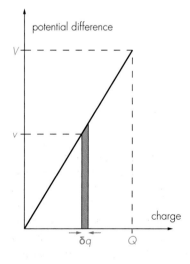

FIG 17.10 The triangular area beneath a *V*–*Q* graph is equivalent to the energy stored by a capacitor.

When a charge of 200 μC is stored on a capacitor with a potential difference across it of 4 V (4 J C^{-1}), it might seem reasonable to assume that the energy stored is 800 μJ. However, this is *not* correct. To see why it is not correct, it is necessary to consider how the capacitor gained its charge in the first place. Initially the capacitor is uncharged. When the first small quantity of charge flows to the capacitor, the potential difference is very small, and so little work is done. As the charge on the capacitor increases, so the potential difference increases, and so more work has to be done to increase the charge on the capacitor. If a graph of potential difference is plotted against charge, it will be a straight line as shown in **Fig 17.10**. If, when the potential difference is v, a small charge δq is added to the capacitor, then the work done will be $v\,\delta q$. This is the area shown shaded on the graph. When the full charge Q has been added, the total work done to charge the capacitor, which will be equal to the energy E stored by the capacitor, will be represented by the total triangular area beneath the graph, and is given by

$$E = \tfrac{1}{2}QV$$

If the capacitor has been charged as a result of a charge Q flowing from a battery of e.m.f. V, then the work done by the battery is QV but the energy stored by the capacitor is only $\tfrac{1}{2}QV$. This implies that energy $\tfrac{1}{2}QV$ has been wasted within the

A mechanical analogy

A mechanical analogy to this situation is a spring being extended a distance x by a force F. The force needs to increase from zero to F as the extension takes place. The work done by the force is the average force multiplied by the extension, i.e.

$$W = \tfrac{1}{2}Fx$$

If a fixed force F is applied to the spring, for instance by dropping a weight on it, then the spring will gain some kinetic energy, stretch beyond its equilibrium extension, and oscillate until it loses energy as internal energy, so that it does store energy of $\tfrac{1}{2}Fx$. The $\tfrac{1}{2}$ in this equation and in the $\tfrac{1}{2}QV$ equation is often forgotten.

resistance of the circuit as internal energy. So the energy stored when a capacitor with a potential difference across it of 4 V carries a charge of 200 µC is 400 µJ, not 800 µJ.

The expressions

$$E = \tfrac{1}{2}QV \quad \text{and} \quad C = Q/V$$

can be combined to give the energy stored in terms of different quantities:

K
$$E = \tfrac{1}{2}QV = \tfrac{1}{2}Q\frac{Q}{C} = \tfrac{1}{2}\frac{Q^2}{C}$$
$$E = \tfrac{1}{2}QV = \tfrac{1}{2}CVV = \tfrac{1}{2}CV^2$$

QUESTIONS

17.4 Find the charge and the energy stored by a 1000 µF capacitor when the potential difference across it is **a** 6.0 V, and **b** 12.0 V.

17.5 A 47 µF capacitor in a flash-gun supplies an average power of 1.0×10^4 W for 2.0×10^{-5} s. What is the potential difference across the capacitor before it is discharged?

17.6 **Fig 17.11** shows a 20 V supply connected to a 100 µF capacitor in parallel with capacitors of capacitance 10 µF and 47 µF in series. Calculate:

a the charge on the 100 µF capacitor;

b the charge on the 10 µF and 47 µF capacitors;

c the energies stored by each capacitor;

d the potential difference across each capacitor.

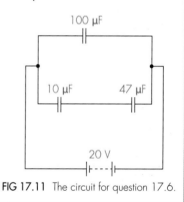

FIG 17.11 The circuit for question 17.6.

17.4 CHARGING AND DISCHARGING A CAPACITOR

A finite time is taken to charge a capacitor from any source of charge. This time has become of increasing importance with the development of computers and the increasing use of digital information. All computer circuits use capacitors to store energy, and there is also stray capacitance between the various parts of any circuit that may have to be taken into account. A problem particularly arises when the frequency of transfer of information is high. If digital information is being handled by a circuit at a frequency of 1000 MHz, then the time for one cycle is a billionth of a second. A capacitor may need to be charged and discharged during this time and, if its value is too high or if the resistance through which it is being charged is too high, it may not be able to gain or lose its charge quickly enough.

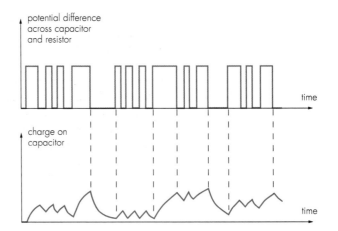

FIG 17.12 When a pulsed signal is applied across a capacitor and a resistor in series, the charge on the capacitor may not have time to reach full charge. The resultant distortion of the signal can be a major problem at high frequency.

INVESTIGATION

The numerical values given for components used in this investigation are for guidance only. If you cannot find precisely these values, use the nearest available ones.

1 Connect a square-wave generator with a frequency of 1 kHz to a 0.47 µF capacitor in series with a 220 Ω resistor. Connect an oscilloscope across the capacitor and adjust the Y-gain and/or the amplitude of the square-wave generator output so that the vertical deflection occupies most of the screen. Then adjust the time-base of the oscilloscope so that approximately two complete cycles of the square wave are shown stationary on the screen. If your trace is not stationary, it will probably be because the synchronisation (synch) needs to be adjusted. Pay particular attention to the earth connections of the instruments you are using. These must be connected together as shown in **Fig 17.13**.

a What is the length of time for one complete cycle?

b Approximately how long does it take for the capacitor to become fully charged?

c Approximately how long does it take for the capacitor to discharge fully?

d Draw a sketch graph showing how the potential difference V_C across the capacitor varies with time.

2 Repeat the experiment with different values of the resistance. First reduce the resistance in convenient steps to about 20 Ω and then increase the resistance in steps to more than 10 000 Ω. For each resistance, make a sketch of the shape of the trace seen on the screen.

3 Reverse the positions of the capacitor and the resistor. Use the 220 Ω resistor again.

a Sketch the pattern shown for the potential difference V_R across the resistor and compare it with the potential difference across the capacitor obtained previously. Do this by drawing graphs of the two potential differences against time, using the same scale on both time axes.

b Find the sum of the two potential differences by adding the values shown on the two sketch graphs together.

4 Disconnect the oscilloscope from the resistor and connect it to X. The sum of V_C and V_R should equal the supply voltage at all times. Why?

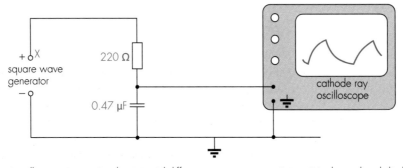

FIG 17.13 A cathode ray oscilloscope measuring the potential difference across a capacitor as it is charged and discharged.

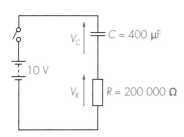

FIG 17.14 The circuit used in the analysis exercise.

Fig 17.12 shows how the potential difference varies across part of a circuit containing a capacitor and a resistor. The corresponding charge on the capacitor is also shown. If the intention is to maintain the shape of the original pulses, it has clearly not succeeded. A great deal of distortion has been introduced as a result of the inability of the capacitor to charge and discharge instantaneously.

A full mathematical analysis of the charging and discharging of a capacitor of capacitance C in series with a resistor of resistance R requires the use of calculus. An understanding of why the curves have the shape they do can be obtained by working through the following data analysis exercise. It can be programmed on a programmable calculator quite easily if you have one available. Do not let the number of significant figures that you record increase beyond four.

ANALYSIS

Charging a capacitor through a resistor

A 400 µF capacitor is to be charged from a 10.0 V battery through a 200 000 Ω resistor as shown in **Fig 17.14**. The switch is turned on at time $t = 0$, so that the sum of the potential differences across the capacitor and the resistor is fixed at 10.0 V.

The capacitor starts in an uncharged state, so the potential difference across it is zero. The potential difference across the resistor must therefore be 10 V. This will only be the case if a current of 50 µA flows through it. The assumption is made that for the next 10 s the current remains constant, so a charge of 500 µC will flow from the battery to charge up the capacitor. Using the notes that are given, copy and complete Table 17.5 by working across and down it line by line. Then plot a graph of the potential difference across the capacitor against time.

TABLE 17.5 Results for the data analysis exercise

TIME /s	V_C /V	V_R /V	I /µA	CHARGE FLOWING IN 10 s/µC	CHARGE ON CAPACITOR 10 s LATER/µC
0	0	10	50	500	500
10	1.25[a]	8.75[b]	43.8[c]	438	938[d]
20	2.35				
⋮					
390					3926[e]
400	9.82	0.18	0.9	9.0	3935
410	9.84	0.16	0.8	8.0	3943
420	9.86	0.14	0.7	7.0	3950
430					
⋮					
∞[f]					

Notes to the table
[a] 500 µC on a capacitor of value 400 µF necessitates 1.25 V across the capacitor.
[b] If there is now 1.25 V across the capacitor, the remainder of the 10 V of the supply must be across the resistor, i.e. 8.75 V.
[c] A potential difference of 8.75 V across a 200 0000 Ω resistor causes a current of 43.8 µA.
[d] 438 µC arrive on the capacitor to be added to the 500 µC already there.
[e] No one is expecting you to have worked through this far manually! It is hoped that you can see the way in which, as the potential difference across the capacitor rises, the potential difference across the resistor falls and the current falls correspondingly. Now proceed a little further from here to see that there is a limit to the quantities involved.
[f] What are these limits at infinite time?

Calculus

The assumption made to do the working in this data analysis exercise is that the current remains constant for a time after each current calculation is made. This assumption is only accurate if the time interval is small compared with a constant called the 'time constant' of the circuit. A time interval of 10 s has been used. It would have been better to have used 1 s, but then there would have been much more arithmetic to do.

The process of needing to use smaller and smaller time intervals is exactly the one which calculus provides. Using calculus, small time intervals are used and the time intervals are then made vanishingly small to get a precise theoretical answer rather than an approximate answer.

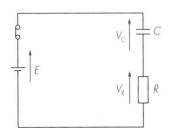

FIG 17.15 Charging a capacitor in series with a resistor.

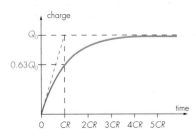

FIG 17.16 After a time equal to one time constant, $t = CR$, the charge on the capacitor is 63% (i.e. $1 - e^{-1}$ as a percentage) of the maximum charge. If the original rate of charging had been maintained, then the capacitor would have been fully charged by this time.

Charging a capacitor

The general equation for charging a capacitor through a resistor can be obtained in the following way. **Fig 17.15** shows a capacitor of capacitance C in series with a resistor of resistance R. The capacitor is initially uncharged and will start to charge at time $t = 0$ when the switch is closed, from a source of e.m.f. E. Let Q be the charge on the capacitor when the time is t.

Using Kirchhoff's second law, that the e.m.f. is the sum of the potential differences, gives

$$E = V_C + V_R$$

$$E = \frac{Q}{C} + IR = \frac{Q}{C} + \frac{dQ}{dt} R$$

The solution to this differential equation is

$$Q = EC(1 - e^{-t/CR})$$

Here E is the e.m.f. of the battery used to charge the capacitor. It is therefore equal to the potential difference across the capacitor at the end of the charging process, because when the current is zero V_R is also zero. Therefore EC is the final charge Q_0 on the capacitor, so the equation can also be written:

$$Q = Q_0(1 - e^{-t/CR})$$

Several mathematical steps have been omitted in this deduction. The final equation, however, is one that you should recognise and be able to use.

Time constant

The **time constant** of a resistor–capacitor circuit is the product CR, which appears in the above equation. It is also the time the capacitor would take to become fully charged if the initial rate of charging is maintained. This is shown in **Fig 17.16**.

Discharging a capacitor

If a charged capacitor is discharged through a resistor, the charge on the capacitor falls exponentially. The equation for discharge, using the same symbols as previously, is:

$$Q = Q_0\, e^{-t/CR}$$

If the capacitor is discharged for a time equal to its time constant, the equation becomes

$$Q = Q_0\, e^{-CR/CR} = Q_0\, e^{-1} = \frac{Q_0}{e} = \frac{Q_0}{2.718}$$

$$Q = 0.368 Q_0$$

After five time constants, the charge is given by

$$Q = Q_0\, e^{-5CR/CR} = Q_0\, e^{-5} = \frac{Q_0}{(2.718)^5}$$

$$Q = 0.0067 Q_0$$

An exponential decay such as this occurs in many branches of physics. The cooling curve for a cup of tea follows approximately the same pattern, and so does the rate of decay of radioactive materials. It is unusual, but possible, to calculate the half-life of a capacitor–resistor circuit (or even of the excess temperature of a cup of tea above its surroundings). If $Q = \frac{1}{2}Q_0$, then

$$\tfrac{1}{2}Q_0 = Q_0\,e^{-t/CR}$$

Q_0 can be cancelled, and taking natural logarithms gives

$$-\ln 2 = -\frac{t}{CR} = -0.693$$
$$t = 0.693CR$$

In other words, a capacitor will discharge by half after 0.693 time constants.

Finding e to a power

Using a calculator to find $e^{-t/CR}$ needs some care. Taking the component values for question 17.7 and a time of 0.005 as an example gives

$$\frac{t}{CR} = \frac{0.005}{0.000\,010 \times 2000} = 0.25$$

Do this first on your calculator, then, *depending on the sequence of working of your calculator*, use

(shift) (ln) (.) (2) (5) (+/−) (=)

to get 0.779.

QUESTIONS

17.7 Use the equation for the charge on a 10 μF capacitor when it is being charged through a 2000 Ω resistor from a source of e.m.f. of 9.0 V to plot a graph of charge against time. You are recommended to do this question by tabulating the values needed for the graph as shown in Table 17.6.

TABLE 17.6 Values for question 17.7

t /s	t/CR (NO UNIT)	$e^{-t/CR}$ (NO UNIT)	$1-e^{-t/CR}$ (NO UNIT)	Q /C
0	0	1	0	0
0.005	0.25	0.779	0.221	
0.010				
0.015				
⋮				
0.10				

17.8 Using the graph that you have drawn from question 17.7, draw a graph of current against time for the same charging process. (Use the fact that the gradient of a charge–time graph is the current $I = dQ/dt$.)

17.9 Using the same numerical values as in questions 17.7 and 17.8, find the following:

  a the current the moment the switch is connected at $t = 0$;

  b the charge on the capacitor when it is fully charged;

  c the length of time it would take the capacitor to charge up fully if this initial rate of flow of charge were continued (this time is called the time constant of the circuit);

  d the value of C multiplied by R;
  [the answers to **c** and **d** should be the same]

  e the unit of CR;

  f the fraction of the total charge that a capacitor has after one time constant and after five time constants.

Charging and discharging compared

When a capacitor discharge occurs, the current flows in the opposite direction to the charging current. The potential difference across the resistor is therefore in the opposite direction for discharge to that for charge. This is shown in the graphs of **Figs 17.17** and **17.18** and their equations, which summarise this work on the charging and discharging of capacitors through resistors.

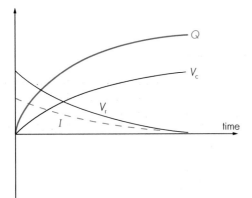

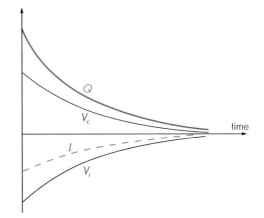

Fig 17.17 Graph showing how Q, V_C, V_R and I vary on *charging* a capacitor through a resistor.

FIG 17.18 Graph showing how Q, V_C, V_R and I vary on *discharging* a capacitor through a resistor.

K

Charge
$$Q = EC(1 - e^{-t/CR}) = Q_0(1 - e^{-t/CR})$$
$$V_C = E(1 - e^{-t/CR})$$
$$V_R = E\,e^{-t/CR}$$
$$I = \frac{E}{R}e^{-t/CR}$$

Discharge
$$Q = Q_0\,e^{-t/CR}$$
$$V_C = E\,e^{-t/CR}$$
$$V_R = -E\,e^{-t/CR}$$
$$I = -\frac{E}{R}e^{-t/CR}$$

When a series of equations such as this is given, it is important not to use them mechanically, but to see how they are related to one another. Here the following points should be noted:

- All the expressions contain an exponential term $e^{-t/CR}$, which governs the rate of rise or fall with time.
- A quantity whose numerical value is rising has a $1 - e^{-t/CR}$ term, whereas one which is falling has just an $e^{-t/CR}$ term.
- V_C and Q are always related by the expression $Q = CV_C$.
- V_R and I are always related by the expression $V_R = IR$.
- Current must always tend to zero as the time increases, since a capacitor can only have a current flowing to or from it when it is charging or discharging. If it is left long enough, no current will flow. The current can be either positive, on charging, or negative, on discharging.
- $V_C + V_R$ equals E during charge, and equals zero during discharge.

EXAMPLE 17.3

A 100 pF capacitor is charged to a potential difference of 50 V and then discharged through a 2.2 kΩ resistor. Find the initial discharge current and how long it will take for the potential difference across the capacitor to fall to 1% of its original value.

At the start, the potential difference across the capacitor is +50 V; and since the only other component in the circuit is the resistor, the potential difference across it must be −50 V. The current is therefore given, using the relation above, by

$$I = \frac{V_R}{R} = \frac{-50\,V}{2200\,\Omega} = -0.023\,A = -23\,mA$$

The equation for I given above under 'Discharge' will of course give the same answer:

$$I = -\frac{E}{R}e^{-t/CR} = -\frac{E}{R}e^{-0}$$
$$= -\frac{E}{R} = -\frac{50\,V}{2200\,\Omega} = -23\,mA$$

Using the equation for the potential difference across the capacitor gives

$$V_C = E\,e^{-t/CR}$$
$$\frac{E}{100} = E\,e^{-t/CR}$$

Cancel E to get

$$0.0100 = e^{-t/CR}$$

At this stage, take natural logarithms of both sides of the equation. [If you are not familiar with this procedure, work carefully through the margin box, as this is a technique that will be used frequently and needs to be understood.] We get

$$-4.605 = -\frac{t}{CR}$$

so

$$t = CR \times 4.605$$
$$= (100 \times 10^{-12}\,F) \times (2200\,\Omega) \times 4.605$$
$$= 1.01 \times 10^{-6}\,s$$
$$t = 1.0\,\mu s$$

A note on logarithms

Since $10^3 = 1000$, we can write $\log_{10} 1000 = 3$, i.e. the logarithm of 1000 to the base 10 is 3. The logarithm of a number is just the power to which the base has to be raised to get that number.

Similarly, since $(2.718)^3 = 20.08$, we write $\log_{2.718} 20.08 = 3$; and since $e^3 = 20.08$, we write $\log_e 20.08 = 3$ (e is a number and is approximately 2.718). Usually $\log_e x$ is written as $\ln x$ and is called a **natural logarithm**.

Scientific calculators have the facility to find the natural logarithm of a number and the reverse of this process, to find the term e^x. You are advised to become familiar with what these keys and the lg and 10^x keys do by practising using them.

QUESTIONS

17.10 What percentage of the original charge remains on a capacitor when it is discharging, after **a** one, **b** two, **c** three, **d** four and **e** five time constants?

17.11 A square-wave signal that is at either 0 V or 500 mV and of frequency 100 kHz (see **Fig 17.19**) is applied to a capacitor of capacitance 0.010 μF in series with a resistor of resistance 100 Ω. Copy **Fig 17.19** and sketch, on the same time axis, graphs of V_C, the potential difference across the capacitor, and V_R, the potential difference across the resistor.
This is called a short-*CR* circuit because the time constant of the circuit is short compared with the period of oscillation.

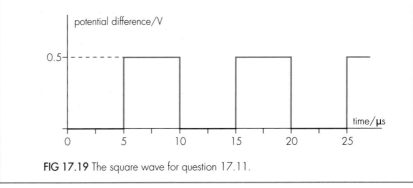

FIG 17.19 The square wave for question 17.11.

17.12 What would the graphs look like in question 17.11 if the resistor used had been 20 000 Ω instead of 100 Ω? Working through this question fully is difficult. You are only expected to show a rough sketch of what happens. This is called a long-*CR* circuit because the time constant of the circuit is long compared with the period of oscillation.

17.13 A 100 μF capacitor carries a charge of 20 mC. It is discharged through an 86 kΩ resistance.

    a Find the time it takes to lose half of its charge (its half-life).

    b Find the time taken for its charge to fall to 1.25 mC.

    c What is the time constant for this circuit?

    d What is the relation between the time constant and the half-life?

SUMMARY

■ Capacitance is charge stored per unit potential difference:

$$C = \frac{Q}{V}$$

■ One farad (1 F) is the capacitance of a capacitor that stores one coulomb (1 C) of charge when a potential difference of one volt (1 V) is across it. Microfarad μF (10^{-6} F) and picofarad pF (10^{-12} F) are more usually used.

■ The capacitance of a parallel-plate capacitor is given by

$$C = \frac{\varepsilon_0 \varepsilon_r A}{d}$$

where $\varepsilon_r = 1$ for a vacuum.

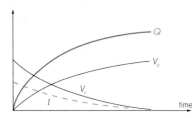

FIG 17.20 Behaviour of Q, V_C, V_R and I when charging a capacitor.

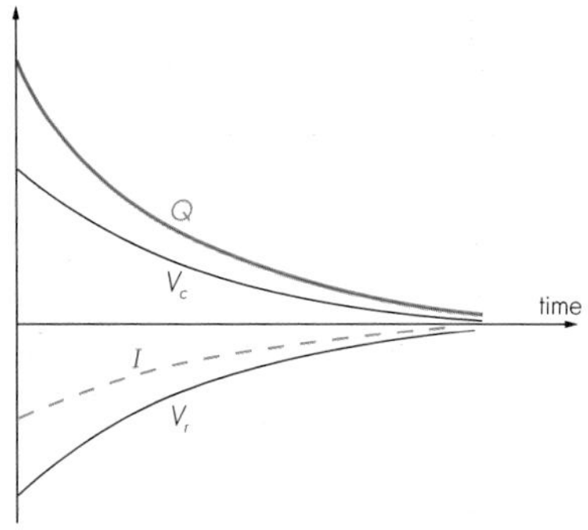

■ Capacitors in series:

$$\frac{1}{C} = \frac{1}{C_1} + \frac{1}{C_2} + \frac{1}{C_3}$$

■ Capacitors in parallel:

$$C = C_1 + C_2 + C_3$$

■ Energy stored in a capacitor:

$$E = \tfrac{1}{2}QV = \tfrac{1}{2}CV^2 = \tfrac{1}{2}Q^2/C$$

■ Charging a capacitor (**Fig 17.21**):

$$Q = Q_0(1 - e^{-t/CR})$$

■ Discharging a capacitor (**Fig 17.22**):

$$Q = Q_0 e^{-t/CR}$$

FIG 17.21 Behaviour of Q, V_C, V_R and I when discharging a capacitor.

ASSESSMENT QUESTIONS

17.14

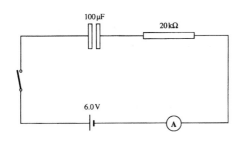

a In the circuit shown, the switch is closed and sometime later the ammeter reads 210 μA. Calculate for that instant
 i the p.d. across the resistor,
 ii the p.d. across the capacitor.
b When the capacitor has become fully charged, calculate
 i the energy stored in the capacitor,
 ii the total energy taken from the battery.

NEAB 1997

17.15 A capacitor of capacitance 0.0015 μF is used in a circuit such that the potential difference across the capacitor is 12 V.
 a Calculate
  i the charge stored by the capacitor,
  ii the energy stored in the capacitor.
 b Outline a simple circuit in which a capacitor is used to store energy.

OCR 1997

17.16 a Define capacitance.
 b An uncharged capacitor of 200 μF is connected in series with a 470 kΩ resistor, a 1.50 V cell and a switch. Draw a circuit diagram of this arrangement.
 c Calculate the maximum current that flows.
 d Sketch a graph of voltage against charge for your capacitor as it charges. Indicate on the graph the energy stored when the capacitor is fully charged.
 e Calculate the energy stored in the fully-charged capacitor.

London 1996

17.17 a Define *potential difference*.
 b When a p.d. is applied to the plates of a capacitor, charges of equal magnitude but opposite in sign appear on the plates.
  i Sketch a graph showing how the numerical charge varies with the applied p.d.
  ii Using this graph and clearly showing your reasoning prove that an amount of energy W is stored in the capacitor where $W = \frac{1}{2}QV$. (Q is the final charge and V the final p.d.)

WJEC 1997

17.18 a An isolated capacitor of capacitance 200 μF has a potential difference across it of 30 V. Calculate
  i the charge stored on one plate of the capacitor,
  ii the energy stored by the capacitor.
 b An uncharged capacitor of capacitance 100 μF is then connected across the charged 200 μF capacitor in **a**. For this combination, state which electrical quantity
  i will have the same total value before and after connection,
  ii will be the same for each of the capacitors after connection.
 c Calculate the total energy stored by the two capacitors in **b** after they have been connected.

OCR 1999

17.19 a In the circuit below, switch A is initially closed and switch B is open. Calculate the energy stored in the 3 μF capacitor when it is fully charged.

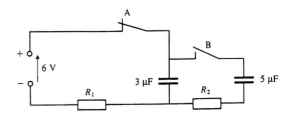

 b Switch A is now opened and switch B is closed. Calculate the final value of the total energy stored in the two capacitors when the 5 μF capacitor is fully charged.
 c State briefly how you would account for the decrease in stored energy.

London 1999

17.20 a Derive a formula for the total capacitance C of two capacitors of capacitance C_1 and C_2 connected in parallel.
 b **i** Sketch a graph to show the variation with voltage V of the magnitude of the charge Q on one plate of a capacitor.
  ii Sketch a graph to show the variation with voltage V of the energy W stored in a capacitor.

Cambridge 1999

17.21 a Derive a formula for the equivalent capacitance of two capacitors in series.
 b A 200 μF capacitor is connected in series with a 1000 μF capacitor and a battery of e.m.f. 9 V. Calculate
  i the total capacitance
  ii the charge that flows from the battery
  iii the final potential difference across each capacitor.

London 1996

17.22 a Copy and complete the circuit below to show the capacitors connected in parallel.

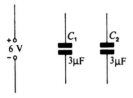

b Copy and complete the same circuit to show the capacitors connected in series.

c Copy and complete the table below using the information in the diagrams.

Capacitors in parallel	Charge on C_1
	Energy stored in C_1 when fully charged
Capacitors in series	Charge on C_2
	Work done by power supply in charging both capacitors

London 1996

17.23 The potential difference between the plates of a 100 μF capacitor is 5.0 V.

a Calculate the charge stored in the capacitor.

b Calculate the energy stored in the capacitor.

c Describe how you would show by experiment that the charge stored in a 100 μF capacitor is proportional to the potential difference across the capacitor for the range of potential differences between 0 V and 10 V.

London 1995

17.24 State whether the following statements are true or false. Give reasons in each case.

a When a battery is connected across a thick wire in series with a thin wire of the same material, electrons move faster through the thick wire.

b When a battery is connected across a high capacitance in series with a low capacitance, the low capacitance stores the larger charge.

c When a battery is connected across a high resistance in parallel with a low resistance, more power is dissipated in the low resistance.

d At the instant when a battery is connected across a high capacitance in parallel with a low capacitance, charge flows faster into the low capacitance.

London 1998

17.25 The figure shows a capacitor of capacitance 100 μF in parallel with a resistor of resistance R, in a circuit which includes an alternating supply and a diode.

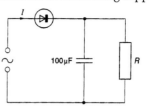

The graph shows the variation with time t of the potential difference V across the capacitor.

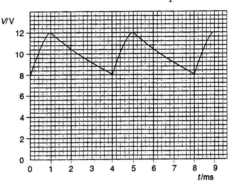

a Sketch a graph to show the variation with time t of the current I in the diode. Calculations are not required.

b Show that

  i the maximum energy stored by the capacitor is 7.2 mJ,

  ii the minimum energy stored by the capacitor is 3.2 mJ.

c **i** Use the values in **b** to calculate the average rate at which electrical energy is dissipated in the resistor while the capacitor is discharging.

  ii Hence, or otherwise, estimate the value of the resistance R.

Cambridge 1997

17.26 A capacitor of capacitance 220 μF is charged so that the potential difference across its plates is 9.0 V.

a Calculate, for the capacitor,

  i the charge stored,

  ii the energy stored.

b A resistor is connected across the terminals of the capacitor. You may assume that the capacitor discharges completely through the resistor in 2.0 ms. Calculate

  i the mean power dissipated in the resistor,

  ii the mean current flowing in the resistor during discharge.

c The capacitor is now charged so that the potential difference across its plates is 18 V. A resistor is connected across its terminals so that it may be assumed that the capacitor discharges completely in the same time, 2.0 ms. What is the mean power dissipated in the resistor?

NEAB 1998

18

MAGNETIC FIELD

LEARNING OBJECTIVES

At the end of this chapter you should be able to:

① define and use the term 'magnetic flux density' and state its unit;

② define and use the term 'magnetic flux' and state its unit;

③ calculate the force on a moving charged particle;

④ use expressions to calculate the magnetic flux density near wires carrying currents;

⑤ calculate the force between parallel wires carrying a current;

⑥ calculate the value of the permeability of free space from the definition of the ampere.

18.1 MAGNETIC PHENOMENA

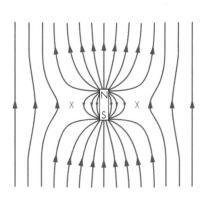

FIG 18.1 The magnetic field surrounding a bar magnet freely suspended in the Earth's magnetic field. Two neutral points, where the magnetic field is zero, are shown by crosses.

Magnets have been discovered from archaeological sites more than two-thousand years old, and have certainly been used for navigation since the eleventh century. Early study of magnetism depended on the properties of the poles of magnets and their behaviour in magnetic fields, such as that of the Earth. If a magnetic field is strong enough, its shape can be found by scattering iron filings in the field. The pattern of weaker fields can be found by using compass needles. **Fig 18.1** shows the magnetic field around a magnet when it is freely suspended in the magnetic field of the Earth. The direction of the magnetic field is taken as the direction in which a force is exerted on the north pole of a magnet.

With the definition of *gravitational field strength* as the force exerted on unit mass, and of *electric field strength* as the force exerted on unit charge, it appears reasonable to define **magnetic field strength** as the force exerted on unit magnetic pole. This was the way in which magnetic field strength was originally defined, but there is a problem with this definition that does not exist with the other two. Whereas it is possible experimentally to use a unit of mass or a single quantity of charge, it is *not* possible to use a single magnetic pole. Because of the way magnetic field is generated, magnetic poles always come in pairs.

In order to avoid this difficulty, the SI definition and the unit of magnetic field strength are approached in a slightly different way from the definitions of gravitational or electrical fields. Instead of considering the force on a magnetic pole, the force on an electric current in a wire is used. The discovery that there is a force acting on a wire carrying a current in a magnetic field was made by Oersted in 1819, but it was not until the 1950s that the effect became the standard way of measuring magnetic field.

18.2 MAGNETIC FLUX DENSITY

A region in which there is a force acting on an electric current in a wire is called a 'magnetic field'. As is always the case with fields in physics, their presence is detected by the force that they exert. The value of the force per unit mass in gravitation is the gravitational field strength in both magnitude and direction. The value of the force per unit positive charge in electrostatics is the electric field strength in both magnitude and direction. Because of the use of electric current to define and measure magnetic field strength, an exactly corresponding statement is *not* made for magnetic field.

To begin with, the force acting on a current-carrying wire is not in the same direction as the field. **Fig 18.2a** shows a wire XY carrying an electric current *I* in a direction at right-angles to a magnetic field *B*. The force acting on the wire is at all places vertically upwards, and depends not only on the current but also on the length of the wire: the longer the wire, the greater the total upward force. Since this is the way magnetic field strengths are measured, however, this is the arrangement that is used for its definition.

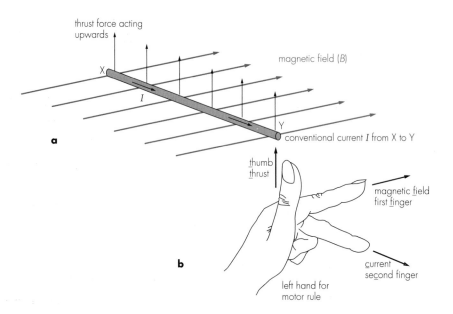

FIG 18.2 **a** The force on a wire carrying a current *I* in a magnetic field *B*. **b** Fleming's left-hand motor rule illustrating how to find the direction of the force.

The quantitative value of magnetic field strength is called the magnetic flux density and it is given the symbol B:

K *Magnetic flux density* **is defined as the force acting per unit current in a wire of unit length at right-angles to the field.**

Put in equation form, B is given by

K $$B = \frac{F}{Il}$$

where F is the total force acting on the wire, l is its length and I is the current. The direction of the force is at right-angles to both the field and the current, and is given by Fleming's left-hand motor rule, as shown in **Fig 18.2b**. This is referred to as Fleming's left-hand motor rule because the force on a wire provides the driving force in an electric motor.

The unit of magnetic flux density is the tesla (T). One **tesla** (1 T) is the magnetic flux density if a wire of length one metre (1 m) carrying a current of one ampere (1 A) has a force exerted on it of one newton (1 N) in a direction at right-angles to both the flux and the current.

Tesla

The tesla is named after Nikola Tesla, an American electrical engineer who designed the power generating station at Niagara Falls and was responsible for many features of high-voltage electrical transmission.

Q **Show that 1 T in base units is $1 \text{ kg A}^{-1}\text{s}^{-2}$.**

If the angle between the current and the field is not 90°, then the force is smaller than that given by the equation $F = BIl$. By resolving I perpendicularly to B, the force becomes

$F = BIl \sin \theta$

where θ is the angle between the field and the current.

EXAMPLE 18.1

In an electric motor a rectangular coil of wire has 150 turns and is 0.20 m long and 0.12 m wide. The coil has a current of 0.26 A through it and is parallel to a field of magnetic flux density 0.36 T, as shown in **Fig 18.5**. Find the torque that is exerted on the coil.

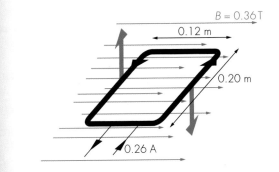

FIG 18.5 The coil for Example 18.1.

Force F on 0.20 m of wire with 0.26 A through it is given by

$F = BIl$
$\quad = 0.36 \text{ T} \times 0.26 \text{ A} \times 0.20 \text{ m}$
$\quad = 0.0187 \text{ N}$

Therefore

turning effect of this force $= 0.0187 \text{ N} \times 0.06 \text{ m}$
$\qquad\qquad = 1.12 \times 10^{-3} \text{ N m}$

torque on a single coil $=$ total turning effect of both forces
$\qquad\qquad = 1.12 \times 10^{-3} \text{ N m} \times 2$
$\qquad\qquad = 2.24 \times 10^{-3} \text{ N m}$

and so

torque on 150 turns $= 2.24 \times 10^{-3} \text{ N m} \times 150$
$\qquad\qquad = 0.337 \text{ N m}$
$\qquad\qquad = 0.34 \text{ N m}$ (to 2 sig. figs)

INVESTIGATION

Find the value of the magnetic flux density inside a solenoid. A possible experimental arrangement is shown in **Fig 18.3**, though several others are available commercially. It is suggested that the solenoid is wound on a cardboard tube of greater than 5 cm diameter and about 40 cm long.

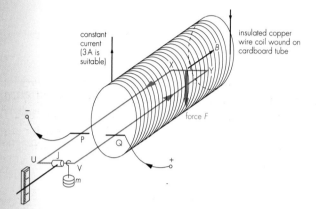

FIG 18.3 An experimental arrangement for measuring the force on a wire carrying a current in a magnetic field.

The wire used should be insulated copper wire of about 24 s.w.g. and the coil should have at least 500 turns in a single layer (s.w.g. stands for 'standard wire gauge' and is a frequently used measure of wire diameter). A double layer containing 1000 turns is better if you have the patience to do the winding! The wire on which the force is to be measured is XY in the diagram. A stiff copper wire rectangular frame UPXYQV is needed, small enough for one end of it (XY) to go into the solenoid, and with an insulator, J, used to join its free ends. The frame is supported on 'knife edges' at P and Q through which current can pass from a d.c. source. It helps if the wire frame is notched at P and Q so that it can easily be returned to its position if it gets displaced. A pointer and a scale showing zero need to be positioned as shown, and a rider, of known mass *m*, placed on UV. An alternative arrangement would be to measure the force on UV by attaching it to the pan of a top-pan balance.

First the frame is balanced by adding Plasticine. Then a steady current (about 3 A is suitable) is passed through the solenoid, and current *I* is passed through XY. In operation, the currents must be in suitable directions for the force that is to be measured to be acting *downwards*. **Fig 18.4** shows a side view of the arrangement, and by applying the principle of moments to the system we get

$$mgd_1 = Fd_2$$

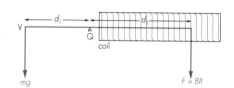

FIG 18.4 Side view of the experimental arrangement in **Fig 18.3**.

This enables the force *F* on the wire to be determined if *m*, d_1 and d_2 are measured. Since $F = BIl$, *B* can be determined once *l*, the distance XY, has been found. Note that any force exerted on PX will be cancelled out by the force on YQ because the currents in these two sections are in opposite directions. In practice, there is little force on either of these sections of the rectangular frame, as the current in these sections is parallel to the magnetic field of the solenoid.

Use the apparatus to plot a graph of magnetic flux density on the axis of the solenoid against distance along the solenoid. This can be found by measuring the distance of the end of the coil from Q for each reading taken. It will almost certainly be easier to move the solenoid rather than the rectangular frame when you do this. Do not stop when XY is outside the solenoid. Take another few readings until the field is too small to be measured. You should also record the number of turns per unit length on the solenoid and the current passing through it. The value of the Earth's magnetic flux density is about 2×10^{-5} T in a horizontal direction, so it should not affect your result significantly.

QUESTIONS

18.1 Calculate the force acting on a wire of length 0.76 m and carrying a current of 3.8 A when placed at right-angles to a magnetic field of flux density 0.065 T.

18.2 A wire of superconducting niobium of radius 0.100 cm can carry a current of 1500 A. The density of niobium is 8600 kg m^{-3}. In what field must the wire be placed so that it levitates, that is, it floats with no visible means of support? [The length of the wire is not needed to solve the problem.]

18.3 A straight wire is placed in a uniform magnetic field of magnetic flux density 0.023 T at an angle of 30° with the magnetic field. The wire carries a current of 8.6 A. Calculate the force on a 3.4 cm length of the wire. Show the direction of the force in a diagram.

18.3 MAGNETIC FLUX

In the investigation above, you were asked to find how the magnetic flux density varied along the axis of a solenoid. At the end of the solenoid, the flux density falls. Diagrammatically the field can be drawn as shown in **Fig 18.6**. The field lines get further apart as the magnetic flux density falls, but a term is needed to show that the total number of field lines is constant. The term used is magnetic flux, which is given the symbol Φ (phi):

 Magnetic flux **is the product of the flux density B and the area A, when the flux is at right-angles to the area.**

In equation form

$\Phi = BA$

The unit of magnetic flux is the weber. One **weber** (1 Wb) is the magnetic flux if a field of flux density one tesla (1 T) exists at right-angles to an area of one square metre (1 m²). Another way of putting this is to say that 1 Wb of magnetic flux per square metre is a magnetic flux density of 1 T.

At first sight there seem to be a lot of terms and units here, but they tend to be used with particular types of problem, and do not usually cause much difficulty. Magnetic flux is a term that is used frequently in the next chapter. These are the terms introduced so far in this chapter:

- **Magnetic field** A name for a region in which a force is exerted on a current-carrying conductor. The pattern of the field can be shown by a field diagram.
- **Magnetic flux density** When it is obvious that the magnetic effect is required, this is often simply called the flux density. The flux density is a vector, it is measured in teslas (T), and it has a numerical value and direction for every point in the magnetic field.
- **Magnetic flux** The magnetic flux density is a flux per unit area. Magnetic flux is the product of magnetic flux density and area. It is measured in webers (Wb). In field diagrams a large value of magnetic flux density is shown by the lines of magnetic flux being drawn close together.

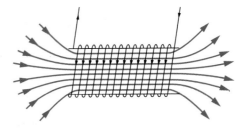

FIG 18.6 The field within a solenoid is uniform apart from near each end.

FORCE ON A MOVING CHARGED PARTICLE IN A MAGNETIC FIELD

18.4

I f a charge moves through a magnetic field at right-angles to the field, it experiences a force on it at right-angles both to its direction of travel and to the direction of the magnetic field. This is because a moving charge is a current, and so Fleming's left-hand motor rule can be used to find the direction of force.

Consider a particle with positive charge q travelling at a constant speed v at right-angles to a magnetic field of flux density B. Assume the particle travels a distance x in time t, so $v = x/t$. The moving charge constitutes a current $I = q/t$. The force F on this current is

$$F = BIl = \frac{Bqx}{t} = Bqv$$

The direction of the force acting on this positively charged particle moving in a uniform magnetic field is shown in **Fig 18.7**. Because the force is of constant magnitude and is always at right-angles to the velocity, the conditions are met for circular motion. Using Newton's second law gives therefore

force = mass × acceleration

$$Bqv = m\frac{v^2}{r}$$

where m is the mass of the charged particle and r is the radius of the circle in which it travels. Reorganising this equation to find the radius of the orbit gives

$$r = \frac{mv}{Bq}$$

showing that the radius is proportional to the velocity of the particle if the charge and mass are constant.

Another feature of the effect of a magnetic field on a moving charged particle is that, since the force acting on the particle is at right-angles to the direction of travel, the work done on the particle is always zero. This must be so, since work done is defined as the product of the force and the distance moved *in the direction of the force*. Here the distance moved in the direction of the force must always be zero, so the work done on the particle is zero. This implies that there can be no increase or decrease in the kinetic energy of the particle, and the motion of a charged particle under the action of a magnetic field must always be at *constant speed*. This is different from the situation in an electric field. The action of an electric field on a charged particle is to change its kinetic energy.

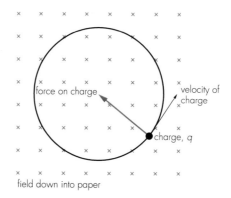

FIG 18.7 A charge moving at right-angles to a magnetic field experiences a force at right-angles to its direction of motion and will travel along a circular path.

EXAMPLE 18.2

An electron travels with a constant velocity through a magnetic field of flux density 0.0076 T and an electric field of electric field strength 56 000 V m^{-1}. Show on a diagram how this is possible, and find the speed of the electron. If the electron then emerges from the electric field, what will be the radius of curvature of the path of the electron in the magnetic field?

Since the electron travels with a constant velocity, the resultant force on it must be zero. This can happen if the force F_B that is exerted on it by the magnetic field is equal and opposite to the force F_E exerted on it by the electric field. This is shown in **Fig 18.8**. The force due to the electric field is in the opposite direction to the electric field because the electron is negatively charged.

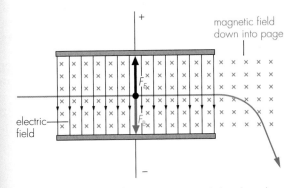

FIG 18.8 A charged particle moving in a straight line through electric and magnetic fields. The force on the charge due to the electric field is equal and opposite to the force due to the magnetic field. When the charged particle leaves the electric field, it travels in a circular path in the magnetic field alone.

In order to find the direction required for the magnetic field, Fleming's left-hand motor rule is used. The current is in the direction from right to left because of the electron's negative charge. In order to obtain a force opposing that due to the electric field, the magnetic field must be directed away from the reader. Since

$$F_E = F_B$$

$$Ee = Bev$$

substituting equations for F_E and F_B previously obtained, the speed v of the electron is given by

$$v = \frac{E}{B} = \frac{56\,000\,\text{V m}^{-1}}{0.0076\,\text{T}}$$

$$= 7.4 \times 10^6\,\text{m s}^{-1}$$

Then using

$$Bev = m\frac{v^2}{r}$$

we can find the radius r of the path in the magnetic field as

$$r = \frac{mv}{Be}$$

$$= \frac{(9.1 \times 10^{-31}\,\text{kg}) \times (7.4 \times 10^6\,\text{m s}^{-1})}{(0.0076\,\text{T}) \times (1.60 \times 10^{-19}\,\text{C})}$$

$$= 5.5 \times 10^{-3}\,\text{m}$$

$$= 5.5\,\text{mm}$$

FIG 18.9 Particle tracks in a bubble chamber. The tracks of positive particles curve anticlockwise and negative particles clockwise in the magnetic field.

QUESTIONS

18.4 An electron travelling through a cathode ray tube with a velocity of 3.2×10^7 m s^{-1} enters a magnetic field of flux density 0.47 mT at a right-angle. What will be the radius of curvature of the electron's path while in the field?

18.5 In a bubble chamber, high-energy particles leave a track of tiny bubbles of gas as they travel through liquid hydrogen at just above its boiling point (see **Fig 18.9**). A positive nuclear particle travels at right-angles to a magnetic field of flux density 2.0 T in a bubble chamber. The speed of the particle is 1.24×10^5 m s^{-1} and the radius of curvature of its track is 1.3 mm.

a Find the ratio of the charge to the mass of the particle.

b Draw a diagram showing the relative directions of velocity, field and force acting, and suggest a possible particle to cause this track.

18.6 A copper wire of cross-sectional area 1.0 mm² carries a current of 5.0 A. [Copper contains 7.8×10^{28} conduction electrons per cubic metre.] The wire is placed at right-angles to a magnetic field of 0.048 T. Find:

a the drift velocity of the conduction electrons;

b the average force exerted by the magnetic field on a single electron as it drifts through the wire;

c the number of free electrons per metre of the wire's length;

d the force on the wire due to the magnetic field.

e Check that the answer that you get for **d** is in agreement with that predicted by the use of the formula $F = BIl$.

18.5 HALL EFFECT

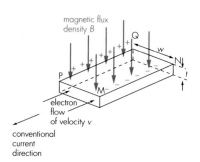

magnetic flux density B

electron flow of velocity v

conventional current direction

FIG 18.10 The Hall effect. A potential difference is set up across a slice of semiconductor when it carries a current at right-angles to a magnetic field.

Question 18.6 asked you to find the force exerted by a magnetic field on an electron flowing through a wire. Because the drift velocity of electrons flowing through wires is small, the force on one electron is correspondingly small. However, the *total* force exerted may be large because of the huge number of electrons that flow.

The Hall effect (**Fig 18.10**) is an effect that arises because of the force that a magnetic field exerts on a moving charge. Electrons cannot drift continually to one side of a wire through which they are flowing because this would raise the potential of one side of the wire unceasingly. In order to prevent sideways movement of electrons, a potential difference is set up between one side of the wire and the other, which counteracts this tendency to drift. This p.d. is called the **Hall potential difference**. It can be shown that the Hall potential difference is larger if the electrons are moving quickly, so in a semiconductor material there is a much greater Hall effect (see the 'in-text question' towards the end of section 14.2).

The Hall potential difference forms the basis of an instrument, called a Hall probe, that is used to measure magnetic flux density. A slice of germanium (a semiconductor) is used in this measuring device. In practice, the Hall probe needs careful adjustment before the Hall potential difference can be measured. **Fig 18.11** shows a suitable circuit. The variable resistance S is adjusted to give a convenient current from the battery, and the milliammeter is used to measure the current through the germanium. The potential divider R is adjusted so that, when there is zero magnetic field, a millivoltmeter connected across XY reads zero. The slice of germanium is then placed at right-angles to the field to be measured, and the milli-voltmeter will record the Hall potential difference, which is proportional to the magnetic flux density.

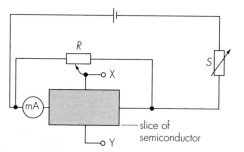

FIG 18.11 An experimental arrangement for measuring a Hall potential difference

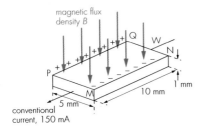

FIG 18.12 The germanium 'slice' for question 18.7.

QUESTIONS

18.7 A piece of germanium has dimensions of 10.0 mm × 5.0 mm × 1.0 mm, and has 4.3×10^{21} charge carriers per cubic metre. A current of 150 mA enters one of its smallest faces and the germanium is placed so that a magnetic field is at right-angles to its largest faces, as shown in **Fig 18.12**. It is found that the side at PQ is 57 mV positive with respect to the side at MN.

a Find the velocity of the elementary charge carriers through the germanium.

b Use $Ee = Bev$ to show that the value of the magnetic flux density is 0.26 T.

c Do the charge carriers have a positive or a negative sign?

18.8 In an experiment on the Hall effect, a current of 0.25 A is passed through a metal strip having thickness 0.20 mm and width 5.0 mm. The Hall potential difference is measured to be 0.15 mV when the magnetic field is 0.20 T.

a What is the electric field across the strip?

b Use the fact that the force due to the electric field balances that due to the magnetic field ($Ee = Bev$) to show that the speed of the charge carriers is 0.15 m s^{-1}.

c Calculate the number of charge carriers per unit volume in the metal.

18.6 ## MAGNETIC FIELD NEAR CURRENT-CARRYING CONDUCTORS

So far in this chapter the measurement of magnetic flux density has been achieved by considering the force that acts on a current in the field. Nothing has been written about the magnitude of the magnetic flux density caused by an electric current or about the factors that affect the magnitude of this flux density. We now look at this aspect.

Field pattern near a current-carrying straight wire

Fig 18.13 shows a wire carrying a current I. One small section of the wire, XY, has been highlighted. The current in XY creates magnetic field in the space all around it. If we choose a point P at random, then we would like to know what field is produced just by this small section. In practice, of course, some leads are required to carry the current to and from XY, but for the moment we will ignore the field created by the current in the leads.

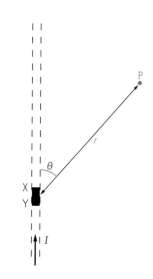

FIG 18.13 A small 'element' of wire carrying a current I.

The field at P is found to be

- inversely proportional to r^2
- proportional to $\sin \theta$
- proportional to l, the small length XY
- proportional to I
- dependent on the material in the space between XY and P

Combining these statements gives

K magnetic flux density at P $\propto \dfrac{Il \sin \theta}{r^2}$

While experiments can be carried out to check this expression, they are difficult and frustrating to do. The real test of such expressions lies in their ability to *predict* magnetic flux densities in many situations over many years.

The expression can be written in the equation form

$$B = \frac{kIl \sin \theta}{r^2}$$

where k is a constant dependent on the material in the space between XY and P. For a vacuum in this space, k, the constant of proportionality, is written as $\mu_0/(4\pi)$, where μ_0 is called the **permeability of free space**. This gives

$$B = \frac{\mu_0 Il \sin \theta}{4\pi r^2}$$

The *numerical* value of μ_0 is found to be $4\pi \times 10^{-7}$. Why this is so will be explained in section 18.7, where its units will also be introduced.

Rationalisation: a reminder

The reason for writing k in this way is to rationalise the equations for magnetic field strength in the same way as was done for electric field strength. The formula for the flux density of any magnetic field showing spherical symmetry will contain a 4π term; one showing cylindrical symmetry will contain a 2π term; and any uniform magnetic field will not contain a π term at all.

Permeability and relative permeability

In the expressions in the previous two subsections, it has been assumed that the space around any current-carrying wire is a vacuum. If this is not so, then the field will be different from the value stated by a factor μ_r. Note that μ_r is called the **relative permeability** of the material and is a plain number. From above, μ_0 is called the **permeability of free space**. The product $\mu_r\mu_0$ is called the **permeability** μ of the material. It is possible for the relative permeability to be slightly less than one: it is 0.999 99 for copper, for instance. Often it is slightly greater than one: it is 1.000 002 for oxygen. For a few materials, called ferromagnetic materials, it is very large. Iron has a value of μ_r that varies with temperature, but is around a value of 1000; and alloys containing iron, cobalt and nickel can have relative permeabilities up to 10 000. This is why placing an iron bar in the centre of a solenoid increases the magnetic flux density to a high value.

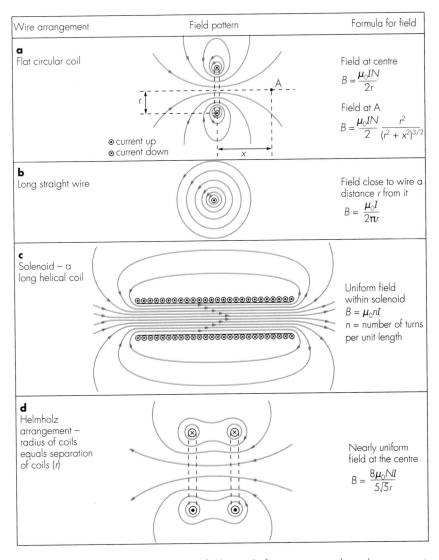

Wire arrangement	Field pattern	Formula for field
a Flat circular coil	⊙ current up ⊗ current down	Field at centre $B = \dfrac{\mu_0 I N}{2r}$ Field at A $B = \dfrac{\mu_0 I N}{2} \dfrac{r^2}{(r^2 + x^2)^{3/2}}$
b Long straight wire		Field close to wire a distance r from it $B = \dfrac{\mu_0 I}{2\pi r}$
c Solenoid – a long helical coil		Uniform field within solenoid $B = \mu_0 n I$ n = number of turns per unit length
d Helmholz arrangement – radius of coils equals separation of coils (r)		Nearly uniform field at the centre $B = \dfrac{8\mu_0 N I}{5\sqrt{5}r}$

FIG 18.14 The magnetic field patterns and field strengths for some commonly used arrangements of current-carrying wires.

Field patterns in other situations

Using the equations on page 322 to find the flux densities B near different configurations of wires involves calculus. Expressions for the flux densities for some situations are given in **Fig 18.14**, without proof, so that flux densities may be calculated when required. Do not try to memorise these equations; they will be given to you if and when they are required.

Cross-Channel electricity

When the cross-Channel power link (Example 18.3) was first established, it was the intention to have a two-way exchange of energy. Peak demand is not at the same time in both countries because of the time difference and differing national habits.

In practice, it has not worked out this way. France generates a great deal of electrical energy from its nuclear power-stations, and always has a surplus, so it imports very little energy from Britain. Britain has saved itself the need to build an extra power-station by taking electrical energy from France on a large scale at a competitive price.

EXAMPLE 18.3

There is a submarine power link between England and France that uses direct current. The cables used for the link are some of the longest, continuous lengths of cable in the world. Each cable consists of a central copper conductor, 10 cm in diameter, insulated with oil-impregnated paper and protected by sheaths of lead and plastic. The outside is 6 mm thick steel armouring, and the whole cable has a mass of 1700 tonnes. Eight of these cables were buried in the sea-bed at a depth of about 1.5 m by an embedding machine, which was able to work at a rate of around 100 metres per hour. Each cable can carry a current of 1850 A and the system is designed to operate at 270 kV.

FIG 18.15 Insulated power cables ready for underground installation. Overhead power cables are only insulated by the air surrounding them.

We need to answer the following questions: **a** What power is transmitted by each cable? **b** What power is transmitted by the whole system of eight cables? **c** At a place where the cable is 60 m below the surface of the sea, what magnetic field is caused by a cable when working at its design power? **d** Bearing in mind that the Earth's magnetic field in the Straits of Dover is 1.9×10^{-5} T, how would you arrange the cables in the sea-bed to avoid affecting ships' compasses?

a We have

> power transmitted by each
> cable = 1850 A × 270 000 V = 500 MW

b Two cables are required for each circuit. Therefore

> total power transmitted = 2000 MW

c The magnetic flux density B at a distance r from a long straight wire is

$$B = \frac{\mu_0 I}{2\pi r}$$

so substituting values gives the magnetic flux density caused by the cable as

$$B = \frac{4\pi \times 10^{-7} \times 1850}{2\pi \times 60}$$

$$= 6.2 \times 10^{-6} \text{ T}$$

d This is about a third of the Earth's natural magnetic flux density, and since it is not in the same direction as the Earth's field, it will significantly alter the direction in which a compass needle will point. At maximum for this depth of water, it would alter the compass direction by an angle whose tangent is 6.2/19. This is an angle of 18°. In shallower water, the deflection could be greater. In order to avoid causing such a large field, the cables are laid in their trenches in pairs. Since the current in one wire of each pair is in the opposite direction to the current in the other wire in the same trench, the magnetic fields cancel one another out.

QUESTIONS

18.9 What is the magnetic flux density 20 cm from the centre of one of the cables linking England and France and carrying a current of 1850 A as in Example 18.3?

18.10 A solenoid gives a very uniform field within it. One particular solenoid is constructed out of a single layer of turns of wire of 1.0 mm diameter with a current of 10 A through it.

a What is the magnetic flux density inside the solenoid?

b If a stronger field is required, then more layers of turns can be added to the solenoid. What problem is likely to arise as more layers are added?

FORCES BETWEEN CURRENT-CARRYING CONDUCTORS

Force between two long parallel wires

If two current-carrying conductors are placed near to one another, then each of the conductors is in the magnetic field that is created by the current in the other. This will cause a force to be exerted on each of the conductors and, by Newton's third law, the forces will be equal and opposite. The simplest situation to consider is when the two conductors are parallel to one another and carry their currents in the same direction. Example 18.4 shows how this can be done.

EXAMPLE 18.4

Find the force per unit length acting on a long straight wire carrying a current of 11.0 A when placed 3 cm away from a similar wire carrying a current of 8.0 A (see **Fig 18.16**).

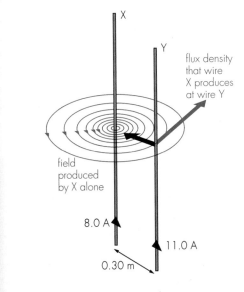
flux density that wire X produces at wire Y

field produced by X alone

8.0 A

11.0 A

0.30 m

FIG 18.16 The wires for Example 18.4.

The flux density that wire X produces at wire Y is given by

$$B = \frac{\mu_0 I_X}{2\pi r} = \frac{\mu_0 \times 8 \text{ A}}{2\pi \times 0.030 \text{ m}} = \frac{4\pi \times 10^{-7} \times 8 \text{ A}}{2\pi \times 0.030 \text{ m}}$$

$$= 5.3 \times 10^{-5} \text{ T}$$

The force that X exerts on Y is given by

$$F = BI_Y l = 5.3 \times 10^{-5} \text{ T} \times 11 \text{ A} \times l$$

The force per unit length on Y is

$$\frac{F}{l} = 5.3 \times 10^{-5} \text{ T} \times 11 \text{ A} = 5.9 \times 10^{-4} \text{ N m}^{-1}$$

The force per unit length on X is of the same magnitude but in the opposite direction:

$$\frac{F}{l} = \frac{\mu_0 \times 11.0 \text{ A}}{2\pi \times 0.030 \text{ m}} \times 8.0 \text{ A}$$

$$= \frac{4\pi \times 10^{-7} \times 11.0 \times 8.0}{2\pi \times 0.030} = 5.9 \times 10^{-4} \text{ N m}^{-1}$$

The wires are pulling each other together.

The force pulling two parallel wires, carrying currents in the same direction, towards one another is usually quite a small force, but it can be important when a current is passing through an ionised gas or plasma. Under these circumstances, there is an effect called the 'pinch effect' that squeezes the current into a very small area of cross-section. This effect is used in a tokamak nuclear fusion reactor (**Fig 18.17**) to achieve temperatures in excess of 50 000 000 K so that fusion of lighter atoms into heavier atoms may take place.

The value of μ_0

Example 18.4 uses the numerical value of μ_0 quoted earlier, but so far nothing has been written about how μ_0 is obtained or about the unit in which it is measured. The definition of the ampere involves a very similar arrangement to that given in Example 18.4. The **ampere** is defined as the constant current which, if maintained in two straight parallel conductors of infinite length placed 1 m apart in vacuum, would produce between these conductors a force of 2×10^{-7} N per metre of length. If we use this definition, then we can calculate the value of μ_0.

In general, the flux density B created by wire X at wire Y is given by

$$B = \frac{\mu_0 I_X}{2\pi r}$$

The force F on Y is

$$F = BI_Y l$$
$$= \frac{\mu_0 I_X}{2\pi r} \times I_Y \times l$$

so the force per unit length on Y is

$$\frac{F}{l} = \frac{\mu_0 I_X I_Y}{2\pi r}$$

Substituting the numerical values from the definition of the ampere gives

$$\frac{2 \times 10^{-7}\ \text{N}}{1\ \text{m}} = \frac{\mu_0 \times 1\ \text{A} \times 1\ \text{A}}{2\pi \times 1\ \text{m}}$$
$$\mu_0 = 2 \times 10^{-7} \times 2\pi \frac{\text{N}}{\text{A}^2}$$
$$= 4\pi \times 10^{-7}\ \text{N A}^{-2}$$

This value of μ_0 comes directly from the arbitrary definition of the ampere. The way the ampere is defined is, in effect, fixing the value of μ_0 at the exact value of $4\pi \times 10^{-7}\ \text{N A}^{-2}$.

FIG 18.17 This 12 tonne field coil is one of several causing a torroidal (doughnut shaped) magnetic field in the Joint European Torus (JET) at Culham in Oxfordshire. The field is used to contain extremely high temperature plasma.

QUESTIONS

18.11 Two long parallel wires separated by a distance of 1.0 cm carry opposite currents of 20 A. Find the magnetic flux density at

a their mid-point and

b a point 1.0 cm from one of the wires and 2.0 cm from the other.

18.12 a Find the force per unit length between two adjacent turns in a transformer if they are separated by a distance of 0.000 50 m and carry a current of 0.46 A. You may assume that the wires behave as long straight wires because their separation is much smaller than their length.

b Use your answer to part **a** to explain why a transformer using mains alternating current can often be heard to buzz.

c What is the frequency of the buzz?

18.8 PRACTICAL APPLICATIONS

There are many domestic and engineering applications of electromagnets and the motor effect. These range from the tokamak, mentioned in section 18.7, and high-energy particle accelerators that are used to probe into the structure of the nucleus, right down to simple electromagnets for lifting scrap iron. Electric motors of a bewildering variety are manufactured by the million every year.

At one time the entire telephone system relied on electromagnetic switches for making connections. Now most of this linking is done using electronic switching systems, but electromagnetic switches are still used in many items of domestic equipment such as central heating controls and washing machines.

Relay

Many electronic control systems, even very complex ones, at the final stage in the system make use of a relay to switch on any mains-operated equipment. A relay is an electromagnetic switch, which has the advantage that it isolates the equipment being switched on from the control system itself. It also has the advantage that it can operate with a small current on the control side and a large current on the mains side. When the controlling current is passing through the electromagnet, the soft-iron armature is attracted to the electromagnet; and as it rocks on the pivot, the lever attached to it pushes the switches open or closed. It is possible to purchase many different types of relays working on different sizes of control current and with an assortment of switches, some normally off, some normally on, and some change-over. **Fig 18.18** shows a relay with two switches. One of them will switch on, and the other will switch off, when a current passes through the electromagnet.

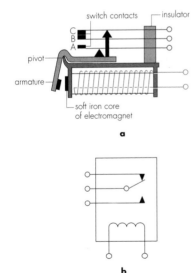
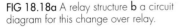

FIG 18.18a A relay structure **b** a circuit diagram for this change over relay.

Moving-coil loudspeaker

In a loudspeaker it is possible, by carefully arranging the shape of the magnetic field, to make a small coil move sideways without turning (in contrast to a motor, where no sideways force acts while the coil does turn). **Fig 18.19** shows a

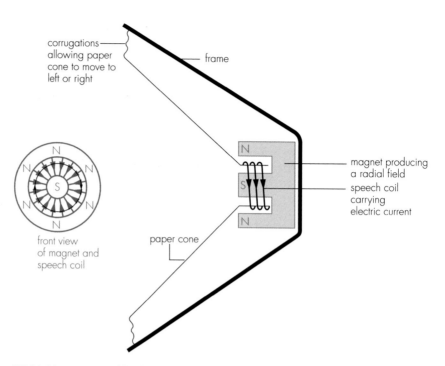

corrugations
allowing paper
cone to move to
left or right

frame

N

N   N

S

N   N

N

front view
of magnet and
speech coil

paper cone

N

S

N

magnet producing
a radial field

speech coil
carrying
electric current

FIG 18.19 A moving-coil loudspeaker.

cylindrical south pole of a permanent magnet surrounded by north poles. The magnetic field is therefore a radial field, which is directed inwards towards the centre. If you use Fleming's left-hand motor rule for any part of the wire carrying the current, you should find that the force on the wire is down into the page, when using the front view diagram. The entire coil therefore will move into the body of the magnet and will pull the paper cone with it. If the direction of the current is now reversed, then the coil will move in the opposite direction and will push the paper cone outwards. If alternating current is passed through the coil, then the coil will oscillate. By varying the frequency, amplitude and pattern of the alternating current, the whole vast range of responses of the cone can be achieved to produce any desired sound.

SUMMARY

- Magnetic flux density is the force acting on unit current in unit length of wire at right-angles to the field:

$$B = \frac{F}{Il}$$

- One tesla (1 T) is the magnetic flux density if a force of one newton (1 N) acts on a current of one ampere (1 A) in a wire of length one metre (1 m).

- If θ is the angle between a magnetic field and a wire carrying a current, then the force on the wire is given by

$$F = BIl \sin \theta$$

- Magnetic flux = magnetic flux density × area:

$$\Phi = BA$$

- The weber (Wb) is the flux if a flux density of one tesla (1 T) exists over an area of one square metre (1 m^2).

■ Magnetic flux density B near a long straight wire:

$$B = \frac{\mu_0 I}{2\pi r}$$

■ Magnetic flux density B at the centre of a flat circular coil:

$$B = \frac{\mu_0 NI}{2r}$$

■ Magnetic flux density B in a long solenoid:

$$B = \mu_0 nI$$

■ Force F between two long parallel wires:

$$F = \frac{\mu_0 I_X I_Y l}{2\pi r}$$

■ The permeability of free space $\mu_0 = 4\pi \times 10^{-7} \, \text{N A}^{-2}$. This figure is exact and comes directly from the arbitrary definition of the ampere. The relative permeability is the factor by which the magnetic flux density is changed, from the value it would have in a vacuum, by the presence of some material.

ASSESSMENT QUESTIONS

18.13 a i The unit of magnetic flux density is the *tesla*. Define the tesla.

ii The diagram represents a uniform horizontal magnetic field.

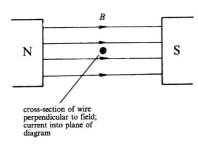

cross-section of wire
perpendicular to field;
current into plane of
diagram

A straight copper wire, perpendicular to the field, carries a current of 5.0 A directed into the plane of the diagram.

1 In what direction is the force on the current-carrying wire?

2 The length of wire in the field is 10 cm, and it experiences a force of magnitude 40 mN. Calculate the magnetic flux density of the field.

b The equation for the magnetic flux density in a long, straight, current-carrying solenoid is $B = \mu_0 nI$.

i Describe experiments to investigate *two* of the factors which govern the magnetic flux density in such a solenoid.

ii What effect has the insertion of a ferrous core on the magnetic field produced by this solenoid?

iii A 500-turn air-cored solenoid has length 800 mm. What current is needed to produce a magnetic flux density of 3.00 mT at the centre of the solenoid?

iv Another solenoid has the same length and number of turns, but is wound so that it has double the cross-sectional area of the solenoid in **iii**. What effect, if any, has this on the magnetic flux density at the centre of the solenoid, if the current is the same as that in the solenoid in **iii**?

c Two long, straight, parallel, thin conductors are placed 50 mm apart in a vacuum. One carries a current of 5.0 A, and the other a current of 4.0 A, in the same direction.

i Draw a clear diagram showing the conductors carrying their currents into the page, and then draw the shape and the direction of the magnetic fields produced separately by each conductor. Mark clearly the directions of the forces they exert on each other, and state whether the conductors attract or repel each other.

ii Calculate the magnitude of the force that a 0.30 m length of each conductor exerts on the other.

CCEA 1998

18.14 a Sketch the magnetic field due to a bar magnet.
  b Use your sketch to explain what is meant by the terms
    i *flux,*
    ii *flux density.*
  c What does the flux density represent?

WJEC 1998

18.15 a Explain what is meant by the term *magnetic field*. Draw a diagram to show the shape of the magnetic field produced by a bar magnet and explain how it is possible to obtain a field of the same shape by using an electric current.
  b Outline the principle of the electromagnet and describe two practical examples of the use of an electromagnet.
  c When a wire carrying an electric current is placed in a magnetic field, it may experience a force due to the field. Describe how to find the direction in which the force acts. Name a device which makes use of this electromagnetic force.
  d An effect further to that described in **c** involves a wire moving in a magnetic field and hence cutting magnetic flux. What physical quantity is produced in the wire as a result of this movement? State the factors which affect the size of the physical quantity produced. Name a device which uses this effect and describe the important features of the device.

Cambridge 1997

18.16 The figure shows two horizontal wires, P and Q, 0.20 m apart, carrying currents of 1.50 A and 3.00 A respectively.

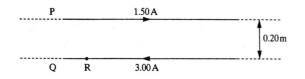

  a On a copy of the outline diagram below draw the *resultant* magnetic field produced by these two currents.

$$\otimes$$

$$\odot$$

$\otimes$ wire P carrying current of 1.50 A into plane of paper
$\odot$ wire Q carrying current of 3.00 A out of plane of paper

b The flux density B of the magnetic field produced at a point by a current I flowing in a long straight wire is given by

$$B = \frac{\mu_0 I}{2\pi a}$$

where a is the perpendicular distance from the wire to the point. Calculate the flux density of the magnetic field at R due to the current in wire P.
  c Hence calculate the force per metre on wire Q due to this field.
  d State the direction of this force.
  e State the direction and magnitude of the force per metre on wire P due to the current in wire Q.

NEAB 1998

18.17 a The SI unit of magnetic flux density is the tesla (T). Express this unit in terms of the newton (N), ampere (A) and the metre (m).
  b The figure illustrates a thin vertical wire carrying a constant current. A Hall probe is used to investigate the magnitude of the magnetic field around the wire.

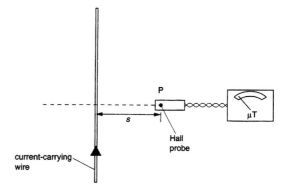

The probe is placed initially at P. Readings of the magnetic flux density B are taken as the probe is placed at various points at distances s from the wire, along a horizontal line passing through the wire. The following readings are obtained.

s/cm	+15	+10	+5	−5	−10	−15
B/μT	10	15	30	30	15	10

  i Plot a graph of the variation of B with s.
  ii It is suggested that the value of B at a point is inversely proportional to the distance of that point from the wire. Discuss whether your graph supports this suggestion.
  iii Assuming that B is proportional to $1/s$, calculate the value of B at a distance of 25 cm from the wire.

Cambridge 1997

18.18 a A child sleeps at an average distance of 30 cm from household wiring. The mains supply is 240 V r.m.s. Calculate the maximum possible magnetic flux density in the region of the child when the wire is transmitting 3.6 kW of power.

b Why might the magnetic field due to the current in the wire pose more of a health risk to the child than the Earth's magnetic field, given that they are of similar magnitudes?

London 1997

18.19 a The magnitude of the force on a current-carrying conductor in a magnetic field is directly proportional to the magnitude of the current in the conductor. With the aid of a diagram describe how you could demonstrate this in a school laboratory.

At a certain point on the Earth's surface the horizontal component of the Earth's magnetic field is 1.8×10^{-5} T. A straight piece of conducting wire 2.0 m long, of mass 1.5 g, lies on a horizontal wooden bench in an East–West direction. When a very large current flows momentarily in the wire it is just sufficient to cause the wire to lift up off the surface of the bench.

b State the direction of the current in the wire.

c Calculate the current.

d What other noticeable effect will this current produce?

London 1997

18.20

PQRS is a section of a circuit, with right-angled corners at Q and at R. It is mounted so that it lies in the same horizontal plane as a uniform magnetic field of flux density B, which is parallel to the sides PQ and RS. When a current, I, flows in the direction shown in the diagram, the wire QR experiences a force which acts vertically upwards.

a i Draw an arrow to show the direction of the magnetic field.

ii Copy and complete the diagram below by sketching the resultant magnetic field pattern around QR, ignoring the effects of the fields due to PQ and RS.

wire QR, carrying current out of plane of paper

iii If the length of QR is 0.15 m, and B is 0.20 T, the force on QR is 0.18 N. Calculate I.

iv Explain why the magnetic forces on sides PQ and RS need not be considered when investigating the equilibrium of PQRS.

b An experiment based on the arrangement described in **a** may be used to determine the magnitude of the flux density of a magnetic field. A current is passed through PQRS, which is a wire frame pivoted about a horizontal axis through PS. The upwards magnetic force on QR is balanced by the weight of a length of ticker tape placed over it. The experiment is repeated for other lengths of tape.

The following results were obtained in such an experiment.

length of tape x/mm	41	83	120	162	208	249
current I/A	0.51	0.98	1.53	2.02	2.55	3.05

The mass of a 2.00 m length of the uniform ticker tape was 1.20×10^{-3} kg.

i Plot a graph of I against x.

ii Show that I and x are related by the equation

$$I = \left(\frac{\mu g}{Bl} \right) x$$

where μ is the mass per unit length of the tape, l is the length of QR and g is the acceleration due to gravity.

iii Hence calculate the magnitude of B from the gradient of the graph.

NEAB 1997

18.21 a The diagram is a plan view of a long straight wire with an electric current going into the page.

Sketch the pattern of the magnetic field due to the current in the wire. Your diagram should indicate the pattern, the direction and the variation in strength of the field.

b i Suggest why a long straight wire is not used as an electromagnet.

ii Describe how a long wire could be made into a useful electromagnet.

c Name one device in the home which includes a small electromagnet, and state the use of this device.

OCR 1999

19 ELECTROMAGNETIC INDUCTION

LEARNING OBJECTIVES

At the end of this chapter you should be able to:

① state the conditions necessary for electromagnetic induction;

② deduce the value of an induced electromotive force;

③ state Faraday's law and Lenz's law;

④ describe the action of a generator;

⑤ understand the importance of the induced electromotive force in the operation of a motor.

19.1 INDUCTION PHENOMENA

Until 1831 electricity was not much more than a novelty. Various scientists had discovered some effects of electric currents, and there were suggestions that perhaps electrical effects were of fundamental significance. But a major restraint on further development was that the sources of electric current were batteries, and at that time batteries were very inefficient. (Even today, batteries are totally unsuitable for large power supplies.) A link between electric current and magnetic field had been established by Oersted in 1819. He had discovered that a magnetic field was produced whenever there was an electric current. In 1822 Ampère showed that there was a force acting between two current-carrying conductors because of their magnetic field.

However, it was not until 1831 when publications by Henry in the USA and Faraday in England announced that an electric current could, in certain circumstances, be induced from a magnetic field. This reverse process is called 'electromagnetic induction' and is of enormous practical importance. Virtually all electrical current is now produced using electromagnetic induction rather than from batteries. Even when batteries *are* used for the supply of electric current, they are often recharged from power supplies which use the effect of electromagnetic induction.

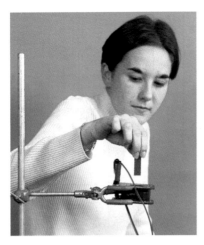

FIG 19.1 Inducing electric current in a coil.

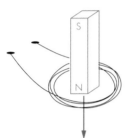

FIG 19.2 Electromagnetic induction is more noticeable if a coil of wire is used.

It is not difficult nowadays to demonstrate electromagnetic induction, because sensitive meters are available to detect even the small current produced in a single wire when a bar magnet is moved near it. You are recommended to try the following demonstration. Use the most sensitive galvanometer you have, and connect a length of wire between its positive and negative terminals. Then move a magnet near the wire, and observe what happens to the galvanometer needle.

The aim is to use the magnet to make the electrons flow in the wire. If a brushing action is used to sweep the electrons *along* the wire, the demonstration is very unconvincing. There is only the occasional current. However, if the magnet is moved sideways *across* the wire, then there is a surge of current in one direction when the magnet passes over the wire. The magnet makes the electrons move in a direction at right-angles to its own direction of movement. If the magnet is moved sideways in the opposite direction, then the current will be induced in the opposite direction. Reversing the polarity of the magnet also causes the current direction to be reversed.

If the wire is now shaped into a circle, then the magnet can move past all of it sideways if it is pushed through the circle. This can be extended, as shown in **Figs 19.1** and **19.2**, by using a coil of many turns in place of a single wire. .

The essential requirements for a current to be produced are that there is a complete circuit and that the wire cuts magnetic flux. Here it has been suggested that the wires are kept stationary and that the magnet moves, but it is equally possible for the magnet to be stationary and for the wires to be moved. It is *relative motion* that is needed. By using larger, stronger magnets, and by using more turns of wire, it is possible to generate as much electrical power as is required. The large generators used by power-stations have fixed coils and the magnets are moved past them. The output of electrical power is huge, but the principle is exactly the same as the above demonstration with a wire, a magnet and a sensitive meter.

19.2 INDUCED ELECTROMOTIVE FORCE

An analogy: a mass in a gravitational field

Before analysing electromagnetic induction and contrasting it with the motor effect, it is worth while comparing a similar pair of mechanical processes, lifting and dropping. In lifting, work is done on the object and it gains potential energy: in dropping, potential energy is lost and kinetic energy is gained. However, consider an object either (a) being lifted on the end of a rope with a constant velocity or (b) being let down with a constant velocity. The force diagrams in **Figs 19.3** and **19.4** must be the same in both cases because, since there is no acceleration, the resultant force on the object must be zero.

The pull on the mass is provided by some external agent. There may be a rope attached to it, with a crane lifting it up or letting it down. Here the important point is that if its velocity is constant, whether it is upwards or downwards movement, the pull of the rope F is equal and opposite to Mg, the weight of the mass. (If you do not believe that this is possible – try it. Use something attached to the end of a piece of string and then lift it or lower it with a constant velocity. While getting it moving requires that the forces are different, keeping it moving with a constant velocity simply requires that the upward force equals the downward force.)

Since the two forces are equal and opposite, we can summarise the two situations as follows:

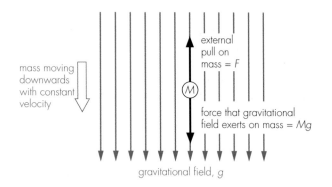

FIG 19.3 The forces on mass *M* when it is moving downwards with constant velocity.

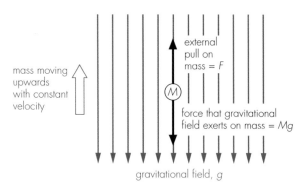

FIG 19.4 The forces on mass *M* when it is moving upwards with constant velocity. They are the same as when it is moving downwards with constant velocity.

■ *Downward motion of mass*

constant velocity	$= v$
time considered	$= t$
distance travelled	$= x$
external force	$= F$
force due to field	$= Mg$
work done by field force on mass	$= Mgx$
energy gained by external agent $= Fx$	$= Mgx$

■ *Upward motion of mass*

constant velocity	$= v$
time considered	$= t$
distance travelled	$= x$
external force	$= F$
force due to field	$= Mg$
work done by external agent $= Fx$	$= Mgx$
energy gained by mass	$= Mgx$

So, going up, the external agent does work *Mgx*, which results in the mass gaining potential energy of *Mgx*; and going down, the gravitational force on the mass does work *Mgx*, which results in a gain of energy *Mgx* by the external agent.

An electric current in a magnetic field

Now consider the corresponding situation when the field is a magnetic field and the mass is replaced by a wire carrying an electric current *I*, as in **Figs 19.5** and **19.6**. Apart from the forces acting at right-angles to the field, the situation is very similar to the one analysed above.

Again, at constant velocity, the two forces in each situation must be equal and opposite. This time the corresponding facts can be summarised as follows:

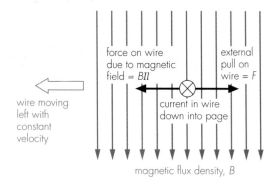

FIG 19.5 The forces on the current-carrying wire when it is moving to the left with constant velocity.

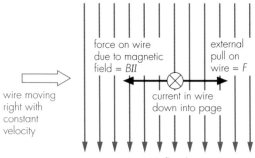

FIG 19.6 The forces on the current-carrying wire when it is moving to the right with constant velocity. They are the same as when it is moving to the left with constant velocity.

■ *Motion of wire to the left*

constant velocity	$= v$
time considered	$= t$
distance travelled	$= x$
external force	$= F$
force due to field	$= BIl$
work done by field force on wire	$= BIlx$
energy supplied to external agent $= Fx$	$= BIlx$

■ *Motion of wire to the right*

constant velocity	$= v$
time considered	$= t$
distance travelled	$= x$
external force	$= F$
force due to field	$= BIl$
work done by external agent $= Fx$	$= BIlx$
energy supplied to current in wire	$= BIlx$

Motion to the left is a motor effect. The wire pulls the external agent and hence uses electrical energy to do work on the wire as it moves along with constant velocity. Motion to the right is a dynamo effect, in which electrical energy is produced as a result of doing work on the wire. A dynamo is an electrical generator. As a result, energy is supplied to the current in the wire in the same way as, when a mass is lifted, energy is supplied to it.

Induced e.m.f.

The energy supplied to the current in the wire was shown to be equal to $Bllx$. The definition of electromotive force E is the energy changed from other forms of energy to electrical energy per unit charge Q. This gives

energy change to electrical energy = EQ = EIt

Here the energy changed to electrical energy is $Bllx$, so we get

$EIt = BIlx$

or

$$E = \frac{Blx}{t}$$

The electromotive force generated by electromagnetic induction when a wire cuts magnetic flux is an **induced electromotive force**.

If **Fig 19.6** is redrawn to show the wire moving from its initial position, a distance x to the right, it appears in plan view as in **Fig 19.7**. Connecting wires to the piece of wire considered are not shown. Here it can be seen that lx is the area swept out by the wire, so that Blx is the flux Φ that the wire cuts. This gives the induced electromotive force E as

$$E = -\frac{\Phi}{t}$$

This quantitative way of finding the induced electromotive force supplied by electromagnetic induction as the flux cut per unit time also shows that an equivalent unit to the weber (Wb) is the volt second (V s). The direction in which the induced current flows can be obtained from Fleming's right-hand dynamo rule. Fleming's left-hand motor rule applies to **Fig 19.5**, and so a mirror image is necessary for **Fig 19.6** as the velocity is in the opposite direction. If you use both hands together on the two diagrams in **Figs 19.5** and **19.6** (and a bit of contortion!), you can see how the motion is given by the left thumb in **Fig 19.5** and the right thumb in **Fig 19.6**.

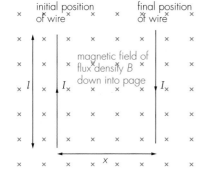

FIG 19.7 The wire moves distance x to the right.

19.3 LAWS OF ELECTROMAGNETIC INDUCTION

Faraday's law

We can make a formal statement of the deduction made in the previous section. It is called **Faraday's law** of electromagnetic induction. It states that:

K **The induced electromotive force across a conductor is equal to the rate at which magnetic flux is cut by the conductor.**

(In Faraday's original statement of the law, when the units used were not SI units, the equality was not established, and the statement was that the induced e.m.f. is proportional to the rate at which flux is cut.)

Lenz's law

The direction in which the e.m.f. is set up is important, as has been shown already. Any current that flows as a result of the e.m.f. must be in such a direction that it opposes the motion. This can be seen by looking at **Fig 19.6**. The current induced in the wire is down into the page. As a result of this current, the wire has a force acting on it because it is in the magnetic field. This force opposes the pull on the wire from the external agent. If the force assisted the external agent, then the wire would be accelerating; and it would be supplying electrical energy and gaining kinetic energy. This would be breaking the law of conservation of energy. If the force assisted motion, it would become possible to use a bicycle dynamo to run a motor on the bicycle, which would eliminate the need for pedalling! This impossibility was formally stated in **Lenz's law**, which is an electrical statement of the law of conservation of energy. It states that:

K **The direction of any induced current is such as to oppose the flux change that causes it.**

Put another way, this says that any electrical energy that can be supplied by a dynamo has to be obtained by doing at least an equivalent amount of mechanical work on the dynamo. To indicate this opposition of the direction of the induced current, the induced e.m.f. that causes it is given a negative sign in the equation that summarises both Faraday's law and Lenz's law. The equation is

$$E = -\frac{\Phi}{t}$$

or more accurately as a rate of change of flux using calculus notation

$$E = -\frac{d\Phi}{dt}$$

EXAMPLE 19.1

A coil of 1000 turns and enclosing an area of 4.0 cm² is rotated from a position where its plane is perpendicular to the Earth's magnetic field, of magnetic flux density 6.0×10^{-5} T, to one where it is parallel to the field (**Fig 19.8**). Calculate (a) the flux cut by the coil in making this rotation, and (b) the average induced e.m.f. when the time taken is 5.0 ms.

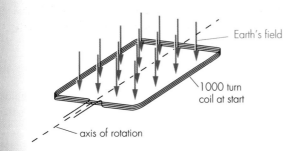

FIG 19.8 The coil for Example 19.1.

a We have

flux through one
turn of the coil = flux density × area
= 6.0×10^{-5} T × 0.0004 m²
= 2.4×10^{-8} Wb

so

flux through 1000 turns = 2.4×10^{-5} Wb

(This term is called the flux linkage: **flux linkage** is the product of the flux through a coil and the number of turns of wire in the coil.)

b Here

average flux cut
per second = $(2.4 \times 10^{-5}$ Wb$)/(0.0050$ s$)$
= 4.8×10^{-3} V

Since the flux cut per unit time is the induced e.m.f., we have

average induced e.m.f. = 4.8×10^{-3} V

19.1 In order to measure the movement of sea-water, an oceanographer immerses two electrodes into the water at a distance apart of 500 m. When she measures the e.m.f. between the two electrodes, she finds that it is 8.3×10^{-3} V. If the vertical component of the Earth's magnetic flux density is 3.7×10^{-5} T, find the velocity of the water in a direction at right-angles to the line joining the electrodes.

[Note: The conductor moving in this case is the sea-water. The problem can easily be solved if the area of sea-water that passes the line joining the electrodes per unit time is found, as the flux through this area is the flux cut per unit time. Note also that different values are often given for the flux density of the Earth's magnetic field. This is because it varies from place to place, and also because sometimes its horizontal component is needed and sometimes, as here, its vertical component is needed.]

19.2 An aeroplane has a distance between its wingtips of 36 m and it is travelling horizontally with a velocity of 280 m s^{-1}.

a If the vertical component of the Earth's magnetic flux density is 3.7×10^{-5} T, find the induced e.m.f. between the plane's wingtips.

b Why would a voltmeter in the plane connected between the wingtips always read zero?

19.4 GENERATORS

A coil mechanically rotated in a magnetic field

If a flat coil of N turns of area A is placed in a magnetic field of flux density B, then the flux through each turn is given by $BA \sin \theta$, where θ is the angle between the plane of the coil and the direction of the magnetic field (see **Fig 19.9**). The total flux cut by all the turns of the coil is therefore given by

$$\Phi = BAN \sin \theta$$

If now the coil is rotated with an angular velocity ω, the angle $\theta = \omega t$, and the flux cut by the coil will vary continually and so induce an e.m.f. across the ends of the coil. The induced e.m.f. can be found by differentiating

$$E = -\frac{\mathrm{d}\Phi}{\mathrm{d}t}$$

But recalling from the previous section that the induced e.m.f. is such that it will oppose the rotation causing it, we will drop the minus sign and just consider the *magnitude* of the induced e.m.f., which is therefore.

$$E = \frac{\mathrm{d}\Phi}{\mathrm{d}t} = \frac{\mathrm{d}(BAN \sin \theta)}{\mathrm{d}t}$$

$$= BAN \frac{\mathrm{d}(\sin \theta)}{\mathrm{d}t}$$

$$= BAN \frac{\mathrm{d}(\sin \omega t)}{\mathrm{d}t} = BAN \omega \cos \omega t$$

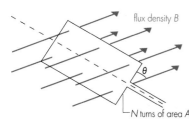
flux density B

θ

N turns of area A

FIG 19.9 A flat coil rotated in a magnetic field.

FIG 19.10 Sinusoidal variation of output e.m.f. from an a.c. generator.

It can be seen from this equation that the output e.m.f. is directly proportional to the flux density, the area of the coil, the number of turns on the coil and the angular velocity. It will also vary sinusoidally as shown in **Fig 19.10**, reaching a maximum value of

$$E_0 = BAN\omega$$

when $\theta = 0$. That is, the maximum output occurs when the plane of the coil is parallel to the magnetic field. This is where flux is cut fastest.

Output: a.c. or d.c.?

This theory forms the basis of all generators. Generators vary enormously in their construction depending on the requirements of their user and on their manufacturer. The output from a generator can be taken through slip rings or it may be taken through a split-ring commutator.

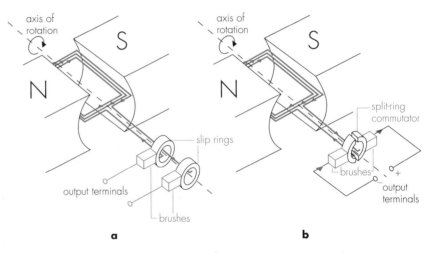

FIG 19.11 **a** An a.c. generator using slip rings. **b** A d.c. generator using a split-ring commutator.

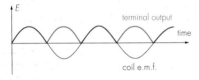

FIG 19.12 The variation of output e.m.f. from a d.c. generator.

Fig 19.11a shows a generator using slip rings. The leads from the ends of the coil are connected to two rings, insulated from one another. Against these rings are pressed carbon brushes through which the a.c. output is taken. A generator like this, which produces an a.c. output, is called an alternator.

Fig 19.11b shows the output being taken through a split-ring commutator, against which two carbon brushes are pressed. The two halves of the commutator rotate with the coil. So a carbon brush will be pressed against one half of the commutator for half a revolution, and against the other half of the commutator for the other half a revolution. In this way, one of the carbon brushes will always be positive and the other carbon brush will always be negative. The output is shown in **Fig 19.12**. It is not a smooth output, but it is d.c. in the sense that the current only ever flows in one direction. The output can be smoothed with smoothing capacitors.

Another way of producing direct current is to put the a.c. output of the alternator through a rectifier. This is standard now for supplying the d.c. power required for a car, although old cars used the system shown in **Fig 19.11a** for their dynamos. For the largest a.c. generators in power-stations, the coils remain fixed and the magnet, which is an electromagnet, rotates. This does away with the necessity for slip rings or commutators, except for the connection to the electromagnet. A photograph of 600 MW generators is shown in **Fig 19.13**.

FIG 19.13 A power station generator.

FIG 19.14 Starter motors for cars are being manufactured. In this photograph the rotating part of the motor, the armature, is being assembled.

A current-carrying coil in a magnetic field

It may seem odd to delay discussion of the electric motor until the end of a chapter on electromagnetic induction, rather than putting the explanation where it seems to fit, namely in a chapter dealing with the motor effect. The reason for doing this is that, when a motor is in operation, it is a current-carrying coil rotating in a magnetic field. One construction for a simple d.c. motor is shown in **Fig 19.11b**. When that particular diagram was discussed earlier, it was described as a d.c. generator; but if the generator, instead of being turned to produce an electric current, has a direct current passed through it, then it behaves as a motor.

The variety in the size, type and construction of electric motors is even greater than the variety of construction of generators, and there is no intention here to cover more than the basic principles of the motor. A simple d.c. motor is the starter motor on a car (**Fig 19.14**). It uses brushes to feed current into a series of coils. By using many coils, there is no possibility that the coils will not rotate in the required direction immediately the current is switched on. Domestic motors, such as those used on a washing machine or vacuum cleaner, need to work on a.c. mains. Their rotation is caused by clever manipulation of the magnetic field within the motor. They are called induction motors.

Balance between various torques

Because it comprises a current-carrying coil rotating in a magnetic field, a motor generates an induced e.m.f. that affects the current through it. The interaction of the input e.m.f. and the induced e.m.f. is fundamental to the operation of any motor. When a motor is switched on, a current passes through the armature and produces a torque on it, which causes rotation. At the same time an induced e.m.f. is produced, which, according to Lenz's law, must be in a direction such as to oppose motion. Also opposing the motion will be the system on which the motor is doing work.

The dynamic balance between these various torques controls the speed of the motor. In many practical situations an electric motor does not need a gearbox to control its speed. The speed is controlled automatically by the motor itself and its response to the system to which it is connected (**Fig 19.15**).

EXAMPLE 19.2

A shunt-wound motor is a motor in which the magnetic field is supplied by an electromagnet connected in parallel with the motor's armature. One such motor has an armature whose resistance is 4.0 Ω and field windings whose resistance is 120 Ω. It runs off-load at a rate of rotation of 2800 revolutions per minute, at which speed it takes a current of 5.0 A from 240 V mains. Find the speed at which the motor runs when the supply current is 20 A, and find how much useful power is then being supplied by the motor.

The circuit diagrams for the motor are shown in **Figs 19.15** and **19.16**. The resistance of the armature is within the armature, but on the circuit diagram it is shown outside and in series with the armature. This is to emphasise that the resistance of the armature causes a quite different effect from the effect caused by it being a coil rotating in a magnetic field. The diagram drawn this way also ought to prevent you from considering the 4 Ω and the 120 Ω resistors as being in parallel with one another. They are not in parallel because, as a result of the induced e.m.f., they do not have the same potential difference across them (see 'Resistors in parallel' in section 15.2).

■ *Off-load*

In this case (**Fig 19.15**) we have:

potential difference across 120 Ω field resistance		= 240 V
current through field resistance	= (240 V)/(120 Ω)	= 2.0 A
supply current		= 5 A
armature current	= 5.0 A − 3.0 A	= 3.0 A
potential difference across 4 Ω armature resistance	= 3.0 A × 4.0 Ω	= 12 V
induced e.m.f.	= 240 V − 12 V	= 228 V

This last line needs more comment. The armature is supplied from a supply of e.m.f. 240 V. That is, 240 J are supplied with every coulomb of electric charge that passes through the armature. We have established that, of these 240 J, 12 J are wasted as heat in the resistance of the armature. This leaves a further 228 J able to do work (against friction and air resistance in this off-load case). If we use our definition of potential difference as the energy

per unit charge converted from electrical energy into other forms of energy, then there is a potential difference of 12 V across the resistor, as 12 J C^{-1} are converted from electrical energy into heat energy, and a potential difference of 228 V across the armature, as 228 J C^{-1} are converted from electrical energy into mechanical energy. This is the induced e.m.f. It is directly proportional to the rate of rotation of the armature. The values for the powers are:

power wasted as heat in field windings	= 2 A × 240 V	= 480 W
power wasted as heat in armature	= 3 A × 12 V	= 36 W
mechanical power output doing work against friction	= 3 A × 228 V	= 684 W
total power input	= 5 A × 240 V	= 1200 W

■ *On-load*

The problem can be repeated when on-load (**Fig 19.16**). The figures have been inserted onto the circuit diagram. Circuit diagrams are intended to be used: they are useful working diagrams, and an extra diagram should be sketched out when changes are made. Since 18 A is now the current through the armature, the induced e.m.f. must be 168 V. The speed drops in the same proportion, so

$$\text{speed} = \frac{168}{228} \times 2800 = 2063 \text{ revolutions per minute}$$

The new values of the powers become:

power wasted as heat in field windings	= 2 A × 240 V	= 480 W
power wasted as heat in armature	= 18 A × 72 V	= 1296 W
mechanical power output doing work	= 18 A × 168 V	= 3024 W
total power input	= 20 A × 240 V	= 4800 W

This large motor has a mechanical power output of 3024 W. The rate at which work was done against friction when off-load was 684 W. It is unlikely that this has dropped very much at the lower speed, so the useful power output when on-load will be about 2400 W. The answer is not reliable to any more than two significant figures.

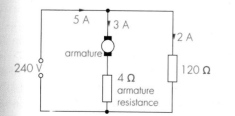

FIG 19.15 Circuit diagram for the motor in Example 19.2: off-load.

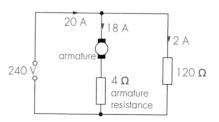

FIG 19.16 Circuit diagram for the motor in Example 19.2: on-load.

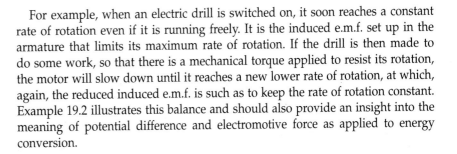

For example, when an electric drill is switched on, it soon reaches a constant rate of rotation even if it is running freely. It is the induced e.m.f. set up in the armature that limits its maximum rate of rotation. If the drill is then made to do some work, so that there is a mechanical torque applied to resist its rotation, the motor will slow down until it reaches a new lower rate of rotation, at which, again, the reduced induced e.m.f. is such as to keep the rate of rotation constant. Example 19.2 illustrates this balance and should also provide an insight into the meaning of potential difference and electromotive force as applied to energy conversion.

Back e.m.f.

The term **back e.m.f.** is often used instead of 'induced e.m.f.' when dealing with motors, but this is a confusing term, for two reasons:

- Back e.m.f. seems to imply opposition, something not wanted, yet in a motor it is the induced e.m.f. that is responsible for the mechanical work output of the motor.
- Back e.m.f. is a potential difference. This is indeed the case here. Energy is being converted from electrical energy into other forms of energy, so perhaps the quantity should be placed on the other side of any energy equation and given a positive sign.

One thing is certain: you will need to be particularly careful with signs in such problems, so you are recommended to check at the end that energy supplied equals energy used.

A term used in the USA for induced e.m.f. is motional e.m.f. It has the advantage that it clearly indicates how the e.m.f. is produced, and it would be recognised if you wished to use it.

QUESTIONS

19.3 A motor in a toy runs off a 6 V battery of negligible internal resistance, and uses a permanent magnet to provide the magnetic field. The armature has a resistance of 12 Ω and the motor uses a current of 0.087 A when it is off-load and is rotating at 2400 revolutions per minute. Find:

    a the current through the motor when it is jammed and the power that is then wasted as heat;

    b the power used to overcome friction and the power wasted as heat when it is off-load;

    c its speed when it is on-load and drawing a current of 0.18 A.

19.4 A 12 V d.c. motor uses a permanent magnet to provide its magnetic field. It has an armature with a resistance of 2 Ω, and when it is off-load it takes a current of 0.50 A and turns at 3000 revolutions per minute. Plot a sketch graph to show how the speed of the motor on-load varies with the current.

Carry out question 19.4 experimentally. That is, plot a graph of rate of rotation against current for a motor. A low-voltage d.c. motor is required, together with a suitable power supply. A small, low-power motor as used in toys is suitable.

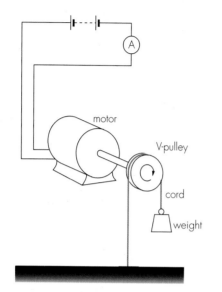

FIG 19.17 An experimental arrangement for finding the power output of a motor.

There is no problem in measuring the current to the motor with a d.c. meter, but it is quite tricky to keep the speed constant when the motor is on-load. You will need to clamp the motor securely and, depending on its physical size and power, fix a brake to it. This can be done in the way shown in **Fig 19.17** with a cord under tension rubbing on a V-belt pulley wheel. Adjusting the weight changes the load on the motor. (This is the way many engines are tested for output power. The power output is often called the 'brake horse-power': it is the power output when under these test conditions.) A rev. counter or a stroboscopic lamp has to be used to measure the speed of the motor.

Be careful of the following, particularly if your motor is quite powerful:

■ The brake getting too hot. Industrial brakes can be very large and are often water-cooled.

■ The motor getting too hot as it is made to do more work. Keep it running only long enough to take a reading.

■ The ammeter having too much current through it. If the motor gets jammed for any reason, not only might the motor burn out, but also the ammeter may be damaged.

SUMMARY

■ Faraday's law of electromagnetic induction states that the e.m.f. produced when a wire cuts a magnetic field is equal to the rate of cutting of magnetic flux.

■ Lenz's law states that any induced current is always in a direction to oppose any flux change that causes it.

■ Faraday's law and Lenz's law can be combined in an equation:

$$E = -\frac{d\Phi}{dt}$$

■ The work done against the induced e.m.f. in a motor is the useful work output of the motor. Work done against the resistance of the wires is wasted as thermal energy.

■ If the e.m.f. generated by a generator causes a current, then work must be done on the generator. Some of this work is changed into electrical energy, its useful output, and the rest is changed into thermal energy in the resistance of the wires of the generator.

■ Output e.m.f. of a generator is

$$E = BAN\omega \cos \omega t$$

ASSESSMENT QUESTIONS

19.5 **a** What is meant by the term *electromagnetic induction?*

b Describe an experiment you could perform in a school laboratory to demonstrate Faraday's law of electromagnetic induction.

c An aircraft has a wing span of 54 m. It is flying horizontally at 860 km h^{-1} in a region where the vertical component of the Earth's magnetic field is 6.0×10^{-5} T. Calculate the potential difference induced between one wing tip and the other.

d What extra information is necessary to establish which wing is positive and which negative?

London 1998

19.6 **a** State Faraday's law of electromagnetic induction.

b The figure shows two adjacent, identical, air-cored coils A and B, arranged coaxially. A is connected to a centre-zero galvanometer G (a sensitive analogue current detector) and a switch S1, and B to a battery and a switch S2. There is no electrical connection between the coils.

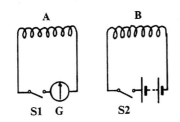

i The experiments listed in the table were performed independently. Possible readings on the galvanometer are suggested. Which result would be observed for each experiment?

Experiment	Details	Galvanometer reading
1	S1 was closed first; S2 was then repeatedly opened and closed	zero steady non-zero oscillatory
2	S2 was closed first; S1 was then closed	zero steady non-zero oscillatory
3	S2 was closed first; S1 was then repeatedly opened and closed	zero steady non-zero oscillatory

ii The battery in the circuit containing coil B is replaced with an a.c. supply of variable frequency. Switches S1 and S2 are both closed. Describe and explain the response of the galvanometer needle for supply frequencies of
1 2 Hz,
2 2 kHz.

CCEA 1998

19.7 **a** Explain why you would expect an electromotive force to be developed between the wing tips of an aircraft flying above the Earth.

b An aircraft, whose wing tips are 20 m apart, flies horizontally at a speed of 600 km h^{-1} in a region where the vertical component of the Earth's magnetic flux density is 4.0×10^{-5} T.
Calculate
i the area swept out by the aircraft's wings in 1 s,
ii the e.m.f. between the wing tips.

NEAB 1997

19.8 An induction microphone converts sound waves into electrical signals which can be amplified.

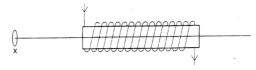

a Describe the stages by which the sound waves are converted into electrical signals. State whether the signals are a.c. or d.c.

b If the alternating output from a signal generator were fed into the microphone, describe and explain what would happen to the diaphragm.

London 1996

19.9 **a** The figure shows a long solenoid with the current in it in the direction shown by the arrows. Sketch the magnetic flux pattern in and around the solenoid.

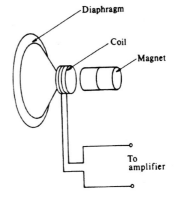

b A small coil is placed at X and is then moved along the centre line of the solenoid to Y. Describe how the magnetic flux linkage through the coil varies as it is moved from X to Y.

c It is found that there is an e.m.f. induced in the coil when it is moved steadily from X to Y. Explain why this e.m.f. occurs and describe how the magnitude and direction of the e.m.f. change during the movement of the coil.

Cambridge 1997

19.10 a State Lenz's law of electromagnetic induction

b A small coil Y is placed inside a long solenoid X. The coils are coaxial, as shown *in section* below.

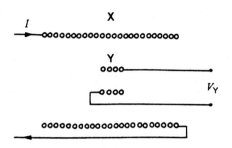

By changing the current I, the magnetic flux density B_X within the solenoid X is varied with time t as shown below. This produces an induced e.m.f. V_Y in the coil Y.

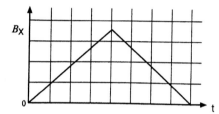

i Sketch the variation of V_Y with t.

ii During the change of magnetic flux density in solenoid X, the maximum induced e.m.f. in coil Y is 5.0 V. The maximum magnetic flux through each turn of coil Y is 0.75 Wb. Coil Y has 20 turns. Calculate the time required for the magnetic flux density in solenoid X to increase from zero to its maximum value.

CCEA 1998

19.11 The coil of a simple a.c. generator rotates in a uniform magnetic field. Part of a diagram of this generator is shown below.

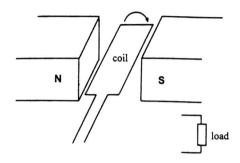

a Show the arrangement by which the e.m.f. generated in the coil is connected to the load.

b i Sketch a graph on appropriate axes to show how the output e.m.f. E from this generator varies with time t when the coil is rotated at a constant rate.

ii Sketch a graph on appropriate axes to show how the peak (maximum) output voltage E_o varies with the angular velocity ω of the coil.

c The coil of the generator is rotated at a constant rate. The output has a peak e.m.f. of 24 V at a frequency of 180 Hz. The speed of rotation of the coil is then reduced to one-third of the initial speed. What will now be the peak e.m.f. and frequency of the output?

CCEA 1998

20 ALTERNATING CURRENT (a.c.) THEORY

LEARNING OBJECTIVES

At the end of this chapter you should be able to:

① describe sinusoidally varying currents and voltages in terms of their amplitude, frequency and phase;

② use a phasor diagram to analyse an a.c. circuit;

③ define and use the terms 'reactance' and 'impedance';

④ calculate the power supplied to an a.c. circuit;

⑤ define and use the terms 'root mean square current' and 'root mean square potential difference'.

20.1 CIRCUIT LAWS

The abbreviation a.c. is short for *alternating current*, but it is often used in a descriptive way just to indicate that the electrical quantities being considered are oscillating. For this reason, the term *a.c. current* is often heard in laboratory discussions despite the fact that strictly this term repeats the word 'current'. Similarly, *a.c. voltage* is heard. This is even more of a nonsense, as you cannot have a current voltage. These abbreviated terms will not be used here, but they do highlight the fact that currents and potential differences do not always have constant values.

Sinusoidal variation

An alternating current is one in which the charge carriers, usually the electrons, oscillate about a fixed point. That is, there is zero drift velocity. The purest oscillation is one of simple harmonic motion in which there is a sinusoidal variation in the current. This oscillation can be caused by an alternator in which the coil rotates in a uniform magnetic field (see section 19.4). The way in which an oscillation takes place may be very complex, as for the speech pattern in **Fig 20.1**.

FIG 20.1 An oscilloscope screen on which is displayed a speech pattern.

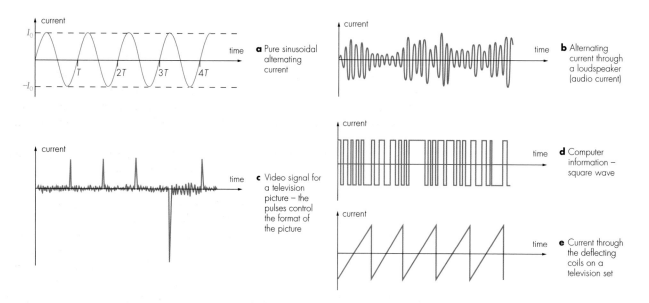

FIG 20.2 Some commonly occurring a.c. signals. Only the first one is a pure sine wave oscillation.

Some examples of alternating currents and their practical uses are shown in **Fig 20.2**. These may be added to direct currents to make matters even more complex. To start with, we will consider the simplest, sinusoidal alternating current and progress to the more complicated situations later. Once the simple situation has been thoroughly understood, there will be no problem with the more complicated one.

A pure alternating current I is shown in **Fig 20.2a**, where the value of the current at any time t is plotted. It has the equation

K $\quad I = I_0 \sin(2\pi f t)$

where I_0 is the peak value of the current and f is the frequency of the oscillation. A term called the *angular frequency* ω (see Chapters 10 and 11) is often used, and is defined by the expression

$\omega = 2\pi f$

As with simple harmonic motion and waves, the frequency is related to the period T by the expression

$f = 1/T$

Note that the angle $2\pi f t$ has the unit radian. If you are unfamiliar with sine waves, you are advised to plot a few to see the effect the different terms have. These are suggested in question 20.1 that follows.

QUESTION

20.1 Tabulate values for the following graphs and plot them on graph paper. You will need to choose small values of t if you are not to get $I = 0$ too frequently, and your calculator must be set for angles in radians.

a $I = I_0 \sin(2\pi f t)$, where $I_0 = 2.0$ A and $f = 3$ Hz

b $I = I_0 \sin(2\pi f t)$, where $I_0 = 2.0$ A and $f = 12$ Hz

c $I = I_0 \sin(2\pi f t)$, where $I_0 = 6.0$ mA and $f = 12\,000$ Hz

d $I = I_0 \cos(2\pi f t)$, where $I_0 = 6.0$ mA and $f = 12\,000$ Hz

Phase difference

Question 20.1(d) asks you to plot current against time when the equation is a cosine variation rather than a sine variation. You should know or have seen that the shape of this curve is the same as the shape of the sine wave (question 20.1(c)) but that it has been displaced by a quarter of a cycle. This is called a **phase difference** of a quarter of a cycle. There is a phase difference of $\pi/2$ between the sine and the cosine curves.

Often, when two pure alternating currents of the same frequency are compared, it is found that they are out of phase with one another. This is illustrated by the graph in **Fig 20.3**, where the two currents have the same peak value I_0. Current I_2 reaches its maximum at a later time than does I_1, and I_2 is said to **lag** behind I_1; I_1 is said to **lead** I_2. Be careful here. It might look as though I_2 is in front of I_1, but I_2 reaches its maximum later than I_1 and it is definitely I_2 that lags behind I_1. Since these two currents have the same frequency, the phase difference between them is constant, and its value may be stated – in radians. If the two currents do not have the same frequency, then the phase difference between them is continually varying, so there is little point in stating it. The value of the phase difference as an angle in radians can be found by realising that one cycle corresponds to 2π radians, so in this case, where I_2 lags behind I_1 by a quarter of a cycle, I_2 has a phase difference of $\pi/2$ with I_1.

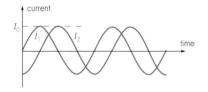

FIG 20.3 Two alternating currents of the same frequency and amplitude but out of phase with one another. I_2 lags behind I_1.

20.2 PHASOR DIAGRAMS

Phasor diagrams for alternating currents

Drawing sine waves is not easy and, in any case, if several of them have to be compared with one another, the diagrams become hopelessly confused. A phasor diagram is a way of representing a pure oscillation with a single straight line. The **phasor** is a rotating vector. In **Fig 20.4** the phasor is shown on the left-hand side of the graph and is assumed to be rotating in an anticlockwise direction, so that it completes one revolution in the period T of the cycle of the current.

The phasor is shown after it has completed an eighth of a revolution. As it rotates, the height of the tip of the phasor marks the value of the current at that instant and hence the height of the sine wave graph. Apart from the ease of drawing phasors compared with drawing the sine wave, they have the advantage that they can be treated as vectors when the addition of alternating currents is necessary.

Consider the part of a circuit shown in **Fig 20.5** in which two alternating currents I_1 and I_2 combine at a junction to produce a total current I_3. Kirchhoff's first law can be applied to this situation, so that at any instant $I_3 = I_1 + I_2$. However, if I_1 and I_2 are not in phase with one another, some surprising results can occur. For instance, if I_1 and I_2 are in **antiphase** (that is, exactly out of phase with one another or out of phase by a phase angle π), then I_3 will be zero because the two currents will cancel one another out. This is shown in **Fig 20.6**.

Also on **Fig 20.6** are shown the phasors that represent each of the currents. If the phasors are added together as if they were vectors, then they give a resultant of zero. This procedure can be carried out with as many different currents as are required, provided that all those plotted on one phasor diagram have the same frequency. The example that follows illustrates this, together with questions 20.2 and 20.3.

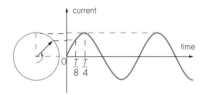

FIG 20.4 A phasor shown (left) with the current variation it represents.

FIG 20.5 Kirchhoff's first law applies to alternating currents.

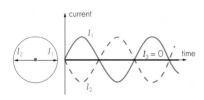

FIG 20.6 Currents in antiphase give a total current of zero, as shown by the phasor diagram (left) or the graph.

EXAMPLE 20.1

An alternating current I_1, of peak value 1.6 A, meets at a junction with an alternating current I_2, of the same frequency and peak value 2.2 A. The two currents are out of phase by a quarter of a cycle. I_2 leads I_1. Find the peak value of the total current I_3 and find its phase relationship to I_1.

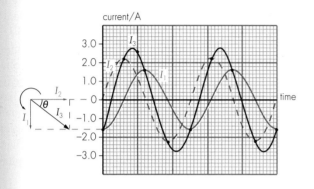

The first thing to notice about this problem is that the peak value of current I_3 is *not* 1.6 + 2.2 = 3.8 A. As an example, the full diagram of the sine wave variation of the currents will be drawn (**Fig 20.7**). Once you are certain you understand what the phasor diagram is doing, the full sine wave diagrams can be omitted. On the wave diagram, the addition of I_1 and I_2 is done here by adding the two currents together point by point at many different times. The peak value of the resultant current is seen to be 2.7 A. This can be obtained from the phasor diagram, using Pythagoras's theorem:

$$I_3{}^2 = I_1{}^2 + I_2{}^2 = 1.6^2 + 2.2^2$$

$$I_3 = \sqrt{2.56 + 4.84} = 2.7 \text{ A}$$

FIG 20.7 The phasor diagram (left) for Example 20.1 is much easier to draw and to work from than the sine wave graphs it represents.

QUESTIONS

20.2 Two pure sine wave alternating currents of the same frequency have the same peak values of 2.8 mA.

  a Use a phasor diagram to find their sum if they are out of phase by an angle of $\frac{1}{2}\pi$.

  b What is the phase relationship of each of the currents to their sum?

20.3 The mains supply uses three phases called the red phase, the blue phase and the yellow phase. These, ideally, are equal currents out of phase with each other by a third of a cycle.

  a Draw a phasor diagram to represent these currents and find their sum.

  b Use your answer to explain why, on power transmission lines (like those shown in **Fig 3.14**), the current for customers is carried along thick cables but the neutral cable at the very top of the pylon can be thin.

Phasor diagrams for alternating potential differences

As stated earlier, a pure alternating current can be produced by an alternator in which the coil rotates in a uniform magnetic field. The alternating current in a resistor connected to the alternator is being caused by an output e.m.f. from the alternator, which oscillates in the same way as the current. Alternating potential

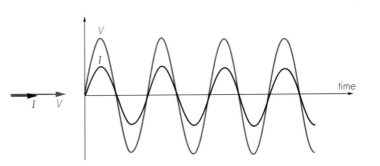

FIG 20.8 How alternating current and potential difference vary with time across a resistor.

differences can be shown on phasor diagrams in the same way as alternating currents. **Fig 20.8** shows how the current I and the potential difference V across a resistance R vary with time. When plotting such variations, the time scale must be the *same* for both quantities; but of course the vertical scales will be different, since for one graph the y-axis will be calibrated in volts and for the other in amperes.

For a resistance R, the current is in phase with the potential difference, as Ohm's law applies. If the potential difference applied is given by

$$V = V_0 \sin(2\pi ft)$$

then the current I at any instant is

$$I = \frac{V}{R} = \frac{V_0 \sin(2\pi ft)}{R} = I_0 \sin(2\pi ft)$$

POWER IN A RESISTIVE CIRCUIT

Power in circuits with alternating current

When an alternating current exists in a resistance, there is a heating effect as there would be when a direct current flows. In the case of alternating current, however, the current is sometimes in one direction and sometimes in the other. The direction makes no difference to the heating effect. If the frequency of the alternating current is f, the power has a frequency $2f$ because maximum current occurs twice per cycle.

Since the frequency of the power is not the same as that of the current or potential difference, power cannot be shown on a phasor diagram. The graph drawn in **Fig 20.9** shows how the power varies with time. Every point on the power curve is obtained by multiplying the current at that moment by the potential difference. Note that when the potential difference is negative, the current is also negative, and so the power is positive. Power cannot be supplied from a resistor back to the supply, so the power cannot be negative.

The relative heights of the three curves are not significant, since these will depend on the scales used on the axes. The shape of the power curve *is* important though. It is of twice the frequency, and the top half is the same shape as the bottom half, so the two areas shaded are equal. The average power, P, is therefore half of the peak power, P_0. We can therefore write for any instant

$$P = VI = V_0 \sin(2\pi ft) \times I_0 \sin(2\pi ft)$$
$$= V_0 I_0 [\sin(2\pi ft)]^2$$

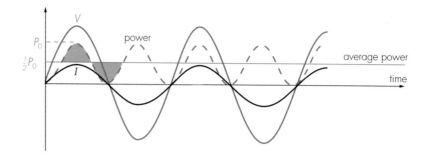

FIG 20.9 The variation of potential difference, current and power with time when the circuit component in use is a resistor.

and the maximum value of the power is given by

$$P_0 = V_0 I_0$$

since the maximum value of the sine of any angle is 1. So

$$\text{average power} = \tfrac{1}{2}P_0 = \tfrac{1}{2}V_0 I_0$$

Since $V_0 = I_0 R$ this gives

$$\text{average power} = \tfrac{1}{2}I_0^2 R$$

Root mean square values

The $\frac{1}{2}$ in this expression for average power is awkward. In electrical work with direct current, the power is given by $P = I^2 R$, and there would be many complications in having one rule for power with alternating current and a different rule with direct current. So, to overcome the problem when working with alternating current, it is not usual to use peak currents. Instead, a current called the **root mean square** (r.m.s.) current is used. This is written $I_{\text{r.m.s.}}$ and it is defined as the *direct current* that produces the same heating effect as the alternating current. This is shown in **Fig 20.10**.

Now we have

$$\text{average power supplied by alternating current} = \tfrac{1}{2}I_0^2 R$$

and

$$\text{power supplied by direct r.m.s. current} = (I_{\text{r.m.s.}})^2 R$$

If these are to be the same then

$$\left(I_{\text{r.m.s.}}\right)^2 R = \tfrac{1}{2}I_0^2 R$$

$$\left(I_{\text{r.m.s.}}\right)^2 = \frac{I_0^2}{2}$$

$$I_{\text{r.m.s.}} = \frac{I_0}{\sqrt{2}} = \frac{I_0}{1.414} = 0.707\, I_0$$

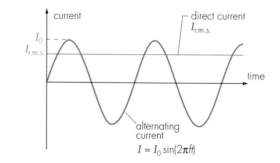

FIG 20.10 The root mean square current.

Since R is constant,

$$V_{\text{r.m.s.}} = 0.707V_0$$

where $V_{\text{r.m.s.}}$ is the direct potential difference that would cause the same heating effect as the alternating potential difference:

$$V = V_0 \sin(2\pi ft)$$

When a mains voltage is said to be 230 V, the value being quoted is the r.m.s. value. The peak value of a 230 V mains is given by

$$V_0 = \sqrt{2} \times V_{\text{r.m.s.}}$$
$$= 1.414 \times 230 \text{ V}$$
$$= 325.3 \text{ V}$$

Its minimum value is -325.3 V, so it has an overall maximum change in value of 650.6 V. This value can be important in some instances. A diode in a mains circuit can be subjected to this peak inverse voltage. That is, it can have the full 650.6 V across it in the opposite direction to that in which it will conduct, and it is important that it is not damaged by it. The peak inverse voltage that can be applied across a diode is normally quoted by the manufacturer.

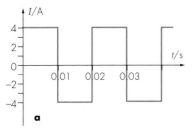

a

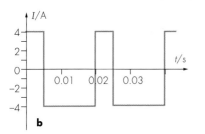

b

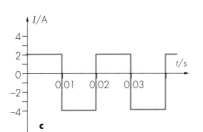

c

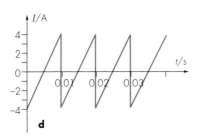

d

FIG 20.11 Alternating currents for question 20.5.

QUESTIONS

20.4 An electric kettle is marked 240 V, 1440 W. Although not stated, these values may be taken to be r.m.s. values. Find:

a the r.m.s. current;

b the peak current;

c the peak potential difference;

d the peak power;

e the average power.

20.5 The r.m.s. current is the direct current with the same heating effect as the alternating current used. For a sinusoidal a.c., the peak current is $1.41 \times$ r.m.s. current, but the value 1.41 only applies for sinusoidal a.c. Consider each of the currents shown in **Fig 20.11**, and calculate its r.m.s. value. [You should be able to answer **a** and **b** easily, but **c** and **d** are more difficult, so follow the guidance given.]

a What is the r.m.s. value for this current?

b What is the r.m.s. value for this current?

c Find the average power in a resistor R for the negative part of the cycle, and the average power for the positive part of the cycle. Use these to find the average power for the whole cycle. Now, what is the r.m.s. value for this current? [The answer is *not* 3.0 A.]

d This calculation requires the use of calculus. But if the question is set as a multiple choice question with answers, you should be able to eliminate all but one of the answers as being impossible. Explain how to eliminate all but 2.3 A, which is the correct answer, from the following answer list:

A 4.0 A   **B** 2.82 A   **C** 2.3 A   **D** 2.0 A

20.4 REACTANCE

Capacitors and alternating current

I f a capacitor of capacitance C has a potential difference V across it, then using the definition of capacitance, the charge Q on the capacitor is given by

$$Q = CV$$

This will apply even when the potential difference is changing, so if the potential difference is alternating and has the value $V = V_0 \sin(2\pi ft)$, the charge on the capacitor at time t will be

$$Q = CV_0 \sin(2\pi ft)$$

In graphical form the charge will therefore vary in the way shown in **Fig 20.12**. The charge on the capacitor increases as long as the current is positive. At the peak of the graph, A, there is no increase in charge, so the current flowing to or from the capacitor is zero. At B also the current will be zero. On the other hand, at D, the charge on the capacitor is changing rapidly. This can only occur if there is a flow of charge: that is, a current in the circuit. The current that is necessary to alter the charge stored in the capacitor is given by the rate of change of charge on the capacitor. This current is the gradient of the charge–time graph, or in calculus terms is given by

$$I = \frac{dQ}{dt}$$

The graph of current–time is shown in **Fig 20.13**, where numerical values have been supplied for both the charge and the corresponding current. Although the maximum charge is only 220 μC, the maximum current is found from the gradient to be 0.28 A. The calculation of the current using calculus gives

$$Q = CV_0 \sin(2\pi ft)$$

$$\frac{dQ}{dt} = I = CV_0 \, 2\pi f \, \cos(2\pi ft)$$

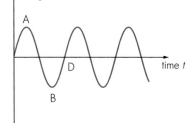

FIG 20.12 Charge–time graph for a capacitor.

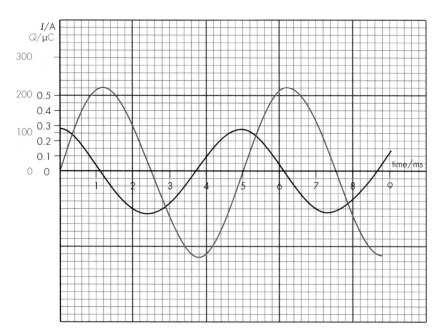

FIG 20.13 Graphs showing how the charge on, and the current to, a capacitor vary with time.

As can be seen from the graph, the current is a cosine curve if the charge, and hence the potential difference, is a sine curve. The current is leading the potential difference by a quarter of a cycle, $\pi/2$ rad.

The maximum value of the current (I_0) must be $CV_0 2\pi f$ since $\cos(2\pi ft)$ has a maximum possible value of 1.

The reactance of the capacitor

As with resistors, it is important to know what the alternating current will be for a given alternating potential difference, so the ratio

$$\frac{\text{maximum potential difference}}{\text{maximum current}}$$

is called the **reactance** of the capacitor, and is given the symbol X_C.

Reactance is measured in ohms because it is a potential difference in volts divided by a current in amperes, and for a capacitor has the value

$$X_C = \frac{V_0}{I_0} = \frac{V_0}{CV_0 2\pi f}$$

$$X_C = \frac{1}{2\pi f C}$$

Since

$$\frac{V_0}{I_0} = \frac{1.41\, V_{\text{r.m.s.}}}{1.41\, I_{\text{r.m.s.}}} = \frac{V_{\text{r.m.s.}}}{I_{\text{r.m.s.}}}$$

we also get the ratio

$$X_C = \frac{\text{r.m.s. potential difference}}{\text{r.m.s. current}} = \frac{1}{2\pi f C}$$

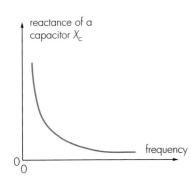

FIG 20.14 The variation for the reactance of a capacitor with frequency.

Note that the reactance of a capacitor depends not only on its capacitance but also on the frequency being used. The reactance cannot therefore be marked on a capacitor. **Fig 20.14** shows the variation of reactance with frequency. The higher the frequency, the lower the reactance.

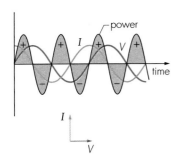

FIG 20.15 The variation of potential difference, current and power with time when the circuit component is a capacitor.

The phasor diagram for a capacitor is shown in **Fig 20.15**. The current leads the potential difference by $\pi/2$ rad. The full graph for this phasor diagram is also shown. It also shows the interesting feature that, if the potential difference and current graphs are multiplied together at each instant to find the instantaneous power, the power is sometimes negative. If the current and the potential difference are in the same direction, then the capacitor is being charged up and extra energy is being stored. There are times, however, when the potential difference and the current exist in opposite directions. Here energy is being supplied from the capacitor back to the a.c. source. These two possibilities give the power graph as shown with alternating positive and negative loops of equal size. The power supplied to a capacitor over a long period of time is therefore zero; and energy supplied to it in one quarter of a cycle is given back to the source during the next quarter of a cycle.

EXAMPLE 20.2

An r.m.s. alternating current of 5.7 mA passes to a capacitor when the applied r.m.s. alternating potential difference is 13.2 V. Find the reactance of the capacitor and the value of capacitance if the frequency of the supply is 250 Hz.

$$\text{reactance} = \frac{\text{peak potential difference}}{\text{peak current}}$$

$$= \frac{V_{\text{r.m.s.}}}{I_{\text{r.m.s.}}} = \frac{13.2 \text{ V}}{5.7 \text{ mA}}$$

$$= 2316 \ \Omega = 2300 \ \Omega \text{ to 2 sig. figs}$$

But

$$\text{reactance} = \frac{1}{2\pi f C}$$

so

$$2316 \ \Omega = \frac{1}{2\pi \times 250 C}$$

$$C = \frac{1}{2\pi \times 250 \times 2316} = 2.75 \times 10^{-7} \text{ F}$$

$$C = 0.275 \ \mu\text{F}$$

QUESTIONS

20.6 Use **Fig 20.13** to find the current from the graph of charge against time, and check your values on the current–time graph. Make the calculations on tangents to the graph at points when $t = 0$, 0.2 ms, 2.5 ms and 4 ms.

20.7 **a** Plot a graph to show how the reactance of a 1.0 μF capacitor varies as the frequency changes from 100 Hz to 3000 Hz.

b Explain why the reactance is smaller with higher frequency.

20.8 A capacitor in a fluorescent lamp is needed to pass an r.m.s. current of 0.50 A when the r.m.s. potential difference across it is 240 V and the frequency of the supply is 50 Hz. Find the reactance of the capacitor and its capacitance.

20.9 A 0.47 μF capacitor has an r.m.s. potential difference of 20.0 V across it at a frequency of 4000 Hz.

a What is its reactance at this frequency?

b What will be the current through it?

20.5 CAPACITOR–RESISTOR CIRCUITS

The phasor diagram

A capacitor is often in a circuit in series or in parallel with a resistor. When a circuit contains both resistive and reactive parts, a phasor diagram is needed to deduce the current in each component. Consider the series circuit shown in **Fig 20.16**.

In this circuit the capacitor of capacitance C and the resistor of resistance R are in series with the a.c. supply, so the current from the supply and through C and R must be the same at all instants. This step is important. It tells us where to start drawing the phasor diagram, which is shown in the stages of its construction in **Fig 20.17**.

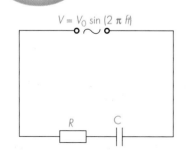

FIG 20.16 A simple capacitor–resistor series circuit.

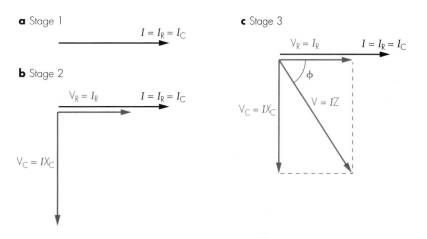

FIG 20.17 Constructing the phasor diagram for a simple capacitor–resistor series circuit.

- *Stage 1* The supply current I, the current through the capacitor I_C and the current through the resistor I_R are identical at all times.
- *Stage 2* This now enables the potential differences across the resistor and the capacitor to be determined. $V_R = IR$ and V_R is in phase with the current through the resistor. The current and potential difference are always in phase with one another for a resistive component. For any capacitor the current leads the potential difference by a right-angle and so the potential difference phasor is behind the current phasor and has a value of IX_C.
- *Stage 3* The total potential difference V is now found by adding V_R and V_C to one another vectorially. The value of the total potential difference can be calculated using Pythagoras's theorem.

The phasor diagram gives

$$V^2 = V_R{}^2 + V_C{}^2 = (IR)^2 + (IX_C)^2 = I^2(R^2 + X_C{}^2)$$

Because this circuit contains both resistive and reactive components, the ratio V/I is not a pure resistance or a pure reactance. Another term is used to indicate this. The ratio is called the **impedance (Z)** and it too is measured in ohms because it is a potential difference in volts divided by a current in amperes.

We have therefore

$$Z = \frac{V_{\text{max}}}{I_{\text{max}}} = \frac{V_0}{I_0} = \frac{V_{\text{r.m.s.}}}{I_{\text{r.m.s.}}}$$

$$= \sqrt{\left[R^2 + \left(\frac{1}{2\pi f C} \right)^2 \right]}$$

which can also be written

$$Z = \sqrt{(R^2 + X_C{}^2)}$$

Phasors: peak or r.m.s. values?

Strictly, phasor diagrams show *peak* values of current and potential difference. In practice, this means that the diagram has all its vectors 1.41 times longer than the r.m.s. values; a diagram that used r.m.s. values throughout would be of the same shape.

No difficulty is therefore likely to arise provided that you are consistent and either always use r.m.s. values or always use peak values.

EXAMPLE 20.3

Find the current through each component of **Fig 20.18** and the power to the resistor.

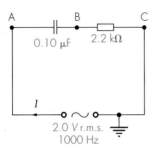

FIG 20.18 Circuit diagram for Example 20.3.

$$\text{reactance of capacitor} = \frac{1}{2\pi f C} = \frac{1}{2\pi \times 1000 \text{ Hz} \times 0.10 \times 10^{-6} \text{ F}}$$

$$= 1590 \text{ C}$$

The phasor diagram is shown in **Fig 20.19**. Make a habit of using a phasor diagram to record useful information as you go along. It is not a diagram to be drawn and then ignored. There is one point to be aware of in their use (see box on previous page). In this example, peak values will be used so that you can see how the 1.41 term cancels out.

From the phasor diagram we get

$$(1590I)^2 + (2200I)^2 = (2 \times 1.41)^2$$

where I is the peak current. Solving for I gives

$$(2.53 \times 10^6)I^2 + (4.84 \times 10^6)I^2 = 8.0$$

$$(7.37 \times 10^6)I^2 = 8.0$$

$$I^2 = \frac{8.0}{7.37 \times 10^6}$$

$$I = 1.04 \times 10^{-3} \text{ A}$$

The r.m.s. value of I is therefore

$$\frac{I}{\sqrt{2}} = \frac{1.04 \times 10^{-3} \text{ A}}{\sqrt{2}} = 0.74 \text{ mA}$$

Note an unexpected feature of a.c. circuits. If a good a.c. voltmeter (one with a high resistance) is placed across the supply terminals, it should read 2.0 V r.m.s. If it is placed across the capacitor, it will read

$$IX_C = \frac{0.74 \text{ A}}{1000} \times 1590 \text{ } \Omega = 1.18 \text{ V}$$

If it is then transferred to measure the p.d. across the resistor, it will record

$$IR = \frac{0.74 \text{ A}}{1000} \times 2200 \text{ } \Omega = 1.63 \text{ V}$$

At first sight it seems as if 1.18 V + 1.63 V = 2.0 V! But you need to remember that V_C and V_R are not in phase, so there is no reason why they should add arithmetically. If you check:

$$1.18^2 + 1.63^2 = 2.01^2 \quad \text{to 2 sig. figs.}$$

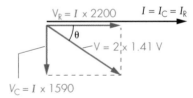

FIG 20.19 Phasor diagram for Example 20.3.

QUESTIONS

20.10 A 0.022 μF capacitor and a 10 kΩ resistor are connected in series with a power supply of frequency 1 kHz and an r.m.s. potential difference of 6.0 V.

a What is the current through the circuit?

b What is the potential difference across the resistor and the capacitor?

c Without making any further detailed calculations, find the approximate current through the circuit if the capacitor is replaced by a 2.2 μF capacitor.

20.11 A 120 V, 60 W bulb is used on 240 V, 50 Hz mains, A capacitor is put in series with the bulb in order to limit the current to the value needed for normal bulb operation. What capacitance is required for the capacitor?

20.6 THE TRANSFORMER

Transmission of electrical power

Alternating current is used for the transmission of electrical power throughout the world for the following two reasons:

First, compared with d.c., it is much easier to switch a.c. on and off. The reason for this is that a.c. is zero many times per second as a result of the variation in current, so a switch in an a.c. circuit can disconnect at one of these instants. Switching large direct currents off is surprisingly difficult, because if a larger current is to be reduced to zero in a short time then the rate of change of current is high. This sets up a large e.m.f. by electromagnetic induction, which can cause sparking across the terminals of the switch. Switching equipment therefore has to be robust and carefully designed if direct current is to be switched off safely.

Secondly, a.c. is used for the commercial transmission of electrical power because of the need for high-voltage transmission. The following example illustrates this essential requirement.

Example 20.4 shows that there is always commercial pressure on the electricity supply industry to use higher and higher potential differences. Local transmission at 230 V can only be through thick cables for distances up to about 500 m. For longer distances, potential differences of 11 000 V, 25 000 V, 125 000 V, 275 000 V and 400 000 V are used. Experimentation is taking place to use even higher values of potential difference for long-distance transmission.

Since these high potential differences are extremely dangerous, it is essential to be able to change the potential difference from one value to another with high efficiency, reliability and safety. Transformers meet these requirements, and so a.c. is used, since a transformer is only able to function on a.c. This is the main reason

EXAMPLE 20.4

A power-station supplies a factory with 1.0 MW of electrical power at a p.d. of 10 000 V (see **Fig 20.20**). The total resistance of both of the cables between the power-station and the factory is 0.50 Ω. Find the percentage of the power-station's output power that is delivered to the factory. How does this figure change if the supply potential difference is only 250 V?

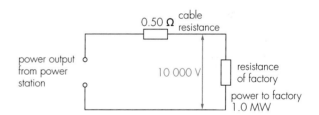

FIG 20.20 Circuit diagram for Example 20.4.

When the supply p.d. is 10 000 V:

current supplied to factory	= (1000 000 W)/(10 000 V)	= 100 A
potential difference across cables	= 100 A × 0.50 Ω	= 50 V
power wasted in cables	= 50 V × 100 A	= 5000 W
power output from power-station		= 1005 000 W
% of power-station's output supplied	= (1000 000/1005 000) × 100	= 99.5%

If the supply p.d. is reduced to 250 V, for the same power:

current supplied to factory	= (1000 000 W)/(250 V)	= 4000 A
potential difference across cables	= 4000 A × 0.50 Ω	= 2000 V
power wasted in cables	= 2000 V × 4000 A	= 8000 000 W
power output from power-station		= 9000 000 W
% of power-station's output supplied	= (1000 000 /9000 000) × 100	= 11%

That is, the percentage of the power supplied falls to only 11%, a totally unacceptable value.

why a.c. is used for the transmission of electrical power.

Types of transformer

A basic transformer consists of two coils, a primary coil and a secondary coil, wound on the same soft-iron core. The design of a transformer and its construction depend on the use to which it is to be put. A transformer is shown in **Fig 20.21**, but it is difficult to illustrate the vast range of different types and powers of commercial transformers. The smallest are for high-frequency work and might use less than a microwatt of power. The largest are mighty constructions capable of handling many megawatts at high voltage. The symbol for a transformer is given in **Fig 20.22a**, together with some variations in **Figs 20.22b** and **c**. Note that there must be at least four connections to a transformer: two primary and two secondary.

We now look at how transformers work. Assume that there is an alternating current through the primary coil. This will set up a varying magnetic field in the core, which links with the secondary coil and by electromagnetic induction sets up an e.m.f. across each turn of wire in the secondary coil. The e.m.f. across the secondary (E_s) is therefore dependent directly on the number of turns in the secondary coil (n_s) and the rate of change of the magnetic flux in the primary coil. This gives

$$\frac{E_s}{E_p} = \frac{n_s}{n_p}$$

The term n_s/n_p is called the turns ratio. If n_s is greater than n_p, then E_s is greater than E_p, and the transformer is called a **step-up transformer**. When n_s is less than n_p, and so E_s is less than E_p, this is called a **step-down transformer**.

Efficiency and power loss

Transformers are usually very efficient. The largest ones can be over 99% efficient. The lost 1% of power can be a problem with high-power transformers as, for example, 1% of the power of a transformer handling 1 MW of power is 10 000 W. This power causes considerable heating of the transformer, so on large transformers some form of cooling system is needed. This is frequently a convected flow of oil, up through the heart of the transformer and down in cooling tubes on the exterior. These are clearly visible in **Fig 20.23**.

Power can be lost in a transformer in different ways. Three important losses are mentioned here.

■ *Copper losses* The wires of the transformer are normally copper, which, although a good conductor, does have some resistance. When there is a current I in the wires, there will be a heating effect. The power loss is I^2R, where R is the resistance of the coil of the transformer through which current I is passing.
■ *Hysteresis losses* Magnetising, de-magnetising and re-magnetising the core of the transformer many times per second requires some power, although modern magnetic materials can reduce the power required to a low value.
■ *Eddy current losses* Not only is the secondary coil in a changing magnetic field, but also the core of the transformer has a changing field within it. An e.m.f. will therefore be induced in the core of the transformer, and this e.m.f. can cause swirls of induced current, called eddy currents, to flow in the core. Most transformer cores are made of sheet steels and built up like plywood, called laminations. If the layers are insulated from one another, it reduces the size of the eddy currents appreciably and hence reduces power losses.

FIG 20.21 A large transformer.

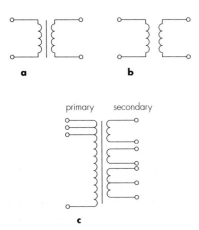

FIG 20.22 Circuit symbols for transformers: **a** transformer with soft-iron core; **b** air-cored transformer; and **c** multi-tap step-down transformer with soft-iron core.

FIG 20.23 The cooling system of a transformer can be seen on this photograph. Oil within the transformer becomes heated and, as the oil rises, it sets up a convection current. The oil cools as it falls through the external tubes.

An *ideal* transformer does not lose any power, so in that case

electrical power input = electrical power output

$$E_p I_p = E_s I_s$$

and the turns ratio is

$$\frac{n_s}{n_p} = \frac{E_s}{E_p} = \frac{I_p}{I_s}$$

where I_p and I_s are the primary and secondary r.m.s. currents respectively. A transformer that steps up the voltage, steps down the current.

QUESTIONS

20.12 A transformer used to step up 240 V to 600 V has a primary coil of 100 turns. What must be the number of turns on the secondary coil?

20.13 One of the largest transformers ever made handles a power of 1.50×10^9 W. This transformer is used to step down 765 kV to 345 kV.

   a What is the r.m.s. current in the primary?

   b What is the r.m.s. current in the secondary?

   c What is the turns ratio?

   d How much power is wasted by the transformer when working at 99.8% efficiency?

20.7 RECTIFICATION

Diodes

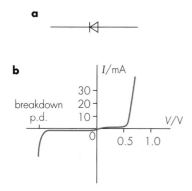

FIG 20.24 a The symbol for a diode. **b** A typical voltage–current characteristic for a diode.

A diode is an electrical component that allows current to pass through itself in one direction only. Diodes are usually manufactured from silicon that has a controlled number of impurity atoms within it. At one end of the diode the impurity atoms make it p-type (positive) semiconductor and at the other end it is n-type (negative). When in use in a circuit, if the p-type end is positive and the n-type end is negative, then there will be a current through the diode. When the polarity is changed, however, there will be very little current. The symbol for a diode is shown in **Fig 20.24a**. The arrow points in the direction of the current. **Fig 20.24b** shows how the current I varies with the voltage V across the diode. It should be noticed that there is a potential difference of about 0.7 V across the diode when it conducts. When reverse biassed, virtually no conduction occurs until the p.d. is high enough for breakdown to occur. This can be when the reverse p.d. is as high as 100 V.

Use can be made of breakdown. A normal diode is irreversibly damaged at breakdown, but some diodes, called **zener diodes**, are used to keep a voltage fixed. These diodes can be purchased with breakdown at specific p.ds (a common one has a breakdown p.d. of 5.1 V). They are used in circuits with reverse polarity and always have the same p.d. across them whatever the current. They do need to be used with a resistor in series to prevent too much current. A fixed output of 5.1 V is obtained from the circuit shown in **Fig 20.25** even if the input p.d. fluctuates. The diagram shows the symbol for a zener diode.

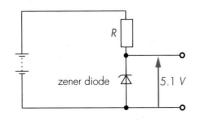

FIG 20.25 A voltage stabilisation circuit using a zener diode.

Rectifiers

If a diode is placed in a circuit with an a.c. power supply in series with a resistor, then the current can be in one direction only. This circuit is shown in **Fig 20.26**, where the potential difference V_r across the resistor is being monitored with an oscilloscope.

The trace on the oscilloscope will be as in **Fig 20.27**, and shows only positive-going peaks. This circuit is called a **half-wave rectifier**. If the potential difference V_d across the diode is monitored, it will have a potential difference across it when it is not conducting, so will show the negative-going peaks. If these voltages are compared, the three traces shown in **Fig 20.28** are obtained. The potential difference V_s of the supply is given by $V_s = V_d + V_r$ at all instants.

There is a small forward potential difference across the diode when it conducts, but if the value of the resistor's resistance is high and the forward resistance of the diode is low, almost all the forward potential difference will be across the resistor. When the negative potential difference is applied, very little current flows, and so there is very little potential difference across the resistance. Although the potential difference across the resistor is in one direction only, it is far from being a constant potential difference.

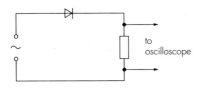

FIG 20.26 Basic rectifier circuit.

Reservoir capacitor

In order to smooth the potential difference, a capacitor can be added in parallel with the resistor. This is called a 'reservoir capacitor' and it performs a similar function to a reservoir for a water supply. A reservoir fills up with water when it rains, so that it can supply water at a steady rate all the time. A capacitor charges up when charge can flow to it, and can discharge steadily all the time. This circuit is shown in **Fig 20.29a**, and the corresponding potential differences and currents in **Fig 20.29b**. The symbol used for the reservoir capacitor in **Fig 20.29a** is that for a polarised capacitor. These are usually electrolytic capacitors, and they must be connected into the circuit the correct way round. They are used because it is possible to make electrolytic capacitors with high values of capacitance (see section 17.1).

Fig 20.29b contains a great deal of information, but it is worth while analysing it properly. The input potential difference V_{input} to the circuit is sinusoidal. The

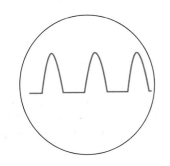

FIG 20.27 Characteristic shape of half-wave-rectified power supply.

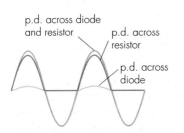

FIG 20.28 Comparison of the three traces.

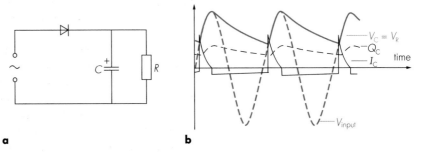

FIG 20.29 A capacitor placed across a load resistor in a rectifier circuit has the effect of smoothing the output voltage, V_R.

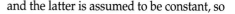

EXAMPLE 20.5

A half-wave-rectified mains supply is used to obtain a p.d. of 340 V d.c., which must not fall below 338 V d.c. when supplying a current of 40 mA. This is said to be a *ripple* of less than 2 V d.c. Find the value of the smoothing capacitor required.

Mains frequency is 50 Hz and the peak value of the mains must be 340 V. **Fig 20.30** shows the problem involved. In the time between two peak values of potential difference, the potential difference must fall by less than 2 V.

time between peaks = 0.020 s
current during this time = 40 mA

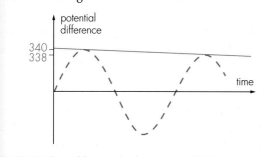

FIG 20.30 The problem involved in Example 20.5.

and the latter is assumed to be constant, so

charge lost from reservoir capacitor = 40 mA × 0.020 s
= 0.8 mC = 800 μC

This causes a drop in potential difference across the capacitor of only 2 V. So since the graph of charge against potential difference for a capacitor is a straight line through the origin

$$C = \frac{Q}{V} = \frac{\text{change in } Q}{\text{change in } V} = \frac{800\,\mu C}{2\,V} = 400\,\mu F$$

This means that the charge stored when the capacitor is fully charged is

400 μF × 340 V = 136 000 μC

and this falls to

400 μF × 338 V = 135 200 μC

during the 0.020 s between peaks – a fall of 800 μC as required.

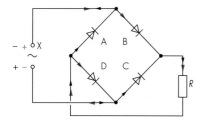

FIG 20.31 A bridge rectifier circuit.

potential difference V_C across the capacitor is equal to the potential difference V_R across the resistance, as they are in parallel. The V_C and V_R graph enables Q_C, the charge on the capacitor, to be plotted, and the gradient of this curve is the current to the capacitor, I_C.

Q **What will the graph of I_R look like?**

The time when current flows to the capacitor is shown as a positive current, and the time it is flowing from the capacitor is shown as negative. The following example shows how the value required for a reservoir capacitor can be estimated.

Other points to note about reservoir capacitors are that they do have a large charge on them, which is gained during the first few cycles after switching-on. The surge of current at switch-on can damage the rectifier, and when switching off care must be taken not to touch them as they may still be charged. For this reason, many television sets have labels on them warning people not to open the back of the set even when it is switched off. It can take several minutes for the unwanted charge to leak away, usually through the resistance of the dielectric of the capacitor.

One way of reducing the problem of ripple is to use both halves of the a.c. cycle instead of just the positive half of the cycle. A bridge rectifier is often used: this is four diodes connected in the way shown in **Fig 20.31**. Whether terminal X is positive or negative, current will pass through R in the same direction. The output is as shown in **Fig 20.32**. The output can be smoothed, as with half-wave rectification, but here the time for which the smoothing capacitor supplies the charge is only half the previous time; for mains, this is therefore a reduction from 0.02 s to 0.01 s, and the capacitor can be reduced to half its previously calculated value.

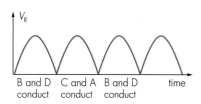

FIG 20.32 The output voltage from a full-wave rectifier.

ANALYSIS

Filter circuits

Modern electronics makes a great deal of use of filter circuits, which are used to allow different frequencies to be passed to different parts of a circuit. To see the principle behind filter circuits, answer the questions below, which refer to the circuit shown in **Fig 20.33** in which two inputs are being used.

a What is the reactance of the 0.32 μF capacitor at both 50 Hz and 5000 Hz?
b Draw a phasor diagram. What is the impedance of the circuit at 50 Hz?
c What are V_C and V_R when the frequency is 50 Hz?
d What are V_C and V_R when the frequency is 5000 Hz?

If you have done this correctly, you will see that for the 50 Hz frequency most of the output voltage is across the

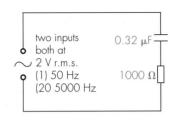

FIG 20.33 Circuit for data analysis exercise.

capacitor, whereas for the 5000 Hz frequency most of the output voltage is across the resistor. An amplifier connected across the resistor will therefore pick up mostly high frequencies, and one across the capacitor will pick up mostly low frequencies. This is a rather unsophisticated filter. In practice, filters can be tuned to different bands of frequencies.

SUMMARY

- A pure alternating current has the equation

  $$I = I_0 \sin(2\pi ft)$$

  where I is the instantaneous current, I_0 is the peak current and f is the frequency. This equation assumes that $I = 0$ when the time $t = 0$.

- The root mean square current, $I_{r.m.s.}$, is the direct current that gives the same heating effect as the alternating current. For sinusoidal currents its value is given by

  $$I_{r.m.s.} = \frac{I_0}{\sqrt{2}} = 0.707 I_0$$

- Instantaneous power has a frequency twice that of the current and has a maximum value that is twice the mean power in a resistive circuit.

- Phasor diagrams for resistors and capacitors:

RESISTOR	CAPACITOR
V and I in phase	I leads V by 90°
resistance $= \dfrac{V_0}{I_0} = R$	resistance $= \dfrac{V_0}{I_0} = \dfrac{1}{2\pi fC}$
power $= V_{r.m.s.} \times I_{r.m.s.}$	power $= 0$

- A phasor is a rotating vector and represents a sinusoidal variation of voltage or current. Impedance can be found from a phasor diagram as the vector sum of resistance and reactance.

- For an ideal transformer

  $$E_p I_p = E_s I_s$$

  and the turns ratio

  $$n_s / n_p = E_s / E_p = I_p / I_s$$

ASSESSMENT QUESTIONS

20.14 Explain, with the aid of a simple labelled sketch, the operating *principle* of a transformer.

WJEC 1998

20.15 A small wind-powered generator is placed on a hill above a campsite. It is connected to a caravan on the site by a pair of cables, each 140 m long. The figure shows how a pair of transformers have been included to raise the efficiency of this transmission system.

a Which is the step-down transformer?.

b The step-down transformer has 1000 turns of wire on its primary coil and 50 turns on its secondary coil, and it can be treated as ideal. The caravan requires an a.c. power supply with an r.m.s. potential difference and current of 12.0 V and 8.00 A respectively. For the primary coil of the step-down transformer, calculate

 i the p.d.,

 ii the current,

 iii the rate at which electrical energy is being delivered.

c Each cable has a resistance per unit length of 0.130 Ω m⁻¹. Calculate the total rate at which electrical energy is transferred to heat energy in the two cables.

d Suggest one advantage and one disadvantage of doubling the number of turns of wire on the secondary of the step-up transformer and the primary of the step-down transformer.

Cambridge 1997

20.16 a A transformer in a power supply for a portable CD player reduces the 240 V mains voltage to 6.0 V and delivers 2.0 W of power to the player.

 i Copy and complete the diagram by inserting the circuit symbol for a transformer. Identify the primary and secondary coils.

 ii Calculate the turns ratio of the transformer.

 iii Find the current in the primary coil and in the secondary coil, assuming the transformer is 100% efficient.

b In practice, transformers are less than 100% efficient. Name three sources of power loss in a transformer.

c Transformers are also used to improve the efficiency of power transmission over large distances from power station to consumer.

 i Name the type of transformer needed

 1 at the power station,

 2 at the consumer's end of the transmission line.

 ii Explain how the use of transformers improves the efficiency of power transmission.

CCEA 1998

20.17

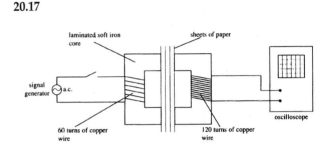

The diagram shows an experimental arrangement used by a student investigating transformers. The signal generator provides a sinusoidal a.c. output of 7.0 V r.m.s.

a Initially the paper is removed and the two C-shaped laminated soft iron cores are pushed firmly together. The switch is then closed.

Stating any assumptions which you make, calculate the approximate *peak-to-peak* potential difference measured by the oscilloscope.

b As part of the investigation the student places sheets of paper, to act as spacers, between the C-shaped cores and connects a 100 Ω resistor across the terminals of the oscilloscope. After each sheet is added, the cores are squeezed together and the peak-to-peak potential difference is measured from the oscilloscope.

Two of the results are recorded in the table below:

number of sheets of paper between cores	peak-to-peak p. d. /V	power delivered / W
1	18	
3	12	

Copy the table and complete the third column to show the power delivered to the resistor. Give your working.

c Explain why the core of a transformer

 i is made from iron,

 ii consists of laminated sheets.

NEAB 1997

20.18 a i Draw a fully-labelled diagram of a simple a.c. generator. Sketch a graph showing how the output voltage from the generator varies with time. Label this graph **1**.

ii The coil of this generator is now rotated at half the rate of that in **i**. On the same axes as your graph in **i**, sketch a graph of the variation of the new output voltage with time. Label this graph **2**.

·b Below is a diagram of a step-up transformer.

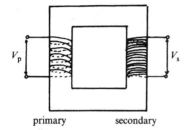

primary   secondary

The primary coil has N_p turns, and the primary voltage is V_p. The secondary has N_s turns and a voltage of V_s.

i Describe briefly the principle of operation of the transformer.

ii Explain what is meant by the term *step-up*.

iii Write down a formula relating N_p, V_p, N_s and V_s.

iv A step-up transformer operates at a primary voltage of 110 V. The transformer supplies a load, connected to the secondary coil, with a current of 2 A. The ratio of the primary and secondary windings is 1:25. Assume the transformer is 100% efficient. Determine the secondary voltage, the primary current and the power output.

c A power station transmits 120 kW of electric power to a factory some distance away. The transmission lines have a total resistance of 0.40 Ω.

i Calculate the power loss in the transmission lines if the power were to be transmitted at 240 V. Express this power loss as a percentage of the power transmitted from the generator.

ii At what minimum voltage should the power be transmitted if the percentage power loss in the transmission lines is to be reduced to less than 0.010%?

iii Your answers to **c i** and **ii** should demonstrate the advantage of transmitting electric power at high voltages. Suggest *two* reasons why it is not practicable to reduce transmission losses effectively to zero by increasing the transmission voltage indefinitely.

CCEA 1998

20.19 The graph shows the variation of potential difference with time across a 2 kW electric kettle connected to an a.c. supply. The water in the kettle is boiling.

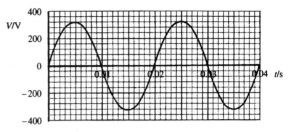

a What is the frequency of the a.c. supply?

b Sketch the current–time graph which would be obtained for the same kettle over the same time interval. Add a scale to the current axis.

c Calculate the resistance of the kettle.

London 1998

20.20 A transformer, to be used in a low voltage power supply, is connected to the 240 V r.m.s., 50 Hz, sinusoidal a.c. mains supply and gives an r.m.s. output voltage of 12 V. There are 1800 turns on the primary coil.

a i Calculate the number of turns on the secondary coil.

ii Assuming that there are no energy losses, calculate the current in the primary coil when the current in the secondary coil is 9.0 A.

iii A 2.5 Ω resistor is connected across the secondary coil. Calculate the rate at which electrical energy is transformed into internal energy in the resistor.

b i Calculate the peak value of the output voltage of the transformer.

ii The output from the transformer is rectified by the use of a diode in series with the load. Sketch the output waveform, showing clearly how the value of the voltage across the load changes with time. Your graph should include suitable voltage and time scales.

iii State and explain briefly how the output can be smoothed.

c State *two* causes of energy loss in the core of a transformer.

AEB 1996

21

ELECTRONICS

LEARNING OBJECTIVES

At the end of this chapter you should be able to:

① distinguish between digital and analogue systems;

② use a variety of different sensors to provide inputs to analogue systems;

③ use a variety of different output systems and appreciate the importance of linking the output of one system to the input of another;

④ understand how a transistor can function as a switch;

⑤ understand the use of an operational amplifier with feedback.

 21.1 INTRODUCTION

The subject of electronics is vast. It is almost entirely a twentieth-century subject, the electron only having been discovered in 1897. Whole degree and further degree courses are planned and taught to give students an insight into its basic principles and to point them in possible directions for further study. Many volumes of books are written on the subject and even more on just some aspect of electronics. This chapter concentrates on input and output devices and the way that these can be linked together through an operational amplifier (op-amp). A more comprehensive account of electronics is given in the related title *Electronics* by Adams and Hutchings, published by Nelson. In studying this chapter, you should be prepared to experiment frequently. The electronic components required are cheap and simple to set up on a breadboard. A battery as a power supply together with occasional use of a multimeter, an oscillator and an oscilloscope complete the apparatus requirements.

21.2 CIRCUIT INPUTS

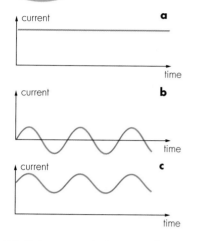

FIG 21.1 **a** A smooth d.c. current; **b** a sinusoidal a.c. current; and **c** a current, which contains both d.c. and a.c. components.

Both inputs to and outputs from electronic circuits can be a.c. or d.c. or a combination of the two, as shown in **Fig 21.1**. The behaviour of a circuit can be very different for a steady current as opposed to a varying one, so it is necessary to be quite clear which type of current is being used. Circuits can be used to monitor almost anything from the vital to the trivial. **Fig 21.2** shows a picture of a baby in an intensive care unit. The baby is wired up to a series of sensors, which monitor such features as heart beat, breathing and temperature. Devices are now available for use in the home so that babies can be monitored while they are asleep to check that they are still breathing. If they stop breathing, an alarm sounds so that the parents can be alerted. It is being found that this is an encouraging way of reducing the relatively large number of distressing cot-deaths that happen each year. (Putting babies down to sleep on their backs rather than on their fronts has also reduced cot deaths considerably.)

Sensors that function on temperature, light intensity and mechanical force are clearly needed, besides straightforward switching and resistive devices. These are incorporated into potential divider circuits as shown in **Fig 21.3**. The sensor can be connected either to switch on, or to switch off, a circuit when a particular low or high value is reached.

Several of these sensors can be used in conjunction with one another or with other sensors, which may respond to any of magnetic field, sound, infra-red, ultra-violet, radiation, humidity, speed, etc.

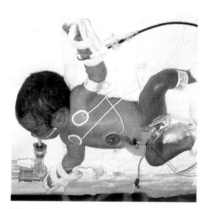

FIG 21.2 A premature baby in an intensive care unit. A ventilator at top right corner maintains the child's breathing. The three electrodes attached to the child's chest are connected to an electro-cardiography machine to monitor heartbeat. The fourth sensor on the baby's abdomen monitors lung movement. Various drips supply antibiotic drugs and vitamin supplements.

Microphone

Some input devices produce an electromotive force, but many do not. For example, microphones are of several different types. In a moving-coil microphone, the changes in air pressure resulting from a sound wave move a small coil within a strong magnetic field. By electromagnetic induction, a small, varying e.m.f. is created, which can be connected across the input of an amplifier. A crystal microphone also creates an e.m.f. When some crystals are squeezed, an e.m.f. is created across them, and use is made of this effect in a crystal microphone, where the sound wave vibrations are used to force a diaphragm against the crystal. Other types of microphone require some external source of potential difference. In a capacitor microphone, for example, the sound waves move one plate of a capacitor and hence adjust its capacitance. If a battery is connected across the capacitor, then a change in the charge on the plates will occur when the capacitance changes. The fluctuation of charge results in a current in a resistor and hence a potential difference across it that varies with the sound oscillations. The circuits into which these components are connected have to take account of their different characteristics.

Thermistor

The negative temperature coefficient (n.t.c.) thermistor is one particularly frequently used input device. It is made of a semiconductor material, such as nickel oxide, and its resistance falls markedly when the temperature is raised. The extra kinetic energy of the atoms at the higher temperature enables more electrons to move from atom to atom. Table 21.1 is a guide to how the resistance of some common thermistors varies with temperature. The resistance of a length of copper wire that happens to be 1 Ω at 20°C and a constantan wire with the same dimensions are shown for comparison.

TABLE 21.1 How the resistance of some common materials and thermistors changes with temperature

	RESISTANCE/Ω	
	20°C	100°C
a copper wire	1.0	1.3
a similar wire of constantan	31	31.1
disc thermistor	25	1
rod thermistor	380	28
bead thermistor	5000	80

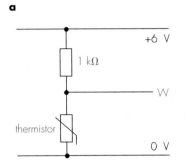

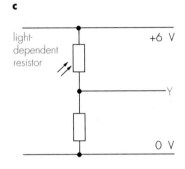

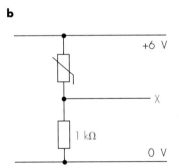

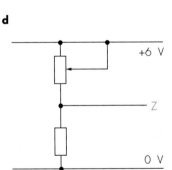

FIG 21.3 Different arrangements of input components.

EXAMPLE 21.1

In **Fig 21.3a** the thermistor has a resistance of 80 Ω when hot and 5000 Ω when cold. What is the potential at W in each case?

Using the equation for a potential divider (section 15.3), we get

■ **Hot**

$$\text{p.d. across thermistor} = \frac{80}{80 + 1000} \times 6 \text{ V}$$

$$= 0.44 \text{ V}$$

potential at W = 0.44 V

■ **Cold**

$$\text{p.d. across thermistor} = \frac{5000}{5000 + 2000} \times 6 \text{ V}$$

$$= 5.0 \text{ V}$$

potential at W = 5.0 V

Note that, as the temperature rises, the potential at W decreases. This change can be used to switch off a circuit when a pre-determined temperature is reached. This could be used in a central heating control system to switch the heating off when a room is warm enough.

Light-dependent resistor (LDR)

The resistance of cadmium sulphide decreases markedly when it is illuminated. This is made use of in an LDR, for which the resistance may change from 1000 kΩ in full moonlight to 30 Ω in full sunlight.

Photodiode

Where a rapid response is required, a photodiode is used rather than an LDR. All diodes are normally light-sensitive, but they are normally placed in light-excluding materials so that when reverse-biassed there is virtually no current through the diode. If a window allows light through to a reverse-biassed diode, then there will be a reverse current that is proportional to the light intensity. The photocurrent for a typical photodiode varies from about 1 nA in the dark to 0.1 mA in the light.

QUESTIONS

21.1 The thermistor referred to in Example 21.1 is used in the circuit in **Fig 21.3b**. How does the potential at X change when the temperature of the thermistor is altered from cold to hot?

21.2 The same thermistor as used in Example 21.1 has a resistance of 80 Ω at 100 °C. What value resistance should be used in place of the 1 kΩ resistor if it is required that the potential at W (**Fig 21.3a**) is to be 0.6 V at this temperature?

21.3 The light-dependent resistor (LDR) in **Fig 21.3c** has a resistance of 1200 Ω when illuminated and a resistance of 10 MΩ when in the dark. Find the value of a suitable resistor that can be placed in series with it so that the potential at Y changes from near zero to 2 V if the LDR is illuminated.

21.3 CIRCUIT OUTPUTS

Outputs from electronic circuits can be used to operate almost anything, but there are a few details that always need to be taken care of. For example, there is no point feeding a d.c. output into a loudspeaker. A lamp, on the other hand, can function on either d.c. or a.c. provided the current to it is not too high and not too low.

Light-emitting diode (LED)

An LED is a common indicator for microelectronic circuits. As with all diodes, the current flows more readily in one direction than in the opposite direction. First, LEDs are diodes; they only allow current through them in one direction, so they must be connected the correct way round. Secondly, LEDs made of gallium phosphide are easily damaged if the current through them is too large. In order to limit the current to around 10 mA, a resistor having a resistance of about 600 Ω is placed in series with the LED. For a.c. operation an LED needs protection against reverse voltages. To do this, a diode needs to be connected in parallel with the LED in order to allow the current to pass in the opposite direction to that in the LED.

EXAMPLE 21.2

Design a circuit that can be used to light an LED when a 6 V_r.m.s. a.c. supply is switched on. Details given for an LED from a manufacturer's data sheet:

Maximum forward current I_F = 20 mA
Maximum forward p.d. V_F = 2.0 V
Maximum reverse p.d. V_R = 5.0 V

A 6 V a.c. supply has a maximum positive voltage of $6 \times 1.41 = 8.5$ V and a maximum negative voltage of 8.5 V.

Since 2 V is the maximum allowable voltage across the LED, there needs to be in series with the LED a resistor across which there must be 6.5 V, when the current through it is 20 mA.

The resistor must have a value greater than 6.5 V / 0.020 A = 325 Ω.

8.5 V is a greater reverse p.d. than the allowable 5.0 V, so a diode must be connected in parallel with the LED to allow current to flow in the opposite direction. The whole circuit is shown in **Fig. 21.4**.

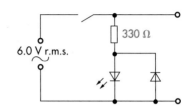

FIG 21.4 Circuit designed so that, when power from an a.c. source is switched on, an LED lights up.

Relay

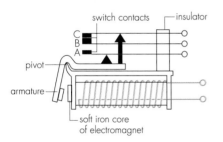

FIG 21.5 The relay.

While LEDs and lamps can be connected directly to an output, there are very many devices that need to be controlled by microelectronic circuits but require a much greater current than the circuit can supply. In these cases, relays are used.

Relays are electromagnetic switches that can be used to allow small currents to switch on or off large currents. **Fig 21.5** shows a relay (see also **Fig 18.18**). For the relay shown in **Fig 21.5**, when there is a current through the coil, it magnetises the core of the electromagnet and this attracts the armature. The armature pivots and the other end of the armature, which has two insulating rods attached to it, moves upwards and pushes through holes against the springy metal supporting the switch contacts. The probe will move contact A so that it touches contact B, and at the same time the other probe will push contact C away from contact B. This type of relay is called a change-over relay, and contacts B and C are normally closed (NC). Contacts B and A are normally open (NO). A diode should always be connected across a relay in a microelectronics circuit because, when the current through the coil of the relay is switched off, there is a surge in potential across the coil as a result of electromagnetic induction. The diode will allow the consequent current to flow through itself for the brief time the surge exists, so preventing any damage to the circuit.

Fig 21.6 shows various configurations that can be used for the output from a point A in the circuit:

- Circuit **a** is used with a.c. The capacitor blocks d.c. from A but allows a.c. to pass to the next stage in the circuit.
- Circuit **b** shows a light-emitting diode, which will be on only when A is at low potential.
- Circuit **c** shows the rearrangement necessary for the LED to be on only when A is at a high potential.
- Circuit **d** shows a relay, with its protective diode, operating a 6 V motor on its switch terminals when A is at low potential.

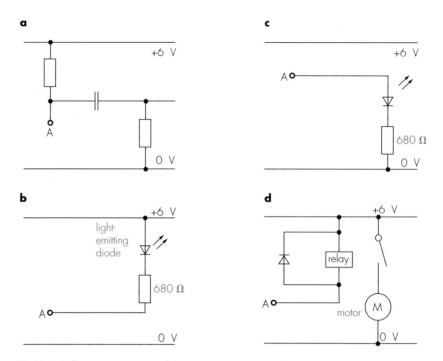

FIG 21.6 Different arrangements of output components.

QUESTION

21.4 For the circuits of **Figs 21.6b** and **21.6c**, the current through the LED is 5 mA. If the potential difference across the LED under these operating conditions is 2.2 V, find the potential at A in each case.

21.4 THE TRANSISTOR AS A SWITCH

Having considered inputs and outputs for electronic circuits, it is now time to consider the **transistor.** The name transistor is a contraction of 'transfer resistor', and transistors were first constructed in 1948 in the Bell Telephone Laboratories in the USA by Shockley, Bardeen and Brattain, who were awarded the Nobel Prize for physics in 1956.

A glance at an electronics catalogue shows how many thousands of types of transistor are manufactured for different purposes, and at this stage it is necessary to consider only the basic function of a transistor and see why it might be called a transfer resistor.

The base current in a transistor controls the collector current. The current through the collector, I_c, is directly proportional to the current through the base, I_b, so with zero base current there will be zero collector current as seen in the investigation.

K The ratio I_c/I_b is called the *d.c. current gain*, h_{FE}. Typically h_{FE} is of the order of 100.

INVESTIGATION

It is recommended that circuits are set up on a circuit board. There are several different types available and the solderless ones are easier to experiment with as components can be interchanged easily. Do not be tempted to use crocodile clips and long trailing wires. They are a source of much frustration. Firm connections are essential.

One of the main problems in setting up electronic circuits is having components that are compatible with one another. Because of the number of different values of components, it is possible to have a circuit functioning perfectly but not to know that it is doing so, simply because a current is not large enough to make the particular bulb you are using light up. Please therefore experiment. Try different components, logically, when things seem not to work – you will not be the first person to damage a transistor by overloading it!

Start with the simplest possible transistor circuit. Apart from the transistor itself, all you will require is a bulb, a 6 V battery and some connecting wire. A suitable transistor is a BC 108, and the bulb should be 6 V, 60 mA. The transistor is an n-p-n silicon transistor and it has three connections to it: emitter, base and collector. The arrangement of these and the circuit diagram are given in **Fig 21.7**.

Keep X and Y apart and nothing happens. No current is possible through the transistor if the base current is zero. Now hold X in one hand and Y in the other hand. If your resistance is low enough, there will be a current of up to 100 μA around the circuit from the battery, through you to the base, emitter and back to the battery.

As soon as this base current flows, the lamp should light, although it will probably be dim. The important feature of the transistor's function is that a current of about 30 mA from the collector to the emitter is possible as a result of a current of 100 μA in the base. In other words, a small current in a circuit of high resistance (you) becomes transferred into a larger current in a circuit of low resistance (the bulb) – a transfer resistor. This is the basis of transistor action, so it is worth while making sure that you have seen it clearly. Any difficulty here will probably be because of one of three things:

① You let X touch Y and thereby overheated the transistor.
Remedy – use a new transistor and do not do the same thing again.

② Your resistance is too high to allow sufficient base current.
Remedy – moisten your hands or hold X and Y, still separated from one another, between the same two fingers, or replace yourself by a 100 kΩ or 22 kΩ or 10 kΩ resistor (or by all three, one at a time).

③ Your bulb has too small a current to make it glow.
Remedy – put a milliammeter in series with the bulb to show the collector current. Then try again. At least you should be able to see the milliammeter needle fall as your hands dry out. This is the basis of some lie detectors

Interplay between the resistance connected between X and Y and the current through the bulb should be observed. The higher the resistance between X and Y, the lower the current through the base. A small base current results in a small collector current.

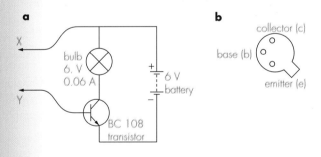

FIG 21.7 a The circuit for the investigation. **b** The connections to the BC 108 transistor.

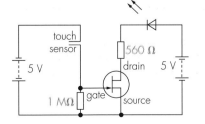

FIG 21.8 The use of a junction transistor (JFET) as a touch switch to light an LED.

Transistors such as the BC 108 are current-controlled devices. Transistors called field effect transistors have their output current controlled by the voltage across their input terminals. A junction field effect transistor (JFET) can be used as a switch. **Fig 21.8** shows a circuit that can be used to switch on an LED using a touch sensor. When no contact is made to the touch sensor, the input, known as the gate and shown with the arrow, is at zero volts. This is the condition for maximum p.d. across the other two terminals of the transistor, known as the source and the drain, and there is maximum current output from the drain, which lights the LED. When the sensor is touched, say by a finger between the two probes, electrical connection is made and the potential of the gate falls to −5 V. This gives zero current output from the drain and the LED goes out.

Q How could you rearrange the circuit so that the LED comes on when the touch sensor is touched?

QUESTIONS

21.5 **Fig 21.9** shows a circuit combining an input potential divider, X and Y, a junction transistor as a switch, and an output Z. Explain how using a thermistor as X, a resistor as Y an LED (plus protective resistor) as Z enables the circuit to switch on a warning light in a greenhouse when the temperature gets too low.

21.6 Design circuits similar to that in **Fig 21.9** using, where necessary, touch sensors, thermistors, LDRs, relays and LEDs to:

   a switch on a fan when it gets hot,

   b switch on a light when it gets dark,

   c switch on a pump when a water level gets low,

   d switch off a heater when it gets hot,

   e switch off a refrigerator when it is cold enough in the food compartment.

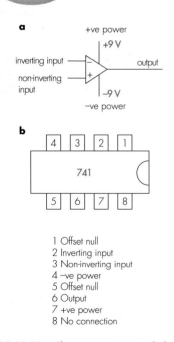

FIG 21.9 Circuit for questions 21.5 and 21.6

THE OPERATIONAL AMPLIFIER (OP-AMP)

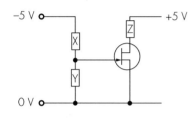

b

4	3	2	1

741

5	6	7	8

1 Offset null
2 Inverting input
3 Non-inverting input
4 −ve power
5 Offset null
6 Output
7 +ve power
8 No connection

FIG 21.10 **a** The op-amp circuit symbol and **b** the pin connections for the 741 op-amp.

The last three sections in this chapter concern amplification of analogue information using a particular circuit item called an operational amplifier (usually abbreviated to op-amp). An op-amp is an integrated circuit amplifier containing many components and having the following characteristics:

■ a high gain (A) of the order of 10^5;
■ a high input impedance of the order of 2 MΩ;
■ a low output impedance of the order of 100 Ω;
■ an ability to accept positive, negative or alternating inputs;
■ two inputs, one of which gives an output of the opposite sign to the input.

The symbol for an op-amp *is* given in **Fig 21.10** together with a plan of the pin configuration of the 741 op-amp. The positive and negative power supplies must be equal and opposite: +9 V and −9 V are convenient values, but any value between 5 V and 15 V can be used. These essential connections are always assumed, but not usually put on to circuit diagrams. The output voltage can never be higher than the voltage of the power supply. Two other connections need to be mentioned. They are marked as 'offset null' on pins 1 and 5, and enable the output to be adjusted to zero when the input is zero. This is only important when high gain is required for a d.c. input, and will not concern us here.

When a basic op-amp circuit is used, inputs V_1 and V_2 can be applied to the inverting and the non-inverting inputs respectively, as shown in **Fig 21.11**, and an output voltage will be produced. It is necessary to be particularly careful with the direction of an applied voltage. The convention used is for the head of the arrow

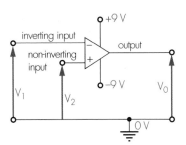

Fig 21.11 A basic op-amp circuit.

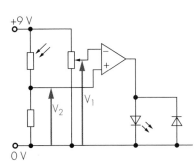

FIG 21.12 The comparator circuit.

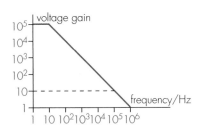

FIG 21.13 The open-loop voltage gain for the 741 op-amp plotted as a function of frequency. The meaning of the dotted line will be explained later in the chapter.

showing the applied voltage to be taken to be positive, and for the voltage to be taken relative to the zero voltage of the power supply. It is usually convenient on a breadboard to have the bottom rail connected to the mid-voltage point between the positive and negative power supplies, here assumed to be +9 V and −9 V.

The amplifier output V_0 under these circumstances is given by

$$V_0 = A (V_2 - V_1)$$

where A is called the open-loop gain. There is, of course, still the limitation that V_0 cannot be greater than +9 V or less than −9 V. Since A is about 100 000, very tiny values of V_2 and V_1 can soon cause saturation, the situation when V_0 has reached 9 V. Note that

- if V_2 is zero, $V_0 = -AV_1$
- if V_1 is zero, $V_0 = +AV_2$
- if V_1 is larger than V_2, the output is negative,
- if $V_2 = V_1$, the output is zero.

The comparator

When an op-amp is used in open-loop mode, it compares the size of two input voltages and gives an output that will be +9 V if V_2 is greater than V_1 and −9 V if V_1 is greater than V_2. This circuit is shown in **Fig 21.12**, in which V_1 and V_2 are being compared. In the light V_2 will be high and higher than V_1, which can be set at any desired value using the variable resistor. This will cause a high positive output and so the LED will be on. In the dark, the output will be negative and the LED will go out. Reversal of the operation can be achieved by swapping over the LDR and the resistor beneath it on the diagram.

Of course, in other circuits, the values of V_1 and V_2 may vary, in which case V_0 will vary correspondingly. An a.c. input will cause an a.c. output. As shown, V_1, the inverting input, causes an output opposite in sign to itself. The reason why this basic circuit is not used much is that the gain, quoted as about 100 000, is very dependent on the frequency of the signal being used. This would mean that an audio signal would be badly distorted, as the high frequencies would be amplified by a different amount from the low frequencies. The variation in gain is shown in **Fig 21.13**. The term **open-loop gain** is used for this situation. It can be seen from the graph that the op-amp amplified 1 kHz signals 1000 times, but 10 kHz signals only 100 times. The way in which an op-amp can be used so that it does not produce distorted signals is by using a technique called negative feedback. This technique drastically reduces the gain of the op-amp but it also enables good-quality amplification to be obtained.

21.6 NEGATIVE FEEDBACK

Negative feedback is not only used with op-amps. Negative feedback is an extremely important control technique. The correct function of the human body is impossible without it. Standing erect puts you in unstable equilibrium. When a minor disturbance occurs, you start to fall over and set in train a series of actions – which you do not even notice unless you have your eyes closed, because the system of feedback is so good. As you move, your eyes detect small changes in the image on the retina, so a signal is sent to the brain. The fluids in the

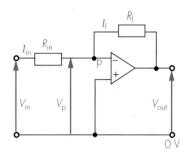

FIG 21.14 The basic negative feedback circuit.

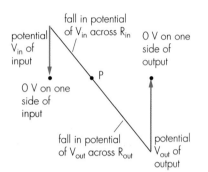

FIG 21.15 A chart showing how the potential changes around the circuit. Three points at which the potential is almost zero make the theory of the op-amp very much simpler than it would otherwise be.

semicircular canals in your ears start to move. That too is detected and messages are sent to your brain. Your brain controls muscles in your feet, which restore you to equilibrium. The foot muscles exert forces to move you back if you fall forward, or to move you forward if you fall back. They exert large forces if you have moved a long way, small forces if you have only moved a short way. This is feedback in action.

The same principle is involved in steering a car or bicycle. If you find yourself going too far to the right when on a straight course you move the steering wheel or handlebars to the left. If you have not corrected enough, then you move them a bit further. It is a sensitive system and is carried out continually and almost imperceptibly.

Feedback in an op-amp is done by connecting a feedback resistor, R_f, from the output back to the inverting input as shown in **Fig 21.14**. The feedback is in antiphase to the input because the inverting input of the op-amp is being used.

The gain in a circuit using negative feedback is called the **closed-loop gain** and to calculate it some approximations need to be made. The approximations are good provided the resistances of circuit components are low compared with the input impedance of the op-amp. The two approximations made are as follows:

1 The current through R_{in} equals the current through R_f. At first sight this seems strange, since perhaps a feedback current is expected in the opposite direction, but remember, the input impedance of the op-amp is of the order of 2 MΩ whereas R_{in} and R_f are usually in the range of 1 kΩ to 100 kΩ. Only a small fraction I_{in} can therefore go into the op-amp inverting input.

2 The potential at P is near to zero. P is called **virtual earth**. This arises because of the very large open-loop gain A of the op-amp. The value of A could well be 100 000 and since

$$V_{out}/V_P = A$$

this makes V_P extremely small, since the maximum possible value of V_{out} is only 9 V or so.

These two approximations now make the theory extremely easy – especially if a chart of potential around the circuit is used (**Fig 21.15**). Because of the virtual earth, any rise in potential caused by the input is balanced by a similar loss in potential across R_{in}. Similarly the fall in potential across R_f must be equal to the rise in potential across the output. This gives

$$V_{in} = I_{in} R_{in}$$

and

$$-V_{out} = I_f R_f$$

But by using the approximation $I_{in} = I_f$ we get

$$\frac{V_{out}}{V_{in}} = \frac{\pm I_f R_f}{I_f R_{in}}$$

K

Closed-loop gain for inverting amplifier $= \dfrac{V_{out}}{V_{in}} = \dfrac{\pm R_f}{R_{in}}$

This shows that the closed-loop gain does not depend upon the open-loop gain, A, at all. It is determined only by the values of R_{in} and R_f. The gain can be altered at will by choosing suitable values for R_{in} and R_f. If the gain required is 10, then R_f could be 10 kΩ and R_{in} would be 1 kΩ. This constant gain can be achieved over a range of frequencies from zero to 10^5 Hz. The dotted line on **Fig 21.13** shows the frequency range for this gain.

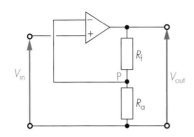

FIG 21.16 Negative feedback when the non-inverting input is used.

This circuit should be set up on a breadboard and, with an input fixed at, say, 0.5 V, the output should be measured for different values of R_{in} and R_f.

If the input signal is to be used with the non-inverting input, the feedback signal is still fed into the inverting input as shown in **Fig 21.16**. As explained in section 21.5, the gain of an op-amp is very large, so for V_{out} to be as small as 9 V the difference between the inverting and the non-inverting inputs must be very small. This implies that the potential at P must be very nearly equal to V_{in}. Also, as is always the case with op-amps, the current to the inverting input is nearly zero. Using these approximations enables the potential divider formula to be used, giving

$$\frac{V_P}{V_{out}} = \frac{R_a}{R_a + R_f}$$

$$\frac{V_{out}}{V_P} = \frac{V_{out}}{V_{in}} = \frac{R_a + R_f}{R_a} = 1 + \frac{R_f}{R_a}$$

$$\text{gain} = \frac{V_{out}}{V_{in}} = 1 + \frac{R_f}{R_a}$$

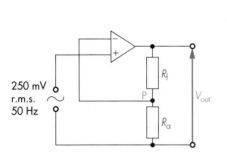

FIG 21.17 Negative feedback when the non-inverting input is used.

QUESTIONS

21.7 Plot a sketch graph to show how the output of an op-amp with open-loop gain of 100 000 varies with the inverting input if the non-inverting input is zero. (See **Fig 21.12**.)

21.8 An op-amp using negative feedback (**Fig 21.15**) has an input voltage of 26 mV and an input resistor (R_1) of 1.2 kΩ.

   a Find the value of the output potential difference if the feedback resistor R_f has a value of 8.6 kΩ.

   b Find also the values of input current I_{in} and feedback current I_f.

21.9 A non-inverting amplifier (**Fig 21.17**) has a 50 Hz a.c. input of 250 mV r.m.s. Its supply potentials are +5 V and −5 V. Sketch a graph showing how the output potential varies with time if $R_f = 29$ kΩ and $R_a = 1.0$ kΩ.

21.7 PRACTICAL CIRCUITS USING THE OP-AMP

The obvious example for use of an op-amp is for producing amplification in an amplifier. The op-amp is, however, much more than just an amplifier. It is better to think of it as a manipulator of electrical signals. It can take any signal and multiply it or add a constant to it, or subtract it from another signal. **Fig 21.18** gives a few examples.

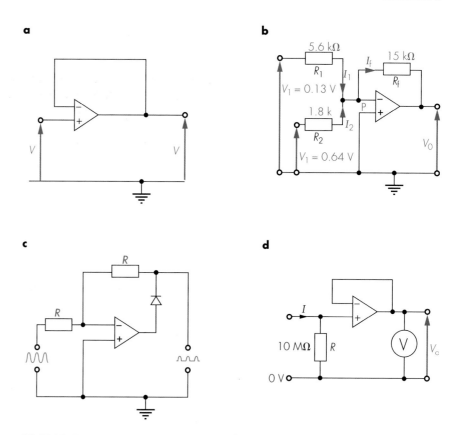

FIG 21.18 Common op-amp circuits – see Example 21.3 and question 21.11: **a** voltage follower; **b** adder; **c** precision rectifier; and **d** nanoammeter.

EXAMPLE 21.3

Find the output from the circuit in **Fig 21.17b** using the details given on the figure.

Since the input impedance of the op-amp inverting input is high, a negligible current enters it at P. Therefore

$$I_f = I_1 + I_2$$

Since the potential at P is a virtual earth:

the p.d. across R_1 is $V_1 = I_1 R_1$

the p.d. across R_2 is $V_2 = I_2 R_2$

the p.d. across R_f is $-V_0 = I_f R_f$

So

$$\frac{\pm V_0}{R_f} = \frac{V_1}{R_1} + \frac{V_2}{R_2}$$

$$\frac{\pm V_0}{15\ \mathrm{k\Omega}} = \frac{0.13\ \mathrm{V}}{5.6\ \mathrm{k\Omega}} + \frac{0.64\ \mathrm{V}}{1.8\ \mathrm{k\Omega}}$$

$$= 0.0232\ \mathrm{mA} + 0.3556\ \mathrm{mA}$$

$$= 0.3788\ \mathrm{mA}$$

giving

$$V_0 = -\ 15\ \mathrm{k\Omega} \times 0.3788\ \mathrm{mA} = -5.68\ \mathrm{V}$$

It should be noted that the output is therefore the sum of the outputs that each input would have given on its own. This is an adder circuit. If the circuit is to be used for direct addition, then R_1, R_2 and R_f are made equal, in which case $V_0 = (V_1 + V_2)$.

This can be extended as required by using an R_3, R_4, R_5, etc. If subtraction is required, then one of the inputs can be connected into the circuit the opposite way round. An adder circuit such as this is used as a mixer, and it would not be unusual to have the inputs from 30 microphones all connected into the same op-amp.

QUESTIONS

21.10 Sketch circuit diagrams showing circuits that will give the following outputs. V_1, V_2, V_3, etc., are inputs.

a $-(V_1 + V_2 + V_3)$

b $-(V_1 + V_2 - V_3)$

c $-(V_1 + \frac{1}{2}V_2)$

d $-(V_1 + 3V_2)$

e $-(2V_1 + 3V_2 + 4V_3)$

f $-6(V_2 - V_1)$

g $-3(V_1 - \frac{1}{2}V_2 + \frac{1}{3}V_3 + \frac{1}{4}V_4)$

21.11 All these questions refer to the circuit diagrams of **Fig 21.18**.

a What is the purpose of circuit **a** where the input and the output potential difference are the same?

b What would be the output from circuit **b** if V_2 were to be -0.64 V rather than 0.64 V?

c Why are the two resistors in circuit **c** the same?

d Circuit **d**, like circuit **a**, is called a voltage follower, in which the input voltage equals the output voltage. Show that the output voltage is 0.36 V when the input current is only 36 nA.

SUMMARY

- The base current for a transistor is small and it controls the collector current. The current gain, $h_{fe} = I_c / I_b$.

- An operational amplifier (op-amp) has very high input impedances both on its inverting input and on its non-inverting input, and a very high voltage gain, called the open-loop gain, A, which is dependent on the frequency. The output is given by $V_0 = A(V_2 - V_1)$.

- By the use of negative feedback, the op-amp's gain can be made to depend on the value of the resistances of the resistors in the circuit. Since the inputs are virtual earths:

  – using the inverting input

  $$\frac{V_{out}}{V_{in}} = \pm \frac{R_f}{R_{in}}$$

  – using the non-inverting input

  $$\frac{V_{out}}{V_{in}} = 1 + \frac{R_f}{R_a}$$

ASSESSMENT QUESTIONS

21.12 a Silicon diodes can be used in *forward bias* mode or *reverse bias* mode. Explain what is meant by forward bias and reverse bias.

b In the circuits **A**, **B** and **C**, S is a silicon diode and Z is a 6.0 V zener diode. For **each** of the three circuits, state the value of the voltage V_d you would expect to find across the diode and hence determine the current I, flowing through the ammeter.

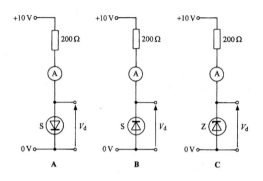

c Draw a circuit diagram, including a transistor, a relay and a diode, to show how a silicon diode can be used to protect a transistor when it is used to switch a relay. Explain how the diode protects the transistor.

NEAB 1998

21.13 Electronic systems may be used to perform mathematical operations on analogue or digital signals.
A circuit designed by a student to carry out a sequence of mathematical operations on the analogue input voltage V_{IN} is shown in the figure.

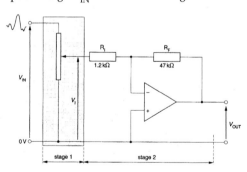

Stage 1 of the circuit consists of a linear potentiometer. Explain what mathematical operation the potentiometer performs on the input voltage V_{IN}.
Stage 2 multiplies the voltage V_I by a certain factor. Derive this factor. With the aid of a suitable sketch graph, show the relationship between the voltages V_{OUT} and V_I.
Determine the output voltage V_{OUT} in terms of the input voltage V_{IN} when the potentiometer is set at its mid-position.

OCR 1999

21.14 A diagram for a temperature-sensing circuit is shown.

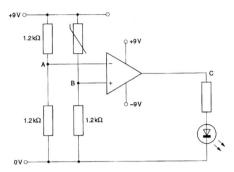

At 15°C, the resistance of the thermistor is 2.2 kΩ and the resistance decreases to 680Ω at 100°C. The light-emitting diode (LED) is on when the output from the operational amplifier (op-amp) is positive and high.
a State the voltage at A.
b The thermistor is immersed in a water bath at 15°C. For the thermistor at this temperature
  i calculate the voltage at B,
  ii state the output voltage from the op-amp,
  iii state whether the LED is on or off.
c Describe and explain what happens to the voltages at B and at C as the temperature of the thermistor is gradually increased from 15°C to 100°C.

OCR 1999

21.15 The bridge circuit shown below contains a negative temperature coefficient thermistor. When the temperature of the thermistor reaches a certain value the LED in the output circuit comes on. This can be used to indicate when the temperature in a greenhouse falls below a certain value.
The graph shows the characteristics of the thermistor.

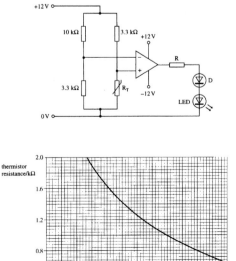

a Calculate the resistance of the thermistor at which the LED comes on and hence state the temperature at which this occurs.

b i Explain briefly the purpose of the diode D.

ii Determine the resistance of R if the current through the LED is not to exceed 15 mA. Assume that the forward voltage drop across D is 0.7 V and that across the LED is 1.8 V.

NEAB 1996

21.16 a The transfer characteristic in figure 1 shows the relationship between the output voltage, V_{out}, and input voltage, V_{in}, for a practical operational amplifier circuit.

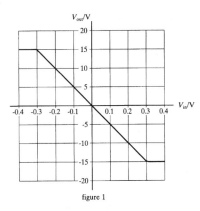

figure 1

i State whether the amplifier is being used in the *inverting* mode or *non-inverting* mode. Explain your reasoning.

ii Determine the voltage gain of the amplifier.

b Draw a circuit diagram which shows how the operational amplifier in part **a** is being used. Include in your diagram the power supply voltages and suggest suitable values for the components used.

c i State what is meant by *open loop gain*.

ii Sketch on a copy of the axes in figure 2 a voltage gain against frequency graph for an operational amplifier in open loop mode. On the same graph draw the equivalent curve for the amplifier described in part **a**. Label which curve represents which mode.

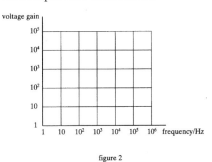

figure 2

iii Give **one** advantage of using an amplifier as in part **b**, rather than in open loop mode.

NEAB 1998

21.17 This question is about operational amplifiers.

a The figure shows the circuit symbol for an op-amp.

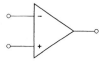

i Copy the diagram. Add extra lines as necessary to label the input voltages, V_+ and V_- and output voltage V_{out}.

ii Explain the terms *input impedance* Z_{in} and *output impedance* Z_{out} for an amplifier.

iii The input and output voltages are related by the formula $V_{out} = A(V_+ - V_-)$. Explain why the values of A and Z_{in} are taken as infinity for an ideal op-amp.

b Resistors of value 10 kΩ and 100 kΩ are connected to the op-amp to construct an inverting amplifier of gain 10.

i Draw a labelled circuit diagram of this amplifier.

ii One point on the circuit diagram is called a *virtual earth* point. Label this point E on your diagram. Explain what is meant by the term and how it is created in this circuit.

O&C 1998

21.18 The figure shows a diagram of a circuit based on an operational amplifier (op-amp)

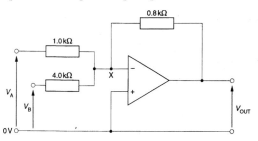

a State the name of the circuit shown.

b Explain why, when the output from the op-amp is not saturated, the voltage at X is approximately zero.

c In a particular case, the input voltages V_A and V_B are 3.0 V and 6.0 V respectively.

i 1. Calculate the currents in the 1.0 kΩ and 4.0 kΩ resistors.

2. Hence explain why the current in the 0.8 kΩ resistor is 4.5 mA.

ii Calculate the output voltage V_{OUT} from the op-amp.

iii Show, by substitution, that the magnitude of the output voltage V_{OUT} is given by the equation

$$V_{OUT} = (0.8V_A + 0.2V_B).$$

d Suggest a practical application of the circuit of the figure.

OCR 1999

21.19

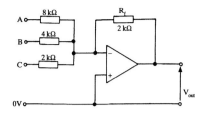

a In the circuit shown, an input of 5 V is applied simultaneously to each of the inputs A, B and C. Calculate the output voltage, V_{out}, of the operational amplifier.

b Calculate the current through the feedback resistor R_f.

c **i** Explain what is meant by *virtual earth*.
  ii State which point is at virtual earth.
  iii State the property of the operational amplifier which makes this point a virtual earth.

NEAB 1997

21.20 This question is about inverting amplifiers.

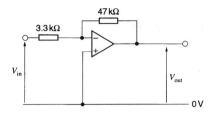

a The figure above shows an op-amp used as an inverting amplifier.
  i State the *input impedance* and *open loop gain* of an ideal op-amp.
  ii Explain the type of feedback in the circuit above and state the gain of the circuit.
  iii Explain why the inverting input of the op-amp in the circuit above is often referred to as a *virtual earth*.

b

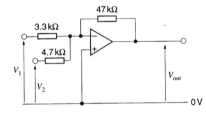

The figure above shows a modified circuit. A voltage of 0.40 V is applied as V_1 and a voltage of 0.30 V as V_2 to the two inputs.
  i Find the currents in the 3.3 kΩ and 4.7 kΩ input resistors.
  ii Calculate the output voltage of the op-amp.
  iii Show that the circuit in **Fig. 4.2** behaves as a summing amplifier whose input and output voltages are related by the equation
$$V_{out} = -(14\,V_1 + 10\,V_2)$$

O&C 1997

MATTER

The structure of matter has intrigued people from our very earliest history until the present day. In the twentieth century, many answers were found to age-old problems, yet new questions are continually being asked. Much remains to be discovered about the solids, liquids, gases and plasmas that together provide a Universe of which we occupy such a tiny part. ■

22 PHASES OF MATTER

LEARNING OBJECTIVES

At the end of this chapter you should be able to:

① define and use the terms 'density' and 'relative density';

② describe the arrangement of molecules within solids, liquids and gases;

③ use the mole and the Avogadro constant correctly;

④ explain the significance of both kinetic and potential energies in determining the phase of matter;

⑤ define and use the term 'specific latent heat';

⑥ define and use the term 'pressure';

⑦ explain how the effect of pressure in a fluid gives rise to a buoyancy force on an object within the fluid, and why bodies float.

22.1 DENSITY

Density is defined as mass per unit volume. An object with a mass m and volume V therefore has a density ρ given by the equation

K $\quad \rho = \dfrac{m}{V}$

The SI unit of density is the kilogram per cubic metre (kg m^{-3}). The densities of various materials are given in Table 22.1. Some of the values given are only estimates because either the density can vary within the material or there is uncertainty in making the measurements. Other values are known more accurately. The density of copper, for example, has the value of 8900 kg m^{-3}, and this value varies only a very small amount with temperature and pressure, and does not vary

TABLE 22.1 Densities of various materials (where given, pressures are quoted in atmospheres, 1 atm = 10^5 Pa).

MATERIAL	DENSITY/kg m^{-3}
intertellar space	$\sim 10^{-20}$
air in 'vacuum' in TV tube	$\sim 10^{-7}$
air at 100 °C and 1 atm	0.95
air at 0 °C and 1 atm	1.29
air at 0 °C and 50 atm	65
balsa wood	150
ethanol	810
ice at 0 °C and 1 atm	920
water at 100 °C and 1 atm	958
water at 0 °C and 1 atm	1000
water at 0 °C and 50 atm	1002
aluminium	2700
iron	7800
copper	8900
lead	11 300
mercury	13 600
gold	19 300
osmium (densest element)	22 500
centre of Sun	160 000
white dwarf stars	$\sim 10^9$
atomic nuclei	$\sim 10^{16}$

within the structure of different pieces of copper. The density of a substance such as wood, however, does vary from place to place within any particular tree. Different parts of the ring system within each growing tree have different densities, depending on the time of year and on whether the growing season was wet or dry. Substances, like copper, that have a uniform density throughout are said to be **homogeneous**. Wood, on the other hand, is non-homogeneous. Average densities can be quoted for non-homogeneous substances, but they are often not very reliable.

In dealing with the density of gases, the temperature and pressure must be quoted, as these have a large effect on the density. For liquids, the density varies with the temperature and, to a lesser extent, with the pressure. Pressure variation makes very little difference to the density of a solid, and temperature has only a small effect.

The term **relative density** is sometimes used. This is a dimensionless quantity defined by the equation

A common mistake

A common mistake in problems dealing with density is to substitute *relative density* into formulae instead of *density*. An error of 1000 is consequently introduced.

$$\text{K} \quad \text{relative density} = \frac{\text{density of substance}}{\text{density of water}}$$

QUESTION

22.1 **a** Using values given in Table 22.1, find the relative density of mercury.

b Find the mass of 1 cm^3 of mercury.

[These two values are numerically the same, but their units are different.]

QUESTIONS

22.2 Calculate the volume occupied by 6.3 kg of iron. Give your answer in both cubic metres and cubic centimetres.

22.3 The density of antifreeze for a car's radiator determines the temperature at which it will freeze. Antifreeze is a mixture of water and ethylene glycol. The density of ethylene glycol is 1120 kg m^{-3}.

    a If the density of the antifreeze in the radiator of a particular car is 1045 kg m^{-3}, find the proportion by volume of ethylene glycol in the antifreeze mixture.

    b What is the proportion of ethylene glycol by mass?

22.2 PHASE

Mention was made in the previous section of the different properties of solids, liquids and gases. The general term used to indicate these three properties is **phase**. It is not only in density measurements that the differences between the three phases of matter are important. In making measurements of elasticity, heat conductivity, electrical conductivity, heat capacity and many more, there is often a startling difference between a property of the solid and the corresponding property of the liquid. The difference is even greater when the substance changes to a gas. In this section, the phase of a substance will be related to the pattern of molecular movement.

Three phases of matter are usually easily recognised as **solid**, **liquid** and **gas**. There are materials, however, that do not fit well into any of these categories — but usually these materials are made up of a wide variety of different molecules. Butter, for example, does not melt at one particular temperature. It gradually gets softer and more liquid as the temperature rises, so it is difficult to know whether to classify soft butter as a solid or a liquid. But then, butter is not a simple substance chemically. It is the complex molecular composition of butter that gives rise to its lack of a unique melting point.

To solid, liquid and gas should be added a fourth phase at extremely high temperatures. This high-temperature phase is called **plasma**, and consists of atomic nuclei and electrons all moving with very high speed but not connected to one another as would be the case at lower temperatures. Since stars are at high temperatures and most of the mass of the Universe is concentrated in the stars, it has been estimated that 90% of all the matter in the Universe is plasma.

Not all substances can exist in all phases, though, by adjusting the temperature and the pressure, many can. Many biological substances, for example, would undergo chemical changes at much lower temperatures than those at which they would be able to melt. Heating sugar will give a sticky black mess before liquid sugar is produced. This is because, on heating, the carbohydrate molecules of the sugar lose water, leaving behind black carbon.

Phase

The word 'phase' as used here has no connection with the use of the word in waves and oscillations.

22.3 THE MOLECULAR STRUCTURE OF MATTER

Many scientists have contributed to the idea of atomic and molecular structure. Early scientists had no direct evidence of atoms or molecules but were working from indirect evidence. Dalton first proposed that an element was composed of many identical atoms. Avogadro's and Gay-Lussac's experiments on gases led them to believe that chemical reactions took place between elementary units of matter. Mendeleyev laid down the basis for the modern Periodic Table of elements.

Certain terms used when dealing with the molecular structure of matter need to be given at this stage.

The mole, and other quantities

The mole, which was defined in section 1.2 as one of the seven base units of the SI system of units, is a measure of the amount of substance:

■ The **mole** is the amount of substance that contains as many entities as there are atoms in 0.012 kg of carbon-12 (^{12}C).

Accurate experimental determination shows that the number of atoms in 0.012 kg of carbon-12 is $(6.022\,17 \pm 0.000\,02) \times 10^{23}$. A mole of hydrogen molecules therefore contains 6.022×10^{23} molecules (to four significant figures). This will be *two* moles of hydrogen *atoms* because each hydrogen molecule contains two hydrogen atoms. These figures enable the mass of a single atom of carbon-12 to be calculated as

$$(0.012\,\text{kg})/(6.022\,17 \times 10^{23}) = 1.992\,64 \times 10^{-26}\,\text{kg}$$

The mass of a single atom of carbon-12 is also, by definition, 12 times the unified atomic mass constant, and so:

■ The **unified atomic mass constant** is one-twelfth of the mass of a single atom of carbon-12, and is therefore

$$(1.992\,64 \times 10^{-26}\,\text{kg})/12 = 1.660\,53 \times 10^{-27}\,\text{kg}$$

This mass is sometimes simply called the unit, u, of atomic mass (or the atomic mass unit):

$$1\,\text{u} = 1.660\,53 \times 10^{-27}\,\text{kg} = 1.66 \times 10^{-27}\,\text{kg (to 3 sig. figs)}$$

It is approximately the mass of a hydrogen atom and also approximately the mass of a proton or neutron.

There may be a time in the future when the unit of mass will be based on, say, the mass of a proton, but at present the link between the mass of sub-atomic particles and the mass of any piece of matter has to be found experimentally.

Other quantities with which you will need to be familiar are as follows:

■ The **molar mass** M_m is the mass per mole of the substance and has the unit kilogram per mole ($kg\,mol^{-1}$).
■ The **molar volume** V_m is the volume per mole of a substance and has the unit cubic metre per mole ($m^3\,mol^{-1}$).
■ The **Avogadro constant** N_A is numerically equal to the number of atoms in 0.012 kg of carbon-12. The value is $N_A = 6.022 \times 10^{23}\,mol^{-1}$.
■ The **relative molecular mass** M_r is defined by the following equation and so has no unit:

$$\text{K} \quad M_r = \frac{\text{mass of a molecule}}{\text{one-twelfth the mass of a carbon-12 atom}}$$

$$= \frac{\text{mass of a molecule}}{1.66 \times 10^{-27}\,\text{kg}}$$

Direct evidence for molecules

Direct evidence for molecules and for their movement came with the discovery of the tiny, jerky motion of minute grains of pollen when suspended in water. The motion is called *Brownian motion* after Robert Brown, who first recorded it in 1827. Einstein first analysed the motion mathematically and was able to deduce a value for the Avogadro constant from his deduction. Agreement between the value so deduced and other experimental values gave strong evidence for the theory. Brownian motion can be seen clearly in the motion of smoke particles in air, and gives direct evidence of collisions between individual molecules and the tiny smoke particles.

QUESTION

22.4 The kinetic energy of a smoke particle of mass 10^{-15} kg is the same as that of a nitrogen molecule of mass 5×10^{-26} kg and speed 1×10^3 m s^{-1}. What is the speed of the smoke particle?

INVESTIGATION

Brownian motion

If you have not previously seen Brownian motion, then you should try to use a microscope to see the effect now. Brownian motion gives direct evidence for perpetual molecular movement. High power is not necessary for the magnification of the microscope, but good illumination of the particles is required.

There are several commercial cells available for viewing the motion of smoke particles in air, and one of these is illustrated in **Fig 22.1**. If you use a cell such as this, it is important to focus the microscope on the region in which the light is also being focussed. The smoke particles are barely visible; they appear as minute spots of light, which are probably much smaller than you would expect. You do not really see a particle, and certainly could not describe its shape. What you can just see is a small amount of light scattered from the surface of a particle whose mass is probably about 10^{-15} kg.

A particle such as this is shown in **Fig 22.2**. It is surrounded by a vast number of invisible molecules that continually strike it. The random nature of these impacts causes just enough irregularity for the movement of the smoke particles to be visible.

If you do not have one of these cells available, then simply place a spot of water onto a microscope slide, add a trace of some fine powder of intense colour (carmine is suitable), put a cover glass on top of the drop and view the powder. Watch a single smoke or carmine particle for a time while it remains in focus, and then draw a sketch showing a typical pattern of its movement.

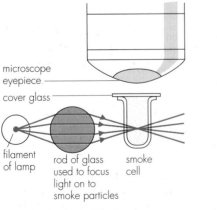

microscope
eyepiece

cover glass

filament
of lamp

rod of glass
used to focus
light on to
smoke particles

smoke
cell

FIG 22.1 The experimental arrangement for viewing Brownian motion. The rod of glass is to focus light from the lamp onto the smoke particles.

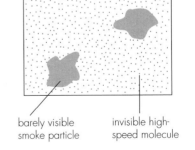

barely visible
smoke particle

invisible high-
speed molecule

FIG 22.2 On a large scale the barely visible smoke particles appear as large patches. They might have a mass of about 10^{-15} kg. The 'dots' represent invisible molecules whose mass may be about 10^{-25} kg and whose speed is of the order of 10^3 m s^{-1}.

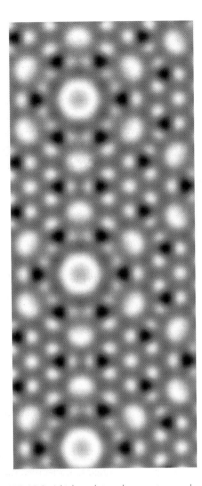

FIG 22.3 A high-resolution electron micrograph of the surface of a silicon crystal. Magnification x 5 000 000.

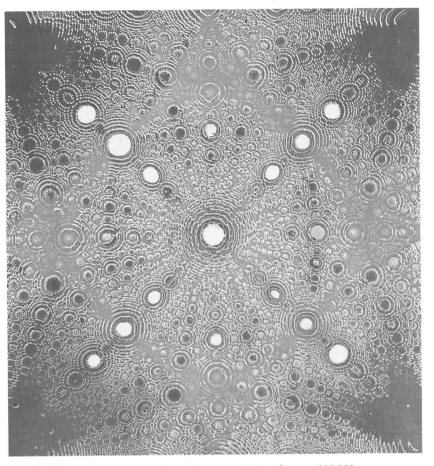

FIG 22.4 A field ion micrograph of atoms in a platinum crystal. Magnification x 200 000

Modern experiments have reinforced earlier ideas that matter consists of particles and that these particles are in rapid motion. The photographs in **Figs 22.3** and **22.4**, taken using a field ion microscope and an electron microscope respectively, both show the pattern of atoms within a crystal.

The forces holding matter together

Any theory of the structure of matter must be able to explain why some substances are solid, some are liquid and some are gaseous. The theory that the particles of matter are in rapid motion can explain why gases are possible, but seems to make the explanation of the solid phase difficult. The atoms photographed in **Figs 22.3** and **22.4** do not appear to be in rapid motion. They do not appear to be particularly blurred, as photographs of fast-moving objects often are.

The key to understanding this paradox lies in realising that there are two conflicting influences on the atoms or molecules in any substance. On the one hand, there is the fact that they are moving and therefore have kinetic energy; on the other hand, they clearly attract one another. This is why the atoms in **Figs 22.3** and **22.4** are so close to one another and so uniformly packed. Because they attract one another, they would need to gain potential energy to become separated. It is the balance between the kinetic energy and the potential energy of atoms and molecules that is all-important.

The appreciable force acting between atoms, molecules and ions is an electrical force. The gravitational force between atomic particles exists, but is far too small to have any important influence on their individual behaviour. In the complex variety of materials, there are many different ways in which the electrical force exists.

■ The force of attraction between ions of sodium and chlorine in crystalline sodium chloride is said to be an *ionic bond*. Each sodium atom loses one electron to become a positive sodium ion. Each chlorine atom gains one electron to become a negative chloride ion. The sodium and chloride ions attract one another to give the ionic bonding in crystalline sodium chloride.

■ A different type of bonding occurs within a hydrogen molecule. This bond is called a *covalent bond*, and it is characterised by electrons being shared between the two hydrogen atoms.

■ Another way atoms can be bonded together is with a *metal bond*. Here electrons are not attached to particular atoms but are free to move through a network of positive metallic ions within the crystalline structure of a metal.

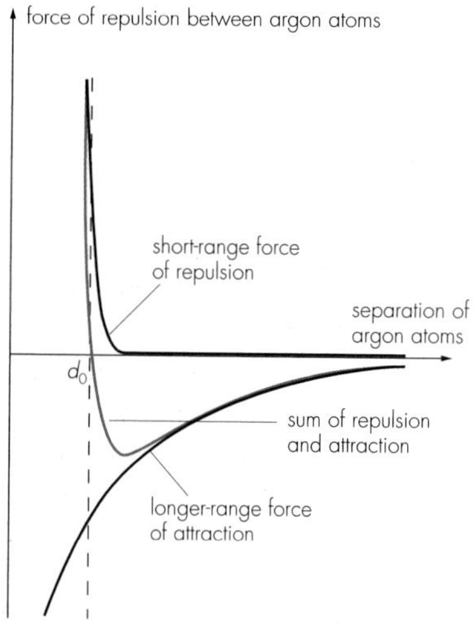

FIG 22.5 The force between two atoms can be one of either attraction or repulsion depending on their separation.

A simplification: argon – a monatomic gas

Inter-atomic forces frequently lead to the formation of molecules. The whole of chemical science is concerned with such inter-atomic bonding and the way in which different molecules can be formed as a result of inter-atomic forces. Here it is useful to be able to discuss inter-atomic forces for a particular example without having the complication of the formation of molecules. For this reason, an inert gas – argon – will be taken as an example.

Argon forms about 1% of the Earth's atmosphere, and it is called a *monatomic* gas because the atoms of argon move around singly and do not form molecules. There are nevertheless weak forces of attraction between argon atoms. These forces are called *van der Waals forces*. Besides forces of attraction, forces of repulsion exist between argon atoms if they are close enough to one another. In **Fig 22.5**, the force of repulsion and the force of attraction between two atoms are both plotted, together with their sum, against the separation d of the atoms. In the absence of any other influence on the atoms, they would come to be in equilibrium a distance d_0 apart, at which the force of attraction equals the force of repulsion. If some other influence then moves them together, they will repel one another; and if they move apart from one another, then they will attract each other. Notice also how steep the graph is for low values of d. This makes it extremely difficult to squash argon atoms much closer than a separation of d_0, and effectively sets a limit to the maximum density of argon.

This graph can also be used to find the potential energy of a pair of atoms. The total force graph is redrawn in **Fig 22.6a**, where it is drawn to scale. If the potential energy of the two atoms is taken to be zero when they are separated by a large distance, then the work done on one of the atoms in moving it to a distance d from the other is given by the (negative) area beneath the force graph as shown. This has been done, and the graph of potential energy against d has also been plotted in **Fig 22.6b**. This graph shows that the atoms have minimum potential energy when in their equilibrium position; they are said to be at the bottom of a potential well.

If some argon atoms are cooled down to near absolute zero, then the argon will be a solid and all of the atoms will have very low total energy. The kinetic energy will be nearly zero and the atoms will have minimum potential energy. An argon atom at the bottom of the potential well would require 1.7×10^{-21} J to escape from one other argon atom. In practice, there are going to be many

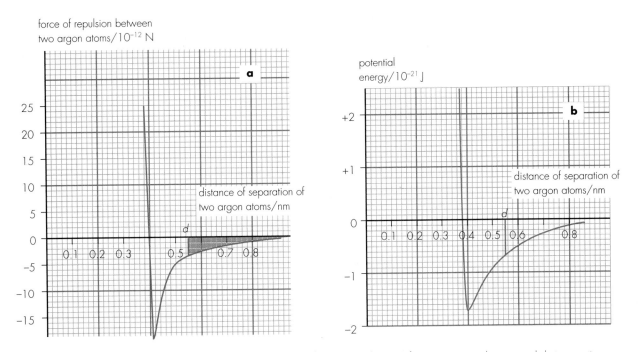

FIG 22.6 a The force between two argon atoms varies with their separation. b The potential energy of two argon atoms also varies with their separation.

argon atoms surrounding it and not just one. To escape from a solid mass of other atoms actually requires about six times this energy. If some heat is then supplied to the argon atoms, the atoms gain kinetic energy. At first, even the fastest atoms have nothing like enough kinetic energy to escape from the potential well that they are in, but eventually there are enough atoms escaping to break down the structure of the solid material. This happens at a particular temperature, the melting point of argon. The graph enables an approximate value of the melting point to be obtained, and the calculation is done in the following example.

Example 22.1 shows how the attraction between atoms is binding them together, whereas the kinetic energy of the atoms is responsible for separating them. At one particular temperature, the melting point, enough bonds are broken so that the atoms do not remain in a fixed pattern. At a higher temperature, the

EXAMPLE 22.1

The mean kinetic energy E_k of any atom at a temperature T is given by $E_k = \frac{3}{2}kT$ where k is the Boltzmann constant, 1.38×10^{-23} J K^{-1} (see section 25.4). A significant number of atoms have kinetic energy $10kT$. Find the melting point of argon using data from the graphs in **Fig 22.6**.

From the graphs,

gain in potential energy required to release an argon atom from one other argon atom $= 1.7 \times 10^{-21}$ J

gain in potential energy required to release an argon atom from solid argon $\quad = 6 \times 1.7 \times 10^{-21}$ J

and from the information at the start of the example,

kinetic energy of high-speed argon atoms
$$= 10kT = 10 \times (1.38 \times 10^{-23} \text{ J K}^{-1}) \times T$$

Enough high-speed atoms break away from the crystal when this kinetic energy is equal to the potential energy gain required, so

$$6 \times 1.7 \times 10^{-21} \text{ J} = 10 \times (1.38 \times 10^{-23} \text{ J K}^{-1}) \times T$$
$$T = 74 \text{ K}$$

Despite the approximations made in making this calculation, it gives a figure surprisingly close to the actual melting point of argon, which is 84 K.

boiling point, the higher value of the kinetic energy of the atoms is sufficient for atoms not only to be removed from their potential well but also to be able to do work against the air molecules surrounding them and to escape from the surface into vapour form. For argon, this occurs at 88 K.

QUESTION

22.5 Use the graphs of **Fig 22.6** to answer the following:

a What force exists between te two atoms if they are at a separation of 0.39 nm? Is this force one of attraction or repulsion?

b How much energy is needed to remove an argon atom from a distance of 0.50 nm from a neighbouring argon atom to a large distance (infinity)?

c What is the maximum kinetic energy gained by an argon atom if it is released a large distance from another argon atom and then falls in towards it?

d Where will the atom in **c** first stop?

e What is the gradient of the energy–distance graph when the separation of the molecules is 0.50 nm? [There is a quick way of finding this.]

22.4 SOLIDS

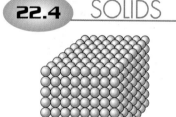

FIG 22.7 Molecules (or atoms) considered as spheres packed in a cubic pattern.

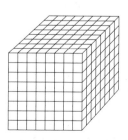

FIG 22.8 Molecules (or atoms) assumed to be cubes packed in a cubic pattern.

A solid normally has a fixed shape; it remains in its shape because there is a fixed pattern of the molecules (or atoms) within it. The molecules vibrate about their mean position, but the amplitude of the vibration is small compared with their separation. The arrangement of the molecules in a mass of a solid is controlled by several factors. One of these factors is the shape of the molecule; another factor is the strength of the attractive force between the molecules. Temperature has an influence on the pattern. Many solids have different molecular arrangements at different temperatures. The stresses that have been applied to the solid also affect its molecular pattern, so the way in which the solid has been treated in the past has an effect on its pattern of molecules and hence on its strength.

The strength of solids is a complex matter and is part of the study of a materials scientist. More detail will be given in Chapter 23 about the effects of applying stress to a solid, but here it is necessary to consider how the pattern of molecules within a solid is related to the density of the solid. To simplify the problem at the start, assume that the molecules behave as solid spheres and that they are packed in a cubic structure as shown in **Fig 22.7**. For ease of calculation, it is often simpler to imagine that each molecule occupies the volume of a cubical box as shown in **Fig 22.8**. If we use the data for argon from the graph in **Fig 22.6**, we can estimate the density of solid argon as shown in the following example.

EXAMPLE 22.2

Find the density of solid argon using data from the potential energy graph (**Fig 22.6**). Explain why the value you obtain is less than the true value. [The relative atomic mass of argon is 40.]

From the previous section

atomic mass unit $= 1.66 \times 10^{-27}$ kg

so

mass of an atom of argon
$= 40 \times 1.66 \times 10^{-27}$ kg $= 6.64 \times 10^{-26}$ kg

From the graph

separation of the centres of two argon atoms in equilibrium $= 0.40$ nm $= 4.0 \times 10^{-10}$ m

so

volume of a cubical box in which there is one argon atom $= (4.0 \times 10^{-10}$ m$)^3$

The density of solid argon is therefore

$$\text{density of argon} = \frac{\text{mass of one atom}}{\text{volume occupied by one atom}}$$

$$= \frac{6.64 \times 10^{-26}\,\text{kg}}{(4.0 \times 10^{-10}\,\text{m})^3}$$

$$= 1040\,\text{kg m}^{-3} \quad \text{(to 3 sig. figs)}$$

This value is lower than the value found by experiment. The reason for this is that argon atoms can and do pack together more tightly than in a simple cubic structure. **Fig 22.9** shows how closer packing of spheres can be achieved. When spherical molecules (or atoms) are packed in three dimensions, they can be close-packed in two ways. **Figs 22.10** and **22.11** show these two arrangements, called hexagonal close-packed and cubic close-packed (sometimes called face-centred cubic).

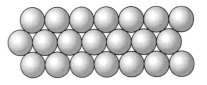

FIG 22.9 Closer packing can be achieved by using a hexagonal pattern. This is called close packing.

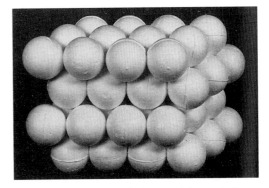

FIG 22.10 A model showing the hexagonal close-packed structure.

FIG 22.11 A model showing the cubic close-packed structure.

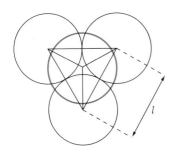

FIG 22.12 The geometry of the atoms in question 22.7.

QUESTIONS

22.6 What is the mean separation of the atoms in argon gas of density 1.4 kg m^{-3}?

22.7 What is the spacing of the atoms in copper of density 8900 kg m^{-3}?

[The relative atomic mass of copper is 63.5. Assume that each atom occupies a cubical box to begin with and then, if your geometry is good enough, find the distance between layers of atoms. **Fig 22.12** shows three atoms in a layer, with a single atom from the next layer, shown coloured green, resting on them. Join together the centres of the four atoms to make a tetrahedron of length of side l. The height of the tetrahedron is the separation of the layers of atoms.]

22.5 FLUIDS

In a liquid, the pattern of the arrangement of a few molecules is, at any instant, very similar to the pattern in a solid. For this reason, the density of liquids is about the same as the density of the corresponding solid. In most cases, the liquid has a lower density than its solid, but water is an exception. The density of ice, as given in Table 22.1, is only $920 \, \text{kg m}^{-3}$, as compared with water of density $1000 \, \text{kg m}^{-3}$. At the molecular level, therefore, molecules in water are packed in more closely than they would be in ice.

The essential difference between molecular behaviour in a solid and in a liquid is that, in the liquid, the pattern of molecules is not fixed; and where pattern does exist, it does not extend far within the liquid. Individual molecules do not maintain contact with the same adjacent molecules for any appreciable time, and the liquid is therefore able to change its shape. **Fig 22.13** is a sketch illustrating this difference. The ability of molecules to flow past one another in a liquid means that a liquid is unable to retain its shape when stretching forces act on it. Nor can it retain its shape when shear forces act on it — shear forces cause an object to be twisted. A liquid can be put under compressive forces, however, in which case its volume will be little changed.

It can be useful to get a visual impression of the differences between molecular behaviour in solids, liquids and gases. Sketches and models of atoms must not be taken too literally, however, and you should not lose sight of the complexity of the actual substance. The huge number of molecules, and the three-dimensional nature of matter, cannot adequately be sketched.

a Solid   **b** Liquid   **c** Gas

FIG 22.13 **a** The atoms in a solid move around, but stay in the same mean position; the atoms are in a regular pattern. **b** In a liquid, there is some pattern, but atoms are able to drift from place to place. **c** In a gas, no pattern of movement is discernible.

22.6 CHANGE OF PHASE

In order to change phase, energy must be supplied. This energy is called **latent heat**.

■ The **specific latent heat of fusion** is the energy per unit mass required to change the substance from the solid phase to the liquid phase.
■ The **specific latent heat of vaporisation** is the energy per unit mass required to change the substance from the liquid phase to the gas phase.

Both of these quantities are given by the following equation:

$$L = \frac{Q}{m}$$

where L is the specific latent heat of fusion or of vaporisation, m is the mass and Q is the quantity of heat energy supplied.

If the change in phase takes place at a fixed temperature, then the energy supplied does not increase the kinetic energy of the molecules of the substance. The average kinetic energy of water molecules at its boiling point of 100 °C is the same whether the water is in the form of liquid water or steam. The energy supplied, as indicated in section 22.3, is used in two ways. First, it increases the potential energy of the system by increasing the distance between the molecules. Secondly, it does work by pushing away molecules in the atmosphere. Some values of melting and boiling points and of specific latent heats of fusion and of vaporisation are given for some typical changes of phase in Table 22.2.

TABLE 22.2 Melting and boiling points, and associated specific latent heats of fusion and vaporisation, for the changes of phase of various substances

SUBSTANCE	NORMAL MELTING POINT/°C	SPECIFIC LATENT HEAT OF FUSION kJ kg^{-1}	NORMAL BOILING POINT/°C	SPECIFIC LATENT HEAT OF VAPORISATION kJ kg^{-1}
helium	−270	5.23	−269	21
hydrogen	−259	58.6	−253	452
oxygen	−219	13.8	−183	213
nitrogen	−210	25.5	−196	201
ethanol	−114	104	78	854
mercury	−39	11.8	357	272
water	0	334	100	2260
sulphur	119	38.1	445	326
silver	961	88.3	2190	2340
copper	1083	134	1187	5070

EXAMPLE 22.3

The specific latent heat of vaporisation of water is 2.26×10^6 J kg^{-1} at a temperature of 100 °C. Steam at a temperature of 100 °C and a pressure of 1.01×10^5 Pa has a density of 0.59 kg m^{-3}. Find the work done against the atmosphere by 1 kg of water molecules escaping from a water surface to form 1 kg of steam, and hence find the increase in the potential energy of the molecules.

From the data given, we have

$$\text{volume occupied by 1 kg of steam at 100 °C} = \frac{1\,\text{kg}}{0.59\,\text{kg m}^{-3}} = 1.695\,\text{m}^3$$

$$\text{volume occupied by 1 kg of water at 100 °C} = \frac{1\,\text{kg}}{1000\,\text{kg m}^{-3}} = 0.001\,\text{m}^3$$

so the change in volume, ΔV, is

$$\Delta V = 1.695\,\text{m}^3 - 0.001\,\text{m}^3 = 1.694\,\text{m}^3$$

From section 7.2,

work done against the atmosphere
$$= p\Delta V$$
$$= 1.01 \times 10^5\,\text{Pa} \times 1.694\,\text{m}^3$$
$$= 0.171 \times 10^6\,\text{J}$$

From the information given at the start of the example

total energy supplied $= 2.26 \times 10^6$ J

so finally

increase in potential energy of the molecules
$$= (2.26 - 0.171) \times 10^6\,\text{J}$$
$$= 2.1 \times 10^6\,\text{J (to 2 sig. figs)}$$

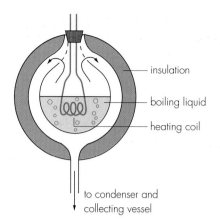

FIG 22.14 Apparatus for measuring the specific latent heat of a liquid.

Note from Example 22.3 that, even in the case of vaporisation, most of the work is done in increasing the potential energy of the molecules. The work needed to be done against the atmosphere to produce the increase in volume is very low, and in the case of a solid melting is negligible.

The numerical values of the specific latent heats vary with the temperature at which the change in phase occurs. To change water into water vapour by evaporation does not require quite as much heat as to heat the same quantity of water to its boiling point and then boil it at the boiling point. This is partly because the density of the vapour depends on the pressure.

To measure the value of the specific latent heat of vaporisation at the boiling point of the liquid, a continuous-flow method can be used. Energy is supplied electrically at a constant rate to the liquid, and the rate at which the liquid boils is found. A suitable apparatus is shown in **Fig 22.14**, where vapour from the boiling liquid provides a jacket surrounding the liquid itself. Outside this is placed another insulating jacket. At first, the heater is left switched on until the temperatures throughout the whole apparatus are steady. At this stage the mass M of liquid that is boiled off in time t is measured. If the potential difference across the heater is V and the current is I then:

power supplied to heater = VI
power supplied to liquid = ML_v/t

where L_v is the specific latent heat of vaporisation. These two expressions are the same if no heat is lost, so

$$VI = \frac{ML_v}{t}$$

$$L_v = \frac{VIt}{M}$$

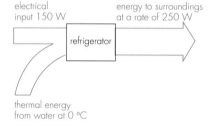

FIG 22.15 The energy flows for the refrigerator in question 22.8.

QUESTIONS

22.8 A refrigerator uses 150 W of electrical power. It extracts thermal energy from water at 0 °C and pumps 250 W of thermal energy into its surroundings, as shown in **Fig 22.15**. Using data from Table 22.2, find the maximum rate at which it can produce ice.

INVESTIGATION

An electric kettle, of the older type which does not automatically switch off when it boils, can be used to give a good estimate of the value of the specific latent heat of vaporisation for water. The power of the kettle will be written on its base. Some water is put in the kettle so that the element is well covered. The full kettle is weighed and the water is then boiled. Once the water has started to boil, the time is noted and the kettle is left switched on until a reasonable amount of water has been boiled away. The kettle is then switched off, the time noted and the energy supplied during the measured time is calculated.

If the kettle and remaining water are now reweighed, then the mass of water boiled away by the energy supplied can be found.

Find what fraction of the energy supplied is wasted by comparing your value with the accurately measured value of the specific latent heat of vaporisation of water, which is given in Table 22.2.

What factors determined the length of time you used for boiling water away?

22.9 The Mediterranean Sea loses a substantial quantity of water by evaporation. This water is replaced by rainfall and by water from the Black Sea, from rivers flowing into it and from the Atlantic Ocean. Using the following data, estimate the rate of evaporation on a fine summer day.

Area of the Mediterranean Sea = 2.9×10^6 km^2
Latent heat of vaporisation of water at sea temperature = 2.45 MJ kg^{-1}
Power of sunlight used for evaporation = 0.9 kW m^{-2}

22.7 PRESSURE

Pressure is **not** a vector

Although pressure is defined in terms of a vector quantity, force, it is not itself a vector. This is because the area on which the force is acting is a vector having an associated direction. In dividing force by area, a vector is being divided by a vector and the result is a scalar quantity.

Pressure

Pressure is defined as the normal force per unit area. Strictly the **average pressure** on a surface is the total force divided by the total area. The **pressure at a point** p is given by the limit

K $p = \left(\dfrac{\delta F}{\delta A} \right)$ as δA **tends to zero**

where δF is the small normal force applied to a small area δA. The SI unit of pressure is the pascal. One **pascal** (1 Pa) is a pressure of one newton per square metre (1 N m^{-2}).

EXAMPLE   22.4

Find the pressure exerted on the foundations by a wall that is 2.0 m high, 20 m long and 0.30 m wide, if the concrete blocks from which it is constructed have a density of 2500 kg m^{-3}.

We have

volume of wall = 2 m $\times$ 20 m $\times$ 0.30 m = 12 m^3
mass of wall = 12 m^3 $\times$ 2500 kg m^{-3} = 30 000 kg
weight of wall = 30 000 kg $\times$ 9.8 N kg^{-1} = 294 000 N
area of base of wall = 20 m $\times$ 0.30 m = 6.0 m^2

and so we get

$$= \frac{\text{total force}}{\text{total area}} = \frac{294\,000\,\text{N}}{6\,\text{m}^2} = 49\,000\,\text{Pa}$$

Note that some unnecessary information has been given in this example. The width of the wall and its length are not needed.

 Work through the problem again using algebraic symbols for width and length and see that they cancel out.

QUESTION

22.10 what is the maximum mass that a house can have if it is built on clay foundations that can have a maximum pressure on them of 220 000 Pa. The area of the foundations is 20 m^2.

Fluid pressure

The word **fluid** means capable of flowing. When a material is said to be a fluid, it must therefore be in its liquid or gas phase. In other words, both liquids and gases are fluids. A solid will transmit a force from one place to another; a fluid transmits a pressure. The pressure exerted by a column of fluid on its base can be found by reference to **Fig 22.16**, where a column of height h and cross-sectional area A is formed from a fluid of density ρ. (The column is shown without any support. In practice, any column of liquid requires support at the sides, possibly from glass tubing.) The pressure on the base is found as follows:

$$\text{volume of fluid in column} = hA$$
$$\text{mass of fluid} = hA\rho$$
$$\text{weight of fluid} = hA\rho g$$

so

$$\text{pressure} = \frac{\text{weight}}{\text{area}} = \frac{hA\rho g}{A} = h\rho g$$

The pressure due to a column of liquid is independent of its area of cross-section. Since the pressure is dependent on the density, care has to be taken when the fluid has a variable density. In finding atmospheric pressure, for instance, the density of the atmosphere must not be taken as having a uniform value. The standard of **atmospheric pressure** is taken as being equivalent to that of a 0.760 m high column of mercury (density 13 600 kg m^{-3}). This gives standard atmospheric pressure p_a as

$$p_a = (0.760 \text{ m}) \times (13\,600 \text{ kg m}^{-3}) \times (9.8 \text{ N kg}^{-1}) = 101\,000 \text{ Pa}$$

Because of variations of g from place to place, this value varies slightly. To overcome this problem, when high accuracy is required, a standard atmosphere is now defined as 101 325 Pa.

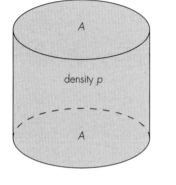

FIG 22.16 A column of fluid exerts a pressure $h\rho g$ on its base.

ANALYSIS

Blood pressure

A doctor measures the blood pressure of a patient with an instrument called a sphygmomanometer. The doctor measures the maximum (systolic) pressure and the minimum (diastolic) pressure and records the systolic/diastolic ratio. A typical, healthy adult at rest will have readings of about 120/80. These readings are the differences in levels in a mercury manometer measured in millimetres and give the pressure excess above atmospheric pressure. Readings above 140/90 are usually regarded as high blood pressure (hypertension).

To take the readings, the doctor forces air into the air sack wrapped around the upper arm at the level of the heart, while at the same time listening to the blood flow through the main artery in the arm (**Fig 22.17**). When the pressure in the air sack is just sufficient to prevent all blood flow, the pressure is the systolic pressure, and the difference in the levels of the mercury is measured. A doctor will normally raise the pressure in the air sack to a value above this pressure and then let it fall gently until blood just starts to flow immediately after the heart has pumped and the pressure is at a maximum. The diastolic pressure can be obtained by lowering the pressure in the air sack still further so that blood is only just being stopped from flowing for a brief interval at the end of the pumping cycle.

In this exercise, assume that a person has blood pressure readings of 120/80, and answer the following questions:

a What would these readings be when measured in pascals?

b What are the maximum and minimum pressures actually present in the blood of the person's arm if atmospheric pressure is 100 000 Pa?

c Show that it is reasonable that the maximum blood pressure in the foot of the person is 130 000 Pa.

d Find by how much the pressure in a person's brain changes when she bends so that her brain, instead of being 0.4 m above her heart, is 0.3 m below her heart.

e [A difficult concept is involved here.] Show that the systolic blood pressure in an astronaut's brain will be zero if he is accelerating in a direction straight upwards towards his head with an acceleration of about 28 m s^{-2}. Assume that his brain is 0.4 m above his heart and that no body mechanisms operate to compensate for such conditions. You may be familiar with a similar effect to this. If you stand up suddenly, you can sometimes feel light-headed as a result of the drop in blood pressure in your brain.

[Relative density of mercury = 13.6. Relative density of blood = 1.06.]

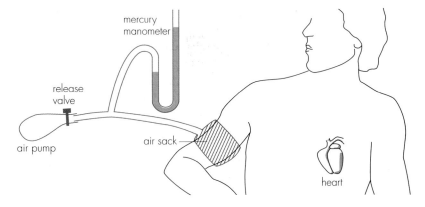

FIG 22.17 A doctor measures the blood pressure of a patient at the level of the heart.

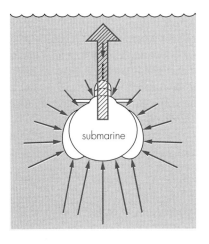

FIG 22.18 The forces acting on a submarine due to water pressure result in a resultant upward force on the submarine. This is the buoyancy force or upthrust.

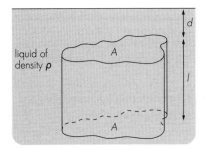

FIG 22.19 There is a buoyancy force $l\rho gA$ on a submerged object.

Buoyancy and Archimedes' principle

The equation given in the previous subsection for the pressure at the base of a column of fluid, i.e. the pressure beneath the surface of a fluid, shows that the pressure increases with depth. Therefore, if an object is submerged in a fluid, the pressure at the bottom of the object is greater than the pressure at the top.

Fig 22.18 shows a submerged submarine. On the submarine are drawn the forces that the water pressure causes, and they can be seen to be acting at right-angles to the surface of the submarine. The sideways forces cancel out, but because there are larger upward forces on the bottom of the submarine than there are downward forces on the top, the resultant of all of these forces is an upward force on the submarine.

The resultant force on a submerged object due to the pressure of the fluid surrounding it is called the **buoyancy force** or **upthrust**. The same reasoning would show that, as a result of the air surrounding a person, everybody has a buoyancy force acting *upwards* on them due to the pressure of the air. The force due to air pressure acting downwards on your head is less than the upward force on your feet because of the higher air pressure near the ground. The buoyancy force on a person is small. Someone with a weight of 800 N usually has a buoyancy force of about 1 N.

To calculate the value of the buoyancy force, consider an object with uniform cross-sectional area A, and which has a horizontal top and bottom. It has a length l and is placed so that its top is a distance d beneath the surface of a liquid of uniform density ρ (**Fig 22.19**). Then

$$\text{pressure on top} = d\rho g$$
$$\text{force downwards on top} = d\rho gA$$
$$\text{pressure on bottom} = (d + l)\rho g$$
$$\text{force upwards on bottom} = (d + l)\rho gA$$

so we get

$$\text{buoyancy force} = (d + l)\rho gA - d\rho gA = l\rho gA$$

FIG 22.20 The supertanker Golar Nikko has a total mass of 105 000 tonnes. It therefore needs to displace 105 000 tonnes of water in order to float.

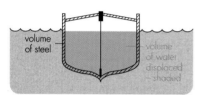

FIG 22.21 The volume of water a ship displaces is very much larger than the volume of steel from which it is constructed. In equilibrium, the upthrust on the ship equals its weight.

Since lA is the volume of the object, $lA\rho$ is the mass of the fluid that the object displaces, and $l\rho gA$ is the weight of fluid displaced. This is a deduction of **Archimedes' principle**, which states that: For an object immersed in a fluid, the buoyancy force is equal to the weight of the fluid displaced. It can be applied generally and not just when the object has a regular shape.

In an object floats, it does so because the buoyancy force acting on it is equal and opposite to its weight. An object that sinks has a buoyancy force acting on it that is less than its weight. The shape of any piece of material is bound to have an effect on the weight of fluid that it displaces. A steel boat can float because the steel is shaped to displace a large volume of water; the volume displaced is clearly much larger than the volume of the steel itself (**Figs 22.20** and **22.21**).

QUESTION

22.11 What is the buoyancy force on a human body of volume $7.4 \times 10^{-2}\,\text{m}^3$ when totally immersed in

a air of density $1.3\,\text{kg m}^{-3}$,

b sea-water of density $1030\,\text{kg m}^{-3}$.

22.12 It is found that 24 m of a supertanker is below the surface when it is in sea-water of density $1030\,\text{kg m}^{-3}$. What depth will be below the surface when it enters fresh water of density $1000\,\text{kg m}^{-3}$? Assume the super-tanker has vertical sides.

22.13 a The Arctic ice-cap is a floating mass of ice. By how much would the sea-level in the oceans rise if the Arctic ice-cap melted as a result of an increased greenhouse effect?

b Why is this answer different from the answer that would be obtained if the question had referred to the Antarctic ice-cap?

 THE BERNOULLI EFFECT

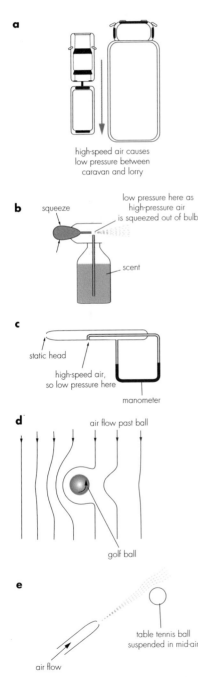

FIG 22.22 A liquid flows at low speed in the wide tube, and speeds up as it enters the narrow tube.

a

high-speed air causes low pressure between caravan and lorry

b squeeze — low pressure here as high-pressure air is squeezed out of bulb — scent

c static head — high-speed air, so low pressure here — manometer

d air flow past ball — golf ball

e table tennis ball suspended in mid-air — air flow

FIG 22.23 Some examples of the use of the Bernoulli effect.

Consider some liquid flowing along a pipe that suddenly narrows, (see **Fig 22.22**). When the volume contained between A and B gets into the narrow section, its shape alters and it fills the space between C and D. Since the distance CD is greater than the distance AB, it follows that the speed of the liquid in the narrow pipe has increased, and consequently its kinetic energy must be higher (its mass has not changed). The increase in kinetic energy can only come about as a result of work being done on the liquid; it must have been pushed from the wide to the narrow part of the tube. This push implies that there is a difference in pressure between the narrow and the wide parts of the tube, and that the pressure is higher when the liquid is flowing in the wide part of the tube. This can be shown by attaching pressure gauges to the tube at different places and seeing that the pressure is indeed lower when the liquid travels fast. The reverse process happens when a liquid slows down if it travels from a narrow tube to a wide one. This effect, of there being low pressure where liquid (or gas) travels at high speed, is known as the **Bernoulli effect**, and it is used in many everyday devices from carburettors to spinning golf balls.

Practical examples of the Bernoulli effect

Fig 22.23 shows five situations that demonstrate the Bernoulli effect. In explaining these effects it is sometimes easier to assume that the object is stationary with the air (or other fluid) moving past it, when in fact it is the object that is moving through the air. The two situations are equivalent.

In **Fig 22.23a** a car towing a caravan is overtaken by a lorry. Air between them flows faster, as it is being funnelled through a narrow gap, so the pressure between them is low. The caravan therefore has higher pressure on one side and lower pressure on the side near the lorry. This results in the caravan being pushed towards the lorry, and it is very disconcerting for the car driver to feel this sideways pull.

A scent spray is shown in **Fig 22.23b**. A rubber bulb is squeezed and air emerges rapidly through a small hole just above a tube leading down into the scent. Low pressure at this point allows atmospheric pressure on the surface of the scent to push scent upwards through a separate small hole and into the air stream, where it atomises (becomes a mass of small droplets). A carburettor uses a very similar arrangement to a scent spray in order to get a petrol–air mixture into car cylinders.

In **Fig 22.23c** a Pitot tube is illustrated. This is a speed-measuring device for the pilot of a glider or a small plane. The hole at the front of the tube, called the static head, has stationary air in front of it. The hole on the side, the dynamic head, has air rushing past it, and so the pressure is lower. A manometer (a pressure gauge) connected between the two gives a pressure difference, which is dependent on the speed of the aircraft and so can be calibrated in metres per second (or knots).

When a golf ball is hit, either by accident or by design, with the club face not at right-angles to the direction of travel, the golf ball will spin. This results in the air flow around the golf ball being greater and faster on one side, as shown in **Fig 22.23d**.

 On which side of the golf ball is the pressure higher? Which way will the golf ball swerve? Explain your answer.

The last example, in **Fig 22.23e**, shows a fairground trick with a table tennis ball. The ball is placed in a high-speed stream of air from a pipe with a nozzle at its end. When the pipe is tilted, the ball will remain in mid-air without any apparent support. Since the high-speed air travels over the top of the ball, the pressure above it is lower than the pressure below it. This pressure difference is sufficient to support the ball. If you do not believe this to be possible – try it. A piece of glass tubing drawn out to make a narrow nozzle works well when connected up to the output of a vacuum pump.

The aerofoil section

When the wing of an aircraft travels through air, it deflects air downwards and so gives some lift to the aircraft. In addition to this lift, however, the shape of the wing, known as an *aerofoil section*, is designed to have the air travelling across its upper surface moving at a greater speed than that travelling beneath the wing (see **Fig. 22.24**). As a result of the Bernoulli effect, this gives a higher pressure beneath the wing and a lower pressure above it, resulting in extra lift. The equation for the lift force F is

$$F = \tfrac{1}{2}SC_L\rho v^2$$

where S is the surface area of the underside of the wing, v is the speed of the wing through the air, ρ is the density of air and C_L is called the lift coefficient. The lift coefficient varies with the attitude of the plane. If it is nose up, then the lift coefficient is higher, provided the nose is not too high up, as at some angle an aircraft will stall.

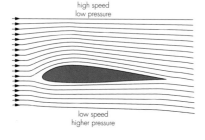

high speed
low pressure

low speed
higher pressure

FIG 22.24 The effect that an aerofoil section has on air flowing past it.

EXAMPLE 22.5

Calculate the speed of take-off that would be required for an aircraft of total mass 2.3×10^4 kg and wing area 65 m^2 if it is dependent solely on the Bernoulli effect for lift. The lift coefficient is 0.83 at take-off and the density of air is 1.3 kg m^{-3}.

We have
lift required = weight of aircraft
$= 2.3 \times 10^4$ kg $\times$ 9.8 N kg^{-1} = 225 400 N and
lift force $= 0.5 \times 65$ m$^2 \times 0.83 \times 1.3$ kg m$^{-3} \times v^2$

Equating these two we get, on rearranging, the equation

$$v^2 = \frac{225\,400\,\text{N}}{0.5 \times 65\,\text{m}^2 \times 0.83 \times 1.3\,\text{kg m}^{\pm 3}}$$

giving $v = 80$ m s$^{\pm 1}$.

QUESTIONS

22.14 Show that the lift coefficient has no unit.

22.15 a Explain why the lift calculated in Example 22.5 will have to be almost the same when the aircraft is cruising at an altitude of 10 000 m.

b Which factors in the lift equation will probably be larger and which smaller when at cruising altitude? Explain your answer.

22.16 Using the data given in Example 22.5, calculate the take-off speed required if the aircraft has no passengers and less fuel aboard and can therefore reduce its total mass to 1.6×10^4 kg.

SUMMARY

- Density is mass per unit volume:

$$\rho = \frac{m}{V}$$

- $$\text{Relative density} = \frac{\text{density of substance}}{\text{density of water}}$$

- The mole is the amount of substance that contains as many entities as there are atoms in 0.012 kg of carbon-12.

- The unified atomic mass constant, u, is one-twelfth of the mass of a single atom of carbon-12:

$$1 \, u = 1.66 \times 10^{-27} \, kg$$

- The Avogadro constant N_A is numerically equal to the number of atoms in 0.012 kg of carbon-12:

$$N_A = 6.022 \times 10^{23} \, mol^{-1}$$

- Properties of the three phases of matter:

	SOLID	LIQUID	GAS
Pattern of atoms	Usually ordered over relatively large distances	Ordered over relatively short distances	No order
Mobility of atoms	Atoms remain in fixed positions relative to one another	Atoms can slide over one another; no fixed structure	Atoms move at random
Density	Atoms closely packed; high density	Atoms closely packed; similar high density to solid	Atoms widely scattered; density of the order of a thousandth the density of the liquid

- Pressure is force acting perpendicularly per unit area; one pascal (1 Pa) is $1 \, Nm^{-2}$.

- Pressure at a depth h in a fluid of density ρ is

$$p = h\rho g$$

- Atmospheric pressure = 0.760 m of mercury
$$= (0.760 \, m) \times (13\,600 \, kg \, m^{-3}) \times (9.8 \, N \, kg^{-1})$$
$$= 101\,000 \, Pa$$

- Archimedes' principle: A body immersed in a fluid has an upthrust on it equal to the weight of the fluid displaced. For a floating body, this is equal and opposite to the weight of the body.

- The Bernoulli effect is that the pressure in a high-speed fluid is low and in a low-speed fluid is high.

ASSESSMENT QUESTIONS

22.17 a A *partially inflated* balloon containing hydrogen carries meteorological equipment up through the atmosphere. Over a short distance its upward speed is almost constant. Draw and label a diagram to show the forces acting on the balloon.

b Explain what can be deduced about the resultant force on the balloon.

c The balloon would not rise if the hydrogen were replaced by an identical volume of air at the same temperature and pressure. Explain why not.

d Explain what happens to each of the following as the balloon and equipment rise a significant distance through the atmosphere.
  i The pressure of the hydrogen
  ii The volume of the hydrogen

London 1996

22.18 The diagram shows a sealed container filled with air and attached to a manometer. The liquid in the manometer is an oil of density 860 kg m^{-3}. The vertical distance between the two oil surfaces is 11.2 cm. The atmospheric pressure is 101.2 kPa.

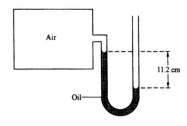

a Calculate the pressure in the sealed container.
b The oil in the left hand arm has more potential energy than the oil in the right hand arm. Identify the source of this potential energy. Explain your answer.
c With reference to kinetic theory, explain why the pressure of the air in the container rises if the temperature rises.

London 1998

22.19 a A physicist inflates a balloon with air to a volume of 1.5 litres and seals it. The density of the surrounding air at the time of the experiment was 1.30 g per litre. Calculate the upthrust on the balloon after the inflation process.

b Sketch and label a free-body force diagram to show the principal forces which act on the inflated balloon at rest on a balance. Given that the mass of air in the balloon is 2.15 g and the mass of the balloon fabric is 4.10 g, show the magnitudes of the forces.

c Calculate the density of the air in the balloon.

d Why is the density of the air in the balloon greater than the density of the air outside the balloon?

London 1998

22.20 The diagram below shows a submarine of total volume 200 m^3 whose depth below the surface of the sea is controlled by the amount of sea water held in a storage tank. The submarine is cruising some distance under the surface. The storage tank contains 10 m^3 of sea water. The density of sea water is 1025 kg m^{-3}.

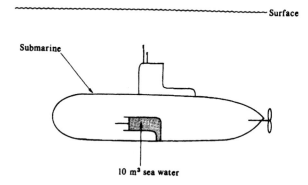

a Calculate the mass of the submarine and its contents.

b If the atmospheric pressure is 1.00×10^3 Pa, calculate the energy needed to empty the water storage tank when the water outlet is 25 m below the sea surface.

The sea water has now been pumped out of the storage tank.

c Calculate the net upward force on the submarine.

d Calculate the maximum value for the upward acceleration of the submarine.

London 1997

22.21 a The atoms of solid sodium are arranged in a body-centred cubic structure. The unit cell of the b.c.c. structure is shown below.

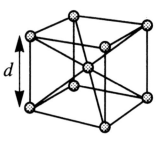

Calculate the length of the cube edge d in this structure.

the density of solid sodium $= 9.7 \times 10^2$ kg m^{-3}
the relative atomic mass of sodium $= 23$

the Avogadro constant = 6.0×10^{23} mol^{-1}

b **i** State a method which could be used to determine *d* experimentally.

ii State *one* other physical property of a polycrystalline specimen which may be determined using the method suggested in part **i**.

NEAB 1997

22.22 The diagram shows an incomplete sphygmomanometer which is a device used for blood pressure measurement.

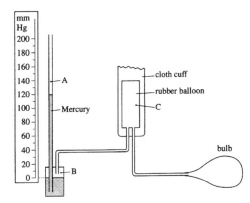

Three valves are required to allow the balloon to be inflated and deflated. Mark the positions of the three valves at suitable places on a sketch diagram. State what each valve is used for.

b If the air pressure is taken as 760 mm of mercury, what is the pressure, in mm of mercury, at points A, B and C on the diagram?

c Describe the procedure for obtaining the systolic blood pressure reading using this instrument.

d A systolic blood pressure reading for a patient is given as 120 mm of mercury. What is this value expressed in pascal?

density of mercury = 1.36×10^4 kg m^{-3}

e This blood pressure reading was, as usual, measured in the arm at the same level as the heart. Estimate the pressure, in pascal, in the foot of this adult patient if she was standing.

density of blood = 1.06×10^3 kg m^{-3}

f Fainting is caused by reduced blood flow to the brain. Explain why one suggested treatment is to lower the patient's head.

NEAB 1997

22.23 In an experiment to estimate the size of an atom, a spherical oil droplet of diameter 0.46 mm is placed on to a clean smooth surface of water. The oil spreads over part of the water surface to form a circular patch of uniform thickness and diameter 78 mm. The thickness of the patch is calculated by using the volume of the droplet and the diameter of the circular patch. The thickness of the patch is taken as an estimate of the size of an atom.

a **i** Using the information above, obtain an estimate of the size of an atom.

ii What is the commonly accepted approximate size of atoms?

iii State *two* reasons why the estimate obtained in part **a i** differs from the commonly accepted approximate size of atoms.

b Name *one* other method of obtaining a value for the size of atoms.

NEAB 1998

22.24 An artery has a cylindrical cross-section of diameter 8 mm. Blood flows through the artery at an average speed of 0.3 m s^{-1}.

a Calculate the average mass of blood flowing per second through the artery.

Density of blood at body temperature, 37 °C = 1060 kg m^{-3}.

b How would you verify experimentally the value of the density of blood quoted above, given a sample of volume approximately 5 cm^3?

London 1996

22.25 a Define *density*.

b The density of water is 1000 kg m^{-3}. Calculate the volume occupied by 0.018 kg of water.

c In water, the mean distance between molecules is 3.10×10^{-10} m. In ice, the mean distance is 3.19×10^{-10} m. Making the simplifying assumption that the arrangement pattern of the molecules is the same in both water and ice, calculate

i the ratio $\dfrac{\text{volume occupied by an ice molecule}}{\text{volume occupied by a water molecule}}$

ii the density of ice.

d Suggest an approximate value for the mean distance between molecules in steam at atmospheric pressure.

Cambridge 1997

22.26 For each of the statements below, say whether the statement is true or false.

a Two conducting objects in contact must be in thermal equilibrium if they have the same temperature.

b Energy is not conserved in oscillating systems unless the speed is constant.

c The specific heat capacity of a substance depends on the amount of substance present

d The specific latent heat (enthalpy) of a substance is associated with changes in molecular potential energy.

e Both melting and vaporisation can proceed at the same time in a mass of substance at constant temperature.

London 1998

23

DEFORMATION OF SOLIDS

LEARNING OBJECTIVES

At the end of this chapter you should be able to:

① define and use the terms 'stress' and 'strain';

② distinguish between elastic and plastic behaviour;

③ explain the behaviour of solids in terms of molecular structure;

④ describe the properties of substances that are ductile, brittle and polymeric.

23.1 STRESS AND STRAIN

Rigidity

Newton's laws state that, when a resultant force is applied to an object, the object will accelerate. On the other hand, a piece of railway track does not seem to accelerate if suddenly a train travels over it. The track seems to have a force applied to it and yet not to accelerate. The track is held rigidly in position and so does not appear to move. The key word here is 'rigidly'. What is meant by a rigid substance? Rigid substances are all solids. Solids have the ability to retain their shape when forces are applied to them in a way that fluids cannot. As explained in section 22.4, this is because if the individual atoms are moved closer to one another they will repel, whereas when they are pulled further apart they will attract. Note, however, that a small movement is necessary to cause a change in the force that exists between the atoms.

This explains the paradox posed at the opening to this section. The force applied to the track by the train *does* cause downwards acceleration of the track. But the additional large forces set up within the parts of the track by atoms being pushed closer together, limit the distance within which acceleration occurs. The vertical distance moved downwards by the track is small, but it is visible if you look at a railway track as a train passes over it.

FIG 23.1 The large acceleration of the ball is caused by the large force acting on it. This large force causes considerable elastic distortion.

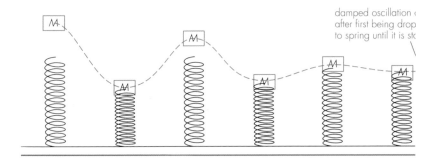

FIG 23.2 Distortion of a spring when a mass falls on to it.

This and similar effects are normally ignored in everyday life, except in such cases as when a person lies on a mattress and the distortion is clearly visible. When we think of a rigid object, we assume that it does not distort when forces are applied to it. But from the point of view of a study of physics, it is essential to realise that all real substances *do* distort when under the influence of an applied force. All substances are to some extent elastic. Elastic distortion of a tennis ball can be seen in **Fig 23.1**. When you drop a piece of steel on to a concrete floor, there might not be much distortion occurring, but there is some. The concrete and the steel both change shape, and large inter-atomic forces are set up, which will stop the movement and, after some bouncing, result in an equilibrium position.

Fig 23.2 illustrates a similar effect when a mass M is allowed to drop on to a spring. Distortion of the spring occurs, the mass stops after some potential energy has been stored by the spring, and oscillation takes place until damped out by friction. The spring and mass reach a different equilibrium position after the mass is added. Putting a book down on a table is physically similar to this, with M being replaced by the book and the spring representing the table. The table is equivalent to a very stiff spring, so the new equilibrium position is established very quickly and the amount of distortion is very little.

Tension

Another important feature of solid behaviour is the way in which solids can be stretched as well as being squashed. When a pair of stretching forces is applied to a solid, the solid is said to be in a state of **tension**. There are some tension effects in liquids but, because of the ease of moving adjacent atoms in liquids, the effect is normally only noticeable as a surface tension.

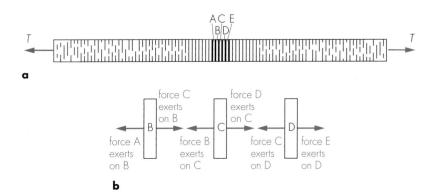

FIG 23.3 A horizontal wire can be considered as a series of slices, for some of which free-body force diagrams are drawn.

In a taut horizontal wire being pulled by equal forces, T, at both ends, each small section of the wire could be considered separately. This is shown in **Fig 23.3a**, where a few of the sections have been labelled A, B, C, D, E. Free-body diagrams for three of the sections (B, C, D) have been drawn in **Fig 23.3b** to show the forces on individual sections. Since all the sections are in equilibrium, the resultant force on each section must be zero. Using Newton's third law gives the force that C exerts on B to be equal and opposite to the force that B exerts on C. This can be continued along the length of the wire, so it follows that throughout the wire the two forces shown acting on each section in **Fig 23.3b** must be equal to T. T is called the *tension* in the wire. Note that tension T is caused by two opposite forces both of magnitude T acting on the ends of the wire. If the wire is held vertically, then the tension in the wire will vary because of the weight of the wire. The tension will be greater nearer the top. A wire undergoing an acceleration must have a different force applied at one end from the other. Very often, however, the assumption is made that the mass of the wire is negligible, and under these circumstances the difference between the forces at each end is negligible.

Stress

The **stress** applied to a wire to stretch it is defined as the force applied per unit area of cross-section. This is sometimes called the *tensile stress*, as forces can be applied in different ways to objects. Under compression, for instance, a *compressive stress* is applied. All stresses are defined as a force per unit area, and therefore they all have the same unit as pressure, the pascal (Pa). If a force T is applied to a wire of cross-sectional area A, then the stress σ is given by the equation

K $\quad \sigma = \dfrac{T}{A}$

EXAMPLE 23.1

The high-tensile steel used to support the roadway of a suspension bridge breaks under a stress of 1.61×10^9 Pa. Find the minimum cross-sectional area for a cable if it is to support a load of mass 2.0×10^5 kg (200 tonnes).

We have

weight of load $= 2.0 \times 10^5$ kg $\times 9.8$ N kg^{-1}
$\qquad\qquad\quad\; = 1.96 \times 10^6$ N

and this must be equal to the tension T in the cable. Now

we can substitute values of stress σ (from the information given) and force (tension T, from above) in $\sigma = T/A$ to find cross-sectional area A:

$$\sigma = 1.61 \times 10^9 \,\text{Pa} = \frac{T}{A}$$

$$A = \frac{1.96 \times 10^6 \,\text{N}}{1.61 \times 10^9 \,\text{Pa}} = 1.22 \times 10^{-3} \,\text{m}^2 = 12.2 \,\text{cm}^2$$

The *minimum* cross-sectional area of the cable must be 12.2 cm^2.

QUESTION

23.1 The calf muscle exerts a tensile force of 1500 N on the Achilles tendon in the lower part of the leg. If the area of cross-section at the centre of the muscle is 70 cm^2 ($= 0.0070$ m^2), and the area of cross-section of the tendon is 1.1 cm^2, find the tensile stress in each of these cross-sections.

Strain

As a result of applying a tensile stress to a wire, a tensile strain is set up within the wire. Stress is the cause, and strain is the effect. The tensile **strain** ε is defined as the extension per unit length. Strain is a ratio of two lengths and therefore does not have a unit. Using the symbols from **Fig 23.4** this can be written in equation form as

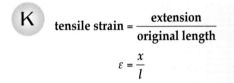

K $$\text{tensile strain} = \frac{\text{extension}}{\text{original length}}$$

$$\varepsilon = \frac{x}{l}$$

Strain is related to stress; and for most materials the strain is proportional to the stress provided the stress is small enough. When strain is proportional to stress, a material is said to be obeying **Hooke's law**, named after Robert Hooke who in the seventeenth century first formulated the law. A graph of stress plotted against strain is straight, up to a certain limit. More detail will be given about stresses beyond this limit in section 23.2. Two stress–strain graphs are shown in **Fig 23.5**. A more rigid material, such as steel, has a steeper slope than a less rigid material, such as aluminium.

The ratio of stress to strain is called a *modulus of elasticity*. If the ratio of stress to strain is large, then the material distorts only a little under the influence of the applied stress, and the material is stiff. For tensile stress and strain, the ratio is called the **Young modulus, E**. This gives

$$\text{Young modulus} = \frac{\text{tensile stress}}{\text{tensile strain}}$$

$$E = \frac{\sigma}{\varepsilon} = \frac{T/A}{x/l}$$

$$E = \frac{Tl}{Ax}$$

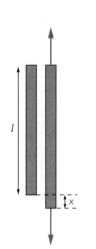

FIG 23.4 The strain ε in a wire is given in terms of length l and extension x as $\varepsilon = x/l$.

Hint

Note that E has a very high numerical value. In order to obtain this, the two numerical values that are large occur on the top of this expression and the two normally small values appear on the bottom.

EXAMPLE 23.2

Use the graphs in **Fig 23.5** to find the Young moduli of steel and aluminium.

Since the graphs are straight lines through the origin, the gradient of each graph is the Young modulus:

$$\text{Young modulus of steel} = \frac{2.0 \times 10^8 \, \text{Pa}}{0.0010} = 2.0 \times 10^{11} \, \text{Pa}$$

$$\text{Young modulus of aluminium} = \frac{1.0 \times 10^8 \, \text{Pa}}{0.0014} = 7.1 \times 10^{10} \, \text{Pa}$$

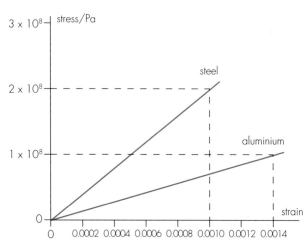

FIG 23.5 Steel requires a larger stress on it for a given strain than does aluminium.

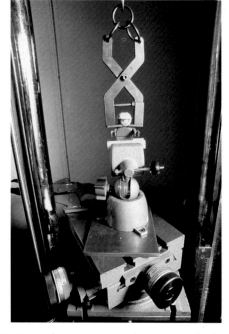

FIG 23.6 Testing to destruction is carried out on many familiar objects. Here a Lego toy is being assessed for safety. The grip will be pulled up until the toy snaps.

A change of convention

Graph axes are normally chosen to have the *cause* of any change plotted on the *x*-axis and the *effect* plotted on the *y*-axis. In the case of stress–strain graphs, however, the axes are, by convention, reversed, so that stress is plotted on the *y*-axis and strain on the *x*-axis.

TABLE 23.1 Readings for question 23.4

LOAD /N	SCALE READING /cm
20	3.021
30	3.070
40	3.118
50	3.167
60	3.215
70	3.264
80	3.314
90	4.270

QUESTIONS

23.2 A wire of diameter 1.0 mm and length 2.3 m is made of copper whose Young modulus is 1.1×10^{11} Pa. Find:

  a the radius and hence the area of cross-section of the wire in m^2;

  b the strain if the wire is stretched by 0.85 mm;

  c the stress;

  d the force necessary to cause this stress.

23.3 **a** The steel cable on a crane has a diameter of 13 mm. What force is necessary to produce a strain of 0.1% in the wire?

  b What is the maximum acceleration with which it can lift a load of mass 1000 kg if the strain in the cable is not to exceed 0.1%. The Young modulus of the high-tension steel used is 2.4×10^{11} Pa. (Be certain to draw a free-body diagram for the load.)

23.4 A copper wire 4.00 m long and 0.96 mm in diameter was used in an experiment to measure the Young modulus of copper. One end of the wire was clamped in a fixed position. The readings shown in Table 23.1 were obtained from the scale of a travelling microscope positioned at the other end of the wire for each load on the wire.

  a Plot a graph of scale reading against load and use the graph to find

    i the Young modulus of copper,

    ii the stress at the limit of proportionality (see next section).

  b This is not a very accurate way of using these data. Why not?

  c How can a more reliable value for the Young modulus of copper be found from the readings?

23.2 ELASTIC AND PLASTIC BEHAVIOUR

Elastic behaviour

In section 23.1 the stress applied to any piece of material was assumed to be sufficiently small that the resultant strain was proportional to the applied stress. If larger stresses are applied, then this ceases to be the case. A graph of tensile stress against tensile strain is shown in **Fig 23.7** for copper. The actual values given on the graph can be considered only as typical values, because they vary from one piece of copper wire to another, depending on the crystalline structure of the copper in the wire and on its previous treatment. Temperature also affects the strain. The end of the straight-line portion of the graph is called the **limit of proportionality**, and is marked as point A. At a point near to the limit of proportionality, the copper ceases to behave elastically and starts to behave plastically. This point is marked as B and is called the **elastic limit** or yield point.

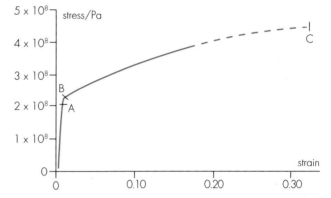

FIG 23.7 A graph of tensile stress plotted against tensile strain for copper.

The distinction between elastic and plastic behaviour is that when a material is elastic it returns to its original shape on removal of the distorting force, whereas if plastic it does not. Plastic behaviour means that the applied force has caused permanent deformation; elastic behaviour means temporary distortion. Most materials are elastic for low stresses and plastic for high stresses. It is important in civil engineering that constructional materials are subjected only to loads that cause elastic deformation. A bridge, distorted by a heavy load moving across it, must spring back into its original shape after the load has passed.

Plastic behaviour

Once a material is plastic, then a small increase in the stress can cause a very large increase in the strain. A material that behaves in this way is said to be **ductile**. It has a large plastic region. Within the plastic region, it is more difficult to measure the strain at a particular stress. There are several reasons for this difficulty.

- In the first case, as the wire is stretched, its area of cross-section reduces. This can be taken into account to begin with; but once the force is sufficiently large, then localised narrowing, called **necking**, takes place at weak points, and the wire will eventually break at one of these points.
- Another problem is that plastic distortion is time-dependent. For an applied stress, the initial strain will have a particular value; but if the strain is measured later, it is often found to have increased. This is known as **creep**.

TABLE 23.2 The Young modulus for different materials

MATERIAL	YOUNG MODULUS /10^{11} Pa
aluminium	0.70
copper	1.1
brass	0.91
iron	1.9
mild steel	2.1
glass	0.55
tungsten	4.1

TABLE 23.3 The breaking stress for these materials

MATERIAL	BREAKING STRESS /10^8 Pa
aluminium	2.2
copper	4.9
brass	4.7
iron	3.0
mild steel	11.0
glass	10
tungsten	20

■ A third problem is that, even if two pieces of wire are taken from the same reel, when stresses are applied it is often found that the strain they reach before breaking can be remarkably different. The microstructure of the two wires cannot be expected to be identical, and neither can the wires be clamped identically where the force is applied. So, whereas one wire might snap with a strain of, say, 10%, the other may not break until the strain is 40%.

The latter part of the curve in **Fig 23.7** is therefore uncertain, and has been shown dashed. The point C is the **breaking point** for the specimen and also shows the **breaking stress**. The breaking stress is the maximum tensile stress of the material (Table 23.2). Approximate values only are given for breaking stress, since it varies markedly from one specimen to another and from alloy to alloy.

Other types of behaviour

Certain other features of different materials can be shown by stress–strain graphs. **Fig 23.8a** is a stress–strain graph for glass. It shows no plastic distortion at all. A material such as glass, which does not behave plastically, is said to be **brittle**. Concrete is another substance that is brittle, as it cannot be permanently distorted. If it does not break under an imposed load, then it will always return to its original shape after the load is removed.

Figs 23.8b and **c** show that the graph can also depend on whether the applied stress is increasing or decreasing. **Fig 23.8b** is the stress–strain graph for copper again, but in this case, once the strain has reached a value of 0.050, the stress has been reduced to zero and the consequent permanent set in the strain can be seen to be 0.036. A similar feature is apparent when rubber is stretched (**Fig 23.8c**). The first thing to notice about this graph is that the strain reaches the high value of 5. This means that the extension of the rubber is five times its original length, whereas in the example given for copper the elastic extension was only one-twentieth of its original length. Quite different values for the strain are obtained for increasing stress as compared with decreasing stress.

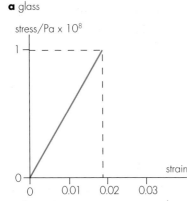

a glass

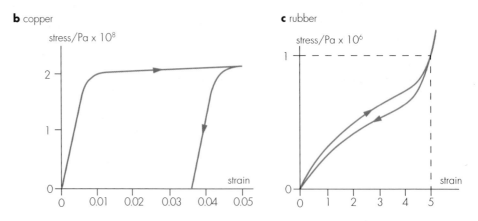

b copper

c rubber

FIG 23.8 a The stress–strain graph for glass is characteristic for a brittle substance. **b** The stress–strain graph for a ductile substance (copper). **c** The stress–strain graph for a polymeric substance (rubber).

Load–extension graphs

All of the graphs discussed so far in this section have been stress–strain graphs plotted for the material in a particular specimen. A similar graph will be obtained if extension is plotted against load. The extension is proportional to the strain, but

Units

The load must have the unit newton (N) and the extension must be measured in metres (m) to give the joule (J) as the unit of work.

the stress is only proportional to the load if the area of cross-section of the wire is constant. In practice, this will affect the shape of the graph only a little for low stress, but more as the load reaches the breaking stress. The area beneath a load–extension graph is the work done in stretching the material (see section 7.2). For the straight-line portion of a load–extension graph this will give:

$$\text{work done} = \tfrac{1}{2} \times (\text{load} \times \text{extension})$$

This is the elastic potential energy stored by the wire, provided the graph for decreasing loads is the same as that for increasing loads.

INVESTIGATION

The creep of copper wire

Set up the arrangement shown in **Fig 23.9**, using bare copper wire of about 28 s.w.g. Pay particular attention to the way that the wire is clamped. It must not slip at the clamp. Holding it firmly between two blocks of hardwood used as jaws is one way of fixing it, and it is then easy to measure the distance between the marker and the jaws to find the original length of the wire. The length of wire used needs to be at least 2 m between the wire clamp and the marker if the extension is to be measured with a ruler, but smaller lengths of wire can be used if a travelling microscope can be used for the measurement of the extension.

If you have not done an experiment such as this before, it is worth measuring the Young modulus for copper first. Plot a graph of load against extension, and for the straight-line portion of the graph find the gradient. Use this gradient, together with measured values of the length and area of cross-section of the wire, to find the Young modulus.

The load must be in newtons, the length of wire used in metres and the area of cross-section, which is πr^2, in square metres. Remember that r is the radius of the wire, not its diameter. Measurement of the diameter must be taken at several places along the wire and the zero reading of the micrometer must be checked. All the travelling microscope readings taken must be recorded. It is standard practice to record values as you go and not to do any arithmetical working or averaging before recording readings. The extension should be calculated from the microscope readings.

Now apply just sufficient load to the wire to make it plastic. Record values of the extension of the wire at intervals over a period of a few minutes. Add a small additional load and repeat the experiment. Carry on adding load until eventually the wire breaks. You will probably break a few wires too quickly to begin with. Plot a series of graphs, one for each load, of extension against time. Summarise your findings.

The investigation can, if desired, be modified to measure breaking stress and to find how breaking stress varies with the diameter of the wire.

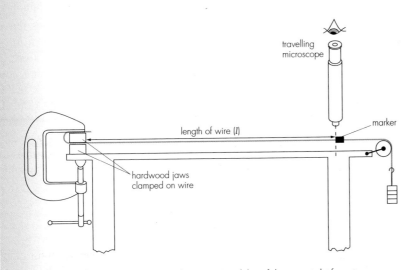

travelling microscope

length of wire (*l*)

marker

hardwood jaws
clamped on wire

FIG 23.9 Apparatus for measuring the Young modulus of the material of a wire.

Warning

Wear eye protection for this investigation in case the wire snaps.

Note

Particular care must be applied in this experiment to get the units correct.

23.5 A copper wire of length 2.30 m and cross-sectional area 8.6×10^{-8} m² is stretched by a tension of 17.3 N. Take the Young modulus for copper to be 1.1×10^{11} Pa.

 a What is the extension of the wire?

 b Find the potential energy stored in the wire as a result of doing work on it.

 c If a force of 173 N were used instead, it is likely that the work done by the force will be much greater than 10 times the value given in **b**, but the potential energy stored will be less than 10 times that value. Why is this?

23.3 MICROSTRUCTURE

The pattern of molecular arrangement in a solid, discussed in sections 22.3 and 22.4, needs to be able to explain the strength, elasticity and plasticity of materials. A full explanation cannot be given here, as the topic of materials science is a huge subject in its own right. The complexity of material structure also makes it difficult to do more than give a few, simplified generalisations.

Crystalline solids

Elastic behaviour does not result in a permanent change in the shape of the material. This implies that the molecules within the material stay in the same basic position in relation to their nearest neighbours. Since elastic strain is usually less than about 1%, it is reasonable to suggest that the individual molecules move a small distance from their equilibrium position on the application of a force and then settle back into their original equilibrium position once the force is removed. **Fig 23.10a** shows a few molecules in their undistorted arrangement, and **Fig 23.10b** shows the pattern after a distorting force has been applied. If the distorting force is now removed, they will return to their original position. If, however, the force is increased, then **slip** occurs and the molecules move into a new position, as in **Fig 23.10c**. The effect of this occurring is shown on a larger scale in **Fig 23.11**. Layers of molecules sliding across one another like this results in visible lines being formed on the surface of the material. Photographs of a metallic surface during the occurrence of slip are shown in **Fig 23.12**.

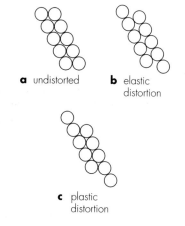

a undistorted **b** elastic distortion

c plastic distortion

FIG 23.10 Elastic distortion results in no permanent change in shape when the distorting force is removed. With plastic distortion, a permanent change in shape occurs.

FIG 23.11 Slip can occur when layers of molecules slide across one another.

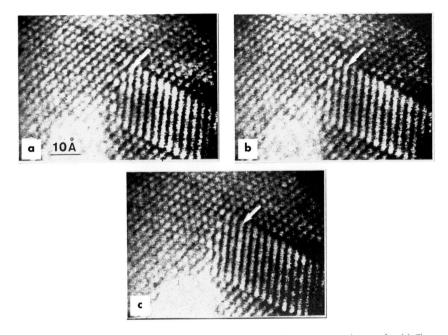

FIG 23.12 These high-resolution electron micrographs show slip in an atomic lattice of gold. The arrows show the movement of whole blocks of atoms along planes of weakness, called slip planes. $10\text{Å} = 10$ angstrom units of length $= 10^{-9}\,\text{m}$

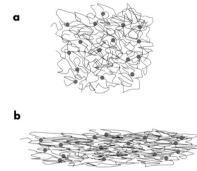

FIG 23.13 A polymeric material such as rubber can undergo a large strain. Its long molecules become stretched to lie parallel to one another.

Polymeric solids

This simple explanation cannot explain how a substance such as rubber can undergo a strain of 500% or more. To have any understanding of the stretching of rubber, it is necessary to know that rubber is a **polymer** with very long chain molecules. The molecules do have an attraction for one another and therefore solid rubber is possible, but they also tangle their molecules to give what can perhaps best be thought of as being like a plateful of cooked spaghetti having occasional cross-links between the strands. If the rubber is then pulled, the strands can be pulled out to lie more nearly parallel to one another. **Fig 23.13a** shows the tangled molecules before stretching and **Fig 23.3b** shows them stretched. Some re-organisation of the cross-links does take place and therefore there can be a permanent distortion of the rubber.

Amorphous solids

Besides crystalline and polymeric solids, there are also solids that are said to be **amorphous**. The word 'amorphous' means lacking in shape. Glass and soot are amorphous solids, and the shape that is lacking in these materials is the crystalline arrangement of the molecules. When pattern does occur, it takes place over only very short distances, of the order of a few molecular diameters.

The strength of a solid

The force that exists between atoms in a solid should give a guide to the strength of that solid. In **Fig 22.6a** the maximum attractive force between two argon atoms was given on the graph as approximately 2×10^{-11} N. The force between the more strongly bonded atoms in metals is greater. In Example 23.3, the strength of a copper wire is calculated after taking the maximum force of attraction between two copper atoms to be 5×10^{-11} N.

EXAMPLE 23.3

Find the maximum load that can be put on a copper wire of cross-sectional area 1 mm², if the maximum force of attraction between two copper atoms is 5×10^{-11} N. The density of copper is 7900 kg m⁻³ and the relative atomic mass of copper is 63.

The Avogadro constant gives 6.022×10^{23} atoms in 0.063 kg of copper. We have

1 kg of copper has volume = 1 kg / (7900 kg m⁻³)
= 0.000 127 m³

0.063 kg of copper has volume = 0.063 × 0.000 127 m³
= 8.00×10^{-6} m³

volume occupied by one atom of copper

$$= \frac{8.00 \times 10^{-6}\,\text{m}^3}{6.022 \times 10^{23}}$$
$$= 1.33 \times 10^{-29}\,\text{m}^3$$

If the atoms are considered to be in a cubical pattern, then the volume of the cube occupied by each atom is the value given above, and the cube root of this value will be the diameter of a copper atom. This is shown in **Fig 23.14**. It should be noted that the distance between the centres of two atoms is the diameter of the atoms. So

separation of molecules $= \sqrt[3]{(1.33 \times 10^{-29}\,\text{m}^3)}$
$= 2.37 \times 10^{-10}$ m

Now consider a layer of copper atoms across the wire at one place. It is easier to imagine that the wire has a square

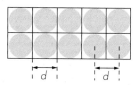

FIG 23.14 In a simple approximation, atoms can be considered to occupy cubes.

cross-section, as shown in **Fig 23.15**, in which the layer being considered has been imagined to be separated from the next layer up and is shaded green. Then

number of atoms in this layer
= (number along a side of 1 mm)²

$$\text{number in a layer} = \left(\frac{0.001\,\text{m}}{2.37 \times 10^{\pm 10}\,\text{m}}\right)^2$$
$$= 1.78 \times 10^{13}$$

Since each of the atoms in this layer pulls on a corresponding atom in the next layer:

total tensile force = $(1.78 \times 10^{13}) \times (5 \times 10^{-11}\,\text{N})$
= 890 N (to 2 sig. figs)

Many approximations have been made in making this calculation, so the only valid deduction that we can make from the calculation is that the maximum tension in a copper wire of area of cross-section 1 mm² should be approximately 900 N.

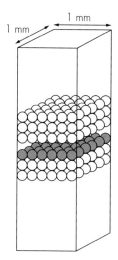

FIG 23.15 A simplification of the atomic layers in a wire.

Imperfections, impurities and strength

The value found in Example 23.3 for the tension in a piece of copper wire would imply that we ought to be able to place a mass of approaching 100 kg on the end of the wire and be able to lift it. When this is tried in practice, the mass is found to be much too large. Something has gone wrong with the theory. If instead of using a normal piece of copper with all its imperfections we use a piece of copper of greater crystalline purity, we find that, instead of getting stronger, as might be expected, the pure crystal of copper is in fact weaker still. The deduction we are forced to make is that a piece of material, while having a maximum strength value

FIG 23.16 The etched surface of a metal, seen here under a microscope, shows the polycrystalline nature of metals.

limited by the size of the inter-molecular attractions, will in practice break at lower stresses. The fundamental flaw in the argument given in the above example is making the assumption that all the molecules need to be pulled apart simultaneously. They can equally well be un-zipped from one another a few at a time. Whether or not this un-zipping can take place depends crucially on how regularly the atoms are arranged. It also depends on the number of impurity atoms present.

The structure and strength of crystalline materials cannot be predicted as easily as might be expected. **Fig 23.16** shows that, in practice, many crystals are present in most metal specimens. The irregularity of this multitude of individual crystals very much affects the strength of any specimen. The boundary between adjacent crystals, the *grain boundary*, can hinder a specimen from breaking. Metal-workers often change the crystal pattern in a piece of metal to increase its hardness. This process is called *work hardening* and it can be achieved by hammering the metal. Rapid changes in the temperature can also be used to alter the strength of a metal.

A fault in the crystal pattern is known as a **dislocation**, and the number and type of dislocations within any material have a marked effect on its strength. Dislocations can be deliberately introduced by using impurity atoms. The strength of steel is much greater than the strength of pure iron. Steel is manufactured by introducing carbon into iron. For more detail on the behaviour of materials, you are advised to use books on the science of materials.

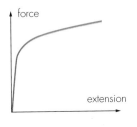

a copper: crystalline and showing ductility

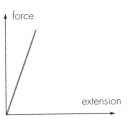

b glass: amorphous and showing brittleness

c rubber: polymeric

FIG 23.17 A summary of the three types of force–extension or stress–strain graph.

SUMMARY

- A solid behaves elastically if no permanent deformation occurs after a distorting force is removed.

- A solid behaves plastically when permanent distortion occurs.

- All solids are elastic when small forces are applied to them; as the force applied to them increases, plastic behaviour may occur before breaking.

- Stress is force per unit area of cross-section; it is measured in pascals.

- Strain is the increase in length per unit length; it has no unit.

- The three types of force–extension or stress–strain diagram (**Fig 23.17**):

- For the straight-line portion of the curves in **Fig 23.17** (up to the limit of proportionality), stress is proportional to strain and Hooke's law is said to apply.

- $$\text{Young modulus} = \frac{\text{stress}}{\text{strain}} = \frac{\text{force per unit area}}{\text{extension per unit length}} = \frac{Fl}{Ax}$$

- Work done in stretching a solid is the area beneath the force–extension graph. If Hooke's law applies, the work done is stored as strain energy. This energy is $\frac{1}{2}Fx$.

- Crystalline solids have long-range order in the pattern of molecules.

- Polymeric solids have long chain molecules, which can cause stiff or flexible solids according to the degree of cross-linking between the chains.

- Amorphous solids have little ordered arrangement of molecules.

ASSESSMENT QUESTIONS

23.6 **a** A rubber cord with a cross-sectional area of 0.025 cm² is stretched by 10% of its original length. The Young modulus of rubber is 4.0×10^7 Pa. Calculate the tension in the rubber cord.

b State *two* assumptions you made in arriving at your answer.

c The stretched rubber cord is attached to a vibration generator.

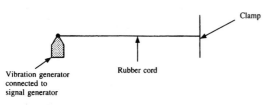

Describe with the aid of diagrams the appearance of the cord as the frequency of operation of the vibration generator is gradually increased.

London 1998

23.7 The table and graph show the properties of *two* materials A and B.

Material	Young modulus/ 10^{10} Pa	Ultimate tensile stress/ 10^8 Pa	Nature
A			
B	0.34	3.2	brittle

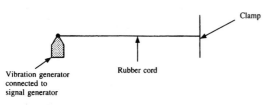

a Use the graph to complete the table for material A.

b Use the table to draw a graph showing the behaviour of material B.

London 1998

23.8 A wire of unstretched length 2.00 m and diameter 2.00 mm hangs vertically from a rigid support and carries a load of mass 4.00 kg at its lower end. The load causes the wire to extend by 0.125 mm. Calculate:

a the tensile stress in the wire;

b the tensile strain;

c the Young modulus for the material of the wire;

d the energy stored in the wire.

WJEC 1997

23.9 The table gives corresponding values of load and extension when masses are hung on a wire of length 1.5 m and diameter 0.30 mm.

load/N	0.0	2.0	4.0	6.0	8.0	10.0	11.0	11.2
extension/mm	0.0	1.0	2.1	3.1	4.2	5.4	7.3	9.0

a **i** Plot a graph of load (vertical axis) against extension (horizontal axis).

ii Indicate on your graph the region over which Hooke's Law is obeyed.

b Use your graph to calculate a value for the Young Modulus of the material from which the wire is made.

NEAB 1996

23.10 The graph shows the strain produced when two different materials are put under increasing stress until they break. One of the materials may be described as *brittle* and the other as *ductile*.

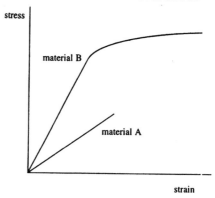

a State which material has the higher Young Modulus, explaining your reasoning.

b Explain the meaning of *brittle* and *ductile*.

c Indicate which material is the brittle material and which is the ductile material.

NEAB 1996

23.11 **a** A materials scientist needs to be able to determine several different properties of a new material. Some of these properties are

i density,

ii the Young modulus,

iii the ultimate tensile stress.

Explain the meaning of each of these terms and outline how the Young modulus and the ultimate tensile stress may be determined.

b The figure shows the force–extension graphs for identically-shaped pieces of steel and a new material.

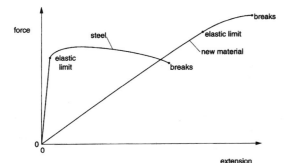

Compare the behaviour of the two specimens. Suggest, with two reasons, which of the two materials would be the more suitable for use in a car bumper.

OCR 1999

23.12 a The Young modulus for leg bone under compression is 9.0×10^9 Pa and the maximum compressive stress which it can undergo before breaking is 1.7×10^8 Pa. Assuming that the force–extension graph is linear up to the point of fracture, calculate

  i the strain at which fracture occurs,

  ii the approximate reduction in length which occurs in a leg bone initially 0.50 m long before it undergoes a compressive fracture.

b Calculate the minimum force required to cause compressive fracture in this bone. (The area of cross-section of the bone in **a** is 6.0×10^{-4} m^2.)

c The weight of a person is considerably less than the force which would cause a compressive fracture. Suggest, with a reason, a situation when a compressive fracture could occur.

OCR 1997

23.13 a i State Hooke's law.

  ii Explain each of the following terms:
     elastic limit   tensile stress   tensile strain

b In an experiment to determine the Young modulus of steel, a long thin steel wire, hanging from the ceiling, is subjected to an increasing load until it breaks.

  i Sketch a typical graph of extension against applied load up to the breaking point.

  ii Label *three* important features of the graph.

NEAB 1997

23.14 This question is about the stretching of rubber.

a Define the Young modulus.

b The Young modulus of rubber is very much smaller than that of steel. Explain this fact in terms of the basic structure of the materials.

c A rubber cord, of Young modulus 2.0×10^8 N m^{-2}, diameter 1.0 mm and natural length 0.80 m, is fixed at its upper end. A mass of 0.60 kg is suspended from the lower end.

  i Calculate the extension of the cord.

  ii Hence determine the spring constant (force per unit extension) of the cord.

d The suspended mass in part **c** is pulled down a short distance and released. Show that the mass oscillates with simple harmonic motion.

O & C 1998

23.15 A steel specimen is made for a certain tensile testing machine. The specimen has a circular cross-section in the shape shown in the diagram.

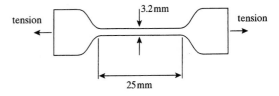

a The large diameter ends of the specimen are clamped in the jaws of the tensile testing machine. The specimen is designed so that when under test the narrow section, of effective length 25 mm, stretches while the large diameter end sections are considered not to stretch.

Explain why the large diameter end sections may be considered not to stretch when the specimen is under test.

b Under test the specimen obeys Hooke's law up to the point where the stress is 2.0×10^8 Pa and the extension is 0.024 mm.

Calculate

  i the extension of the specimen when the stress is 1.0×10^8 Pa,

  ii a value for the Young modulus of the steel,

  iii the cross-sectional area of the narrow section of the specimen,

  iv the tension when the stress is 2.0×10^8 Pa.

NEAB 1998

23.16 The diagram shows curves of the force, F, against separation, r, for pairs of atoms of two crystalline solids X and Y. The solids have the same crystal structure.

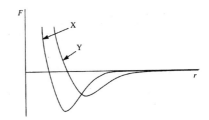

State and explain which solid has the

a larger atoms,

b greater Young modulus,

c greater breaking stress.

NEAB 1997

24 TEMPERATURE AND THERMAL ENERGY TRANSFER

LEARNING OBJECTIVES

At the end of this chapter you should be able to:

① define the ideal-gas temperature and know that it is the same as the thermodynamic temperature;

② use the Kelvin and Celsius temperature scales;

③ explain the two mechanisms of thermal conduction;

④ explain why convection occurs;

⑤ explain how radiation is important in establishing the temperature of any object;

⑥ use empirical temperature scales;

⑦ suggest suitable thermometers for any use.

24.1 THERMAL EQUILIBRIUM

Robert Boyle

Robert Boyle, in the seventeenth century, realised the difficulty of quantifying 'hotness' when he wrote:

'We are greatly at a loss for a standard whereby to measure cold.'

Temperature is a word that will be familiar to all readers, but it is also a word that poses scientific difficulty. We know that if it is a hot day the temperature is high, and that on a cold day the temperature is low (**Fig 24.1**). We know what it feels like to touch a cold or a hot object. Children quickly learn that hot objects can cause pain or burning. Temperature is directly associated with the physiological sense of hotness. But one cannot measure temperature using a physiological effect, because people are differently sensitive to hotness, and because how hot or cold an object feels depends on the materials from which it is made. The air inside an oven is at approximately the same temperature as the shelves in the oven, yet it is possible to put your hands into the hot oven provided you are careful not to touch the shelves. The shelves seem to be much hotter because they conduct heat well and much more energy can flow from them into your hands than can flow from the hot air, air being a bad conductor of heat. Hot and cold are therefore imprecise terms on which it is difficult to base a scientific definition.

Various attempts have been made to overcome this problem. All of the early solutions involved attaching certain values to particular temperatures and then fixing a scale between them. A familiar proposal was suggested by Fahrenheit, who gave 0 °F as the lowest temperature he could obtain with an ice–salt mixture and 96 °F as the temperature of the human body. Neither of these temperatures is sufficiently precise to be used as a standard. Newton also set up a temperature scale. He divided the temperature difference between body temperature and that of melting snow into 12 equal parts. In another way of tackling the problem, the Royal Society thermometer was taken as the standard in the years from 1663 to 1730.

Temperature is a quantity that cannot be defined in terms of other quantities. Like time and mass, it is one of the fundamental quantities. An important statement concerning temperature is the zeroth law of thermodynamics – so called because, by the time it was seen to be necessary to have such a basic law, the first and second laws of thermodynamics had already been stated. The **zeroth law of thermodynamics** concerns thermal equilibrium and states that: if two bodies are in thermal equilibrium with a third body, then they must be in thermal equilibrium with each other. Put in less formal language, if two objects are placed in contact with one another and allowed to reach thermal equilibrium, then no heat will flow from one to the other. If one of these objects is also in thermal equilibrium with a third object, then all three objects will be in thermal equilibrium with one another.

The property that they then all have in common is said to be their temperature. **Temperature** is the property of an object that determines which way heat will flow from it to another object. If you put a hand into some water and heat energy flows from your hand to the water, then the water is at a lower temperature than your hand. If heat energy flows from the water to your hand, then the water is at a higher temperature than your hand. If no heat energy flows, then there is thermal equilibrium between your hand and the water, and they must be at the same temperature. This is further illustrated in **Fig 24.2**. The relationship between heat flow and temperature is used to define the current standard of temperature, the thermodynamic temperature.

FIG 24.1 While everyone is familiar with the effect of being cold, measurement of temperature is not as straightforward as might be expected.

zero   zero
heat flow   heat flow
from A to B   from C to B

FIG 24.2 Here A is in thermal equilibrium with B; and C is in thermal equilibrium with B. The zeroth law of thermodynamics states that A must be in thermal equilibrium with C. The property that is the same for A, B and C is their temperature.

THERMODYNAMIC TEMPERATURE

Thermodynamic scale

The full definition of thermodynamic temperature is complex and is not dealt with at A-level, but it can be shown that it is identical to the ideal-gas scale, which is considered later in this section. At this stage it is just necessary to realise that there exists an absolute thermodynamic scale of temperature. The symbol used for thermodynamic temperature is T, and since it was proposed by Lord Kelvin, the unit of thermodynamic temperature has been named the **kelvin** (K). The kelvin is the fraction $1/273.16$ of the thermodynamic temperature of the triple point of water. This is the temperature inside a triple-point cell as shown in **Fig 24.3**. The **triple point** of any substance is the single temperature at which the solid, liquid and vapour phases of the substance co-exist.

Note that referring to the triple point of water in this way simply attaches a numerical value to the scale of temperature. It does not contradict the statement made above about the scale being independent of the way a property changes with

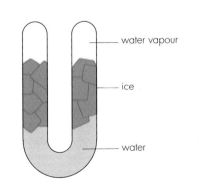

water vapour

ice

water

FIG 24.3 A triple-point cell for water. The temperature in the cell is 273.16 K.

temperature. We are not concerned with the way any property of water changes with temperature, but a single reference temperature *is* needed. The choice of the exact number 273.16 seems unusual at first sight. In the absence of historical precedence, no doubt a number such as 1000 would have been chosen, but the size of a unit of temperature would then have been different from units customarily used. An International Conference on units made the choice of 273.16 in order to make the temperature interval of one kelvin very nearly equal to the old interval of one centigrade degree.

It is still difficult to measure thermodynamic temperature to a high degree of accuracy. The measurement of the thermodynamic temperature of well-defined fixed temperatures is carried out by Standards Laboratories around the world. A list of some of their results is given in Table 24.1 to the nearest 0.01 K, although the uncertainty in the readings is often greater than 0.01 K.

TABLE 24.1 Temperature reference points

DEFINING FIXED POINT	THERMODYNAMIC TEMPERATURE/K
absolute zero	0 exactly
triple point of hydrogen	13.81
boiling point of hydrogen	20.28
boiling point of neon	27.10
triple point of oxygen	54.36
boiling point of oxygen	90.19
freezing point of water	273.15
triple point of water	273.16 exactly
boiling point of water	373.15
freezing point of lead	600.65
boiling point of sulphur	717.82
freezing point of aluminium	933.52
freezing point of silver	1235.08
freezing point of gold	1337.58

Celsius scale

The Celsius scale of temperature is merely an arithmetical adjustment to the thermodynamic scale measured in kelvin. Its value is given by subtracting 273.15 from the thermodynamic temperature measured in kelvin. The symbol *t* is used for Celsius temperatures. We therefore have the defining equation for Celsius thermodynamic temperature:

K $\quad t/°C = T/K - 273.15$

Since the interval of one kelvin is very nearly the same as the old centigrade degree, the effect of subtracting 273.15 is to make the Celsius scale very nearly the same as the old centigrade scale. The change to thermodynamic temperatures has made the centigrade scale redundant, but the International Conference that set up the new scales accepted that, since changing the custom of people is difficult, they would make the new scale agree with the old scale if no great accuracy is required. Table 24.2 should make clear that there *was* a fundamental change in the basis on which temperature is measured with the introduction of thermodynamic temperatures. In the table, temperature values shown in **_bold italics_** are exact by definition. The other temperatures are experimental values, and improvements in experimental techniques in the future may make additional significant figures possible. When more significant figures are able to be given, it will be apparent that centigrade and Celsius temperatures are not quite the same.

TABLE 24.2 A few temperatures shown on different scales. The temperatures that are **exact** by definition are shown in **bold italics**

	KELVIN	CELSIUS	CENTIGRADE
absolute zero	*0*	−*273.15*	−273.15
freezing point of water	273.15	0.00	*0*
triple point of water	*273.16*	*0.01*	0.01
boiling point of water	373.15	100.00	*100.00*
freezing point of silver	1235.08	961.93	961.93

ANALYSIS

Temperature scales

You have decided that the value 273.16 is too awkward to use in a basic definition, and that the International Conference that set up the Kelvin scale should have been bold and set up an entirely new thermodynamic temperature scale. You would like the Conference to use the number 500 for the value of the triple point of water on the new scale instead of 273.16. (And you would not object if it decided to call the new scale of temperature after you!)

Prepare a table showing some temperatures, taken from Table 24.1, on your new scale.

Someone else at the Conference, whose name began with the letter F, suggests that you subtract 468 from your scale of values. He says that this makes things familiar.

Do you think this helps? What is it similar to? What would be the temperature on his or her scale on a nice warm summer day?

Ideal-gas scale

The meaning of the term *ideal gas* will be dealt with more fully in Chapter 25. At this stage an ideal gas can be stated to obey the gas law $pV \propto T$ exactly, where p is the pressure of the gas, V is its volume and T is its temperature on the ideal-gas scale. Since we have already defined pressure and volume, this equation effectively defines an unknown temperature from the equation

K $\quad \dfrac{pV}{T} = \text{constant}$

This also holds at the triple point of water, and so we can write

$$\frac{(pV)_{\text{at unknown temperature}}}{\text{unknown temperature}} = \frac{(pV)_{\text{at triple point}}}{273.16\,\text{K}}$$

In symbols this is written

$$T = \frac{(pV)_T}{(pV)_{\text{tr}}} \times 273.16\,\text{K}$$

where T is the unknown temperature on the ideal-gas scale. It can be shown theoretically that the value of temperature given by the ideal-gas scale is identical with the thermodynamic temperature. This is why the symbol T has been used for temperature on the ideal-gas scale. It also provides a practical method for measuring a thermodynamic temperature, though the method is, in practice, very time-consuming and normally only done in Standards Laboratories.

Conduction electrons

Any solid metal has a crystalline structure in which the positive metallic ions form a three-dimensional lattice. Within the spaces provided by the lattice there exists a sea of free electrons. Most of the electrons in an atom of a metal are associated with a particular atom, but the outer electrons are not attached to particular atoms and are therefore able to drift from place to place within the lattice. These are the electrons that are responsible for electrical conduction, and they are called the *conduction electrons*.

As discussed in section 24.1, a difference in temperature between two bodies results in a flow of heat energy from the one at a higher temperature to the one at a lower temperature. Similarly, if there is a temperature gradient within any object, then heat will flow from the region of higher temperature to the region of lower temperature. If a saucepan is placed on a ring on a cooker, the surface of the bottom of the saucepan in contact with the ring reaches a higher temperature than the surface in contact with the food. Heat energy will therefore flow through the metal from the ring into the food. Heat ceases to flow when there is no temperature gradient. This process of thermal conduction takes place in materials as a result of two quite different methods of energy transfer.

The first mechanism responsible for heat flow depends on the conduction electrons in a metal (see box). When there is a temperature gradient within a metal, the conduction electrons gain energy at the hotter part of the metal and can diffuse to the colder part of the metal, resulting in a transfer of thermal energy. This process can only take place in electrical conductors, because only electrical conductors have conduction electrons.

However, thermal energy can be transferred through electrical insulators, so there must be some other process that allows heat to flow besides that using conduction electrons.

The second mechanism responsible for heat flow depends on coupling between adjacent molecules. The atoms in any solid (not just metals) have vibrational kinetic energy. At a hot region within the solid, the vibrational energy is high. Heat transfer will take place from a region where the vibrational energy is high to an adjacent region where the vibrational energy is lower.

As a result of these two effects, we can state that any good conductor of electricity will be a good conductor of heat. The reverse statement is not true. Some good electrical insulators are good conductors of heat. Quartz, for example, has an electrical resistance about 10^{20} times that of copper, but it conducts heat only a little less well than copper.

All materials change in size as they are heated. In almost all cases, the volume of an object increases as its temperature increases, and the increase is nearly linear with thermodynamic temperature. For ideal gases, the volume, at constant pressure, is directly proportional to the thermodynamic temperature; but for solids and liquids, the volume is not proportional to the thermodynamic temperature. In graphical terms, a straight-line graph only shows proportion if it passes through the origin. Graphs showing the expansion of a solid and a liquid are nearly straight-line graphs not passing through the origin (**Fig 24.4**).

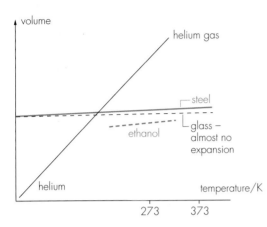

FIG 24.4 Compared with gases, solids and liquids expand only a little when heated.

FIG 24.5 Because hot air is less dense than cold air, it rises. A hot-air balloon depends on this effect for lift.

Increase in the volume of a substance must result in a decrease in its density because its mass is not changed by heating it. This is demonstrated very clearly by a hot-air balloon (**Fig 24.5**). The air inside the balloon is hotter than the air outside and so is less dense. Because of its lower density, it rises.

Most fluids expand when heated and so have a lower density when hot than when cold. The hot fluid floats upwards, with the colder fluid falling, as a result of the buoyancy force exerted on the hot fluid by the surrounding cold fluid being greater than its weight. This movement of the hot fluid upwards and the cold fluid downwards is called a **convection current**. Convection currents can be set up in any fluid, so are possible in gases as well as in liquids.

Convection currents are responsible for the familiar land and sea breezes. They are also responsible for the less familiar *katabatic* winds illustrated in **Fig 24.6**. In mountain areas, air over high land cools and flows down through valleys,

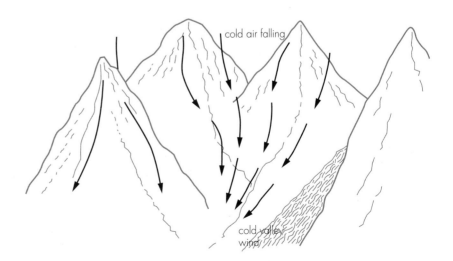

FIG 24.6 Cold air falling into a valley by convection and possibly giving rise to damaging frosts in the valley.

FIG 24.7 In order to rise, a hang-glider must find rising air. This can occur when wind blows towards a ridge or when hot air rises in a convection current. The tiny droplets of water that make up a cloud also rise in upward convection currents.

sometimes causing damaging frosts. The Mistral in southern France is a wind of this type. **Fig 24.7** is a photograph of a hang-glider, which can rise in a convection current of warm air. Hang-glider pilots rely on rising air to gain height and they become very skilled at finding it. They look for indications of rising air. These include corn fields, over which air tends to be hotter than over grass fields, bird movement, cumulus clouds, which are rising water droplets, and other hang-gliders.

The rate of heat transfer by convection from a hot object depends on such factors as whether or not there is a wind blowing, the shape and the surface area of the object. These factors are difficult to quantify, so that although it is possible to state that the rate of loss of heat depends on the temperature difference between the object and its surroundings, it is difficult to state a mathematical relationship for that dependence.

24.5 RADIATION

All objects emit some electromagnetic radiation. The quantity and wavelength of this radiation depend on several factors. In some cases it is clear that electromagnetic radiation is being emitted because it is possible to see and feel the effect of the radiation, whereas in other cases this is not so. The Sun, for example, emits a great deal of radiation. The light that we see is only a small part of the total emitted radiation, which includes ultra-violet, infra-red and microwave radiation. If a graph is plotted of intensity of radiation against wavelength for the Sun, then a curve similar to that in **Fig 24.8** is obtained. Only about 20% of the Sun's radiation is emitted as visible light.

The most important factor controlling radiation is temperature. The power radiated by an object is proportional to the fourth power of its temperature, so an object at a temperature of 2000 K radiates 16 times as much power per unit area as the same object would at 1000 K.

EXAMPLE 24.1

The total energy arriving from the Sun and falling on the upper atmosphere of the Earth is 1400 W m^{-2}. The distance of the Earth from the Sun is 1.5×10^{11} m. Find the power output of the Sun.

The Sun cannot lose energy by conduction or convection because of the vacuum of space; it must therefore lose all its energy by radiation.

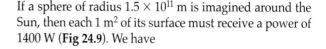

FIG 24.9 Part of an imaginary sphere around the Sun.

If a sphere of radius 1.5×10^{11} m is imagined around the Sun, then each 1 m^2 of its surface must receive a power of 1400 W (**Fig 24.9**). We have

area of surface of this
imaginary sphere $= 4\pi r^2$
$= 4\pi \times (1.5 \times 10^{11} \text{ m})^2$
$= 2.8 \times 10^{23}$ m^2

so that we get

power output of Sun $= 2.8 \times 10^{23}$ m$^2 \times 1400$ W m^{-2}
$= 4.0 \times 10^{26}$ W

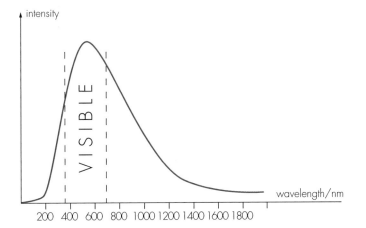

FIG 24.8 Intensity of radiation plotted against wavelength for a hot body, like the Sun.

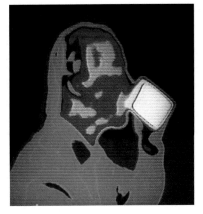

FIG 24.10 Thermograph of a person drinking a hot drink. There is an obvious difference between the cup and the person, who also shows relatively warmer areas.

The temperature of the surface of the Sun is controlled by the figure found in Example 24.1. Because any changes in the power output of the Sun take place only slowly, the Sun can be considered to be in a dynamic power equilibrium. This is not the same as the temperature equilibrium referred to at the start of the chapter, where no heat flowed from one body to another at the same temperature. This is certainly not true for the Sun, because it is losing heat to its surroundings, which are at a lower temperature than itself. The power equilibrium implies that a balance has been reached between the power output of the nuclear reactions going on inside the Sun and the power being radiated away from the Sun. When these two quantities are equal, then there is no net gain or loss of energy by the Sun, and its temperature will therefore remain constant. This happens for the Sun at its surface temperature of about 6000 K.

The radiation loss from an electric fire is something that can be felt readily on your skin. Your skin is a good detector of infra-red radiation in that it will experience a temperature rise which you can interpret as being due to the radiation. You can use your hand to detect the infra-red radiation from a radiator containing hot water. If you place your hand about 3 cm from a domestic radiator, you will be away from the convection currents around the radiator but you can still feel the heat energy reaching you as a result of the radiation. Clearly there is not as much radiation as from a radiant electric fire, but there is some radiation although it is not visible. The radiator does not glow in the dark; it is not giving out light, but it *is* emitting infra-red radiation, and by using infra-red film and special detectors it is possible to take photographs using infra-red radiation (**Fig. 24.10**). This principle is made use of medically. A person, like everything else, emits infra-red radiation at a rate dependent on temperature. A patient with joint inflammation caused by rheumatoid arthritis has a higher temperature at the joint than on surrounding skin. This will show up on a thermograph, where the extra radiation emitted will appear as an obvious 'patch'. Early diagnosis of this condition can help in delaying the onset of the arthritis.

Not only do objects emit radiation, they also absorb it from their surroundings. Where the surroundings are at a higher temperature than the body, then the body will have a net gain in energy; whereas there will be a net loss of energy from the body if it has its surroundings at a lower temperature. This is a direct consequence of the zeroth law of thermodynamics (section 24.1), which applies equally well to transfers of energy by radiation as it does to conduction.

The ability to radiate and absorb radiation is a complex matter, but the two are related. Any good absorber of a particular wavelength is also a good radiator of that wavelength. The colour and texture of a surface are also important. A lemon is yellow because it absorbs blue and violet light well but reflects the red/yellow/green region of the spectrum.

24.6 THERMOMETRIC PROPERTIES

Many other properties of substances vary with temperature besides the pressure and volume of a gas. Any property that varies with temperature may be used as a basis for constructing a thermometer. All of the following properties vary with temperature, and provide the principle on which a thermometer is based:

- Volume of a liquid, such as mercury or ethanol
- Resistance of a metal, such as platinum
- Thermocouple effect, between two metals such as copper and nickel
- Length of a metal rod, such as brass
- Resistance of a semiconductor, such as silicon
- Magnetic properties of a crystal, such as chromium potassium alum
- Vapour pressure of a gas, such as helium
- Radiation from a hot body, such as a lamp filament

Different properties are used over different ranges of temperatures, as will be explained in the next section. Here it is necessary to consider what constitutes a good property to use in making a thermometer.

1 *The change in the property must be large enough to measure accurately or the sensitivity of the thermometer will be low.* This can be done with a normal mercury-in-glass thermometer by having a very narrow capillary tube to exaggerate the expansion of the mercury, and by using a glass that itself expands very little. Some semiconductors change their resistance by a very large factor, 100 times or more, for comparatively small changes in temperature. A device that makes use of this property is called a **thermistor**.

> **Q** What would happen if a thermometer were constructed out of a glass with large expansion containing a liquid with small expansion?

2 *The value of the temperature recorded must be reproducible.* The thermometric property is useless if sometimes one value is given for, say, the melting point of tin, yet on other occasions the value is different. Some thermometric properties are particularly good on this account. The resistance of a platinum resistance thermometer can, if well used, give reproducible results in temperature measurement up to six significant figures. This creates a difficulty because often the thermodynamic temperature itself is not known to that degree of accuracy.

3 *The property being used must be suitable over the temperature range being measured.* The magnetic property of crystals is used for measuring temperatures right down to near absolute zero, but is of no use at high temperatures. The radiation of light from a filament can only be used at high temperatures. Some properties do not change appreciably over a range of temperature. The volume of water cannot be used as a thermometric property for this reason. Between 0 °C and 4 °C water contracts, and from 4 °C upwards it expands, as shown in **Fig 24.11**. This implies that if a water-in-glass thermometer were to be constructed, then its scale would be so bunched up, and doubled up on itself, that it would be useless in the range of 0 °C to 8 °C (**Fig 24.12**). Many thermocouples have ranges of temperature within which they cannot be used, as the thermoelectric e.m.f. does not vary sufficiently.

> **Q** What would happen with the thermometer in Fig 24.12 when the temperature is −1 °C?

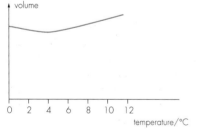

FIG 24.11 A graph showing qualitatively how the volume of water varies with the temperature. It shows that water has a maximum density at 4.0 °C.

FIG 24.12 This unlikely thermometer uses coloured water as its thermometric liquid. It is shown recording a temperature of 10.5 °C, but when the temperature falls, two possible temperatures are given for a single reading.

4 *The thermometers must be able to be calibrated without too much difficulty.* If the property changes uniformly with thermodynamic temperature, then, once two fixed points are calibrated, a linear scale can easily be added. If the property varies irregularly, then calibration is more involved. In the case of a thermistor, which is a piece of suitably prepared semiconductor material, the resistance varies in a far-from-linear manner. This necessitates many calibration points or an equation relating temperature with resistance.

Empirical scales

Calibration equations made up to fit observed facts are called **empirical equations** and are much used in practical thermometry. An equation with which you may be familiar can be considered as an empirical equation. It is the one traditionally used for defining the centigrade scale of temperature. But since the Celsius scale is so similar to the centigrade scale, it is useful as an empirical scale for the Celsius scale. As an empirical scale, it makes the assumption that the length of the mercury column in a mercury-in-glass thermometer increases linearly with the thermodynamic temperature. At the freezing point of pure water, the length of column (l_0) is measured. At the boiling point of pure water, the length of column (l_{100}) is also measured, and a straight-line graph plotted of length against temperature. The graph (**Fig 24.13**), or the equation

$$\frac{t}{100} = \frac{l_t - l_0}{l_{100} - l_0}$$

which is derived directly from the graph, can then be used to find an unknown temperature t if the length l_t of the mercury thread is found at t.

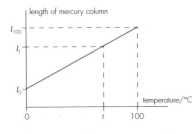

FIG 24.13 An unknown temperature can be found by measuring the length l_t of the mercury thread, and then using the graph to read off the temperature t.

THERMOMETERS

Choosing a suitable thermometer

The choice of thermometer to be used for a particular temperature measurement depends on many factors, such as accuracy, sensitivity, convenience, size, nature of read-out required, cost, range of temperatures being measured, speed of response and availability. Some of these factors will be dealt with here, but, since there is a huge variety of different types of thermometer the list of thermometers will by no means be complete. Before dealing with individual thermometers, it is worth while making some general points about choice of thermometer.

Perhaps the first point to make is the obvious one that the thermometer must be capable of recording temperatures in the range required. Standard laboratory liquid-in-glass thermometers have clearly defined ranges. A common temperature range is $-10\,°C$ to $+110\,°C$. Unless the thermometer is longer than 300 mm, this limits the distance for each degree to not more than about 2 mm. These thermometers cannot be relied on therefore to measure to much greater sensitivity than the nearest half-degree. If the thermometer is shorter – and many are only 150 mm in length – then the sensitivity is bound to fall. The sensitivity will also fall if the range is increased. A mercury-in-glass thermometer of 150 mm length, that reads from $-10\,°C$ to $+350\,°C$ will probably have a distance of only about 3 mm for each $10\,°C$ rise in temperature. A suitable length of scale should therefore be chosen for the range of temperatures to be measured. Wasted scale range increases the uncertainty of measurement.

Another important factor is to be clear what reading is required. In using a clinical thermometer, the temperature required is the maximum temperature reached when the thermometer is placed under the patient's tongue. The thermometer's construction allows this maximum temperature to be recorded. A greenhouse thermometer is normally used to record maximum and minimum temperatures. A meteorological thermometer normally is required to produce a graphical record of the variation in temperature over a period of time. Many thermometers are used to control temperature. That is, they need to be able to operate electrical switches at pre-set temperatures. These switching thermometers are called **thermostats** and may use expansion of metals for their operation, as in an oven, or may use electronic circuitry involving thermistors. In other situations it may be important to have remote sensing. The thermometers in a nuclear reactor, for example, cannot be read directly, so an electrical output is required from them. Hundreds of thermocouples are used throughout the reactor, and the electrical output that each gives enables the temperature at all points to be monitored continuously.

The heat required to raise the temperature of the thermometer itself is negligible for the thermocouples within a nuclear reactor. The thermocouple is permanently in a high-temperature environment and the temperature does not change very much, nor very quickly. In other situations, this may not be the case. If a temperature is changing rapidly, then a thermometer can only respond quickly to the change in temperature if it has a low heat capacity. A low heat capacity is also needed when the temperature of a small object is required. To take an extreme example, if a fly lands on a mercury thermometer, the thermometer will continue to record its own temperature, not the temperature of the fly, which could be above room temperature as a result of muscular activity.

Thermometers do, of course, normally record their own temperature. When in use they are placed in thermal contact with the body whose temperature is required. (Manufacturers give details of the depth of immersion required and whether the thermometer is to be used vertically or horizontally.) If given enough time to come into thermal equilibrium with their surroundings, they have the same temperature as their surroundings, as stated by the zeroth law of thermodynamics. Usually the thermal energy required to heat up the thermometer itself is negligible. But it must be considered that the final temperature recorded for a system is the temperature it acquires after its temperature has been changed by the insertion of the thermometer.

Constant-volume gas thermometer

This thermometer is used to obtain temperature standards on the thermodynamic scale. Basically it consists of a flask containing a fixed mass of gas whose volume can be maintained at a fixed value by controlling the pressure. The pressure of the gas is measured using a manometer and barometer. The value of the pressure is then used to work out the unknown temperature.

Platinum resistance thermometer

FIG 24.14 A platinum resistance thermometer. Two of the terminals are connected to the coil of platinum; the other two are connected to dummy leads.

The advantage of using the resistance of platinum as a thermometer is that resistance may be measured electrically to a high degree of accuracy and over a wide range of temperature. The readings so obtained are difficult to relate to the thermodynamic scale of temperature, although tables are available to help with this conversion, but the readings themselves are extremely reliable and reproducible. This makes comparison of temperature possible even if the exact thermodynamic temperature is not known. The thermometer itself (**Fig 24.14**) is usually a coil of pure platinum wire of diameter about 0.1 mm and of resistance about 25 Ω. To compensate for any change in the resistance of the leads, a pair of dummy leads can be connected in the circuit used to measure resistance.

24.1 Find the Celsius temperature *t* given by a mercury-in-glass thermometer if the length of the mercury column is 40 mm at the ice point, 240 mm at the boiling point of water, and 178 mm at *t*. Answer this question first by using common sense, and then by using the equation above. You should, needless to say, get the same answer in both cases.

24.2 Find the Celsius temperature *t* on the platinum resistance thermometer if an empirical temperature scale for the thermometer is

$$\frac{t}{100} = \frac{R_t - R_0}{R_{100} - R_0}$$

The resistance at the ice point is 8.452 Ω, the resistance at the steam point is 11.550 Ω, and the resistance at *t* is 6.707 Ω.

24.3 The resistance of a platinum resistance thermometer at 90.19 K is 7.525 Ω. Use a straight-line empirical equation to find the temperature when the resistance is 28.60 Ω. [Assume $R = 0\ \Omega$ at 0 K.]

Thermocouple thermometers

When a junction is made between two different metals, a small e.m.f. is set up between them. If any attempt is made to measure this e.m.f., then a continuous circuit has to be made and there will necessarily be other junctions to consider as well. These other junctions usually nullify the effect at the first junction. However, if two junctions are used and they are at different temperatures, then a measurable e.m.f. is set up that depends on the temperature difference between the two junctions. This effect is called the **Seebeck effect** and use is made of it in all thermocouple thermometers.

The Seebeck effect can be demonstrated quite simply by connecting up the circuit shown in **Fig 24.15** using copper and constantan wires twisted firmly together at the two junctions. The e.m.f. generated is of the order of a millivolt, so a very sensitive meter must be used. In practice, the thermometer can be used without an obvious cold junction. The cold junction is kept away from the hot junction and is simply at room temperature.

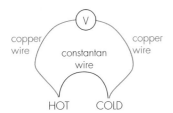
FIG 24.15 Circuit to show the Seebeck effect used in thermocouples.

Make a thermocouple and carefully attach the hot junction to the bulb of a standard laboratory thermometer that reads up to 300 °C or so (**Fig 24.16**). Place the cold junction in a beaker of melting ice, and heat the hot junction of the thermocouple and the thermometer in a sand bath. Measure the e.m.f. of the thermocouple at different temperatures.

Plot a calibration graph for the thermocouple. Why is it preferable to take the measurements as the temperature falls from its maximum value, rather than take them during the course of heating?

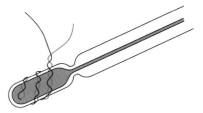
FIG 24.16 The hot junction of a thermocouple wrapped around the bulb of a thermometer.

One particular use of thermocouple thermometers is in the measurement of rapidly fluctuating temperatures. The junction itself can be made very small, so that it has very low heat capacity. It will therefore quickly reach thermal equilibrium with its surroundings.

The e.m.f. of common thermocouples varies in a non-linear way with temperature, so again an empirical equation needs to be used for accuracy. Over small changes of temperature, however, the variation can often be regarded as linear.

Radiation thermometers

Above temperatures of about 1700 K, all of the thermometers mentioned so far become useless. Even with careful choice of metal, they will be about at their melting point. Above this temperature, therefore, a totally different method has to be used. In practice, above the temperature of the melting point of gold, the concentration of radiant energy emitted by the hot object is used to measure its temperature. Detailed analysis of how this is done will not be given here, but it depends on the brilliance and colour of the hot object. Table 24.3 merely gives a guide to the temperature and comparative colour of a few commonly observed hot bodies.

TABLE 24.3 Colours and temperatures of some typical hot bodies

EXAMPLE	COLOUR	APPROXIMATE TEMPERATURE/K
simmering hot-plate (just visible)	dull red	900
fully-on hot-plate	red	1000
element of electric fire	orange	1100
filament of electric lamp	yellow	2500
surface of red giant star (Betelgeuse)	yellow	3300
surface of the Sun	white	6000
surface of white dwarf star (Sirius)	white	11 000
surface of white giant star (Rigel)	blue/white	25 000

Comparison of thermometer readings

Table 24.4 shows how certain empirical temperature values differ from the thermodynamic temperature if it is assumed that there is a straight-line variation of the thermometric property with temperature between 0 °C and 100 °C. The variation in values should make it quite clear why care is necessary when accurate values of temperature are required.

TABLE 24.4 Showing how near different thermometers approach the thermodynamic temperature

THERMODYNAMIC TEMPERATURE /K	CELSIUS TEMPERATURE /°C	CHROMEL–ALUMEL THERMOCOUPLE /°C	PLATINUM RESISTANCE THERMOMETER /°C	CONSTANT-VOLUME HYDROGEN THERMOMETER /°C
273.15	0	0.00	0.00	0.00
293.15	20	19.5	20.25	19.97
313.15	40	39.3	40.35	39.94
333.15	60	59.5	60.35	59.94
353.15	80	79.8	80.20	79.97
373.15	100	100.00	100.00	100.00

SUMMARY

■ The *zeroth law of thermodynamics* states that, if two bodies are in thermal equilibrium with a third body, then they must be in thermal equilibrium with each other. It introduces temperature as a useful concept.

■ A temperature T on the ideal-gas scale is defined by the equation

$$T = \frac{(pV)_{\text{at unknown temperature}}}{(pV)_{\text{at triple point}}} \times 273.16\,\text{K}$$

when the pressure in each bracket approaches zero. It can be proved that this temperature is identical to the absolute thermodynamic temperature.

■ The Celsius temperature is based on the thermodynamic temperature but with a numerical shift of 273.15, so if t is the Celsius temperature

$$t/°\text{C} = T/\text{K} - 273.15$$

■ Thermal conduction takes place in electrical conductors and insulators by lattice vibration. In electrical conductors it can take place additionally by electron diffusion.

■ Convection takes place as a result of hotter, less dense, fluid rising and colder, denser, fluid falling.

■ All objects emit electromagnetic radiation. The quantity of power radiated by an object depends on the nature of the surface of the object and on its temperature.

■ Empirical equations are often used with thermometers. Frequently they assume a uniform increase in a property between two temperatures. The two temperatures are often 0 °C and 100 °C, in which case

$$\frac{t}{100} = \frac{X_t - X_0}{X_{100} - X_0}$$

where X is the property being used.

■ Properties of various thermometers:

THERMOMETER	RANGE/°C	ACCURACY	SENSITIVITY	OTHER COMMENT
mercury-in-glass	−39 to 500	poor	moderate	direct reading
constant-volume gas	−270 to 1500	excellent	good	slow response: used as standard
platinum resistance	−200 to 1200	good	poor	large heat capacity
thermo-couple	−250 to 1500	fair	fair	good for varying temperatures
thermistor	−50 to 200	poor	very high	low heat capacity; good for varying temperatures
radiation (pyrometer)	over 1000	poor	poor	usable over 1500 °C

ASSESSMENT QUESTIONS

24.4 **a** State one similarity and one difference between conduction and convection of thermal energy.

  b By reference to thermal energy transfer, explain what is meant by
  i two bodies having the same temperature,
  ii body H having a higher temperature than body C.

  c **i** Briefly describe how a physical property may be used to measure temperature on its empirical centigrade scale.
  ii Hence explain why two thermometers measuring temperature on their empirical centigrade scales do not agree at all temperatures.

  d The table shows data for ethanol.

density	0.79 g cm^{-3}
specific heat capacity of liquid ethanol	2.4 J g^{-1} K^{-1}
specific latent heat of fusion	110 J g^{-1}
specific latent heat of vaporisation	840 J g^{-1}
melting point	$-120\,°C$
boiling point	78 °C

  Use the data above to calculate the thermal energy required to convert 1.0 cm^3 of ethanol at 20 °C into vapour at its normal boiling point.

  e **i** State the *first law of thermodynamics*.
  ii Suggest why there is a considerable difference in magnitude between the specific latent heats of fusion and vaporisation.

OCR 1999

24.5 The temperature of the water in a hot bath is measured with a thermometer.

  a For each of the statements below indicate whether the statement is true or false.
  i The thermometer will *not* give the correct temperature unless it is in thermal contact with the water.
  ii The thermometer will *not* give the correct temperature unless it is in thermal equilibrium with the water.
  iii The thermometer will *not* give the correct temperature unless it, itself, is in a thermal equilibrium state.

  b A thermocouple has the cold junction immersed in an ice–water mix at 0 °C. When the hot junction is in boiling water, the e.m.f. is 1.65 mV. Estimate the temperature of the hot junction when the e.m.f. is 1.47 mV.

London 1996

24.6 **a** Sketch and label the diagram of a thermocouple thermometer shown below.

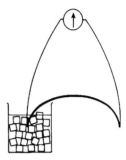

  b A thermocouple operates as a simple heat engine. Explain where the energy comes from, where it goes to and whether the energy transfers are heating or working.

  c Explain one way of increasing the efficiency of a thermocouple when considered as a heat engine.

London 1998

24.7 This question is about the concept and meaning of temperature.
Write a paragraph about each of the following aspects of temperature:
  a how ideal gases can be used to define a temperature scale,
  b the absolute zero of temperature,
  c the conditions under which real gases in practice behave as ideal gases,
  d how you would determine experimentally whether two bodies were at the same temperature. Include a statement of the zeroth law of thermodynamics

O & C 1998

24.8 This question is about energy transfer and the melting of ice.
  a Define the terms *specific latent heat* and *thermal conductivity*. In each case give a formula and carefully define all the symbols.
  b A cubical box of inside dimension 0.30 m is made of polystyrene sheet of thickness 10 mm. It is filled with ice at 0 °C. The exterior of the box is maintained at 25 °C. Using the data below, estimate how long it takes for all the ice to melt. State any assumptions made.
  [Specific latent heat of ice = 3.3×10^5 J kg^{-1}, density of ice = 920 kg m^{-3}, thermal conductivity of polystyrene = 0.020 J s^{-1} m^{-1} K^{-1}]

O & C 1998

24.9 The figure illustrates a cross-section through a double-glazed window fitted in a roof.

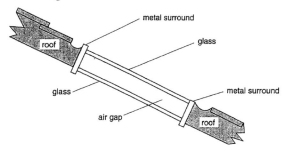

The exterior temperature is much less than the interior temperature.

a Thermal energy passes through the panes of glass and the metal surrounds by conduction.

   i Describe, on a molecular basis, the process of conduction through

      1 glass,

      2 metal.

   ii Hence suggest why metal is a much better conductor than glass.

b Some thermal energy is transferred through the air gap by convection.

   i Sketch the direction of a convection current in the gap.

   ii Describe the physical processes involved in the formation of convection currents.

   iii Suggest a design feature of double-glazed windows which reduces convection currents.

Cambridge 1997

24.10 a Heat transfer from a hot object to its surroundings can be by **i** thermal conduction, **ii** convection or **iii** radiation. Briefly describe the mechanisms by which each of these processes can take place.

   b The figure shows the intensity of the radiation emitted by two light bulbs operating at temperatures of 1500 K and 2000 K.

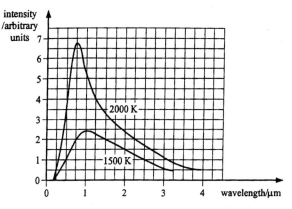

One theory suggests that in some circumstances the product $\lambda_{max}T$ is constant. λ_{max} is the wavelength at which the intensity is at a maximum and T is the absolute temperature of the filament.

Do the data in the figure support this suggestion? Justify your answer.

   c State clearly how, other than by thermal radiation, the light bulb is likely to transfer its energy to the surroundings.

AEB 1996

24.11 a Describe the processes that may contribute to the transfer of thermal energy from a region of higher temperature to one of lower temperature in solids, liquids and gases and in a vacuum.

   b The diagram is a cross-section of a type of vacuum flask.

This flask may be used to keep hot drinks hot or to keep cold drinks cold. Making reference to your answers to **a**, explain how the design features of the flask help to achieve these two functions.

OCR 1999

25

GASES

LEARNING OBJECTIVES

At the end of this chapter you should be able to:

① state and use Boyle's law;

② use the universal gas law equation for an ideal gas;

③ describe the kinetic theory of an ideal gas;

④ relate the kinetic energy of molecules to the temperature;

⑤ describe qualitatively the distribution of molecular speeds within a gas.

25.1 BOYLE'S LAW

FIG 25.1 The atmosphere seems more real when it can be photographed. In fact, light is being reflected from dust or water droplets in the atmosphere.

The behaviour of gases was referred to in Chapter 24, where the variation in the pressure and the volume of an ideal gas was used as a basis for measuring the thermodynamic temperature. In this chapter the properties of a gas will again be considered, but this time the properties will also be related to the behaviour of the molecules within the gas.

■ When the properties under consideration are large-scale properties that can be felt or measured with instruments, such as the pressure of a gas, they are called **macroscopic** quantities.

■ Properties that cannot be sensed directly, such as the velocities of the individual molecules, are called **microscopic** quantities.

Early experiments on gases were done using macroscopic quantities only. In the seventeenth century, Robert Boyle measured the volume of a fixed mass of gas at different pressures, while keeping the temperature constant. He found that the volume was inversely proportional to the pressure. The apparatus shown in **Fig 25.2** can be used to repeat this experiment.

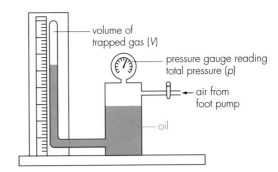

FIG 25.2 Apparatus for measuring how the volume of a gas varies with the pressure exerted on it.

The volume V of trapped gas is measured for different pressures p. Care must be taken to leave enough time in between taking one reading and the next to allow the gas to be in thermal equilibrium with its surroundings. If this is not done, then the rise in temperature that takes place when the pressure of the gas is increased will affect the readings of the volume. If a graph is plotted of p against V, then it is difficult to establish whether or not there is inverse proportion between the two quantities, because the graph is not a straight line. In order to establish inverse proportion, either plot a graph of p against $1/V$ or plot pV against V.

A formal statement of **Boyle's law** is:

 The volume of a fixed mass of gas is inversely proportional to the pressure applied to it if the temperature is kept constant.

Expressed mathematically this is:

$$p \propto 1/V \quad \text{or} \quad pV = \text{constant}$$

The law is an experimental law and has been found to be reliable provided the gas pressure is not too high.

 ## THE IDEAL GAS

The theoretical model of an ideal gas was introduced in Chapter 24. By the definition of an ideal gas, it must obey Boyle's law. Since the equation used to define the ideal-gas scale of temperature is

$$T = \frac{(pV)_T}{(pV)_{tr}} \times 273.16 \, \text{K}$$

it follows that, because 273.16 K and (for a fixed mass of gas) p and V at the triple point of water are all constants, for an *ideal gas*:

$$T \propto pV$$

In other words, because temperature is defined in this way, it follows that the volume of a fixed mass of an ideal gas at constant pressure *must* be proportional to the temperature. If the measurements of an experiment carried out at constant pressure show that the volume of a particular gas is proportional to the ideal-gas temperature, the conclusion that can be drawn from the experiment is that the gas is behaving ideally.

Since for the ideal gas $pV \propto T$, it follows that

$$\frac{p_1 V_1}{T_1} = \frac{p_2 V_2}{T_2}$$

Hint

When you need to solve gas law problems, it is helpful to start with a table of the information given.

where the subscript 1 refers to the initial conditions in the state of a gas and the subscript 2 refers to the final conditions. Note that in all gas law problems the temperature is the *absolute* temperature measured in kelvin and not the Celsius temperature.

Additionally, an ideal gas must not cool down when allowed to undergo a free expansion into a vacuum. This implies that the molecules in an ideal gas do not attract one another, so an ideal gas would never be able to be turned into a liquid. More detail will be given about this when considering the internal energy of gases in section 25.3.

EXAMPLE 25.1

A fixed mass of gas, in passing through a jet engine, has its pressure increased from 3.0×10^5 Pa to 1.3×10^6 Pa, while its temperature rises from 80 °C to 1500 °C. By what factor does the volume of the gas change?

As suggested in the box, we start by drawing up a table of information given (Table 25.1).

TABLE 25.1 Information given in Example 25.1

	INITIAL CONDITIONS (1)	FINAL CONDITIONS (2)
pressure	3.0×10^5 Pa	13×10^5 Pa
volume	V_1	V_2
temperature	80 °C	1500 °C
	= 80 + 273	= 1500 + 273
	= 353 K	= 1773 K

Then we can substitute known values into the previous equation. In this case we get

$$\frac{p_1 V_1}{T_1} = \frac{p_2 V_2}{T_2}$$

$$\frac{(3.0 \times 10^5 \, \text{Pa}) \times V_1}{353 \, \text{K}} = \frac{(13.0 \times 10^5 \, \text{Pa}) \times V_2}{1773 \, \text{K}}$$

$$\frac{V_2}{V_1} = \frac{1773 \times 3.0}{353 \times 13} = 1.16$$

The volume of the gas has increased by a factor of 1.2 (to 2 sig. figs) in passing through the engine. The assumption has been made that the gas behaves as an ideal gas. For the information that is given here to two significant figures, this is a reasonable assumption, since two significant figures clearly does not give a value capable of high accuracy.

Whether or not gases can be considered as having ideal behaviour depends on the pressures being used and on the accuracy required. With nitrogen, for example, at atmospheric pressure and below, calculations using the ideal-gas law give results that are within 0.1% of the experimental value. If the pressure is raised to 100 atmospheres, then the error introduced by using the ideal-gas law is still only of the order of 5% provided the nitrogen is not cooled down to near its boiling point.

The value of the term pV/T is directly proportional to the amount of gas, n (in moles):

$$\frac{pV}{T} \propto n$$

This therefore gives an equation relating the pressure of the gas, p, the volume of the gas, V, the temperature of the gas, T, and the amount of gas, n:

$$\frac{pV}{T} = nR$$

S.t.p.

Standard atmospheric pressure and temperature (s.t.p.) are 1.013×10^5 Pa and 0 °C.

where R is a constant called the **molar gas constant**. If 1 mol of an ideal gas is considered, then its volume at standard temperature and pressure is 0.0224 m³. Substituting in the numerical values gives:

$$(1.013 \times 10^5 \text{ Pa}) \times (0.0224 \text{ m}^3) = 1 \text{ mol} \times R \times 273.15 \text{ K}$$
$$R = 8.31 \text{ J K}^{-1} \text{ mol}^{-1}$$

The volume that a mole of gas occupies is called its **molar volume** and is given the symbol V_m. So $V_m = V/n$. The gas equation, which is called the **equation of state** of the gas, can therefore be written in either of two ways:

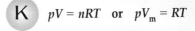

$$\boxed{K} \quad pV = nRT \quad \text{or} \quad pV_m = RT$$

This equation is known as the **universal gas equation**.

QUESTIONS

25.1 During the compression stroke of an internal combustion engine, the volume of the gas in the cylinder is reduced from 4.0×10^{-4} m³ to 5.0×10^{-5} m³. During the stroke, the pressure changes from 1.0×10^5 Pa to 18×10^5 Pa. If the temperature of the gas at the start of the stroke is 22 °C, what will be its temperature at the end of the stroke?

25.2 Find the volume at s.t.p. of a mass of air that has a volume of 3.6×10^{-5} m³ at a pressure of 5.7×10^4 Pa and a temperature of 400 °C.

25.3 The pressure in a car tyre when at a temperature of 10 °C is 1.9×10^5 Pa above atmospheric pressure. If the volume of the tyre does not change, what will be the pressure in the tyre at a temperature of 33 °C? [Atmospheric pressure is 1.0×10^5 Pa.]

25.3 THE KINETIC THEORY OF GASES

Pressure variation in a gas

Since the gravitational force on each molecule acts downwards, there is not quite random motion. Within a small volume, however, this effect is very small: the pressure on the top of a flask is very nearly the same as the pressure on the bottom. It is only when dealing with, say, the Earth's atmosphere that the lack of uniformity of the pressure would need to be taken into account.

The behaviour of a gas on a macroscopic scale concerns the variation of its temperature, volume and pressure. But to obtain a clearer understanding of gas behaviour, we must be able to relate the microscopic movement of the individual gas molecules to these macroscopic quantities. It is the *kinetic theory* that enables this to be done. As usual, any theory must be able to be checked by experiment and must provide greater understanding of the physical problem. The kinetic theory of gases applies the laws of mechanics to molecular movement, so that expressions for the pressure and temperature of a gas are obtained in terms of molecular speed, mass and number.

Brownian motion, which was discussed in section 22.3, gives clear evidence that particles of matter are in perpetual movement. The kinetic theory uses the idea that movement of gas molecules is responsible for the pressure of a gas. Certain assumptions are made about the properties of gas molecules:

1 Any gas consists of a very large number of molecules.
2 The molecules of the gas are in rapid, random motion (but see box).
3 Collisions between gas molecules are elastic.
4 Collisions between gas molecules and the walls of the container are elastic.

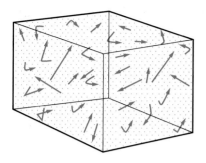

FIG 25.3 The model of a gas consists of a very large number of molecules in rapid, haphazard motion.

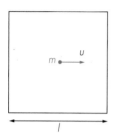

FIG 25.4 A single molecule travelling directly towards a wall of a cubical box.

5 There are no intermolecular attractive forces.

6 Intermolecular forces of repulsion act only during collisions between molecules; the duration of collisions is negligible compared with the time interval between collisions.

7 The volume of the gas molecules themselves is negligible compared with the volume of the container; that is, almost all the gas is empty space.

Several of these assumptions have the effect of defining the microscopic properties of an ideal gas. Assumption 7, for example, implies that the gas must not be at high pressure, because if the pressure is high then the concentration of molecules within the available space will be high. The densities of gases and liquids give a guide to how much space is actually occupied by the molecules. The density of a liquid is roughly 1000 times the density of the corresponding gas at atmospheric pressure, so, of the total volume of the container, the volume of the molecules themselves is only 0.1% of the total volume.

The model of a gas that we are using therefore is one of a huge number of point masses, moving around in completely random zig-zag fashion as they bounce perfectly off one another and off the walls of the container (**Fig 25.3**).

We shall now find an expression for the pressure caused by the continual bombardment of the walls of the container by the molecules. Consider first a single molecule in a cubical box with sides of length l. If the mass of the molecule is m and it is travelling with velocity u directly towards the right-hand wall of the box (**Fig 25.4**), then it has

momentum to the right before collision with wall $= mu$
momentum immediately after a perfect collision $= -mu$
change in momentum (to the left) $= 2mu$

The time interval before the molecule makes another collision with the same wall will be the time taken to travel across the box and back. So

time between collisions with the same wall $= 2l/u$
number of collisions with this wall per unit time $= u/2l$
rate of change of momentum of the molecule $= 2mu \times u/2l = mu^2/l$

By Newton's second law, the rate of change of momentum of the molecule is the force the wall exerts on the molecule. This is equal in size to the force on the wall. So

$$\text{force on the wall} = \frac{mu^2}{l}$$

The force here is an average force over a period of time.

Now if, instead of one molecule, we consider N molecules, all going backwards and forwards (**Fig 25.5**), then the total force F on a wall will be

$$F = \frac{mu_1^2}{l} + \frac{mu_2^2}{l} + \frac{mu_3^2}{l} + \dots + \frac{mu_N^2}{l} = \frac{m}{l}(u_1^2 + u_2^2 + u_3^2 + \dots + u_N^2)$$

where u_1 is the velocity of the first molecule, u_2 is the velocity of the second molecule, etc. If the average value of u^2 is written as $\overline{u^2}$, then this expression can be written more simply as

$$F = \frac{m}{l}(N \times \overline{u^2}) = \frac{mN\overline{u^2}}{l}$$

In terms of V, the volume of the box, this gives the pressure p on the wall as

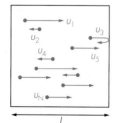

FIG 25.5 A number of molecules travelling towards and away from a wall of a cubical box.

$$p = \frac{\text{force on wall}}{\text{area of wall}} = \frac{mN\overline{u^2}}{l \times l^2} = \frac{mN\overline{u^2}}{V}$$

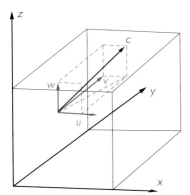

FIG 25.6 A molecule will move at random in three dimensions.

Note

Sometimes $\overline{c^2}$ is written as $\langle c^2 \rangle$.

Also mN is the total mass of the gas molecules, so mN/V is the density ρ of the gas. The pressure p exerted by all these molecules moving back and forth is given by

$$p = \rho \overline{u^2}$$

In an ideal gas, however, the molecules will not just move backwards and forwards in one dimension. They will move in three dimensions, as shown in **Fig 25.6**. Using Pythagoras's theorem for the velocity c_1 of the first molecule gives

$$c_1^2 = u_1^2 + v_1^2 + w_1^2$$

If you imagine doing this for all the molecules of the gas, and then adding all the equations together, you would get

$$N\overline{c^2} = N\overline{u^2} + N\overline{v^2} + N\overline{w^2}$$
$$\overline{c^2} = \overline{u^2} + \overline{v^2} + \overline{w^2}$$

where $\overline{c^2}$ is the average value of c^2, and is called the **mean square velocity**. But since there is no tendency for the gas to exert a greater pressure in one direction rather than any other, it follows that

$$\overline{u^2} = \overline{v^2} = \overline{w^2}$$
$$\overline{c^2} = \overline{u^2} + \overline{u^2} + \overline{u^2} = 3\overline{u^2}$$
$$\overline{u^2} = \tfrac{1}{3}\overline{c^2}$$

The pressure exerted on the wall was shown to be given by $p = \rho \overline{u^2}$. This can therefore be written

K $\quad p = \tfrac{1}{3}\rho \overline{c^2}$

This deduction has not involved collisions between molecules at all. In practice, they will collide with one another, but since both momentum and kinetic energy are being conserved, any gain in these quantities by one molecule will result in another molecule losing an equal amount of that quantity. There is therefore no overall effect on the pressure as a result of collisions between molecules.

EXAMPLE 25.2

The density of nitrogen at s.t.p. is 1.25 kg m^{-3}. Find the mean square speed of a nitrogen molecule in air at s.t.p., and hence deduce the mean kinetic energy of a nitrogen molecule. [Relative molecular mass of nitrogen = 28.]

We have standard atmospheric pressure = 1.013×10^5 Pa.

Substituting into $p = \tfrac{1}{3}\rho \overline{c^2}$ gives

$$1.013 \times 10^5 \text{ Pa} = \tfrac{1}{3} \times 1.25 \text{ kg m}^{\pm 3} \times \overline{c^2}$$
$$\overline{c^2} = 3.039 \times 10^5 / 1.25 = 243\,100 \text{ m}^2 \text{ s}^{\pm 2}$$

1 mol of nitrogen molecules has a mass of 0.028 kg, so using the Avogadro constant, N_A, gives the mass m of a molecule of nitrogen as

$$m = (0.028 \text{ kg})/(6.022 \times 10^{23}) = 4.65 \times 10^{-26} \text{ kg}$$

The mean kinetic energy E_k of a nitrogen molecule is

$$E_k = \tfrac{1}{2} m \overline{c^2}$$
$$= \tfrac{1}{2} \times 4.65 \times 10^{\pm 26} \text{ kg} \times 243\,100 \text{ m}^2 \text{ s}^{\pm 2}$$
$$= 5.65 \times 10^{\pm 21} \text{ J}$$

ANALYSIS

Kinetic theory

The pressure exerted by the gas does not depend on the shape of the container. This exercise enables you to work out the equation for the pressure exerted by a gas, $p = \frac{1}{3}\rho c^2$, for a spherical container of radius r. You are asked to copy out and complete each of the lines of working with an algebraic expression, using the terms given on **Fig 25.7**. A single molecule, of mass m, and travelling with velocity c, hits the wall of the container at an angle of incidence θ.

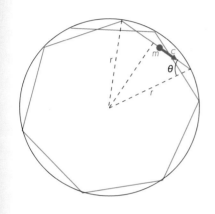

FIG 25.7 The diagram for the analysis exercise.

Component of velocity in a direction towards the wall	=
Momentum towards wall before collision	=
Momentum away from wall after collision	=
Change in momentum at a collision	=
Distance travelled between collisions with wall	=
Time between collisions with wall	=
Number of collisions per unit time	=
Rate of change of momentum (see note)	=

Note

The angle θ should cancel out at this stage; many glancing collisions have the same effect on the wall as fewer direct collisions.

Surface area of a sphere	$= 4\pi r^2$
Total force on wall	=
Pressure on wall due to one molecule	=
Total pressure on wall	=
Volume of sphere	$= \frac{4}{3}\pi r^3$
Total pressure in terms of volume, N, m and mean square velocity	=
Total pressure in terms of mean square velocity and density	$= p = \frac{1}{3}\rho\overline{c^2}$

QUESTIONS

25.4 a Repeat Example 25.2 but for oxygen instead of nitrogen. The density of oxygen at s.t.p. is 1.43 kg m^{-3} and the relative molecular mass of oxygen is 32.

b Compare your answer with Example 25.2. Can you explain why they should be the same?

25.5 Seven molecules have speeds of 200, 300, 400, 500, 600, 700 and 800 m s^{-1} respectively. Find:

a their mean speed;

b their mean square speed (this is different from their mean speed squared);

c the square root of their mean square speed (r.m.s. speed).

When using the kinetic theory, the term that keeps arising is the mean square speed of all the molecules. In question 25.5 above, it was indicated that this term cannot be obtained by squaring the mean speed of the molecules. To overcome this difficulty, the term 'root mean square speed' of the molecules is used. Be careful, when using this term, to *work in the correct order*. The **root mean square speed** of molecules (denoted $c_{\text{r.m.s.}}$) is the square root of the mean (average) of the squared speeds of the molecules (denoted $\overline{c^2}$). It is *not* the square root of the mean speed squared. That is

$$c_{\text{r.m.s.}} = \sqrt{\overline{c^2}}$$

25.4 APPLICATION OF THE KINETIC THEORY OF GASES

Problem-solving

In solving problems using the kinetic theory of gases, extra care must be taken with algebraic symbols and units.

- Temperature will always be in kelvin.
- Relative molecular mass, M_r, has no units.
- Molar mass is the mass of a mole and will be quoted in kg mol^{-1}.

For example, for oxygen: an oxygen atom, O, has a relative atomic mass of 16; an oxygen molecule, O_2, has a relative molecular mass of 32. Since the molar mass of oxygen atoms is 0.016 kg mol^{-1} = 16 g mol^{-1}, the molar mass of oxygen molecules is 0.032 kg mol^{-1} = 32 g mol^{-1}. Then the mass of an oxygen atom is $0.016/(6.022 \times 10^{23})$ kg.

A frequent cause of mistakes is to use 16 instead of 0.016 here. Another common error is not to distinguish carefully between a mole and a molecule: n is used to represent the number of moles of gas; N is likely to be a huge number, as it is the number of molecules.

If $p = \frac{1}{3}\rho\overline{c^2}$ is compared with the ideal-gas law, we find that, whereas at a microscopic level

$$p = \frac{1}{3}\frac{Nm}{V}\overline{c^2}$$

at a macroscopic level we have

$$p = \frac{n}{V}RT$$

where n is the amount of gas in moles. Equating these two expressions gives

$$\tfrac{1}{3}Nm\overline{c^2} = nRT$$

The total kinetic energy E_k of *all* the molecules is

$$E_k = \tfrac{1}{2}Nm\overline{c^2}$$

and therefore

$$Nm\overline{c^2} = 3nRT = 2E_k$$

This finally gives

$$E_k = \tfrac{3}{2}nRT$$

The kinetic energy of the molecules is therefore directly proportional to the temperature.

For an ideal gas the total kinetic energy of the molecules as they move around in their container is the **internal energy** U of the gas. This is given by

K $U = \tfrac{3}{2}nRT$

Note that this depends on the amount of ideal gas, n. It is also directly proportional to temperature. Internal energy will be dealt with in more detail in section 26.2.

Always state exactly what amount of gas is being used and then, as always, look at answers critically to see if they make sense. An answer such as

No. of molecules = 4.1×10^{-12}

is nonsensical.

This expression not only gives the total kinetic energy of the gas molecules, but also enables the average kinetic energy of one molecule to be obtained. Since a mole of molecules contains N_A molecules, where N_A is the Avogadro constant, n moles contain nN_A molecules. The average kinetic energy E_k of *one* molecule is therefore

$$E_k = \frac{3nRT}{2nN_A} = \frac{3RT}{2N_A} = \frac{3kT}{2} = \frac{3}{2}kT$$

Here k is a constant called the **Boltzmann constant** and is the gas constant for an individual molecule; and R is the molar gas constant, so $R/N_A = k$. The numerical value of k is 1.38×10^{-23} J K^{-1}.

The equation of state of the ideal gas can be expressed in terms of k rather than R, giving:

$$pV = NkT$$

where N is the total number of molecules.

QUESTIONS

25.6 Find the average kinetic energy of a molecule in air at 300 K.

25.7 Sketch two graphs on the same temperature axis, to show how

a the average kinetic energy, and

b the root mean square speed

of a gas molecule in air vary with the temperature.

25.5 DISTRIBUTION OF MOLECULAR SPEEDS

In any gas, the molecules are not all travelling with the same speed. As a result of collisions taking place, some molecules gain energy and other molecules lose energy. In between collisions, the ideal-gas molecules have a constant speed. A graph, plotted to show the assortment of different speeds that exist, is shown in **Fig 25.8** for nitrogen gas at a temperature of 1000 K. The graph shows the variation from molecule to molecule of the constant speeds between collisions. It does not deal with speeds during the collisions; these are assumed to take zero time in an ideal gas.

The vertical axis of the graph needs some explanation. Since the number of molecules in a gas is likely to be very high, the unit used on the vertical axis has been adjusted to make the total area beneath the curve equal to one. That is, the whole of the gas is represented by the area under the graph. One-tenth of the total number of molecules making up the gas has been coloured green on the graph to illustrate this; the area covered by the colouring is one-tenth of the total area beneath the graph. The molecules in this 10% of the gas are the fastest 10% of all the molecules. The graph is not symmetrical, and so the mean speed of the molecules, $\bar{c}$, is slightly higher than the most probable speed, $\hat{c}$. The root mean

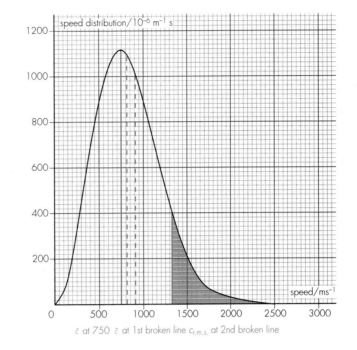

FIG 25.8 A typical molecular speed distribution (for nitrogen gas at 1000 K). More molecules are travelling at 740 m s^{-1} than at any other speed.

square speed, $c_{\text{r.m.s.}}$, is different from the mean speed and the most probable speed. For the molecules in a gas, the approximate relations between them are:

$$\hat{c} \approx 0.8 c_{\text{r.m.s.}} \quad \text{and} \quad \bar{c} \approx 0.9 c_{\text{r.m.s.}}$$

The following example should be worked through to see how the graph may be used.

Use the graph in **Fig 25.8** to answer the following questions: **a** What is the most probable speed? **b** What fraction of the total number of molecules travel with speeds between 700 and 800 m s^{-1}? **c** What fraction of the molecules of the gas travel at more than twice the most probable speed?

a The peak of the curve occurs at a speed of 750 m s^{-1}. More molecules have this speed than any other speed.

b The height of the curve between 700 and 800 m s^{-1} has a nearly constant value of 0.001 15 m^{-1} s. The area beneath the curve between 700 and 800 m s^{-1} is

$$0.001\ 15\ \text{m}^{-1}\ \text{s} \times 100\ \text{m s}^{-1} = 0.115$$

The fraction of the total number of molecules travelling with speeds between 700 and 800 m s^{-1} is therefore 0.115.

c Twice the most probable speed $2 \times 750 = 1500$ m s^{-1}. The fraction required is the area under the curve for values of v above 1500 m s^{-1}. This is difficult to find accurately from the graph, but there are very few molecules travelling at more than 2500 m s^{-1} so they can be neglected. The area found by counting squares is approximately 0.05. In other words, only one molecule in 20 travels at more than twice the average speed.

The figure that one in 20 molecules travels at more than twice the average speed is worth remembering. It can be used to give rough values for even higher speeds, for example:

■ One molecule in 20 travels at twice the average speed.
■ One molecule in 20^2 (one in 400) travels at four times the average speed.
■ One molecule in 20^3 (one in 8000) travels at eight times the average speed, etc.

FIG 25.9 A computer graphic representing evaporation from the surface of a liquid (in this case, water).

Extending this scale it is possible to find approximately that:

- One molecule in 20 000 travels at 10 times the average speed.
- One molecule in 400 000 000 travels at more than 100 times the average speed.

The reason these values are important is that in many situations it is the high-speed molecules that matter. For instance, evaporation takes place as a result of molecules leaving the surface of a liquid (**Fig 25.9**). The faster molecules are more able to leave than the slower molecules. The rate at which a chemical reaction proceeds is often dependent on the temperature. Since high-speed molecules are effectively at a higher temperature than slow molecules, the high-speed molecules are important in activating the reaction.

The molecular composition of the atmosphere of a star or planet depends on its escape velocity. The escape velocity of the Earth is 11 000 m s^{-1}; the average velocity of a hydrogen molecule in the Earth's atmosphere is about 1700 m s^{-1}. These values seem to indicate that hydrogen gas will not be able to escape from the Earth. But, using the figures from above, since one in 8000 hydrogen molecules travels at eight times the average speed, there is one in 8000 hydrogen molecules travelling at more than 13 600 m s^{-1}. This is greater than the escape velocity, and so some hydrogen molecules can escape from the Earth. Given a long enough time, virtually all of them will escape, and so the atmosphere will contain very little hydrogen. The corresponding values for oxygen and nitrogen would show that very few of these gas molecules can escape, and that is why our atmosphere contains them.

The graph showing variation of molecular speed within a gas is very temperature-dependent. As the temperature rises, the number of low-speed molecules decreases and the number of high-speed molecules increases, as shown by the curves of **Fig 25.10**. The total number of molecules remains the same, and so the area under all the curves is constant. The area beneath all of the curves is 1 (indicating all of the molecules in the gas).

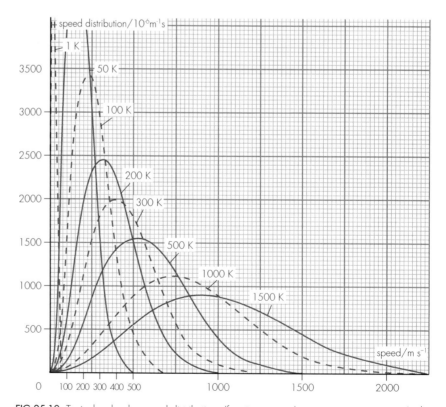

FIG 25.10 Typical molecular speed distributions (for nitrogen gas) at various temperatures. As the temperature of the gas is increased, the average speed of the molecules increases. The gas has a higher internal energy.

SUMMARY

- Boyle's law: at a constant temperature, the pressure of a gas is inversely proportional to its volume.

- From the definition of thermodynamic temperature, $pV \propto T$ for an ideal gas.

- Equation of state for an ideal gas:

  $$pV = nRT \quad \text{or} \quad pV_m = RT$$

- Kinetic theory of gases – basic assumptions:

  - a gas consists of a large number of molecules;

  - the molecules are in rapid, random motion;

  - the molecules collide with each other and with the walls of the container.

- Kinetic theory of gases – simplifying assumptions:

  - gravitational forces are negligible;

  - all collisions are elastic;

  - intermolecular forces are negligible except during a collision;

  - the volume of the molecules themselves is negligible compared with the volume of the container.

- The pressure exerted by a gas is given by

  $$p = \tfrac{1}{3}\rho \overline{c^2}$$

- The Boltzmann constant, k, is defined as R/N_A hence

  $$pV = NkT$$

- When solving problems using the kinetic theory of gases:

  - temperature will always be in kelvin;

  - relative molecular mass, M_r, has no units;

  - molar mass is the mass of a mole and will be quoted in $kg\ mol^{-1}$.

ASSESSMENT QUESTIONS

25.8 **a** Sketch a graph to show the relationship between the volume V of a fixed mass of an ideal gas under constant pressure and the Celsius temperature θ of the gas.

    b **i** Draw a labelled sketch of apparatus which could be used to obtain experimental values of V and θ to enable you to plot the graph in part **a**. Assume that air behaves as an ideal gas.

       ii Explain how the values for V would be obtained from readings taken with the apparatus you have sketched in **b i**.

CCEA 1998

25.9 **a** **i** Write down the equation of state for n moles of an ideal gas.

       ii What is meant by an ideal gas?

    b Calculate the mass of argon gas filling an electric light bulb of volume $8.2 \times 10^{-5}\ m^3$ if the pressure inside the bulb is 90 kPa and the temperature of the gas is 340 K.

       density of argon at standard temperature and pressure = $1.56\ kg\ m^{-3}$
       standard pressure = 100 kPa
       standard temperature = 273 K

NEAB 1996

25.10 This question is about ideal gases.

   a **i** State the equation of state for an ideal gas.

      ii State the relationship between the Kelvin and Celsius temperature scales.

   b A bicycle tyre has a volume (assumed to be constant) of 2.1×10^{-3} m^3. On a day when the temperature was 15 °C the tyre was filled with air to a total pressure of 280 kPa. Assume that the air behaves as an ideal gas.

      i Calculate the amount of air, in mol, in the tyre.

      ii The next day was much warmer. The total pressure in the tyre was found to be 290 kPa. Assume that no air was lost from the tyre overnight. Calculate the temperature.

      iii Explain, in terms of the kinetic theory of gases, why the pressure in the tyre rises when the temperature rises.

O & C 1997

25.11 a **i** What is meant by the *internal energy* of a system?

      ii State two ways in which the internal energy of an ideal gas may be increased.

   b The pressure p of an ideal gas is given by the equation

$$p = \frac{1}{3}\frac{Nm}{V}\overline{c^2}$$

      i State what is meant by $\overline{c^2}$.

      ii Use the equation to show that the total kinetic energy E_k of the N molecules in the gas is given by

$$E_k = 3pV/2$$

      iii Hence show that E_k is proportional to the absolute temperature of the gas.

Cambridge 1997

25.12 Ideal gas behaviour can be predicted using an equation relating pressure, volume and temperature, or explained in terms of molecular motion.

   a Write down an equation of state for an ideal gas, explaining all the symbols used. Hence, by drawing a suitable graph, explain what is meant by the absolute zero of temperature.

   b Using the equation

$$pV = \frac{1}{3}Nm\overline{c^2},$$

    deduce a relationship between absolute temperature and the mean kinetic energy of the individual molecules of an ideal gas.

   c A sample of an ideal gas is contained in a cylinder with a moveable piston. The piston is suddenly withdrawn so as to increase the volume of the gas. By considering collisions of gas molecules with the piston, suggest what change occurs to the root-mean-square speed of gas molecules, and hence the temperature of the gas.

   d By reference to your answer to **c**, suggest reasons for the following observations:

      i rising air currents, in the Earth's atmosphere, cool rapidly as they displace the existing air,

      ii a metal cylinder for compressed air, when being filled to a high pressure, gets hot.

Cambridge 1997

25.13 a A gas cylinder, of constant volume 5.0×10^{-2} m^3, contains 20 mol of argon (considered as an ideal gas). Calculate the pressure of the argon in the cylinder at a temperature of 273 K.

   b The figure shows axes for the pressure p and the Kelvin temperature T for the argon in the cylinder.

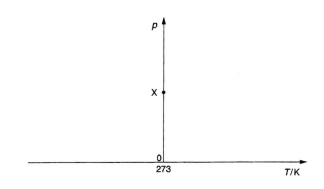

    The point X on the vertical axis represents the pressure calculated in **a**.

      i Copy the figure and draw a line through point X to show the variation with T of p for the argon.

      ii State and explain the value of T at which the line drawn in **i** intersects the T-axis.

OCR 1999

25.14 Two gas containers, A and B, have equal volumes and contain different gases at the same temperature and pressure.

   a Use the ideal gas equation to show that there are equal numbers of molecules in the two containers.

   b The molecules in container A have mass four times as great as the molecules in container B. Sketch and label graphs to show how the molecular speeds are distributed in the two containers, given that the speed distribution curve for the molecules in container A peaks at 200 m s^{-1}.

   c How does kinetic theory account for the equal pressures at the same temperature and volume?

London 1997

25.15 In this question the symbols used have their usual meanings.

a An ideal gas is one that obeys the equation of state

$$pV = nRT \qquad \text{(A)}$$

What single unit could be used for either pV or nRT?

b A sample of ideal gas is trapped in an enclosure, as shown in the diagram. The temperature of the gas is controlled by a water bath around the enclosure, so that the pressure of the gas can be measured at any temperature.

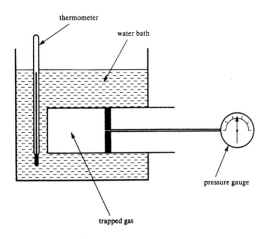

i In the first experiment, the volume of the gas is kept constant as measurements of the pressure are recorded for various temperatures. The sample contains *one mole* of ideal gas in this experiment.

Sketch a graph of the product of pressure and volume, pV, against temperature in °C.

Label this graph F and show numerical values on it for both the temperature axis intercept and the pV axis intercept.

ii A second experiment is then performed, *using 0.5 mole* of ideal gas. Once again the volume is kept constant whilst the pressure is measured at various temperatures.

Add a second graph to show the results expected in this second experiment. Label this second graph S.

c A laboratory vacuum pump will produce a pressure of 1.0×10^{-9} Pa. Calculate the number of molecules per cubic metre in such a vacuum at a temperature of 27 °C.

d From kinetic theory, the product of pressure and volume of an ideal gas is given by

$$pV = \tfrac{1}{3} N m \overline{c^2} \qquad \text{(B)}$$

i State two assumptions which are made about the molecules of the gas when deriving equation B.

ii By applying equations A and B to one mole of gas, show that the temperature of the gas is given by

$$T = \frac{2N_A}{3R} E_k$$

where E_k is the mean kinetic energy of a molecule.

iii If two gases are at the same temperature, what property will their molecules have in common?

NEAB 1998

25.16 a In the kinetic theory of gases, the pressure exerted by a gas is explained by considering the collisions of the molecules with the walls of the container. The figure illustrates the impact of one molecule with a wall.

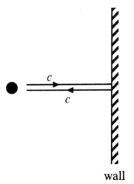

wall

The molecule, of mass m, strikes the wall normally with speed c, rebounding with speed c in the opposite direction.

i Write down an expression for the magnitude of the change of momentum of the molecule as a result of this impact.

ii The collision of the molecule with the wall is elastic. Explain what is meant by an *elastic collision*.

iii One of the assumptions of the kinetic theory is that the gas is made up of a very large number of molecules. Why is this assumption necessary?

b The ideal gas equation may be written

$$pV = nRT$$

i In this equation, what does n represent?

ii Give a consistent set of SI units for the quantities p, V and T in this equation.

c One mole of nitrogen has a mass of 0.028 kg.

i Find the mass of one nitrogen molecule.

ii Calculate the root-mean-square speed of a nitrogen molecule at room temperature (17 °C).

iii Hence calculate the total translational kinetic energy of all the molecules in one mole of nitrogen at room temperature.

CCEA 1998

THE FIRST LAW OF THERMODYNAMICS

LEARNING OBJECTIVES

At the end of this chapter you should be able to:

① define and use the term 'internal energy';

② state a general form of the first law of thermodynamics;

③ apply the first law of thermodynamics to a gas;

④ relate the specific heat capacity at constant volume for a gas to its specific heat capacity at constant pressure;

⑤ use the terms 'isothermal', 'isobaric' and 'adiabatic' correctly;

⑥ use an indicator diagram for a heat engine;

⑦ find the efficiency of a heat engine;

⑧ see the need for the second law of thermodynamics.

26.1 INTERNAL ENERGY

A cup of tea on Concorde

The internal energy of a cup of tea might be 40 000 J. An identical cup of tea at the same temperature but travelling on Concorde at 500 m s^{-1} still has an internal energy of 40 000 J. It might, in addition, have a further 10 000 J of energy due to its movement with the aeroplane.

In any substance, the molecules have kinetic energy because they are moving. They may also have potential energy because of the attractive forces between them. This has been referred to in section 25.4, where the term 'internal energy' was introduced. The **internal energy** U of any object is defined as the sum of all the microscopic kinetic and potential energies of the molecules within the object. It is measured in joules. The word 'microscopic' is introduced here to indicate that it is the kinetic and potential energies of the random movement of the molecules that is the internal energy, and *not* any large-scale movement of the whole object (see box).

One problem associated with internal energy is the need to choose an arbitrary zero of potential energy. This problem is always present when potential energy is being used. In most circumstances, however, it is *changes* in potential energy that are required, in which case the position taken to have zero potential energy does

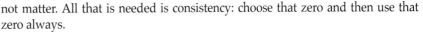

not matter. All that is needed is consistency: choose that zero and then use that zero always.

In the particular case of an ideal gas, this difficulty does not arise. An ideal gas obeys the gas laws precisely, and there is no inter-molecular attraction. The molecules in an ideal gas cannot therefore change their potential energy, so the reference is chosen to give them zero potential energy. This means that the internal energy of an ideal gas is just the kinetic energy of the molecules, and the internal energy can only be changed if the kinetic energy changes. Since a change in the kinetic energy of the molecules implies a change in temperature, it follows that the internal energy of an ideal gas is constant at a constant temperature. Its pressure and volume may be changed, but the internal energy remains fixed. In many cases, only a small error is made by assuming that a real gas is an ideal gas.

26.2 THE FIRST LAW OF THERMODYNAMICS

The laws of thermodynamics are concerned with energy transfer, and they may be applied to all systems. They can be applied to the energy changes taking place within a wire being stretched or to a battery supplying an electric current. They can be applied to the tension in the surface of a liquid or to a chemical reaction. In particular, they enable theoretical calculations to be made concerning the efficiency of heat engines. A **heat engine** is a device that converts thermal energy into work. Steam engines were the earliest types of heat engine.

In this chapter, the laws of thermodynamics will be applied only to the behaviour of gases. The first law of thermodynamics is, in part, a statement of the law of conservation of energy (section 7.7). It goes beyond the law of conservation of energy, however, by stating also that the internal energy of a system depends only on the state of the system.

FIG 26.1 Boulton and Watt's rotative beam engine. The efficiency of such early steam engines was less than 1%.

If we regard all the oxygen in an oxygen cylinder to be the system, then for the oxygen in the cylinder the first law states that its internal energy depends on factors such as the amount of oxygen in the cylinder, the pressure of the oxygen in the cylinder and its temperature. It does not depend on how the oxygen came to be in the cylinder or on the previous history of those particular oxygen molecules. Another cylinder full of oxygen under the same conditions, that is, in the same state, will have the same internal energy. This may seem to be obvious, but note that one could not make the same statement about the work done on the oxygen molecules to put them into the cylinder. That can be done in different ways, involving different amounts of work. It is reasonable to ask how much internal energy there is in the oxygen in the cylinder, but not to ask how much work there is in it.

A general statement of the **first law of thermodynamics** is that:

> **K** The internal energy of a system depends only on its state; any increase in the internal energy of a system is the sum of the heat supplied to the system and the work done on the system.

As usually happens when a basic law is stated in formal language, its meaning and uses are not at first clear, so some examples need to be given. As indicated above, the examples used in this chapter will be concerned with gas behaviour. This will enable you to see how to apply the law to the gases in heat engines such as a car or motor-bike engine (**Fig 26.2**), or a refrigerator or a steam turbine in a power-station.

In all of these cases it is a gas on which, or by which, work is done. The gas is known as the *working substance*. The two ways referred to for increasing the internal energy of a gas are illustrated in **Figs 26.3** and **26.4**:

FIG 26.2 A modern small petrol engine.

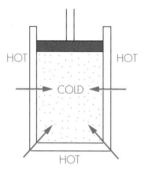

FIG 26.3 Heating a gas by having the surroundings of a cylinder at a higher temperature than the gas within the cylinder.

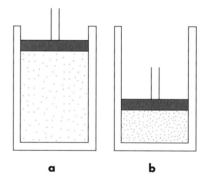

FIG 26.4 Work being done on a gas by a piston.

■ Heating the gas is shown in **Fig 26.3**. Here the space surrounding the cylinder is at a higher temperature than the gas within the cylinder, and so the molecules gain internal energy as a result of heat flow through the walls. Heating also takes place if a fuel is burnt within the cylinder.

■ Working on the gas is shown in **Fig 26.4**. By moving the piston downwards from the position shown in **Fig 26.4a**, molecules are struck by the moving piston and hence gain internal energy in the same way that a tennis ball gains kinetic energy when struck by a moving racket.

Both heating and working are processes that can change the internal energy of the gas, so the first law can be written as an equation in the convenient short-hand form:

> **K** *increase* in internal energy of gas = heat supplied *to* gas + work done *on* gas

EXAMPLE 26.1

A fixed mass of an ideal gas has its state changed from state A to state B as shown in Table 26.1.

TABLE 26.1 Change-of-state data for Example 26.1

STATE	PRESSURE/Pa	VOLUME/m^3	TEMPERATURE/K	INTERNAL ENERGY/J
A	200 000	0.0010	100	300
B	300 000	0.0030	450	1350

Find how much heat has been supplied to the gas, and how much work has been done on it, if the change of state is carried out in the three different ways, (i) to (iii), as shown on the pressure–volume graph in **Fig 26.5**.

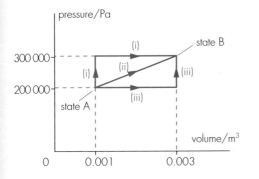

FIG 26.5 A diagram showing change in pressure with volume for three different ways of changing the state of a gas. This is called an 'indicator diagram' (see section 26.4).

Note from the data that the internal energy can be given for each state of the gas, in accordance with the part of the statement of the first law that the internal energy of the gas depends only on its state.

The three ways being considered for changing the state of the gas from state A to state B are as follows:

i a change of pressure at constant volume followed by a change of volume at constant pressure – in both of these sections there will be an increase in temperature;

ii a straight-line change of pressure and volume – this would be difficult to carry out experimentally;

iii a change of volume at constant pressure followed by a change of pressure at constant volume – again both of these sections will involve a rise in temperature.

The area beneath a pressure–volume graph is the work done (see section 7.2). Here we must be careful about whether work is being done *on* the gas to compress it or

by the gas in an expansion. Vertical lines on the graph imply no change in volume and therefore no work being done. In all three cases, expansion of the gas is taking place, so work is being done *by* the gas. The value we expect to get for the work done *on* the gas is therefore negative.

Many problems on the first law of thermodynamics can conveniently be done by completing a table, with the first law statement making the table headings. That is what we do here in Table 26.2.

TABLE 26.2 Initial version of the table

	INCREASE IN INTERNAL ENERGY OF GAS	= HEAT SUPPLIED TO GAS	+ WORK DONE ON GAS
(i)	1350 − 300 = 1050 J		−300 000 Pa × 0.0020 m^3 = −600 J
(ii)	1050 J		−250 000 Pa × 0.0020 m^3 = −500 J
(iii)	1050 J		−200 000 Pa × 0.0020 m^3 = −400 J

So far the table has included in it data given in the question and calculations made of the area beneath the graphs. We can now complete the table by using the statement of the first law of thermodynamics to find how much heat needs to be supplied in each case This is shown in the revised table (Table 26.3).

TABLE 26.3 Updated version of the table

	INCREASE IN INTERNAL ENERGY OF GAS	= HEAT SUPPLIED TO GAS	+ WORK DONE ON GAS
(i)	1050 J	1650 J	−600 J
(ii)	1050 J	1550 J	−500 J
(iii)	1050 J	1450 J	−400 J

A remarkable result that can be seen from these calculations is that the heat supplied to a gas to change its temperature from 100 K to 450 K is different for the three cases. The importance of the first law can be seen from this. The heat supplied to a system and the work done on a system can have different values and are dependent on how the changes are carried out. The increase in the internal energy does not depend on how the changes are carried out, only on the initial and final states of the gas.

A note on signs

Different books use different algebraic notations when stating the first law of thermodynamics. Some use the work done *by* the gas; some use the work done *on* the gas. This introduces a minus sign or the term appearing on the other side of the equation. This problem can be overcome by a clear understanding of the first law. The law is stating that the internal energy of a gas may be increased either by heating it or by working on it.

You are advised to write the equation at the bottom of p.448 using words and in three columns rather than using algebraic symbols. It is important when applying the first law of thermodynamics to get the signs correct (see box). In this equation you should note the emphasis on the words *increase, to* and *on*.

It has been assumed up to this stage that the term 'heat energy' is familiar to the reader, but it is a term that needs some explanation. In everyday speech the word 'heat' is used to mean a variety of different things. It is used as a noun in *heat rises*, as a verb in *heat the kettle* and as an adjective in *a heat pump*. It is very often used incorrectly in such statements as *the heat of the oven is 220 °C*. You can probably remember at some time in your physics lessons being warned to distinguish carefully between 'heat' and 'temperature', heat being measured in joules and temperature being measured in °C. A similar warning is now being given to be careful to distinguish between 'internal energy' and 'heat'. Both of these quantities are measured in joules, and probably in the past you would have used the word 'heat' to cover both of them. In future, be careful to use the words in the correct way, as follows:

> **K**   *Internal energy* **is the sum of all the microscopic kinetic and potential energies of the molecules in an object.**

> **K**   *Heat* **is energy transferred as a result of a temperature gradient.**

It is therefore correct to say that heat escapes by conduction through a window. If it is warmer inside than outside, there is a temperature gradient across the window which causes heat to flow to the outside, where it will increase the internal energy of the atmosphere. Heating is a process in which energy is transferred from a region of high temperature to a region of low temperature. As stated earlier, it is possible to answer the question 'How much internal energy is possessed by an object?' It is not possible or sensible to try to answer the question 'How much heat or work is possessed by an object?'

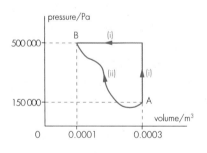

FIG 26.6 Diagram for question 26.1.

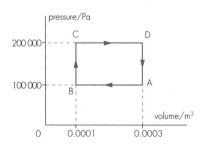

FIG 26.7 Diagram for question 26.2.

QUESTIONS

26.1 A gas has its state changed from A to B by two different paths, as shown in **Fig 26.6**. If path (i) requires the input of 60 J of heat, find:

   a the work done on the gas in changing its state along path (i);

   b the increase in internal energy between A and B.

   c What can be deduced about the work done on the gas, the heat supplied to the gas and the increase in the internal energy of the gas if path (ii) is used?

26.2 An ideal gas undergoes a cycle of changes A → B → C → D → A, as shown in **Fig 26.7**. Copy and complete Table 26.4 for the cycle.

TABLE 26.4 Copy and complete this table in question 26.2

	INCREASE IN INTERNAL ENERGY OF GAS/J	HEAT SUPPLIED TO GAS/J	WORK DONE ON GAS/J
A → B	−50		
B → C	25		
C → D			140
D → A			

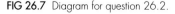

26.3 HEAT CAPACITY

The **heat capacity**, C, of an object is the quantity of heat that must be supplied to it to give it unit rise in temperature. The SI unit for heat capacity is the joule per kelvin ($J\,K^{-1}$). In equation form this definition becomes

$$C = \frac{\Delta Q}{\Delta T} \quad \text{or} \quad \Delta Q = C\,\Delta T$$

where ΔQ is the quantity of heat energy supplied and ΔT is the temperature rise.

The term 'specific heat capacity' is also used. In physics the word 'specific' means *per unit mass*. The **specific heat capacity**, c, therefore is the heat capacity per unit mass. It has the SI unit joule per kilogram per kelvin ($J\,kg^{-1}\,K^{-1}$) and in equation form is

$$c = C/m$$

where m is the mass of the object. Combining the last two equations gives

K $\quad \Delta Q = mc\Delta T$

an equation that you will need frequently.

The specific heat capacity is given for a material, whereas the heat capacity is given for a particular object. This is illustrated in the following example.

EXAMPLE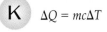

A hot water tank for a house contains 120 kg of water at 15 °C. The tank itself has a heat capacity of 6000 J K^{-1}. Find how long it will take an immersion heater to raise the temperature of the water to 50 °C if the tank is well insulated and the power of the heater is 2500 W. [Specific heat capacity of water is 4200 J kg^{-1} K^{-1}.]

From the equation $\Delta Q = mc\Delta T$, for the water we have

heat supplied to the water
$$= mc\Delta T$$
$$= 120\ \text{kg} \times 4200\ \text{J kg}^{-1}\ \text{K}^{-1} \times 35\ \text{K}$$
$$= 17\,640\,000\ \text{J}$$

Note that a difference in temperature of $(50-15)$°C is, by definition, the same as a difference in temperature of 35 K.

From the equation $\Delta Q = C\Delta T$, for the tank itself we get

heat supplied to the tank $= C\Delta T$
$$= 6000\ \text{J K}^{-1} \times 35\ \text{K}$$
$$= 210\,000\ \text{J}$$

This assumes that the tank reaches the same final temperature as the water. This is a reasonable assumption, since the tank is well insulated and is a good conductor of heat. Note that the quantity of heat required to heat the tank is very small compared with that required to heat the water.

Adding together these two quantities of heat gives

total quantity of heat required
$$= 17\,640\,000\ \text{J} + 210\,000\ \text{J} = 17\,850\,000\ \text{J}$$

The power supplied by the heater is 2500 W = 2500 J s^{-1}, so the total time needed to supply the above quantity of heat is

time required $= (17\,850\,000\ \text{J})/(2500\ \text{J s}^{-1}) = 7140\ \text{s}$

The assumptions made in answering this question make it unreliable to quote three or more significant figures so, having quoted the values that the data give, it is sensible to say that the time required is approaching two hours.

A problem arises with the heat capacity of gases. If you look again at the answers in Example 26.1, you will see that the heat required to change the temperature of the gas from 100 K to 450 K in that example was:

  i   1650 J for the first way of changing the temperature;
 ii 1550 J for the second way of changing the temperature;
iii 1450 J for the third way of obtaining the same temperature change.

This means that the heat capacity of the gas involved can be 4.71 J K^{-1}, 4.43 J K^{-1} or 4.14 J K^{-1}, depending on how the change in temperature takes place. This lack of constancy of heat capacity means that, whenever the heat capacity of a gas is given, it is essential also to say *under what conditions* the value was obtained. For example

- specific heat capacity of air at constant pressure, c_p is 1010 J kg^{-1} K^{-1};
- specific heat capacity of air at constant volume, c_v is 722 J kg^{-1} K^{-1}.

Specific heat capacities of gases can have any value – positive, zero or negative – so the term is meaningless unless the conditions are stated. Values of heat capacity for gases are usually given *per mole* rather than per unit mass; so the **molar heat capacity** is the heat capacity per mole. It is therefore essential to be very careful and know exactly what value is being used, and to make it agree with the quantity of gas actually present. The molar heat capacity of a gas at constant pressure is given the symbol $C_{p,m}$ and the molar heat capacity of a gas at constant volume is $C_{v,m}$. Some values for these constants are given in Table 26.5.

If you look at the values given in Table 26.5, you may notice that there is a nearly constant difference between the two columns of figures. The difference has the value of 8.3 J K^{-1} mol^{-1}, and it is not a coincidence that this is the same numerical value and the same unit as the molar gas constant, R. For an ideal gas,

$$C_{p,m} - C_{v,m} = R$$

You may need to remember this equation. The table shows that sulphur dioxide does not behave as an ideal gas.

TABLE 26.5 Molar heat capacities of various gases at constant pressure ($C_{p,m}$) and at constant volume ($C_{v,m}$)

	$C_{p,m}$/J K^{-1} mol^{-1}	$C_{v,m}$/J K^{-1} mol^{-1}
helium	20.8	12.5
air	29.1	20.8
oxygen	29.4	21.1
carbon monoxide	29.2	20.9
sulphur dioxide	40.4	31.4

26.4 INDICATOR DIAGRAMS

When the pressure, volume and temperature of a gas change, it is useful to be able to show those changes on a graph. However, a single two-dimensional graph cannot display three variables. Instead, it is usual to plot a whole series of graphs of pressure against volume, each at a different temperature, and then to superimpose on these graphs a line to show how the state of a gas varies when certain changes are made. The lines on these graphs are called **indicator diagrams**, and they have been used already in **Figs 26.5** to **26.7**. Some features of indicator diagrams need to be understood in order to make use of them when considering different types of heat engine.

1 An indicator diagram needs to be drawn for a fixed mass of gas. This implies that, for an ideal gas, pV/T will be constant for all points drawn on the diagram.
2 The area beneath the curve is a measure of work done. If the gas is expanding, then work is being done *by* the gas; if the gas is being compressed, then work is being done *on* the gas, usually by a piston compressing it in a cylinder.

3 A vertical line on an indicator diagram indicates a change in pressure at constant volume. For an ideal gas, the pressure will be proportional to the temperature. No work is being done since there is no movement if the volume is constant. This change is called an **isovolumic** change in the state of the gas.

4 A horizontal line indicates a change in volume at constant pressure. For an ideal gas, the volume will be proportional to the temperature. The area beneath the curve is the work done. This change is called an **isobaric** change in the state of the gas.

5 An **isothermal** change in the state of the gas is one in which there is no change in the temperature of a gas. For an ideal gas, the volume will be inversely proportional to the pressure and there will be no change in the internal energy of the gas.

6 A change that takes place in the state of a gas without any heat being supplied to it, or taken from it, is called an **adiabatic** change. For an ideal gas, an adiabatic expansion always results in a fall in the temperature, as the work that the gas does can only be done by using some of its internal energy. This results in a curve of steeper gradient than the isothermal on the indicator diagram.

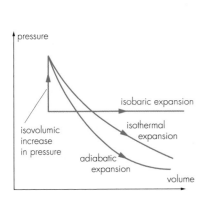

FIG 26.8 Some typical lines and curves on an indicator diagram: Isovolumic – no change in volume. Isobaric – no change in pressure. Isothermal – no change in temperature. Adiabatic – no heat supplied.

These changes are illustrated in **Fig 26.8** and are given in a table in the summary at the end of this chapter. The adiabatic change is always one in which no heat is supplied to the system. Mathematically, the changes that take place in the pressure, volume and temperature are rather complex and are not dealt with in this book. You may, however, find it necessary to use the expression relating p (the pressure of an ideal gas) with V (its volume) while undergoing an adiabatic change. The relationship is

$$pV^\gamma = k$$

where k is a constant and γ is the ratio $C_{p,m}/C_{v,m}$. For air, γ has the value 1.4.

In section 26.3 it was stated that the equation for ΔQ, the heat supplied to a substance in terms of its specific heat capacity, would be needed frequently. The equation

$$\Delta Q = mc\Delta T$$

becomes

$$\Delta Q = nC_{v,m}\Delta T$$

when dealing with n moles of a gas whose molar heat capacity at constant volume is $C_{v,m}$ and which is undergoing a rise in temperature ΔT at constant volume. Using the first law of thermodynamics on a change at constant volume gives this same expression for the increase in the internal energy of the gas since, at constant volume, no work is done on or by the gas. That is

increase in internal energy of gas	=	heat supplied ***to*** gas	+	work done ***on*** gas
$nC_{v,m}\Delta T$	=	$nC_{v,m}\Delta T$	+	0

The internal energy of an ideal gas, however, depends only on its temperature and not on the pressure or volume of the gas. So the change in the internal energy of a gas is *always* $nC_{v,m}\Delta T$ whatever the conditions for the change. If a change in state takes place at constant pressure, then the change in the internal energy of the gas is still $nC_{v,m}\Delta T$.

26.3 The curves in **Fig 26.9** show isotherms for 1 mol of an ideal gas. At all points on the curves, $pV/T = R$, where R is the molar gas constant having a value of $8.31 \, \text{J K}^{-1} \, \text{mol}^{-1}$. For the ideal gas, heat capacity at constant volume $C_{v,m} = 20.8 \, \text{J K}^{-1} \, \text{mol}^{-1}$; heat capacity at constant pressure $C_{p,m} = 29.1 \, \text{J K}^{-1} \, \text{mol}^{-1}$.

a Show, by considering at least three points on the curves, that pV/T does have a constant value of 8.3 (to 2 sig. figs) and show that its unit is $\text{J K}^{-1} \, \text{mol}^{-1}$.

b What gas law is being obeyed if the state of the gas changes along

   i the isotherm AB,

   ii a straight line AC,

   iii a straight line AD?

c How much heat needs to be supplied to the gas to change its state from

   i A to C in a straight line,

   ii A to E in a straight line,

   iii A to B along the isotherm?

d How much work needs to be done on the gas to change its state from

   i A to E in a straight line,

   ii A to C in a straight line? [Be careful with the sign.]

e Estimate, from the area under the curve, the work done on the gas for the state of the gas to change along the isotherm from A to B.

f Use the first law of thermodynamics to find the increase in internal energy if the state of the gas changes from

   i A to C,

   ii A to E.

Why are these two values the same? [Note that, whereas the path taken for finding the heat supplied and the work done always needs to be stated, this is not necessary for the change in internal energy. This is because internal energy of a gas is a function of its state.]

g Apply the first law of thermodynamics to the isothermal change A → B. Check your answer to **c iii**. It should not be zero. How is it possible for a gas to be supplied with heat and yet for its temperature not to rise?

h What is the heat capacity of a gas at constant temperature?

i [*Difficult*] Assume a gas starts at point A and is then allowed to expand to a volume of $0.50 \, \text{m}^3$ without any heat being supplied to it. Its temperature will fall. Find a possible final temperature so that the work done by the gas during the expansion equals its loss in internal energy. State what assumptions you make about the path taken for the change in state.

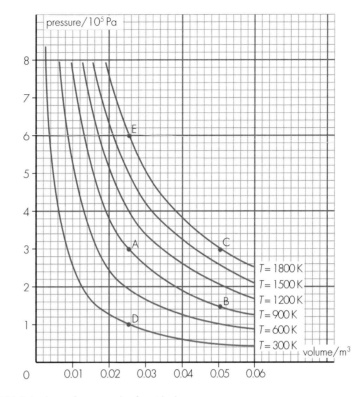

FIG 26.9 Isotherms for one mole of an ideal gas.

HEAT ENGINES

As explained earlier, heat engines are devices that convert thermal energy into work. Heat engines absorb heat from a source, perform some mechanical work, and lose some heat to an exhaust. The internal combustion engine in a car does just this. As a result of burning a small quantity of petrol inside the cylinder, a quantity of heat is supplied to the gas in the cylinder. The pressure in the cylinder rises and the piston is pushed downwards, doing work in the process. The hot gases remaining then have to be ejected from the car through the exhaust pipe. A mechanical engineer, when designing an engine, needs to apply the laws of thermodynamics to the processes taking place in order to try to get the maximum possible amount of work done from a minimum quantity of petrol used. The **thermal efficiency** ε of a heat engine is defined by the equation:

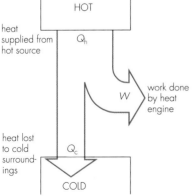

FIG 26.10 A generalised diagram showing the function of a heat engine. This is called a 'Sankey diagram'.

K $$\varepsilon = \frac{\text{work done by engine}}{\text{heat supplied to engine}}$$

This is illustrated on a diagram called a *Sankey diagram* (**Fig 26.10**), where the heat supplied to the engine from the hot source, Q_h, is converted into work, W, and an amount of heat, Q_c, lost to the cold surroundings. Applying the law of conservation of energy to this process gives $W = Q_h - Q_c$, so the efficiency can also be written

$$\varepsilon = \frac{Q_h \pm Q_c}{Q_h}$$

As can be seen from **Fig 26.10**, or from the equation, the efficiency of the heat engine can be increased if more of the energy supplied from the hot source can be changed into work so that less is wasted to the cold surroundings. There are both practical and theoretical reasons why Q_c cannot be zero. In practice, efficiencies are typically 40%.

QUESTION

26.4 The thermal energy supplied by 1 kg of petrol is 48 000 kJ. Find the efficiency of a car that is producing mechanical energy at a rate of 40 kW if it is using petrol at a rate of 8 kg per hour.

Any heat engine needs to be able to do work continuously. After expansion of the burning gases takes place in a petrol engine, the waste gases have to be ejected through the exhaust, and more petrol vapour and air have to be supplied to get the engine ready for the next power stroke. This cyclical pattern of changes can conveniently be shown on an indicator diagram. The *Otto cycle* is an idealised sequence of changes that take place on a fixed amount of gas in a petrol engine (**Fig 26.11**). It consists of the following separate parts.

1   During an intake stroke, air and petrol vapour are transferred from outside the cylinder to inside the cylinder, ideally without change in volume, temperature or pressure. This all occurs at point A on the indicator diagram.
2   From A to B, the piston squashes the gas to a small volume. No heat is supplied to the gas in this stage. This is called the compression stroke, and it is an adiabatic compression.
3   From B to C, as a result of a spark from the sparking plug igniting the petrol, heat is supplied to the gas in a brief moment of time and this causes a large rise in the pressure at constant volume.
4   From C to D is the power stroke of the engine. Work is done by the expanding gases on the piston, and their temperature and pressure fall. No heat is being supplied during this stage, so the expansion is an adiabatic expansion.
5   The stage from D back to A, ready to start again, is complex. It can ideally be considered as a cooling down to the starting temperature, during which time heat is lost, and then an exhaust stroke at constant pressure, volume and temperature just to move the waste gases from the cylinder to the outside.

Note that work is done by the gases in the cylinder on the piston during the power stroke, and that work has to be done on the gases by the cylinder during the compression stroke. The area within the loop is therefore the net work output of the engine in a cycle.

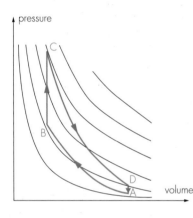

FIG 26.11 Indicator diagram for an idealised petrol engine.

 **Q** **Why is the last sentence true?**

EXAMPLE 25.3

A car engine has a cylinder whose maximum volume is 0.000 50 m³ and minimum volume 0.000 050 m³. It goes through one Otto cycle. Atmospheric pressure is 1.0×10^5 Pa and atmospheric temperature is 300 K. The engine has 0.0125 g of petrol injected during each stroke, which gives 600 J of heat on burning. Calculate the maximum possible efficiency of the engine. [It will not be 100%.] Assume that the molar heat capacity $C_{v,m}$ for the gas in the engine is 21 J K^{-1} mol^{-1}.

The procedure used in solving such problems is to assume that the engine behaves *ideally*. The efficiency found will then be the maximum possible efficiency.

The first stage in solving the problem is to find the amount of air being used in each cycle. At the start

$$pV = nRT$$

so

$$1.0 \times 10^5 \text{ Pa} \times 0.000\,50 \text{ m}^3 = n \times 8.3 \text{ J K}^{-1} \text{ mol}^{-1} \times 300 \text{ K}$$
$$n = 0.020 \text{ mol}$$

The second stage is to find the temperature rise that takes place when the petrol is burnt. This is given by

$$\Delta T = \frac{Q}{nC_{v,m}} = \frac{600 \text{ J}}{0.20 \text{ mol} \times 21 \text{ J K}^{\pm 1} \text{ mol}^{\pm 1}} = 1430 \text{ K}$$

The third stage involves applying the gas equation to each of the four sections of the cycle. As before, it is easiest to use a table. In Table 26.6, some of this information can be put in directly and is shown in normal type. *Italic* type is used to show subsequent working, with **bold** type showing the more difficult steps in the adiabatic expansions. Letters refer to **Fig 26.11**. At all sections in the cycle the value of pV/T has the constant numerical value of 0.020×8.3, which is 0.166. Note that the temperature rise of 1430 K has taken place between B and C.

The fourth stage concerns the first law of thermodynamics particularly. All the information in Table 26.6 is information that, here, is obtained by using the gas equation. It could equally well have been obtained by experiment. It is interesting to note that the temperature reached by the gases in an engine is above the melting point of the metal used in the construction of the engine. This is one reason why the engine always needs to be well

TABLE 26.6 The gas equation applied to the four sections of the cycle in Example 26.3

SECTION OF CYCLE	p/kPa	V/m³	T/K
start A	100	0.000 50	300
after compression B	**2510**	0.000 050	*753*
after ignition C	*7270*	0.000 050	*2180*
after power stroke D	**290**	0.000 50	*870*
after exhaust A	100	0.000 50	300

TABLE 26.7 The first law of thermodynamics applied to the four sections of the cycle in Example 26.3

	INCREASE IN INTERNAL ENERGY OF GAS/J	HEAT SUPPLIED TO GAS/J	WORK DONE ON GAS/J
A to B	190	0	190
B to C	600	600	0
C to D	−550	0	−550
D to A	−240	−240	0

cooled. In applying the first law to the changes of state taking place, it is worth while again to use a table. For all the changes, the change in internal energy is

$$nC_{v,m}\Delta T = 0.020 \text{ mol} \times 21 \text{ J K}^{-1} \text{ mol}^{-1} \times \Delta T$$
$$= 0.42 \text{ J K}^{-1} \times \Delta T$$

The ΔT values needed can be found from the differences of the values in the last column of Table 26.6. So, inserting these ΔT values in the above equation, we can draw up Table 26.7 for the sections of the cycle. Note that there is no overall change in the internal energy of the gas.

The final stage is to examine these data to see that the efficiency of the engine is

$$\varepsilon = \frac{\text{net work done by gas in engine}}{\text{heat supplied to gas in engine}} = \frac{500 \text{ J} \pm 190 \text{ J}}{600 \text{ J}}$$
$$= 0.60 = 60\%$$

This value of 60% gives the maximum possible efficiency that an engine such as this could have even under such idealised conditions as have been assumed.

In practice, there are many factors that will reduce the actual efficiency to well below the maximum possible efficiency. These factors include friction, the gas not behaving ideally, ignition not taking place instantly, heat being lost during the adiabatic compression, and losses of energy on transferring the gas into and out of the cylinder. The indicator diagram for a real engine is more likely to be as shown in **Fig 26.12**. Deviations from idealised behaviour do always reduce the efficiency, but the important point for the operation of a heat engine is that, even if it were working ideally, it would still not be 100% efficient. A final point to note about this problem on a car engine is that the car's heater makes use of some of the wasted 40% of energy.

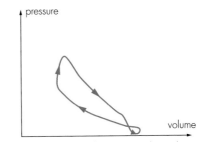

FIG 26.12 Indicator diagram for a practical petrol engine.

The following question refers to a different type of heat engine. Much of the tedious calculation of pressures and volumes has already been done for you, so that the question can concentrate on the thermodynamic properties of the engine.

QUESTION

26.5 In a large diesel engine, air is adiabatically compressed (A to B, **Fig 26.13**) by the piston in a cylinder until the air reaches a very high pressure and temperature. A small amount of diesel oil is then injected into the cylinder and it burns in the hot air (B to C). Expansion continues adiabatically until the piston is pushed down its full amount (C to D), and then cooling and exhaust take place (D to A), before the cycle starts again. The compression ratio of this engine is 25:1. That is, the volume occupied by the gas in the engine has a value at maximum volume that is 25 times its minimum volume. Compression ratios have a considerable influence on efficiencies and are typically much higher in diesel engines than in petrol engines.

TABLE 26.8 Data for question 26.5

	P/kPa	V/cm^3	T/K
start A	100	2250	300
after compression B	9060	90	1090
after injection C	9060	173	2090
after power stroke D	250	2250	750
after exhaust A	100	2250	300

Use the data given in **Fig 26.13** and in Table 26.8, and the molar heat capacities $C_{v,m} = 20.8\,\text{J K}^{-1}\,\text{mol}^{-1}$ and $C_{p,m} = 29.1\,\text{J K}^{-1}\,\text{mol}^{-1}$ for the gas in the cylinder, to answer the questions that follow.

a Use the gas law to find how many moles of gas are used in each cycle.

b Copy and complete Table 26.9 using the first law of thermodynamics. It is suggested that you start by putting in terms known to be zero and then work vertically down the column headed 'Increase in internal energy…'.

TABLE 26.9 Copy and complete this table for question 26.5(b)

	INCREASE IN INTERNAL ENERGY OF GAS/J	HEAT SUPPLIED TO GAS/J	WORK DONE ON GAS/J
A to B			
B to C			
C to D			
D to A			

c Use Table 26.9 to show that the efficiency of this diesel engine is 68%.

d What is the net power output of the engine if it completes 6000 power strokes each minute?

e Suggest a possible use for such an engine.

f If the fuel supplies 45 000 kJ of heat energy per kilogram, how fast is fuel being used when running at 6000 power strokes per minute?

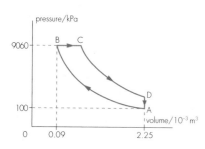

FIG 26.13 Indicator diagram for a large diesel engine.

FIG 26.14 A Rolls Royce RB211 Turbofan engine.

ANALYSIS

The behaviour of gases in a Turbofan jet engine (Rolls Royce type RB211)

The Rolls Royce Turbofan jet engine is shown in **Fig 26.14.** Four of these engines are fitted to a Boeing 747. The thermodynamics of the engine is illustrated by **Fig 26.15**. Air enters the engine at a rate of $420 \, m^3 \, s^{-1}$ when the aircraft is cruising at $250 \, m \, s^{-1}$ at an altitude of 9000 m.

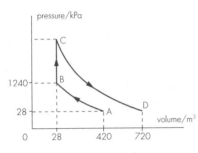

FIG 26.15 Indicator diagram for a Turbofan engine.

The air at this altitude has a pressure of 28 000 Pa and a temperature of 225 K. The air first passes through a compressor. This reduces the volume of gas to one-fifteenth of its original volume. The compressed air then has fuel continuously sprayed into it. The fuel burns, creating a very high-pressure, high-temperature gas that expands rapidly out of the rear of the engine. In doing so it does work on a turbine.

Throughout this exercise, consider the operation of the engine for 1 s. Many approximations are made in the calculations, so the answers are only a guide to the actual values. In reality, the engine is less efficient than expected. It is worth while making intermediate calculations to three significant figures so that two significant figures are reliable for the idealised engine. If you get stuck with a part of this problem, just go on to the next part. The problem has been arranged so that later parts can be done even if earlier parts cannot. Information required can be found in the 'Answers' section on p.548.

a Find the number of moles of air that enter the engine in 1 s using

$$pV = nRT$$

b A $\rightarrow$ B is an adiabatic compression. No heat is supplied to the air during the compression. For the compression, $pV^{1.4}$ is constant. Use the equation

$$p_A V_A^{1.4} = p_B V_B^{1.4}$$

to show that the pressure after compression is $1.24 \times 10^6 \, Pa$.

c What is the temperature at B?

d Fuel is burnt to raise the temperature of the air at C to

1400 K. What is the pressure at C? Assume that there is a constant amount of air throughout the cycle; that is, ignore the chemical processes that can alter the number of moles of gas present.

e How much heat is supplied by the fuel per second? The molar heat capacity of the air at constant volume, $C_{v,m}$, is $20.8 \, J \, K^{-1} \, mol^{-1}$.

f How much aviation fuel is burnt per second? Aviation fuel gives $53 \, 000 \, kJ \, kg^{-1}$.

g How much fuel is used while cruising for 6 h during a flight (four engines)?

h The exhaust gases then expand adiabatically to atmospheric pressure. What is their temperature at D if they have then expanded to $720 \, m^3$?

i Why do large jets cause vapour trails?

j It appears at first sight as though this indicator diagram is not a complete loop. However, while a continuous supply of new air is available at A, the gas supplied to the atmosphere at D occupies a larger volume at D than it originally did at A. Work has to be done against the atmosphere to create this space. How much work has to be done? This is equivalent to completing the indicator diagram with a horizontal line from D to A.

k Copy and complete Table 26.10, as follows:
  i Use the molar heat capacity at constant volume to find the increase in internal energy for each of the four sections of the cycle.
  ii Insert zeros in the table for the sections in which no work is done on the air, or no heat is supplied to, or taken from, the air.
  iii Complete the table.

TABLE 26.10 Copy and complete this table in part **k**

	INCREASE IN INTERNAL ENERGY OF GAS/J	HEAT SUPPLIED TO GAS/J	WORK DONE ON GAS/J
A to B			
B to C			
C to D			
D to A			

l Use Table 26.10 to find the theoretical efficiency of the engine.

m The thrust of the engine was stated to be 220 000 N and the speed of the aircraft is $250 \, m \, s^{-1}$. Find the actual power of the engine.

n What is the efficiency of the engine in practice?

o Why is the efficiency of the engine less at low altitude?

26.6 THE SECOND LAW OF THERMODYNAMICS

FIG 26.16 An early steam engine such as this could be expected to have an efficiency of as little as 1%.

FIG 26.17 A high-efficiency modern steam turbine. The laws of thermodynamics do not permit 100% efficiency. Turbines of this type are usually about 40% efficient.

All the examples given in section 26.5 on heat engines have been used to find the efficiency of engines. Even when these engines are operating under idealised conditions, their efficiencies are not 100%. The question immediately arises therefore as to what factors control the maximum possible efficiency of a heat engine.

Early steam engines (**Fig 26.16**) were extremely inefficient, less than 1%. The most efficient steam engine that British Rail ever had was only about 11% efficient. The heat engines in power-stations, steam turbines (**Fig 26.17**), are usually not more than 40% efficient. Car engines typically are about 30% efficient.

The second law of thermodynamics deals with the question of heat engine efficiency and states that it is impossible, in principle, for any heat engine to have a thermal efficiency of 100%. The problem stems from the fact that heat engines have to work in a cycle, and that the end of one expansion stroke *cannot* take the gas back to a state at which it is ready to start a new cycle. A 100% efficient engine would have an indicator diagram as shown in **Fig 26.18a**, but the second law says that this is impossible. **Fig 26.18b** shows a cycle that can be shown to be the most efficient cycle operating between any two given temperatures. **Fig 26.18c** shows the indicator diagram of a 100% efficient engine, which could only be achieved if the gas could start and finish at a temperature of absolute zero! Since we have to run engines in a world where the outside temperature is in the approximate range of 250–300 K, we have no possibility of achieving 100% efficient heat engines.

The second law is also related to the definition of energy as the *stored* ability to do work. The internal energy of the atmosphere includes the kinetic energy of all the molecules in the atmosphere – energy that is stored in their movement. However, it is extremely difficult to make use of this energy to do work. The energy is present, stored in the movement of the molecules, but the second law results in it being impossible to get this energy out of store to do work unless there is some lower-temperature system available. A temperature difference is necessary before work can be done. The Earth's atmosphere can be considered as a huge heat

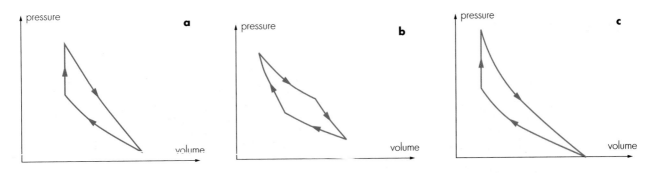

FIG 26.18 a This indicator diagram, in which no heat is lost, breaks the second law of thermodynamics. **b** The indicator diagram of an engine with the maximum possible efficiency for the temperatures used. **c** An indicator diagram in which the efficiency is 100%, but this is only possible with the surroundings at a temperature of 0 K.

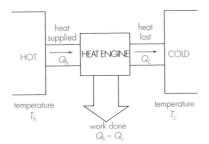

FIG 26.19 The efficiency of any heat engine is limited by the temperatures between which it is operating. All heat engines give up heat to a low-temperature reservoir, often the atmosphere.

engine using hot air near the equator as the source of heat and cold air at the poles as the low-temperature reservoir. Work is done in creating wind movement. One statement of the second law of thermodynamics is that it is impossible for any process to have as its sole result the transfer of heat from a colder to a hotter body.

It can be shown that, when an ideal heat engine takes a quantity of heat Q_h from a hot source at a temperature T_h and wastes a quantity of heat Q_c to its cold surroundings at a temperature T_c (**Fig 26.19**), then these quantities are related by the equation

K $$\frac{Q_c}{Q_h} = \frac{T_c}{T_h}$$

This equation is, in fact, the fundamental equation for the definition of thermodynamic temperature, but here it is needed to calculate the maximum possible efficiency ε_{max} of any ideal engine. Using **Fig 26.19** we get:

$$\varepsilon_{max} = \frac{\text{work done by ideal engine}}{\text{heat supplied from hot source}} = \frac{Q_h \pm Q_c}{Q_h} = 1 \pm \frac{Q_c}{Q_h} = 1 \pm \frac{T_c}{T_h}$$

QUESTIONS

26.6 What is the maximum possible efficiency for a steam engine using steam at a temperature of 100 °C on a day when the temperature is 24 °C? [The answer is *not* 76%.]

26.7 Magnox nuclear reactors, which operate at a temperature of 370 °C, use uranium metal as a fuel, and this is contained in magnesium alloy cans. Advanced gas-cooled reactors use a uranium oxide ceramic as a fuel, contained in stainless steel cans. These materials allow the higher temperature of 645 °C to be used. What improvement can theoretically be made in the efficiency of a nuclear power-station when the output temperature of the reactor is raised from 370 °C to 645 °C, and the cooling water for the reactor is at a temperature of 15 °C?

The implications of the formula for the efficiency of heat engines are enormous, particularly in the electrical supply industry. At present, if a power-station operates at 40% efficiency, the remaining 60% of the heat energy supplied by the fuel is wasted, probably in heating up a river or the atmosphere. The laws of thermodynamics always put upper limits on the efficiency of power-stations, as they do on all heat engines. Since the value of T_c cannot be below the temperature of the outside air for an engine, the only practical way of improving the efficiency of heat engines is by raising the temperature T_h. Problems associated with this are the fact that the materials used in the engine may start to melt or become less strong. The high pressures involved with using steam at high temperatures also cause difficulties. If more people lived near power-stations, they could be supplied with warm water from the cooling system of the power-station just as a car heater uses the cooling water of the engine to heat the inside of the car. Such systems do exist, and they will probably become more popular as fuel costs rise. They are called CHP schemes – the initials stand for combined heat and power.

The problems associated with the large-scale supply of energy and its environmental implications are too large a topic to be dealt with in this book, but the physics of any problem cannot be overlooked. It is always necessary to consider numerical values of physical quantities alongside those of economics. Too many plausible suggestions are made for which neither the economics nor the physics is correct. Given a free choice, the general public want cheap power supplies that they cannot see. But the 'NIMBY' effect is real – people want power supplies but 'not in my back yard'.

The public also want to use power from renewable resources so that fossil fuels are not wasted. As the ultimate source of energy, the Sun's power is able to be harnessed more effectively than it is at present. Solar panels that give modest amounts of tepid water are never likely to be used in great quantity in the UK, but solar cells, which produce electricity, are becoming cheaper and more efficient and offer the prospect of good reliable energy supplies for the future. At present the manufacture of a solar cell uses up as much energy as it can produce from sunlight over a considerable time, but already these cells make commercial sense in out-of-the-way places (**Fig 26.20**).

Other sources of renewable power, such as tidal power, wind power and wave power, are promising on a small scale, but large-scale production of electrical energy using these methods will involve much unsightly machinery, often in places of outstanding natural beauty. In those parts of the world with geysers and volcanoes, useful harnessing of their power is already being achieved. The island of St Lucia in the West Indies, for example, has a scheme which uses hot mud under construction at present to replace an already successful pilot scheme.

FIG 26.20 This roadside emergency telephone in Hawaii is powered by photovoltaic cells.

It will use hot mud in avolcano to provide superheated steam for a power-station. The aim is to produce enough electricity for the island's own needs and to be able to export electrical energy to neighbouring islands.

HEAT PUMPS

An interesting application of the laws of thermodynamics is in the operation of a refrigerator. A refrigerator can be considered as a heat engine working in reverse. One particular refrigerator uses a gas that undergoes a cycle of changes of state, as shown by the indicator diagram in **Fig 26.21**. Notice that the arrows are in the opposite direction to those in indicator diagrams for heat engines. More details of the cycle are shown in Table 26.11.

TABLE 26.11 Details of the heat pump cycle

	INCREASE IN INTERNAL ENERGY OF GAS	HEAT SUPPLIED TO GAS	WORK DONE ON GAS
A to B	47 J gas temperature increasing	47 J heat taken from food	0 J no volume change
B to C	42 J gas temperature still increasing	0 J adiabatic compression	42 J area under curve BC
C to D	−63 J gas temperature falls	−63 J heat lost to atmosphere	0 J no volume change
D to A	−26 J gas temperature falls to starting temperature	0 J adiabatic expansion	−26 J area under curve DA

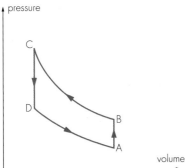

FIG 26.21 Indicator diagram of a heat pump.

The net work done to run this refrigerator for one cycle is 42 J − 26 J = 16 J. This will allow 47 J to be extracted from the food in the refrigerator and 63 J to be released into the atmosphere. This is why the back of a refrigerator is warm. Heat is being extracted from the food and being added to the work done to supply the heat to the atmosphere. The refrigerator is acting as a heat pump to move heat from inside the refrigerator to the outside, as is shown in **Fig 26.22**.

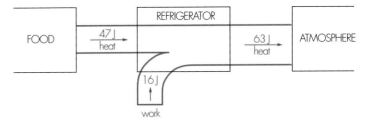

FIG 26.22 A refrigerator considered as the reverse of a heat engine, a heat pump.

This principle can be used for heating systems, and is particularly effective if there is something readily available outside a house that can be cooled down. The atmosphere is a possibility, but a stream or river would be much better. A heat pump uses electrical energy to operate, and it extracts heat from the stream or river and pumps it into the house, as illustrated diagrammatically in **Fig 26.23**.

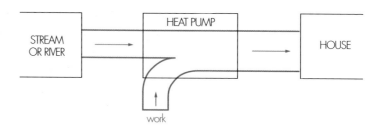

FIG 26.23 A heat pump being used for central heating.

A heat pump system that uses the atmosphere as its source of heat can supply thermal energy at a rate of 6.2 kW for an electrical power output of only 3.3 kW if the outside temperature is 8 °C. When the outside temperature falls to −1 °C, the output falls to 4.5 kW for the same electrical power input.

Inevitably there are some snags with heat pumps. They operate much better in the summer than in the winter because it is easier to cool down the outside source of heat if it is at a high temperature. A heating system has severe limitations, however, if it works less and less well as the outside temperature drops. As mentioned earlier, they do not work particularly well if only the atmosphere is available for cooling down, because the air passed through the heat pump has too little mass. Running water is much better, but not many houses have running water flowing past them. The capital cost of installing heat pumps can also be large because of all the pipework necessary within and outside the pump.

SUMMARY

- The internal energy of a gas is the sum of all the microscopic kinetic and potential energies of the molecules within the gas.

- The first law of thermodynamics states that the internal energy of a system depends only on its present state; the internal energy may be increased either by doing work on the system or by heating it. That is

$$\begin{matrix} \textit{increase} \text{ in internal} \\ \text{energy of a gas} \end{matrix} = \begin{matrix} \text{heat supplied} \\ \textit{to} \text{ the gas} \end{matrix} + \begin{matrix} \text{work done} \\ \textit{on} \text{ the gas} \end{matrix}$$

- Indicator diagrams for a gas are graphs of pressure against volume and show how the state of a gas changes.

- Properties of various changes of state:

	ISOTHERMAL	ADIABATIC	ISOBARIC	ISOVOLUMIC
definition	no change in temperature	no heat supplied	no change in pressure	no change in volume
work done on gas	area under curve	area under curve	$p\Delta V$	zero
heat supplied to gas	*not* zero	zero	$C_p\Delta T$	$C_v\Delta T$
increase in internal energy	zero	$C_v\Delta T$	$C_v\Delta T$	$C_v\Delta T$
law applying*	$pV = k$	$pV^\gamma = k$	$V/T = k$	$p/T = k$

* $pV = nRT$ applies additionally in all cases.

■ For an ideal gas, $C_{p,m} - C_{v,m} = R$.

■ Since the internal energy of a fixed mass of an ideal gas depends only on its temperature, the change in internal energy can always be calculated from $C_v\Delta T$, no matter how the pressure and volume change.

■ The second law of thermodynamics makes the maximum possible efficiency of a heat engine to be

$$\varepsilon = 1 \pm \frac{T_c}{T_h}$$

ASSESSMENT QUESTIONS

26.8 **a** State the first law of thermodynamics as applied to a fixed mass of gas when heat energy is supplied to it so that its temperature rises and it is allowed to expand.
Define any symbols you use.

b When 5.0×10^{-3} m³ of water is boiled away at an atmospheric pressure of 1.0×10^5 Pa, the volume of steam produced is 8.3 m³.
The work done against a constant external pressure when a gas expands is the product of pressure and change in volume. The specific latent heat of vaporisation of water is 2.3 MJ kg⁻¹. Assume that the density of water is 1000 kg m⁻³.
Calculate

  i the work done when the water expands to become steam,

  ii the increase in the internal energy of the water molecules.

NEAB 1996

26.9 This question is about a refrigerator.

a State the condition for a body to be in thermal equilibrium with its surroundings.

b **i** What is meant by *internal energy*?

  ii State the *first law of thermodynamics*.

  iii Explain why the temperature of a gas rises when compressed in a pump such as a tyre pump.

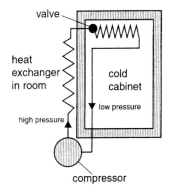

c In a domestic refrigerator a gas is compressed outside the cold cabinet. It is then allowed to expand through a valve into pipes within the cold cabinet (see diagram). With reference to the first law of thermodynamics, explain

  i why the gas expanding in the pipes in the cold cabinet is cooled,

  ii how the refrigerator transfers thermal energy from the contents of the cold cabinet to the room outside.

O & C 1997

26.10 a **i** Derive, from the definitions of work done and pressure, the expression

$$W = p\Delta V$$

for the work W done by a gas which is expanding against a constant external pressure p.

ii Suggest how the expansion in **i** might be carried out without the gas suffering a loss of internal energy.

b The figure shows a hot-air balloon.

The volume of the hot air in the balloon when fully inflated is 2.0×10^3 m^3. Atmospheric pressure is 100 kPa.
Calculate the work done against the atmosphere during the inflation of the balloon.

c The air in the balloon in **b** is heated by using gas burners which supply energy at a cost of 30 pence per kilowatt hour.
The work done in **b** represents 25% of the energy supplied by these burners.

i Express 1 kilowatt hour in terms of joules.

ii Calculate the cost of the inflation of the balloon.

Cambridge 1997

26.11 a What is a biofuel?

b Give an example of a biofuel.

c Why is it very inefficient to use biofuels to generate electricity for home heating rather than using them to heat homes directly?

London 1996

26.12 a A 24 W filament lamp has been switched on for some time. In this situation the first law of thermodynamics, represented by the equation $\Delta U = \Delta Q + \Delta W$, may be applied to the lamp. State and explain the value of each of the terms in the equation during a period of *two* seconds of the lamp's operation.

b Typically, filament lamps have an efficiency of only a few percent. Explain what this means and how it is consistent with the law of conservation of energy.

London 1996

26.13 a Pure water is boiled in a kettle to produce steam at 100 °C.

i Describe the motion of a typical water molecule in the steam.

ii Compare the mean kinetic energy of the water molecules in the steam with those in the boiling water.

b A mass of 0.20 kg of water at 100 °C was poured into a copper container which was at a temperature of 20 °C. The temperature was monitored with time and the results are given in the table. The water and the copper container rapidly reached a temperature of 90 °C, after which the temperature of the water dropped slowly.

specific heat capacity of water = 4190 J kg^{-1} K^{-1}
specific heat capacity of copper = 385 J kg^{-1} K^{-1}

time/s	temperature/°C
0	100
6.0	90
60	87

Calculate the amount of energy given to the copper by the water during the first 6.0 s.

NEAB 1998

26.14 A large hotel and sports complex uses a heat pump to provide hot water for central heating by transferring energy from a nearby river. The maximum power output from the heat pump installation is 920 kW and, when delivering this output, the electrical power input to the system is 368 kW.

a Determine the rate at which energy must be extracted from the river when the installation operates at maximum output.

b Explain briefly why the river is unlikely to turn to ice when the heat pump extracts energy from it.

c Estimate the annual fuel cost of running the heat pump if the heating requirement is equivalent to maximum output for 9 hours per day over 260 days each year. 1 kWh of electrical energy costs 8.0 pence.

d Calculate the annual saving in fuel costs when using the heat pump compared to an electrical heating installation of the same output using heaters operating directly from the mains electricity supply.

NEAB 1997

26.15 a State what is meant by the *internal energy* of a system.

b In a car, the driver's safety air-bag is suddenly inflated from a capsule producing a gas at high pressure. The volume of the bag increases from zero to 0.50 m^3, displacing atmospheric air which is at a constant pressure of 105 kPa.
Calculate the work done against atmospheric pressure during this inflation.

c Many of the gas molecules within the air-bag in **b** collide with the inner surface of the bag as it expands.

State and explain what happens to
  i   the speeds of these molecules,
  ii the internal energy of the gas.

OCR 1999

26.16 a State what is meant by the *internal energy* of a system.
  b An ideal gas is contained in a flask. Explain qualitatively how the motion of the molecules of the gas results in forces on the walls of the flask.
  c The flask in **b** is heated gently. Gas is allowed to escape so that the pressure in the flask remains unchanged.
    i   State what happens to the r.m.s. speed of the gas molecules in the flask.
    ii State and explain what happens to the number of collisions per unit time made by the gas molecules with the walls of the flask.

OCR 1999

26.17 a Define *specific heat capacity*.
  b An open-air swimming pool of surface area 60 m² has a uniform depth of 1.5 m and is heated by the Sun. The rate at which energy arrives from the Sun per unit area of water surface is 800 W m⁻². Of this energy, 20% is reflected and the rest is absorbed by the water. Water has specific heat capacity 4200 J kg⁻¹ K⁻¹ and density 1000 kg m⁻³.
  Calculate
    i   the rate at which energy is absorbed by the pool,
    ii   the mass of water to be heated,
    iii the mean rate of rise of temperature of the water,
    iv   the time taken for the temperature of the water to rise by 3.0 K.
  c Suggest three reasons why, in cool climates, it is difficult to maintain a high water temperature in a swimming pool if it is only heated directly from the Sun.

OCR 1999

26.18 a What is meant by a *heat engine*?
  b Explain why there is a constant search for materials to make turbine blades that will operate at higher temperatures to improve the efficiency of thermal power stations.

London 1997

26.19 A sample of a substance, initially in the solid state, is heated electrically at a constant rate. Energy transfer to the container and surroundings may be neglected. The variation with time t of the temperature θ of the sample is shown in the graph.

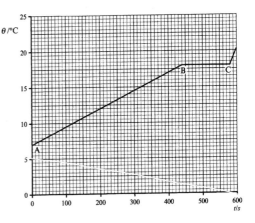

a Explain why the graph has the form that it does in the regions **i** AB and **ii** BC.
b Use the graph to deduce the melting point of the substance.
c The mass of the sample is 0.025 kg, and the rate of supply of electrical energy is 150 W. Making use of information from the graph, calculate
  i   the specific heat capacity of the solid substance,
  ii the specific latent heat of fusion of the substance.

CCEA 1998

26.20 This question is about heating and cooling a fixed mass of ideal gas.
An ideal gas is contained in a hollow cylinder, sealed at one end, with a frictionless piston at the other. See diagram. The volume of the trapped gas is V_0 when its pressure is p_0 and its temperature T_0.

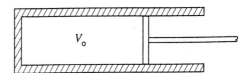

The system is used as a heat (Stirling) engine. The following cycle of changes ABCDA occurs to the gas as the piston moves in and out. The gas is
from A to B   compressed at constant temperature to half of its initial volume V_0,
from B to C   heated at constant volume to 1.5 times its initial temperature T_0,
from C to D   expanded at constant temperature back to its initial volume,
from D to A   cooled at constant volume back to its initial temperature.
a State which of the gas laws is obeyed during the process from A to B.
b Find the values of the pressure p of the gas at the end of each of the stages of the cycle in terms of the initial pressure p_0. Give your reasoning.
  i   at B,
  ii   at C,
  iii at D.

c Draw a *pV* diagram showing the cycle of operation of the engine. Mark the points A, B, C and D on your diagram. Use axes like those below.

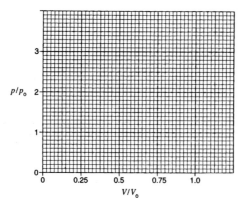

d Consider the stage A to B of the cycle. Is work *W* done on or by the gas? Is thermal energy *Q* supplied to the gas or extracted from it? How are *Q* and *W* related? Explain your reasoning.

O & C 1998

26.21 The diagram shows the theoretical pressure–volume diagram of an engine in which a fixed mass of gas is taken through a closed cycle of changes. The cycle consists of two isothermal processes A and C, separated by two constant pressure processes B and D. Process A occurs at a temperature of 300 K.

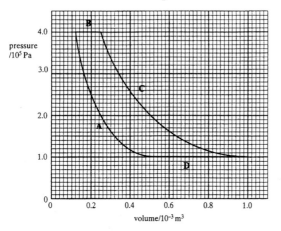

a In which two processes is work done by the gas?
b Calculate the number of moles of gas contained in the cylinder.
  $R = 8.3\,\text{J mol}^{-1}\,\text{K}^{-1}$
c Calculate the temperature of process C.
d Use the diagram to show that the net work done by the gas in one complete cycle is 70 J.

NEAB 1998

26.22 An inventor has designed a heat engine for a small combined heat and power plant which will operate between temperatures of 900 °C and 80 °C. The inventor makes two claims about the performance of the engine:
  claim 1 When the engine consumes fuel of calorific value 44 MJ kg^{-1} at a rate of 8.6 kg h^{-1}, it will deliver a useful mechanical output power of 80 kW.
  claim 2 At the same time, the engine will also provide energy at the rate of 25 kW for heating purposes.
a Use the calorific value of the fuel and the fuel flow rate to calculate the input power to the engine.
b Calculate the maximum possible efficiency of any heat engine which operates between temperatures of 900 °C and 80 °C.
c Using the result of your calculation in **b**, and any other necessary calculation, explain whether either or both of the inventor's claims are justified.

NEAB 1997

27 CHARGED PARTICLES

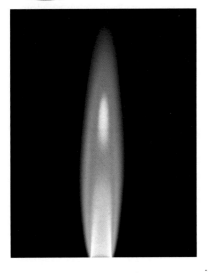

LEARNING OBJECTIVES

At the end of this chapter you should be able to:

① understand the significance of Thomson's and Millikan's experiments in finding the elementary charge;

② calculate the paths of beams of charged particles in both electric and magnetic fields;

③ find the specific charge of an electron or ion;

④ use and describe the principles of a cathode ray oscilloscope;

⑤ describe a mass spectrometer.

27.1 ELECTRONS

During the nineteenth century, sources of electrical energy became more powerful and more reliable. The development of dynamos and induction coils in place of batteries meant that higher voltages could be used. At the same time, vacuum pumps using electric motors were becoming more efficient. By combining these two developments, the conduction of electricity through gases could be investigated in glass tubes called *discharge tubes*. It was discovered that if a high voltage was applied across a gas at low pressure then the gas glowed with a colour characteristic of the gas at that pressure. The glowing gas conducts electricity and indicates that there must be some charged particles within the gas.

It is now known that any light being emitted from a gas shows that the gas is ionised. This phenomenon is used in many modern examples of lighting technology. Street-lights frequently use sodium vapour discharge lamps, which emit a characteristic orange glow. Higher-pressure sodium vapour lamps are now increasingly being used because they produce additional colours besides orange, and this gives a more pleasing, whiter, light. These lamps have high efficiency of conversion of electrical energy into light energy in comparison with ordinary filament lamps. Another example of a discharge tube is the common fluorescent tube. This contains a mixture of argon and mercury vapour. Examples of emission of light from ionised gases are shown in the photographs in **Figs 27.1** and **27.2**.

FIG 27.1 If a gas is visible, then ionisation is occurring, as in this propane gas flame from a Bunsen burner.

FIG 27.2 The atmosphere glows when it is bombarded by charged particles from the Sun. The particles collect in the strong magnetic field near the North Pole, causing the *aurora borealis* or 'northern lights'. This picture was taken in Alaska. The exposure was several seconds — long enough to expose the film with starlight, but not long enough to notice the rotation of the Earth.

INVESTIGATION

The resistance of a Bunsen flame

Use two steel rod probes about 3 mm thick and place them in retort stands so that they are about 3 mm apart (**Fig 27.3**). Use crocodile clips on the probes to connect them to an EHT d.c. power supply through a 100 μA ammeter (EHT stands for extra high tension).

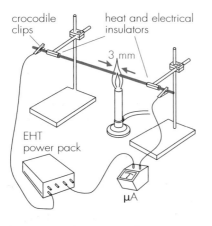

FIG 27.3 The experimental arrangement for the investigation.

There are several commercially available EHT power units capable of supplying a voltage of a few kilovolts. For safety, all of them have very high-value resistors in series with their output, and they often have built-in voltmeters. Some preliminary experimentation will probably be necessary to obtain a suitable potential difference across the probes for the range of meter being used. It has been found that 2000 V is suitable when using a 100 μA ammeter.

No current will flow when the Bunsen flame is not present. By supporting the Bunsen at different heights, find the current through the flame at different places within and around the flame. Use the value of the potential difference given on the EHT meter or scale to calculate the resistance of the flame. The values obtained for resistance will not be very accurate because of the internal resistance in the power supply; however, the variation of resistance from place to place can be found. Draw a plan of the flame and mark on it the resistance of 3 mm of the flame at different places.

If the pressure in a discharge tube is reduced, it is found that, even though little visible light comes from the gas itself, a glow can still arise from the glass of the tube, and that the glow can be affected by the presence of a magnetic field. It is also found that there is a small electric current through the tube. This suggests that a flow of charged particles through the tube is the cause of the current and the glow, and that the particles come from the material of the cathode itself. It was soon realised that the properties of these particles were the same no matter what metal was used for the cathode, so the particles seem to be present in all metals. These beams were originally called *cathode rays* and are now known to be beams of electrons.

27.2 THE SPECIFIC CHARGE OF AN ELECTRON

At the end of the nineteenth century, several fundamental experiments were done. One of these was performed by J.J. Thomson in Cambridge using cathode rays. He deflected a beam of cathode rays in an evacuated tube and was able to find the ratio of the charge to the mass for an individual particle in the beam of rays. The ratio of charge to mass is called the **specific charge** for any particle.

Although Thomson's cathode ray tube looked very different from a modern cathode ray tube, it did have electrodes and a deflecting system, which were very similar in several respects to the modern tube. However, the practical application of the cathode ray tube was not the only important feature of Thomson's experiment. To see the fundamental significance of Thomson's work, it is first necessary to examine the theory of his experiment.

If an electron is accelerated in a vacuum through a potential difference V, then its loss in potential energy is equal to its gain in kinetic energy. Since the potential difference is the potential energy per unit charge, we get

$$Ve = \tfrac{1}{2}mv^2$$

where e is the charge on the electron, m is the mass of the electron and v is the electron's final velocity if it is assumed to start from rest. Note that the final velocity depends on the potential difference between the plates, but not on how far apart the plates are placed. The equation can be rearranged to give the speed of an electron as

$$v = \sqrt{\frac{2Ve}{m}}$$

The principle of the experiment is illustrated by **Fig 27.5**. Electrons are emitted from the cathode and are accelerated in a vacuum towards an anode, which is at a potential $+V$ with respect to the cathode. Here most of the electrons collide with the anode, but some pass through a small hole in the anode and proceed through another anode at the same potential into the right-hand part of the apparatus, which is also all at potential $+V$ with respect to the cathode.

If no field is applied, the electrons coast on in a straight line to the end of the tube, where they create a small spot of fluorescence on the glass. However, if a magnetic field is applied over a region of the tube, as shown, then the electrons experience a force Bev acting on them in a direction at right-angles to their path, as shown in **Fig 27.5**. This causes them to travel in an arc of a circle and to hit the glass tube at P. (Section 18.4 explains why the force is Bev.) If the geometry of the

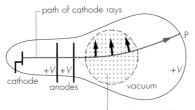

FIG 27.5 Cathode rays are deflected in a circular path when they travel through a uniform magnetic field.

magnetic field and deflection is used to find r, the radius of curvature of the path, then we can apply Newton's second law to the curved part of the electron path:

force = mass × acceleration

$$Bev = m \times \frac{v^2}{r}$$

This equation can be combined with the equation for v found above. It is impossible to determine v, e and m from these two equations as, although we can measure B, r and V, this still gives only two equations for three unknowns. What *can* be found though is the specific charge e/m by eliminating v. From the previous equation we can write

$$Be = \frac{mv}{r} = \frac{m}{r} \times \sqrt{\frac{2Ve}{m}}$$

Squaring this gives

$$B^2 e^2 = \frac{m^2}{r^2} \times \frac{2Ve}{m}$$

and cancelling and rearranging leads to

$$\frac{e}{m} = \frac{2V}{B^2 r^2}$$

TV tubes and monitors

A modern television set uses the same principle in producing its picture as J.J. Thomson's cathode ray tube. Thomson did not die until 1940, so he was aware before his death of the use to which his important idea had been put. Though no doubt he would have been staggered to realise just how many television tubes (and also computer monitors) are now produced annually.

When J.J. Thomson substituted his readings into this equation, he found that the value he obtained for e/m was over 1800 times greater than for any previously known particle or substance. The value of the specific charge of an electron is now known accurately to be

$(1.758\,803 \pm 0.000\,003) \times 10^{11} \, \text{C kg}^{-1}$

This is a huge figure, when compared with the value of $9.65 \times 10^7 \, \text{C kg}^{-1}$ for hydrogen ions – a value known to Thomson. The implications of this number were not lost on him. He deduced either that there existed a particle of about the mass of a hydrogen ion but with 1800 times the charge, or that this was a particle with a negative charge of about the same size as the positive charge on the hydrogen ion but with a mass only $1/1800$ that of the hydrogen ion.

Since at that time atoms were *defined* as the smallest possible particles, and the atom of hydrogen was the smallest atom, you should be able to see how revolutionary was the suggestion that Thomson made in 1897 that he had discovered a sub-atomic particle with mass $1/1800$ of the mass of the hydrogen atom. This discovery, of what is now called the *electron*, over a hundred years ago, not only revolutionised academic ideas on matter but has led to the entire electronics industry and much more besides. It is the high value of the specific charge on the electron that makes it so useful, because it is so easily manipulated with electric and magnetic fields. It can be given enormous accelerations using only modest voltages. Oscillation frequencies for electrons of more than $1000\,000\,000$ Hz are commonplace in radio aerials.

While J.J. Thomson was certain that he had discovered a sub-atomic particle, final proof of its existence depended on being able to find the charge and mass separately rather than as a ratio. It took another 15 years before that was achieved.

27.3 THE ELEMENTARY CHARGE

Millikan's oil-drop experiment: principles

One important feature of Thomson's work was that the nature of the gas used did not affect the value for e/m obtained. Until the time of Thomson it had been assumed that there were about 80 different elements and that the atoms of each of these elements were different. Thomson's experiment implied that all these atoms contained within them identical 'cathodic corpuscles' – as Thomson called electrons. Once Thomson had shown that atoms were not the smallest particles of matter, the hunt was on for other sub-atomic particles. That hunt is still going on today.

If cathode rays consist of sub-atomic particles of known fixed specific charge, then finding the charge on one of these particles enables the mass of the particle to be found. The Millikan oil-drop experiment enabled this to be done. The apparatus used was not complicated, and many schools and colleges have copies of the original apparatus that you may be able to use. Essentially the apparatus consists of a pair of parallel metal plates across which a potential difference may be applied. A negatively charged oil droplet is held in the space between the plates so that the upward force on it due to the electric field is equal and opposite to its weight. **Fig 27.6** shows this in principle.

If the potential difference between the plates is V and their separation is x, then the electric field E in the space between the plates is given by

$$E = \frac{V}{x}$$

Since electric field is defined as the force per unit charge, the force F that the field exerts on a charge q is given by

$$F = qE = q\frac{V}{x}$$

This is an upward force in the arrangement shown in **Fig 27.6** if q is negative.

The weight of the droplet is

$$\text{weight} = mg = \tfrac{4}{3}\pi r^3 \rho g$$

where r is the radius of the droplet, ρ is the density of the oil and g is the Earth's gravitational field (that is, the gravitational force per unit mass).

When the oil drop is held stationary, these two forces are equal and opposite, which gives

$$\frac{qV}{x} = \tfrac{4}{3}\pi r^3 \rho g$$

$$q = \frac{4\pi r^3 \rho g x}{3V}$$

but for V positive, q must be negative. Hence q can be calculated if r, ρ, g, x and V can be measured.

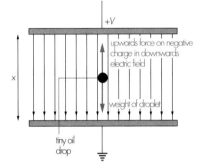

+V

upwards force on negative charge in downwards electric field

x

weight of droplet

tiny oil drop

FIG 27.6 The principle of Millikan's experiment to measure the fundamental charge.

Millikan's oil-drop experiment: practical problems

This concludes the principle of establishing the charge on the oil drop, but there are some practical problems. A large oil drop cannot be used because its weight will be much greater than any possible upward electrical force on the elementary charge. For this reason an atomiser is used and this produces a cloud of suitable tiny oil droplets. However, these are difficult to see unless they are strongly illuminated and are viewed through a microscope. It is then impossible to measure their radius directly because they are so small.

A suitable practical arrangement is shown in **Fig 27.7a**. The plates are separated by about 5 mm and V is typically about 300 V. Once the microscope is focussed on the illuminated space under the hole through which oil droplets can enter, and the p.d. is applied across the plates, a squeeze on the atomiser will cause a shower of droplets to be seen, as shown in **Fig 27.7b**. A few droplets may be seen to be moving upwards. These are the ones which have somehow acquired a negative charge and will therefore have an upward electrical force on them. The p.d. V is adjusted to keep a particular drop stationary and its value is measured. If the electric field is then switched off and the time t taken for the drop to fall across the scale a measured distance y is found then the terminal velocity v of this drop is given by $v = y/t$.

When falling with its terminal velocity, the drop is not accelerating, so the viscous force of the air upwards equals its weight. The viscous force is given by an equation, known as Stokes' equation, with which you are probably unfamiliar. It is

$$\text{viscous force} = 6\pi\eta rv$$

where η is called the coefficient of viscosity of air and has the value $1.8 \times 10^{-5}\,\text{N s m}^{-2}$ at 20 °C. This gives

$$\text{weight} = \text{viscous force}$$
$$\tfrac{4}{3}\pi r^3 \rho g = 6\pi\eta rv$$

from which r can be determined. This value of r can then be used to find q by substitution into the original equation.

Further practical points to note about this experiment are as follows:

- The microscope inverts the image, so everything falling is seen to be rising.
- The microscope can most easily be focussed on the correct region of the space between the plates if bright illumination is used and if it is initially focussed on a thin wire placed in the hole.
- Particles can go sideways as a result of draughts and they move out of the field of view of the microscope. If a particle is lost this way, that reading has to be abandoned. The light pipe is used to minimise the heating effect of the lamp to try to reduce unwanted movement of the air inside the apparatus.
- This experimental work was started using water droplets but they evaporated too much, so Millikan changed to using oil. He also took account of the buoyancy effect of the air, but this is only a 0.15% correction and would be unlikely to be one of your larger uncertainties.

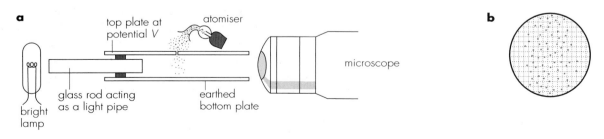

FIG 27.7 **a** Apparatus for Millikan's experiment to measure the fundamental charge. (The microscope is offset, towards the reader as drawn here, so as not to be looking directly towards the light source.) **b** The field of view as seen through the microscope.

■ Millikan repeated his readings on a particular drop after irradiating it with a burst of X-rays. This can change the charge of the drop, so different values of q are obtained for the same value of r.

Quantisation of charge

The experiment needs to be repeated many times to show what is meant by the term *quantisation of charge*. Typical results from the readings with a few droplets are given in Table 27.1, which were taken using oil of density 970 kg m^{-3}, $x = 5$ mm = 0.005 m and $y = 0.0025$ m.

As can be seen from Table 27.1, the charge on a droplet is not constant, but is frequently around 1.6×10^{-19} C. When it does not have this value, it has, in this case, values of approximately either twice or three times 1.6×10^{-19} C. If an average is taken, therefore, we must assume two charges for the fifth and eleventh

TABLE 27.1 Typical results from Millikan's oil-drop experiment

V /V	t /s	r /10^{-7} m	q /10^{-19} C
288	60.0	6.29	1.54
140	65.2	6.03	1.64
200	73.6	5.68	1.64
120	109.2	4.66	1.51
265	38.4	7.86	3.28
210	70.0	5.82	1.68
336	51.2	6.81	1.68
370	50.4	6.86	1.56
273	60.0	6.29	1.63
48	92.0	5.08	4.88
285	37.6	7.94	3.14
212	71.3	5.77	1.62

readings and three charges for the tenth reading. The average is therefore the total of the last column divided by 16 (not 12):

$$25.80/16 = 1.61$$

This gives the charge on an electron as -1.61×10^{-19} C, though the spread of results here, and the two significant figures value used for the viscosity of air, does *not* permit three significant figures to be quoted. These results therefore give

electronic charge = -1.6×10^{-19} C

The fundamental unit of charge, e, is taken as the same, but positive, value, namely 1.6×10^{-19} C, and is now known to be the charge on the proton. A more accurate value is 1.602×10^{-19} C.

This experiment has by now been performed accurately many times and in practical work no one has found a charge less than this fundamental unit of charge. This is the meaning of the term **quantisation**. Charge is not infinitely divisible. It is apparently not possible to have a free charge less than the fundamental charge and all charges are whole-number multiples of this charge.

Putting this value for the electron's charge as -1.602×10^{-19} C together with its specific charge e/m of -1.759×10^{11} C kg^{-1}, gives the mass m of the electron as

$$m = \frac{1.602 \times 10^{-19}\,\text{C}}{1.759 \times 10^{11}\,\text{C kg}^{-1}} = 9.11 \times 10^{-31}\,\text{kg}$$

This is 1/1837 of the mass of the hydrogen atom.

EXAMPLE 27.1

Find the kinetic energy that an electron acquires if it falls through a potential difference of 1 V, and its charge is -1.60×10^{-19} C.

This can be found by recalling that

one volt = one joule per coulomb = 1 J C^{-1}

Therefore for a charge of -1.60×10^{-19} C, the kinetic energy E_k acquired as the potential difference rises by one volt is

$$E_k = 1 \text{ J C}^{-1} \times 1.60 \times 10^{-19} \text{ C} = 1.60 \times 10^{-19} \text{ J}$$

The quantity of energy found in Example 27.1 is called an **electron-volt** (eV):

K **One electron-volt = 1 eV = 1.60×10^{-19} J**
1 MeV = 1.60×10^{-13} J

EXAMPLE 27.2

A beam of electrons travelling at 1.35×10^7 m s^{-1} enters a uniform electric field between two plates of length l, separated by a distance d. One plate is at a potential of +50 V and the other is at −50 V, as shown in **Fig 27.8**. If $l = 0.060$ m and $d = 0.020$ m, find θ, the angular deflection of the beam.

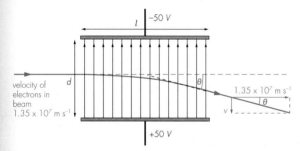

FIG 27.8 Finding the deflection of the electron beam in Example 27.2.

The electric field, E, between the plates is the potential gradient

$$E = \frac{100 \text{ V}}{0.02 \text{ m}} = 5000 \text{ V m}^{-1}$$

The field is in a direction towards the top of the page, so an electron, which has a negative charge, will have a force on it acting downwards towards the bottom of the page. Since this force is always downwards, there is no horizontal acceleration, and the horizontal velocity is therefore constant. This gives the time taken to travel between the plates as

$$\text{time taken} = \frac{\text{distance}}{\text{velocity}} = \frac{0.060 \text{ m}}{1.35 \times 10^7 \text{ m s}^{-1}} = 4.44 \times 10^{-9} \text{ s}$$

The downward force on an electron is given by

$$\begin{aligned} \text{force downwards} &= Ee \\ &= 5000 \text{ V m}^{-1} \times 1.60 \times 10^{-19} \text{ C} \\ &= 8.0 \times 10^{-16} \text{ N} \end{aligned}$$

While between the plates, the downward acceleration is therefore given by

$$\frac{\text{acceleration}}{\text{downwards}} = \frac{\text{force downwards}}{\text{mass}} = \frac{8.0 \times 10^{-16} \text{ N}}{9.11 \times 10^{-31} \text{ kg}}$$
$$= 8.78 \times 10^{14} \text{ m s}^{-2}$$

(Note how huge this acceleration is. This is possible because of the very small mass of the electron.)

For the downward direction, it is now possible to use $v = u + at$ to get

$$\begin{aligned} v &= 0 + (8.78 \times 10^{14} \text{ m s}^{-2} \times 4.44 \times 10^{-9} \text{ s}) \\ &= 3.90 \times 10^6 \text{ m s}^{-1} \end{aligned}$$

The forward velocity is unchanged, so using the vector diagram shown in the bottom right part of **Fig 27.8** gives

$$\tan \theta = \frac{v}{1.35 \times 10^7 \text{ m s}^{-1}} = \frac{3.90 \times 10^6 \text{ m s}^{-1}}{1.35 \times 10^7 \text{ m s}^{-1}}$$

$$= 0.289$$

$$\theta = 16° \quad \text{to 2 sig. figs}$$

In questions similar to Example 27.2, the normal laws of electricity and mechanics can be applied, provided speeds do not become so large that relativistic effects become important. You will need to pay particular attention to the following:

■ Keep all quantities in SI units. Distances must be in metres, not centimetres; energies must be in joules, not in electron-volts; etc.

■ Use powers of 10 correctly. Remember that 10^7 is put into a calculator as 1 exp 7 and *not* as 10 exp 7.

■ Check answers. Think at each stage whether the figure you have found is reasonable: velocities cannot be larger than the speed of light, 3×10^8 m s^{-1}; accelerations are likely to be high, masses to be tiny, and charges to be tiny. If you find a number like $m = 3.23 \times 10^{26}$ kg or $q = 1.6 \times 10^{19}$, you have probably omitted the minus sign on a power of 10.

■ Work from first principles and not from a formula.

QUESTIONS

27.1 **a** What kinetic energy does an electron acquire if it leaves a cathode at zero potential and travels through a vacuum to an anode at +200 V?

b With what speed does it reach the anode?

27.2 A uniform electric field exists in the space between two uniformly charged parallel plates a distance of 0.014 m apart. An electron travels at right-angles to both plates from rest at the negative plate and takes 3.7×10^{-9} s to reach the positive plate. Calculate:

a the acceleration of the electron;

b the force on the electron;

c the electric field;

d the potential difference between the plates;

e the velocity of the electron when it reaches the positive plate.

27.3 An electron beam is generated in a magnetic field of flux density 0.88 mT in a direction at right-angles to the field. The beam forms a circular path of radius 6.5 cm. Find the speed of the electrons.

27.4 THE CATHODE RAY TUBE

In section 27.2 Thomson's work on cathode rays was discussed, and it was stated that the rays existed in a gas at low pressure when a high potential difference is applied across the tube. This is not the only way of producing electron beams. If a plentiful supply of electrons is required, they can be obtained more readily if the cathode is heated. Emission of electrons from a hot surface is called *thermionic emission*. It is the process used to produce all the necessary electron beams in a television set. Some materials emit electrons more readily than others, and many metals are near their melting points before an appreciable flow of electrons from the surface takes place. Oxides of strontium and barium emit electrons readily at temperatures around 1000 °C, and these are used in television sets, where they are heated by special heater elements placed in contact with them.

Once a beam of these highly charged, extremely mobile and controllable particles has been produced in a vacuum, it can be directed very readily to any required target. This principle is used not only in the cathode ray tubes of oscilloscopes and televisions but also in electron microscopes and X-ray tubes. **Fig 27.9** shows a beam of electrons passing through an evacuated tube until some of them hit a Maltese cross. The remaining electrons cause a shadow of the cross to be seen.

The control of movement of the electrons is sometimes done using magnetic fields and sometimes using electric fields. In most applications, it is first necessary to produce a focussed beam of electrons. This itself poses problems because electrons repel one another strongly, so a beam that starts off as a fine beam tends to broaden, as shown in **Fig 27.10**. This shows electrons from a heated cathode passing through two small holes lined up on two anodes. The electrons pass through the holes but then spread out because of mutual repulsion.

The focussing is done using an electron gun, which consists of a heated cathode, a grid that is a nearly closed cylinder surrounding the cathode, and a series of several focussing and accelerating anodes. **Fig 27.11** shows three anodes, two of which are circular plates with holes in their centre and one of which is cylindrical. The shape of the electric field in the region of the anodes has the effect of deflecting those electrons which are not going straight along the axis of the electron gun. An assembly like this is behaving as an electronic lens and focussing the cathode rays. The amount of focussing is controlled by adjusting the potential of an anode. A high degree of precision is possible nowadays in electron gun technology, so a narrow beam of electrons can be directed at a point less than 0.1 mm wide on the screen. This is useful for high-resolution graphics.

Control of the rate at which electrons pass through the tube is achieved using the grid. The grid is at a potential slightly less than that of the cathode. If it is made more negative, it reduces the number of electrons that can pass it to reach the anodes. Making it less negative allows more electrons to pass, so it controls the electron current. This in turn controls the brightness at any point on the screen.

The commercial production of cathode ray tubes is big business, and most of them are colour tubes with three electron guns, one for each of the three primary colours, red, green and blue. There are commercial pressures that lead to there being a big choice of style and characteristics. Most television tubes have magnetic deflection. For oscilloscopes, however, electric deflection is more usual because the current requirement for the deflection is much smaller. A sketch of a typical cathode ray tube in an oscilloscope is shown in **Fig 27.12**. The electrical connections with the tube and supports to keep anodes and deflection assembly in position are not shown.

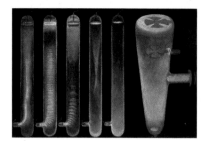

FIG 27.9 In an early classic demonstration of cathode rays, a Maltese cross was placed in their path. A clear shadow of the cross was produced, showing that the cathode rays travelled in straight lines in the absence of electric and magnetic fields. When one of these fields is present, the shadow is displaced.

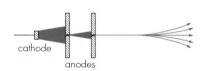

FIG 27.10 A beam of electrons tends to spread out because electrons repel one another.

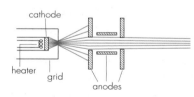

FIG 27.11 An electron gun.

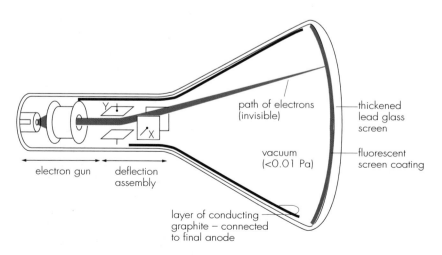

FIG 27.12 A cathode ray tube with electrostatic deflection.

27.5 THE CATHODE RAY OSCILLOSCOPE

The cathode ray oscilloscope has been referred to in Chapter 11, where it was suggested as a versatile instrument for measuring potential difference and time. It was seen to be particularly useful for examining potential differences that vary regularly and rapidly. These are the type of potential differences that occur, for example, as the output from a microphone or within electronic circuits. The cathode ray oscilloscope is able to display on its screen a steady pattern showing how the potential difference is changing. Because of the stability of the pattern on the screen, measurements can be made on it. Often the accuracy of these measurements is not particularly good because the focussing of the trace is poor and the square grid usually placed in front of the screen is rather coarse. However, if more accurate measurements are required, better-quality oscilloscopes can be purchased.

Time measurement with an oscilloscope can be done very accurately if the instrument itself is accurately calibrated. No instrument can be better than its calibration.

The basic principle of the cathode ray oscilloscope involves using a cathode ray tube in which the X-plates are used with a sawtooth potential difference called a time-base. Its purpose is to move the spot from left to right across the screen at a controllable, constant speed of known value, and then to return the spot to the left-hand side of the screen suddenly, the *flyback*. Calibrated speeds can vary on different oscilloscopes from around 0.1 cm s^{-1} to 10^6 cm s^{-1}. The potential difference required to cause this spot movement is generated internally within the oscilloscope and is shown in **Fig 27.13a**. This voltage causes no vertical movement of the spot. A negative value at the start means that the spot is on the left-hand side of the screen. As the voltage rises to zero, the spot moves uniformly to the centre of the screen and then continues to the right until flyback occurs. An accurately linear variation is needed for this increasing voltage if measurements are to be made.

It is usual to suppress the spot while flyback occurs. This is done by applying the voltage shown in **Fig 27.13c**. The grid is made more than usually negative so that there is no electron beam through the tube during flyback. The voltages shown as (a) and (c) on **Fig 27.13** produce, by themselves, on the screen of the oscilloscope, what appears to be a succession of spots appearing on the left-hand side of the screen and moving across the screen before disappearing on the right-hand side.

You are recommended to set up an oscilloscope to see this if you are unfamiliar with its operation. No external input is required. Adjust the spot for focus and for brightness, and switch the time-base so that the spot sweeps across the screen at a slow rate. If you then increase the time-base speed, you will notice that at high speed the moving spot appears as a horizontal line. This is due to persistence of vision. Your eye is no longer able to follow an individual spot. Persistence of vision is essential for watching television. It would be most unpleasant if the horizontal motion of the spot could be seen when a television picture is formed. The same time-base principle is used for picture formation. The spot moves across the screen 625 times per picture, each sweep being a little lower down the picture than the previous one. Since 25 complete pictures are displayed each second, the time-base on a television set needs a frequency of $625 \times 25 \text{ s}^{-1} = 15\,625 \text{ Hz}$.

With an oscilloscope, the voltage variation that it is desired to examine, the *signal*, is applied to the Y-plates. If the signal is sinusoidal, as shown in **Fig 27.13b**, then as the spot moves across the screen it will also be moved up and down to give an appearance on the screen as shown in **Fig 27.14**. Note that nearly two cycles are displayed, but before the end of the second cycle is reached the flyback has started so that the spot is ready to begin its next sweep across the screen. For the pattern

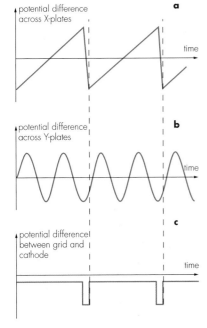

FIG 27.13 The time-base on X-plates is synchronised with the signal on the Y-plates. The flyback is suppressed by switching the beam off during flyback.

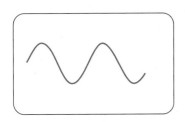

FIG 27.14 The spot on the oscilloscope screen can be moved up and down as it moves across the screen, to give the appearance of a sine wave.

on the screen to appear stationary, it is essential that each sweep of the time-base begins at the same point on the signal pattern. This is achieved by the signal in **Fig 27.13** having a frequency exactly twice the time-base frequency.

Keeping a fixed pattern on the screen of the oscilloscope is awkward to do manually because frequencies often drift slightly. To overcome this problem, synchronisation of the time-base and signal frequencies is achieved by introducing a pause after flyback. The pause keeps the sweep from starting until a particular Y-plate voltage is reached as it increases. This enables different frequencies to be used on the X- and Y-plates, yet for the trace on the screen still to be stationary.

Different oscilloscopes from different manufacturers have different additional facilities. Ranges of time-base frequencies are different, so are the outputs and inputs that can be used. Some oscilloscopes are double-beam, some have long-persistence screens. All oscilloscopes have amplifiers of variable gain connected to the Y-input. This enables signals as small as a few microvolts or as large as several hundreds of volts to be displayed on the screen.

In addition, most oscilloscopes have the option of putting the signal in through a capacitor as shown in **Fig 27.15**. If this is done, then any d.c. input is blocked and only a.c. signals can be displayed. It often happens that a signal has an a.c. and a d.c. component, as shown in **Fig 27.16**. By opening the a.c./d.c. switch to the a.c. position, only the a.c. component of the signal is displayed on the screen, and a larger gain can then be used.

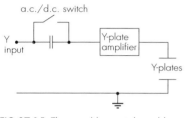

FIG 27.15 The a.c./d.c. switch enables d.c. to be passed to the Y-plates or not, as required.

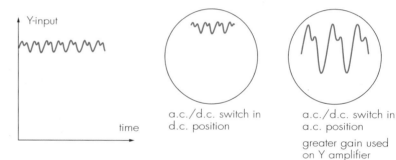

FIG 27.16 The effect of the a.c./d.c. switch.

ANALYSIS

Ships' radar

A very popular display for ships' radar uses a large-screen cathode ray tube as shown in **Fig 27.17**. The screen is coated with a phosphorescent material, so that it glows for a few seconds after it has been bombarded by electrons. In use, a trace on the screen rotates in a way similar to a hand on a watch. One end is fixed at the centre of the screen and the other end rotates in a circle every two seconds. The trace rotates with the same angular velocity as the radar transmitter/receiver aerial on the ship's mast.

The voltages applied to the X- and Y-plates to achieve this rotation are shown in **Fig 27.18**, in which it has been assumed that 16 traces are made per revolution. In practice it is many more, so the trace seems to rotate smoothly, not in 16 jerks per revolution. At the beginning of each radial

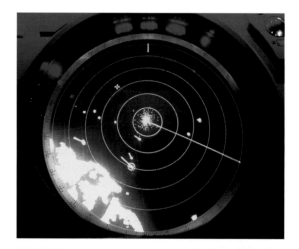

FIG 27.17 A sweep radar screen in use. Reflections of radar waves from targets are shown as dots on the screen.

trace, a short burst of high-frequency radio waves is emitted from the aerial. If any ship, or the coast, is close enough, a reflected echo can be received by the aerial before the end of that trace. The echo is amplified and applied to the grid of the cathode ray tube to make the trace particularly bright at that point. The brightness lasts until the trace sweeps round on its next revolution. The trace itself has to be visible but is not so bright that its afterglow lasts too long.

The data needed to answer the following questions are given below for a particular radar and are illustrated in **Fig 27.19**:

Frequency of radio waves	7.23 GHz
Duration of a pulse	0.5 μs
Radius of cathode ray tube screen	10 cm
Speed of radio waves	3.0×10^8 m s^{-1}
Pulse repetition frequency*	5000 Hz

(*This is the frequency with which pulses are transmitted. It is therefore the number of radial traces in one second, half a revolution.)

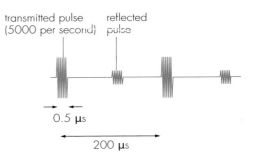

FIG 27.19 A radar system sends out a succession of strong pulses of short-wavelength electromagnetic waves and receives weak reflected echoes.

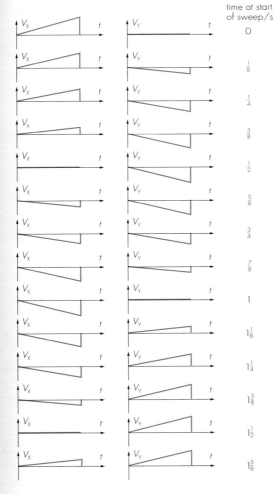

FIG 27.18 The voltages applied to the X- and Y-plates to achieve a rotating trace.

a What is the wavelength of the radio waves used?
b How many oscillations take place in each 0.5 μs pulse?
c What is the period of a wave of frequency 5000 Hz?
d How far can a wave travel in 200 μs?
e What is the maximum range of this radar?
f What speed is needed for the spot as it moves outwards along each radius? [Ignore the flyback time.]
g A boat is at a distance of 15 km from the radar. What will be the distance from the centre of the screen of its echo?
h If an echo is at a distance of 1 cm from the centre of the screen, how far is the ship causing the echo away from the radar?
i What is the minimum distance for an object to provide an echo separate from the emitted pulse? To what screen distance does this correspond?
j Explain how the graphs of **Fig 27.18** give a rotating trace.
k The ship carrying the radar changes direction. Explain what effect this will have on the screen display.
l How would the display be changed if the sensitivity of the cathode ray tube is altered so that the spot moves at 5000 m s^{-1}? How can this help to overcome a difficulty that part **i** highlights?

QUESTIONS

27.4 [This can be answered either by reasoning or, preferably, by using an oscilloscope and signal generator and making the required changes.]

Draw the appearance of the screen of an oscilloscope when the following changes are made, one at a time, to the potential differences applied to its plates. Assume that you start with potential differences that vary in the way shown in **Fig 27.13**.

a The peak value of the potential difference across the Y-plates is doubled.

b The frequency applied to the Y-plates is doubled.

c The frequency applied to the Y-plates is halved.

d The frequency of the potential difference applied to the Y-plates is reduced to one-eighth of the value shown.

e The frequency of the time-base is doubled.

f The frequency of the time-base is halved.

g The amplitude of the time-base is halved.

27.5 a Sketch the trace seen on an oscilloscope screen when a 50 Hz a.c. input of 2.0 V r.m.s. is applied. The gain on the Y-plates is set at 1 V cm^{-1} and the time-base is set to 10 ms cm^{-1}.

b How will the pattern alter if the time-base control is changed to 100 ms cm^{-1}?

27.6 The current through a cathode ray tube is 4.8 μA and the electrons are accelerated by a potential difference of 2000 V. Calculate:

a the speed of the electrons through the tube;

b the number of electrons passing a point in the tube per second;

c the number of electrons per metre of beam.

Positive rays

I n the search for sub-atomic particles, which started with J.J. Thomson's discovery of the electron, it was quickly realised that, since interesting particles were emitted from the cathode of a discharge tube, it was sensible to look for positively charged particles that may be emitted from the anode. J.J. Thomson himself did this, but came up against two particular difficulties. In the first place, any particles emitted had far smaller charge-to-mass ratio and so could not be deflected so easily; and, secondly, there was a huge variety of different particles. Whereas cathode rays consisted of one type of particle whatever gas or electrodes were used, any particles emitted from the anode did depend very much on the gas within the tube. It was therefore some years before the charge-to-mass ratios for these positive rays, as they were called, could be measured to any great accuracy. Aston, a student of Thomson's, developed a piece of apparatus called a mass spectrometer by 1921 which could be used to determine charge-to-mass ratios.

The mass spectrometer

Modern mass spectrometers are capable of measuring charge-to-mass ratios to a high degree of accuracy. Data concerning the mass of atoms are almost always now determined using mass spectrometers.

A diagram of the Bainbridge mass spectrometer is shown in **Fig 27.20**. It consists of a positive ion source, which may be an ionised gas or a heated crystal of the substance under investigation, a collimator, a velocity selector, a deflecting chamber and a photographic film detector. The collimator is a pair of slits, S_1 and S_2, which allow only a narrow beam of positive ions into the space between the plates. In this region both strong electric and strong magnetic fields exist, arranged so that they exert opposite forces on positive ions.

If an ion, of charge $+q$, is travelling with speed v in this region, then

force exerted on ion by electric field $= qE$
force exerted on ion by magnetic field $= B_1qv$

where E and B_1 are the electric and magnetic field strengths respectively. For a certain velocity, the forces are not only opposite but also equal in magnitude.

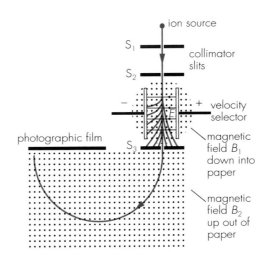

FIG 27.20 The principle of the Bainbridge mass spectrometer.

Charged particles of this velocity are therefore not deflected. This occurs when

$$qE = B_1qv \quad \text{so} \quad v = E/B_1$$

The charged particles then pass through slit S_3 into a region of uniform magnetic field strength B_2. In this field the force acting on them is B_2qv at right-angles to their direction of travel. This causes them to move in a circle of radius r, and since

$$\text{force} = \text{mass} \times \text{acceleration}$$

we get

$$B_2qv = m \times \frac{v^2}{r}$$

where m is the mass of the charged particle. This gives

$$B_2q = \frac{mv}{r} = \frac{m}{r} \times \frac{E}{B_1}$$

$$\frac{q}{m} = \frac{E}{rB_1B_2}$$

This equation shows that the radius of the path of a particle depends on its charge-to-mass ratio, and **Fig 27.21** shows how a photographic film appears when placed in the path of the particles. The darkest lines indicate large numbers of particles of a particular charge-to-mass ratio.

EXAMPLE  27.3

Work on electrolysis, the conduction of electricity through liquids, shows that a hydrogen ion has a charge-to-mass ratio of $q/m = 9.65 \times 10^7$ C kg^{-1}. A certain mass spectrometer of the type described has $B_1 = 0.93$ T, $B_2 = 0.61$ T and $E = 3.7 \times 10^6$ V m^{-1}. Calculate the radius of the paths of each of the following ions in the mass spectrometer: **a** H$^+$, **b** H$_2^+$, **c** He$^+$ and **d** He^{2+}.

a For H$^+$: Rearranging the last equation gives

$$r = \frac{E}{B_1B_2} \frac{1}{(q/m)}$$

Substituting q/m for the hydrogen ion and the other values stated for the mass spectrometer gives

$$r = \frac{(3.7 \times 10^6 \,\text{V m}^{-1})}{(0.93\,\text{T}) \times (0.61\,\text{T})} \times \frac{1}{(9.65 \times 10^7 \,\text{C kg}^{-1})}$$

$$= \frac{3.7 \times 10^6}{0.93 \times 0.61 \times 9.65 \times 10^7}\,\text{m} = 0.0676\,\text{m}$$

b For H$_2^+$: An H$_2^+$ ion is a molecule of hydrogen that has lost one electron. It therefore has the same charge as an H$^+$ ion but twice the mass. Its charge-to-mass ratio is therefore half that for H$^+$ and so, for the same spectrometer conditions, the radius of its path is doubled, which gives $r = 0.135$ m.

c For He$^+$: In the same way, He$^+$ has the same charge as H$^+$ but four times the mass. Therefore its charge-to-mass ratio is a quarter that for H$^+$ and so the radius of its path is four times as big, giving $r = 0.270$ m.

d For He^{2+}: Similarly, He^{2+} has twice the charge and four times the mass of H$^+$. Its charge-to-mass ratio is therefore half that for H$^+$, and the radius of its path is twice as big, giving $r = 0.135$ m.

Note that H$_2^+$ and He^{2+} would not be separated.

Before mass spectrometers were used, the relative atomic masses of atoms had been obtained by chemical or electrolytic methods and many atoms were known to have relative atomic masses that were approximately whole numbers. Chlorine, however, seemed an exception with its relative atomic mass of 35.5. If high enough energy is used with a chlorine source in a mass spectrometer, it is possible to create Cl$^+$ ions and to pass them through the instrument. When this is done, no line is found that corresponds to a relative atomic mass of 35.5. Instead, a line corresponding to a relative atomic mass of 35 is found, together with another line, of one-third the intensity, corresponding to a relative atomic mass of 37. It was seen, therefore, that chlorine consists of two types of atom of relative atomic masses 35 and 37. The weighted average of these masses is 35.5. More detail about *isotopes*, as they are called, will be given in Chapter 28.

QUESTIONS

27.7 a What is the speed of a positive ion that is undeflected when passing through a region in which the magnetic field strength is 370 mT and the electric field is 8.1×10^4 V m^{-1}.

 b How must the fields be arranged to make zero deflection possible?

27.8 In a Bainbridge mass spectrometer, the electric field in the velocity selector is 1.19×10^5 V m^{-1} and the magnetic flux density over the whole of the path followed by the ions is 0.53 T. The radii of the paths of several different types of singly charged ions of oxygen were found to be 7.08, 7.53 and 7.97 cm.

 a What is the mass of each of the oxygen ions? The charge on each ion is 1.60×10^{-19} C.

 b What is the relative atomic mass of each of these ions?

 c Another line was found corresponding to a path having a radius of 14.16 cm. What ion caused this line?

SUMMARY

■ Millikan's experiment:

 – for oil drop moving at terminal velocity

 $\frac{4}{3}\pi r^3 \rho g = 6\pi \eta r v$ (to find r)

 – for oil drop stationary in electric field

 $\frac{4}{3}\pi r^3 \rho g = Ee$

 Hence, using r, find e. It is found that e has discrete values – that is, charge is quantised.

■ Force on a charge in electric and magnetic fields:

	ELECTRIC FIELD STRENGTH E	MAGNETIC FIELD STRENGTH B
stationary charge	Eq	0
charge moving at velocity v	Eq in the direction of the electric field	Bqv at right-angles to both velocity and field

■ A charge q accelerated through a potential difference V in a vacuum gains kinetic energy equal to the work done on it, so $qV = \frac{1}{2}mv^2$.

■ The previous equation enables the speed of electrons to be found. Deflection experiments enable the speed and the specific charge (q/m) to be found for particles.

■ Cathode ray oscilloscope: Y-plate sensitivity is quoted in volts per centimetre; X-plates have time-base quoted in milliseconds per centimetre.

■ Mass spectrometer: used to find the charge-to-mass ratio for ions by using deflection techniques.

ASSESSMENT QUESTIONS

27.9 In an experiment to measure the charge of the electron, a charged oil droplet of mass 6.1×10^{-15} kg was observed between two horizontal parallel metal plates 6.0 mm apart, as shown in the diagram. The droplet was held stationary when the top plate was at a potential of +750 V relative to the lower plate.

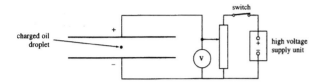

a Calculate the magnitude of the charge carried by the droplet and state the sign of charge carried by the droplet.

b Outline why measurements of the charge carried by charged oil droplets led Millikan to the conclusion that electric charge is quantised.

NEAB 1996

27.10 An oil drop of mass m and charge $-Q$ is between two horizontal conducting plates. An electric field of variable strength E may be applied between the plates.

a The electric field is adjusted until the drop is stationary. The diagram illustrates the drop and the plates.

● oil drop

i On a copy of the diagram, draw labelled arrows to represent the forces acting on the drop.

ii Explain which plate is positive, so that the drop remains stationary.

iii Write down the equation relating the forces on the drop when it is stationary.

iv Explain why the plates must be horizontal.

b The electric field is now switched off and the drop, of weight 8.2×10^{-14} N, descends. Eventually, the drop reaches a terminal velocity v of 1.8×10^{-4} m s^{-1}. The viscous force F on the drop is given by

$$F = kv,$$

where k is a constant for this drop. Determine the constant k.

OCR 1999

27.11

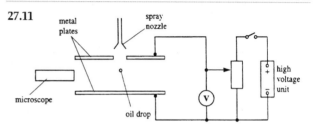

a A charged oil droplet of mass 5.8×10^{-16} kg was observed between two oppositely charged horizontal metal plates spaced 30 mm apart. The droplet was held stationary when the top plate was at a potential of +530 V, relative to the lower plate, as shown in the diagram. Determine the magnitude and sign of the charge of this droplet.

b The droplet suddenly moved upwards when beta radiation was briefly directed between the two plates. It was brought to rest again by reducing the potential of the top plate to +355 V.

  i Assuming no change of the mass of the droplet, calculate its new charge.

  ii Hence explain why the droplet moved upwards suddenly.

NEAB 1997

27.12 a A stationary negatively-charged particle experiences a force in the direction of the field in which it is placed. State, with a reason in each case, whether or not the field is
  i magnetic,
  ii electric,
  iii gravitational.
  b i Calculate the magnitude of the electric field strength required to maintain an electron in a fixed position in the gravitational field of the Earth, near its surface.
  ii Hence explain why gravitational effects are ignored when considering the motion of electrons in electric fields.
  c Atoms of neon-20 are ionised by the removal of one electron from each atom. For a neon-20 ion,
  i state the charge on the ion,
  ii calculate its mass.
  d The ions in **c** are accelerated from rest in a vacuum through a potential difference of 1400 V. They are then injected into a region of space where there are uniform electric and magnetic fields acting at right angles to the original direction of motion of the ions, as shown.

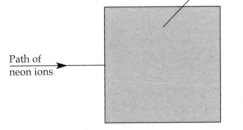

The electric field has field strength E and the flux density of the magnetic field is B.
  i Copy the diagram and on your diagram indicate clearly the directions of the electric and magnetic fields so that the ions pass undeflected through the region.
  ii Calculate the speed of the accelerated ions on entry into the region of the electric and magnetic fields.
  iii The electric field strength E is 6.2×10^3 V m^{-1}. Calculate the magnitude of the magnetic flux density so that the ions are not deflected in the region of the fields.
  e The mechanism by which the neon atoms in **c** are ionised is changed so that each atom loses two electrons. State what change occurs in
  i the speed of the ions entering the region of the electric and magnetic fields in **d**,
  ii the path of the ions in the two fields.

OCR 1999

27.13 a Write down an equation for the force F experienced by a charged particle travelling at an angle θ to a magnetic field. Identify all other symbols appearing in your equation.
  b Explain in words why a beam of charged particles may be regarded as an electric current.
  c The figure shows the path of a beam of identical charged particles in a vacuum. The particles enter, pass through, and leave a uniform magnetic field.

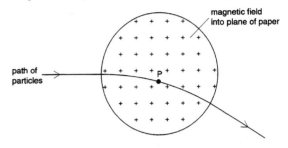

The directions of the initial path of the beam and of the magnetic field are at right angles, with the field directed into the plane of the paper.
  i Deduce the sign of the charge on the particles.
  ii Draw an arrow to show the direction of the force acting on the particles in the beam at point P.
  iii Explain why the speed of the beam, when leaving the field, has the same magnitude as that when entering the field.
  d A beam of charged particles is projected along the direction of a magnetic field. Describe and explain the subsequent motion of this beam.

Cambridge 1997

27.14 a In a linear accelerator, electrons are accelerated through a series of electrodes along a straight line, as represented below.

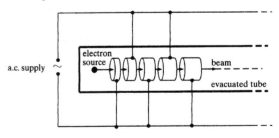

  i Describe how an alternating voltage of constant frequency applied between adjacent pairs of electrodes accelerates the electrons.
  ii Explain why electrodes shown on the diagram are of different lengths.
  b The Stanford Linear Accelerator can accelerate electrons to an energy of 50 GeV.
  i Calculate the work done, in J, on an electron accelerated through 50 GV from rest.
  ii Calculate the mass gained by an electron that has been accelerated through 50 GV from rest.

NEAB 1998

27.15 a Write down expressions for the force acting on a particle of mass m carrying a charge q when it is moving with speed v at right angles to

   i an electric field of intensity E,

   ii a magnetic field of flux density B.

b

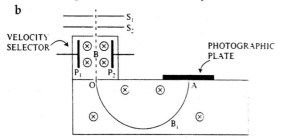

A *mass spectrometer* is an instrument for measuring nuclear masses. The diagram shows some basic features of such an instrument which, when working, is evacuated.

Ions carrying a charge e^- (i.e. they are singly ionised atoms) pass through fine slits S_1 and S_2 and enter a region called a *velocity selector* with various velocities. They then pass between a pair of parallel charged plates P_1 and P_2 which produce a uniform electric field of strength E perpendicular to the

initial direction of the motion. A magnetic field of uniform flux density B is also applied into the plane of the paper and at right angles to E and the ion path.

   i Obtain the condition for the ions to pass undeviated through the velocity selector and emerge with velocity v through the further slit O.

   ii Explain why the term *velocity selector* is a good description of the function of this part of the spectrometer.

c The ions passing through O now enter a region where a different magnetic field of uniform flux density B_1 is applied, again at right angles to the motion and into the plane of the paper. As a result the ions travel along a semi-circular path of radius r and strike a photographic plate at A, where they are recorded.

   i Explain why the ions follow a circular path.

   ii Show that the distance OA is proportional to the mass m of the ions.

   iii What is the significance of the result of **c ii** for a beam consisting of ions of different masses but all having the same charge?

WJEC 1998

28 ATOMIC STRUCTURE

LEARNING OBJECTIVES

At the end of this chapter you should be able to:

1. use the term 'nuclear atom' correctly;

2. use nuclear equations;

3. appreciate the random nature of radioactivity;

4. understand the term 'mass defect';

5. describe the processes of fission and fusion;

6. describe several uses of radioactive nuclides.

28.1 ATOMIC PARTICLES

The discovery of the electron as a sub-atomic particle by J.J. Thomson in 1897 led many people to question what else existed within an atom. It was known that an atom was neutral, so if it was possible for a negatively charged particle to be ejected from the atom, then there must be some positive charge left behind. Thomson himself carried out a long series of experiments on the positive particles that are present in a discharge tube. He came to the conclusion that, whereas many of the negative particles present in such a tube are all the same and are independent of the nature of the electrodes and the gas in the tube, there is a huge variety of different positive particles and these *do* depend on the nature of the gas present. Thomson incorrectly concluded that the positive charge remaining on the atom was distributed evenly within the atom (see box), together with any electrons not ejected.

In 1896, just before the discovery of the electron, Becquerel had discovered the phenomenon that came to be called 'radioactivity'. He found that a photographic plate placed under some crystals containing uranium became exposed even though no light had reached it. Further study of the phenomenon by such people

'Plum-pudding' model

This view of the atom became known as the 'plum-pudding' model of the atom, because it was assumed that the positive and negative charges were randomly distributed much like the 'plums' or currants in a pudding.

as Marie Curie and Rutherford led to the discovery of other radioactive materials, and the isolation of three types of naturally occurring radioactive emissions called alpha, beta and gamma by Rutherford in 1899. More detail on radioactivity is given later in this chapter.

It was soon realised by Rutherford that alpha radiation was capable of creating intense ionisation of the air through which it passes, owing to its charge and large kinetic energy. It can be detected by allowing it to collide with a fluorescent screen such as one coated with zinc sulphide. On impact with the screen, a tiny flash of light, called a *scintillation*, is produced, which can be seen by the eye if a magnifying lens is used. Alpha radiation can therefore be useful in investigating the structure of the atom. By 1908 Rutherford had established that alpha radiation was a stream of positively charged particles identical to helium atoms that have lost two electrons. He did this by collecting a vast number of alpha particles in a discharge tube and then observing the light radiated from the tube. The spectrum he found contained traces of helium, although none had been present before the experiment started. Question 28.1 illustrates how, even with strong sources of alpha particles, only a very small mass of helium is produced.

QUESTIONS

28.1 A radioactive source is emitting 8000 000 alpha particles per second at an approximately constant rate for 50 days.

   a What mass of helium gas is produced?

   b What will be its volume at s.t.p.?

   Assume values for the Avogadro constant and the molar gas constant. The mass of a helium atom is 6.7×10^{-27} kg.

28.2 An alpha particle travels a distance of 7 cm through air and causes an increase in the ability of the air to conduct electricity. Measurements indicate that the alpha particle causes 10 000 electrons per centimetre to be knocked away from the atom to which they had been attached. Find the initial speed of the alpha particle if it requires 30 electron-volts to remove the electron. The mass of a helium atom is 6.7×10^{-27} kg; 1 eV = 1.6×10^{-19} J.

Beta rays were also found to be high-energy particles. They have a negative charge and can be deflected to a much greater extent by a magnetic field than alpha particles. Their charge-to-mass ratio can be determined from their deflection in magnetic and electric fields, and is found to be identical to that for the electron. These particles are emitted from the *nucleus* of an atom during radioactive decay and are electrons. Beta particles have much less mass than alpha particles and therefore cause much less ionisation. This also means that they can penetrate greater thicknesses of materials than alpha particles.

Gamma rays are uncharged, so cannot be deflected by electric or magnetic fields. They do not interact with matter very strongly, and so can penetrate considerable distances through matter before being absorbed. Gamma rays are not particles. They are electromagnetic radiation similar to X-rays of short wavelength.

Experiments done by Marie Curie in 1899 on the amount of ionisation caused by alpha particles gave values for the speed of alpha particles, in the way shown by question 28.2. It was these high-speed particles that were used by Rutherford to penetrate what he thought of as a 'plum-pudding' atom. He was using the alpha particle as a bullet to break open an atom, and he hoped that he could identify any particle that emerged. In no way did Rutherford expect what he and his research students, Geiger and Marsden, actually found.

The experiment he suggested that Geiger and Marsden perform was to take a thin layer of massive atoms in a vacuum and to fire alpha particles at them. A thin layer was necessary so that there was less chance of repeated collisions, and the hope was that an atom could be burst open. In practice, the thin layer was a gold foil only a few hundred atoms thick, and the alpha particles usually went straight through. Occasionally, however, an alpha particle was deflected through a large angle or even rebounded from a gold atom. The apparatus is shown diagrammatically in **Fig 28.1**, and the pattern of alpha particle scattering is shown in **Fig 28.2** in which thousands of undeflected alpha particles have not been drawn. The rigorous mathematics of why this occasional scattering was so surprising is too complex to be dealt with here, but the following example gives some idea of the problem facing Rutherford.

EXAMPLE 28.1

An alpha particle of mass 6.7×10^{-27} kg and speed 1.0×10^7 m s^{-1} is to be stopped in a distance equal to the radius of a gold atom, 1.3×10^{-10} m. Find the deceleration of the alpha particle, assumed constant, and the force necessary to cause this deceleration. What electrostatic force exists between the positive charge in a gold atom and an alpha particle when separated by a distance equal to the radius of the gold atom? [The total positive charge within a gold atom is 79 times the charge on a hydrogen ion; the total positive charge within a helium atom is twice the charge on a hydrogen ion.]

Using the equation relating initial velocity u, final velocity v, distance travelled s and constant acceleration a gives

$v^2 = u^2 + 2as$
$0 = (1.0 \times 10^7 \text{ m s}^{-1})^2 + 2a(1.3 \times 10^{-10} \text{ m})$
$-2a(1.3 \times 10^{-10} \text{ m}) = 1.0 \times 10^{14} \text{ m}^2 \text{ s}^{-2}$

So

$$a = \frac{\pm 1.0 \times 10^{14} \text{ m}^2 \text{ s}^{\pm 2}}{2 \times 1.3 \times 10^{\pm 10} \text{ m}}$$

$$= \pm 3.85 \times 10^{23} \text{ m s}^{\pm 2}$$

The force F necessary to cause this deceleration is

$F = ma$
$= 6.7 \times 10^{-27} \text{ kg} \times 3.85 \times 10^{23} \text{ m s}^{-2}$
$= 0.0026 \text{ N} = 2.6 \times 10^{-3} \text{ N}$

The deceleration of the alpha particle is enormous. Also a force of 0.0026 N is extremely large, when the mass of the particle on which it is acting is considered.

The positive charge on a gold atom is $79 \times 1.6 \times 10^{-19}$ C. The positive charge on an alpha particle is $2 \times 1.6 \times 10^{-19}$ C. So when they are 1.3×10^{10} m apart, the electrostatic force of repulsion F' is

$$F' = \frac{q_1 q_2}{4\pi\varepsilon_0 r^2}$$

$$= \frac{(79 \times 1.6 \times 10^{\pm 19} \text{ C}) \times (2 \times 1.6 \times 10^{\pm 19} \text{ C})}{4\pi \times (8.85 \times 10^{\pm 12} \text{ F m}^{\pm 1}) \times (1.3 \times 10^{\pm 10} \text{ m})^2}$$

$$= 2.2 \times 10^{\pm 6} \text{ N}$$

When the size of this electrostatic force F' is compared with the size of the force F necessary to cause the alpha particle to rebound, it is seen that the two have completely different orders of magnitude. The electrostatic force is too small by a factor of about 1000.

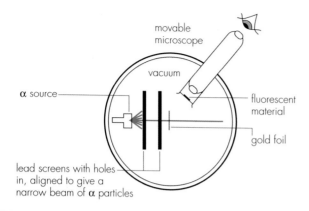

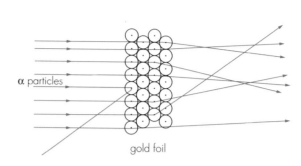

FIG 28.1 An experimental arrangement to detect alpha particle deflections.

FIG 28.2 Alpha particle deflections by the atoms in a gold foil (note that, for clarity, the paths of thousands of undeflected alpha particles have not been drawn).

The large discrepancy between the two figures quoted in Example 28.1 startled Rutherford. There was no way in which his 'plum-pudding' concept of the atom could agree with the results of Geiger and Marsden's experiment. He thought they had been mistaken in taking their readings. But, having established that the results were genuine, a complete re-think on the internal structure of the atom was needed. There was no doubt that the force of repulsion had to be at least as large as the figure of 0.0026 N calculated in Example 28.1. This implies that the electrostatic force is too small. The only way in which this force can become larger is for the distance between the two charges to be smaller. This is illustrated in **Fig 28.3**, where it is shown that, if the mass and the positive charge on the gold atom are concentrated into a smaller and smaller particle, then the alpha particle can approach closer to the charge, and hence a much larger force of repulsion can be achieved. It was the gold foil experiment, therefore, that resulted in the idea of a nucleus within an atom that contained most of the mass of the atom and all of its positive charge.

More detailed analysis of the results of this experiment shows that the diameter of the atom is of the order of 10 000 times the diameter of the nucleus. The nucleus of an atom occupies a very small fraction of the atom's total volume. This has the effect of making the density of nuclear matter extremely high. Its value was given in Table 22.1 as being of the order of 10^{16} kg m^{-3}. The density of all nuclear matter is approximately constant, so a nucleus of twice the diameter of a second nucleus will have $2^3 = 8$ times its mass. The electrons orbit the nucleus but because of their very small mass they do not appreciably deflect alpha particles. Alpha particles do

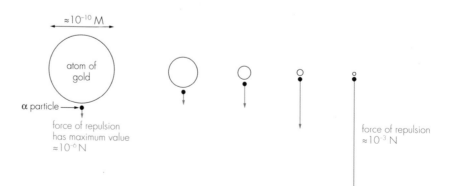

FIG 28.3 A force large enough to deflect alpha particles appreciably as they pass through a gold foil can only be supplied if the alpha particle is very close to all the positive charge in an atom. The diagram shows how the force increases as the volume occupied by the positive charge decreases.

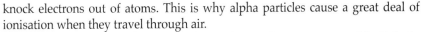

knock electrons out of atoms. This is why alpha particles cause a great deal of ionisation when they travel through air.

The nuclear model of the atom with its orbiting electrons was used by Bohr in 1913 as a basis of his theory of the emission of line spectra. More detail will be given on line spectra in Chapter 29.

QUESTION

28.3 The nucleus of uranium has a radius of 9×10^{-15} m. Find the radius of a nucleus (beryllium) that has a mass of $1/27$ of the mass of the uranium nucleus.

The proton

Having established that all the positive charge in an atom resides in its nucleus, there still remains the question of what particles exist within the nucleus. Moseley, also in 1913, was working on the frequency of X-rays from many elements, and he found the proof that within the atom there is a quantity that increases in regular steps from one atom to the next. He was convinced that this fundamental quantity was the positive charge on the nucleus. His work led directly to the idea that positive charge is quantised, and that the number of these positive charges within the nucleus of an atom determines the nature of the atom. Until the work of Moseley, atomic masses had been considered to be the critical quantity to determine the nature of a substance, but from this time onwards the atomic number was seen to be the factor of prime importance. This provided strong support for the periodic table of Mendeleev. The order in which the elements were placed by Mendeleev was in atomic number sequence.

The **proton number** Z of any element is the number of elementary units of positive charge on the nucleus of the atom. Each of these units of positive charge is associated with a particle called a **proton**. The proton number used to be called the atomic number.

Table 28.1 lists the first few elements to illustrate the point that the number of protons within an atom determines which element it is. All hydrogen atoms must have one proton in their nucleus. If there are two protons in the nucleus of a substance, then the substance must be helium; six protons in a nucleus must be a carbon nucleus; etc. Included in Table 28.1 are the atomic masses. At the beginning of the century these would have been found by chemical analysis. The pattern of numerical values for atomic masses clearly shows that there are a large number of values that are very close to being whole numbers, but the whole numbers are not equal to the atomic number.

TABLE 28.1 The proton number for the first 10 elements together with the average mass of an atom. These figures have significant differences from those in Table 28.2

ELEMENT	PROTON NUMBER	AVERAGE ATOMIC MASS/u
hydrogen	1	1.008
helium	2	4.003
lithium	3	6.939
beryllium	4	9.012
boron	5	10.811
carbon	6	12.011
nitrogen	7	14.007
oxygen	8	15.999
fluorine	9	18.998
neon	10	20.183

The neutron

The resolution of the problem of what particles exist within a nucleus was not solved satisfactorily until 1932 with Chadwick's discovery of a second nuclear particle, the neutron. The **neutron** is a particle of slightly greater mass than the proton, but it is uncharged. Since the mass of the proton and the mass of the neutron are very nearly equal to one another, it is not surprising that many nuclear masses are integral multiples of this mass. On first consideration it would seem sensible to regard the mass of the proton as the fundamental unit of mass. Indeed, this may become the standard of mass at some time in the future. But at present there are practical difficulties in relating this to the mass of large-scale bodies.

A small unit of mass is needed, however, and it is called the **atomic mass unit** (u). One atomic mass unit (1 u) is one-twelfth of the mass of an atom of carbon, which contains six protons, six neutrons and six electrons. The conversion factor between the atomic mass unit and the kilogram is found experimentally to be

$$1\,u = 1.660\,566 \times 10^{-27}\,kg$$

In these units the masses of the proton, neutron and electron are

$$m_p = 1.007\,276\,u$$
$$m_n = 1.008\,665\,u$$
$$m_e = 0.000\,549\,u$$

Nuclear particles, both protons and neutrons, are called **nucleons**. The number of protons plus neutrons in the nucleus of an atom is called the **nucleon number** A of the atom. The nucleon number used to be called the mass number. In symbol form the structure of an atom is given by

K — nucleon number
Symbol
proton number

A
Symbol
Z

^{65}
Cu
29

The example $^{65}_{29}$Cu gives the information that the atom of copper (Cu) contains 29 protons. It must therefore contain 29 electrons if it is a neutral atom. Since this particular copper atom contains 65 nucleons altogether, and 29 of them are protons, then the other 36 nucleons must be neutrons.

An atom of a particular nuclear structure is called a **nuclide**. **Fig 28.4** shows diagrammatically the structure of five different nuclides. Atoms of three different hydrogen nuclides are shown, together with an atom of helium and an atom of bismuth. The bismuth nuclide is $^{209}_{83}$Bi and is the largest stable nuclide. Diagrams such as **Fig 28.4** are useful for showing certain features of atomic structure. But there are some drawbacks to such diagrams: in the first place, they are only two-dimensional diagrams of a three-dimensional object; secondly, it is impossible to draw them to scale since the nucleus is so small; and, thirdly, the representation of an electron, even when shown as a cloud of charge surrounding the nucleus, is only a small part of the whole story. Think of the figure purely as a convenient diagram and *not* as a photograph of an atom.

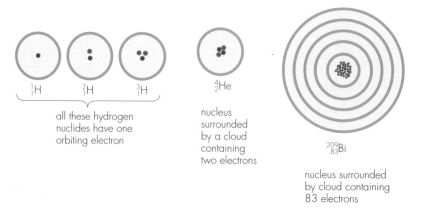

$^{1}_{1}$H $^{2}_{1}$H $^{3}_{1}$H

all these hydrogen nuclides have one orbiting electron

$^{4}_{2}$He

nucleus surrounded by a cloud containing two electrons

$^{209}_{83}$Bi

nucleus surrounded by cloud containing 83 electrons

FIG 28.4 Diagrammatic representation of some stable atoms from the smallest $^{1}_{1}$H to the largest $^{209}_{83}$Bi.

ISOTOPES

As described in Chapter 27, working on the positive particles present in discharge tubes, Thomson found that the atoms of each element were not necessarily all the same. At the time he could not explain why this was, but the development of the mass spectrometer showed, for example, that three-quarters of the atoms of chlorine had a mass of about 35 u, whereas the other quarter had a mass of 37 u. This gives the average mass of a chlorine atom as 35.5 u, a figure that had been something of a mystery to chemists for years, since they were unable to explain why, with so many atoms having an atomic mass that is nearly a whole number, chlorine should have an atomic mass definitely not a whole number. The two different types of chlorine atoms are said to be isotopes. **Isotopes** are atoms with the same atomic number but different mass numbers. The discovery of the neutron explained that isotopes were atoms with the same number of protons but with different numbers of neutrons. Modern mass spectrometer determinations enable the mass of about 2500 different nuclides to be measured. Most of these nuclides are unstable, but there are about 270 known stable nuclides. Each coloured square on **Fig 28.5** represents a stable nuclide.

There are some elements that have a large number of isotopes. Tin, for example, has an atomic number of 50. All tin atoms therefore have 50 protons. Together with the 50 protons, stable tin nuclides may have 62, 64, 65, 66, 67, 68, 69, 70, 72 or 74 neutrons. Each of the following nuclides is therefore an isotope of tin (Sn):

Isotope	$^{112}_{50}Sn$	$^{114}_{50}Sn$	$^{115}_{50}Sn$	$^{116}_{50}Sn$	$^{117}_{50}Sn$
Abundance (%)	1.0	0.6	0.3	14.2	7.6
Isotope	$^{118}_{50}Sn$	$^{119}_{50}Sn$	$^{120}_{50}Sn$	$^{122}_{50}Sn$	$^{124}_{50}Sn$
Abundance (%)	24.0	8.6	33.0	4.7	6.0

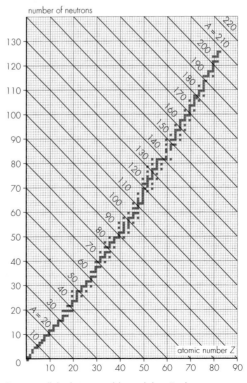

FIG 28.5 A chart showing all the known stable nuclides. Each green square represents a stable nuclide.

Percentage abundances of each isotope are also given above. In tin mined on the Earth, the percentage abundance of each isotope is remarkably constant whether the tin is mined in Cornwall, Bolivia, Malaysia, Zaire, Zambia, China or Nigeria. This is true for any element. Variation in the abundance of any nuclide is always small.

Detailed information about all the stable nuclides can be found in reference books, but sufficient information about the common nuclides is given in Table 28.2 to be able to see the pattern of atomic structure and for your needs here. The corresponding data for the elementary particles is also given. An asterisk (*) indicates that the nuclide is unstable. Percentage abundance cannot normally be given for unstable nuclides, but in the case of uranium an exception is made because uranium when mined has the proportion stated. This is partly because uranium-238 decays very slowly.

28.4 MASS DEFECT

Table 28.2 can be used to find the mass of any particular nuclide. It can also be used to find the total mass of the particles within that nuclide. An important discrepancy appears between those two values, as is shown in Example 28.2.

EXAMPLE 28.2

Find the mass of the particles contained within the $^{40}_{20}$Ca atom and compare it with the atomic mass.

An atom of $^{40}_{20}$Ca consists of 20 protons, 20 neutrons and 20 electrons, so the total mass of these particles is:

$$
\begin{aligned}
20 \times 1.007\ 28\ \text{u} &= 20.145\ 60\ \text{u} \\
20 \times 1.008\ 67\ \text{u} &= 20.173\ 40\ \text{u} \\
20 \times 0.000\ 55\ \text{u} &= \ \ 0.011\ 00\ \text{u} \\
\hline
\text{total mass} &= 40.330\ 00\ \text{u}
\end{aligned}
$$

From Table 28.2, however, the atomic mass of $^{40}_{20}$Ca is only 39.962 59 u. There is thus a difference in mass of

$$\Delta m = 40.330\ 00\ \text{u} - 39.962\ 59\ \text{u} = 0.367\ 41\ \text{u}$$

to be accounted for!

This is a startling result. The law of conservation of mass appears to be broken. The mass of the individual particles is *greater* than their mass when they are all fitted together into an atom. This would be totally inexplicable were it not for Einstein's *theory of relativity*, in which he equates energy E with an equivalent amount of mass m by the equation

K $E = mc^2$

where c is the speed of light in free space. Using the theory of relativity, we therefore have

$$
\begin{array}{ccc}
\text{mass of} & \text{mass equivalence of the energy} & \text{mass of} \\
\text{whole} \quad + & \text{required to separate the atom} \quad = & \text{individual} \\
\text{atom} & \text{into its individual particles} & \text{particles}
\end{array}
$$

which can be rewritten as

$$\Delta m = \text{mass of individual particles} - \text{mass of whole atom}$$

TABLE 28.2 Numerical data for some nuclides

ELEMENT	SYMBOL	PROTON NUMBER Z	NEUTRON NUMBER	NUCLEON NUMBER A	ATOMIC MASS /u	ABUNDANCE (%)
proton	p	1	0	1	1.007 28	
neutron	n	0	1	1	1.008 67	
electron	e	−1	0	0	0.000 55	
hydrogen	H	1	0	1	1.007 83	99.99
(deuterium)	(D)	1	1	2	2.014 10	0.01
helium	He	2	1	3	3.016 03	0.000 13
			2	4	4.002 60	~100
lithium	Li	3	3	6	6.015 13	7.4
			4	7	7.016 01	92.6
beryllium	Be	4	5	9	9.012 19	100
boron	B	5	5	10	10.012 94	19.6
			6	11	11.009 31	80.4
carbon	C	6	6	12	12 exactly	98.9
			7	13	13.003 35	1.1
nitrogen	N	7	7	14	14.003 07	99.6
			8	15	15.000 11	0.4
oxygen	O	8	8	16	15.994 92	99.76
			9	17	16.999 13	0.04
			10	18	17.999 16	0.20
fluorine	F	9	10	19	18.998 41	100
neon	Ne	10	10	20	19.992 44	90.9
			11	21	20.993 84	0.3
			12	22	21.991 38	8.8
sodium	Na	11	12	23	22.989 77	100
magnesium	Mg	12	12	24	23.985 04	78.7
			13	25	24.985 84	10.2
			14	26	25.982 59	11.1
aluminium	Al	13	14	27	26.981 54	100
calcium	Ca	20	20	40	39.962 59	97
			22	42	41.958 63	0.6
			23	43	42.958 78	0.1
			24	44	43.955 49	2.1
			26	46	45.953 69	0.003
			28	48	47.952 52	0.2
iron	Fe	26	28	54	53.939 62	5.8
			30	56	55.934 93	91.7
			31	57	56.935 39	2.2
			32	58	57.933 27	0.3
technetium	Tc	43		no naturally occurring nuclides		
lead	Pb	82	122	204	203.973 07	1.4
			124	206	205.974 46	25.2
			125	207	206.975 90	21.7
			126	208	207.976 64	51.7
bismuth	Bi	83	126	209	208.980 42	100
radium	Ra	88	135	223*	223.018 57	
			136	224*	224.020 22	
			138	226*	226.025 40	
			140	228*	228.031 23	
uranium	U	92	142	234*	234.034 90	trace
			143	235*	235.043 93	0.7
			146	238*	238.050 80	99.3

* Unstable nuclides.

The difference between the mass of its constituent particles taken separately and the mass of the whole atom itself is called the **mass defect**, Δm. The mass defect is small compared with the total mass of the atom. It is only when working to a large number of significant figures that the mass defect becomes apparent, but it is a real and crucial difference. The energy ΔE released when an atom is formed from its constituent particles is given by

$$\Delta E = \Delta mc^2$$

In Example 28.2 the mass defect is 0.367 41 u and ΔE, the energy equivalence of this mass, is given by

$$\Delta E = \Delta mc^2 = (0.367\ 41 \times 1.660\ 56 \times 10^{-27}\ \text{kg}) \times (2.998 \times 10^8\ \text{m s}^{-1})^2$$
$$= 5.484 \times 10^{-11}\ \text{J}$$

Note that the mass must be converted from a value in atomic mass units (u) into a mass in kilograms before being substituted into the equation.

The energy of nuclear reactions is often given in electron-volts (eV) or in MeV (see section 27.3), where

$$1\ \text{eV} = 1.602 \times 10^{-19}\ \text{J}$$
$$1\ \text{MeV} = 1.602 \times 10^{-13}\ \text{J}$$

The energy equivalence of the mass defect here is therefore

$$\Delta E = 342.3\ \text{MeV}$$

QUESTION

28.4 Show that the energy equivalence of a mass of 1 u is 932 MeV.

If the mass defect is calculated for all nuclides, then it is found that, with the exception of the H nuclide, all nuclides have a mass defect. Some of these figures are given in Table 28.3.

TABLE 28.3 Mass defect of selected nuclides

NUCLIDE	MASS DEFECT/u	MASS DEFECT PER NUCLEON/u
^1_1H	0	0
^2_1H	0.002 40	0.001 20
^4_2He	0.030 40	0.007 60
^7_3Li	0.042 16	0.006 02
^9_4Be	0.062 48	0.006 94
$^{11}_5\text{B}$	0.081 86	0.007 44
$^{12}_6\text{C}$	0.099 00	0.008 25
$^{14}_7\text{N}$	0.112 43	0.008 03
$^{16}_8\text{O}$	0.137 08	0.008 57
$^{24}_{12}\text{Mg}$	0.212 96	0.008 87
$^{40}_{20}\text{Ca}$	0.367 41	0.009 19
$^{56}_{26}\text{Fe}$	0.528 75	0.009 44
$^{110}_{46}\text{Pd}$	1.010 56	0.009 19
$^{208}_{82}\text{Pb}$	1.757 84	0.008 45
$^{209}_{83}\text{Bi}$	1.761 89	0.008 43
$^{238}_{92}\text{U}$	1.935 38	0.008 13

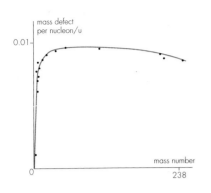

FIG 28.6 A plot of mass defect per nucleon against mass number.

The mass defect increases with the number of particles within the nucleus, so an additional column has been added to Table 28.3 that gives the mass defect per nucleon. This is therefore giving the amount of mass defect for each of the nuclear particles, and this figure shows an interesting sharp rise at the start to a maximum for the iron nuclide followed by a gradual fall. **Fig 28.6** shows the values of the mass defect per nucleon ($\Delta m / A$) plotted against mass number. The fact that the mass defect per nucleon is not the same for all nuclides has important practical applications. In practice, it is not possible to make atoms by taking protons, electrons and neutrons and forming them into atoms. The fact that there is a very strong repulsion between protons makes union between them and the required number of neutrons impossible at ordinary temperatures. What *is* possible is a rearrangement of the protons, electrons and neutrons in a particular nuclide to form different nuclides. In questions 28.5, 28.6 and 28.7 you are asked to use the data in Table 28.3 to find how the mass defect changes when certain nuclear changes take place.

Some of these changes are impossible practically, but they do illustrate the fact that, if the light atoms can be joined together to make heavier atoms, then the mass defect increases. This can only happen if energy is released. This process is called **fusion** and is the source of the energy of the Sun and the stars.

If the heavy atoms are broken into two roughly equal parts, then again energy is released. This process is called **fission** and is used in all nuclear power-stations. On **Fig 28.6** any alteration in nuclear structure that results in moving towards the maximum of the graph has the effect of releasing energy. One of the reasons why there is so much iron in the Universe is that the iron atom is the atom with the greatest mass defect. Note that an atom with a high mass defect needs a great deal of work to be done on it in order to separate it into its constituent nucleons. It is therefore in a low energy state. The term **binding energy** is used for an atom to indicate how much energy would be released if it were formed from separate protons, electrons and neutrons. The binding energy is therefore the work that would need to be done to separate the atom into individual protons, electrons and neutrons. It is *not* the energy holding the atom together. In practice, the energy required to remove the electrons from an atom is extremely small compared with the energy required to separate the nucleons. For this reason, often only the nucleons are considered.

QUESTIONS

28.5 a What is the change in the mass defect if a $^2_1$H atom joins with a similar atom to make a single atom of the nuclide $^4_2$He?

    b What energy is released in this process?

28.6 a What is the change in the mass defect if four $^4_2$He atoms join to make a single $^{16}_8$O atom?

    b What energy is released in this process?

    c What energy would be required to separate a single $^{16}_8$O atom into four $^4_2$He atoms?

28.7 a What change in mass defect takes place if a single atom of $^{238}_{92}$U breaks up into two $^{110}_{46}$Pd atoms and several neutrons?

    b What energy is released in this process?

28.5 NUCLEAR REACTIONS

In the previous section, various reorganisations of atomic nuclei were suggested. Some of these transformations can be carried out in practice, but others cannot. It was not until 1896, when Becquerel discovered radioactivity, that any possibility existed of changing one element into another. Until that time, creating valuable elements from base elements was merely a dream of the alchemists.

Even after the discovery of radioactivity, it took several more years before the first nuclear transformation was observed in the laboratory. This took place in Rutherford's laboratory in 1919. He used a zinc sulphide screen to detect the particles, but use of a cloud chamber later confirmed his results and enabled pictures to be taken of the particles before and after the reaction. **Fig 28.7** shows alpha (α) particles being produced from a source, and most of them simply removing electrons from atoms and hence causing ionisation of the nitrogen gas through which they are passing until they have lost all their energy and stop. One of the α particles, however, enters the nucleus of a nitrogen atom and causes a nuclear reaction to occur. The forked track shows that in the reaction two particles are produced. Measurement of the angles of the tracks and knowledge of the energy of the α particles enabled Rutherford to show that the long track was a proton. This gives a nuclear reaction that can be expressed in equation form:

$$^4_2He + {}^{14}_7N \rightarrow {}^{17}_8O + {}^1_1H$$

which is sometimes written

$$^{14}_7N(\alpha,p)^{17}_8O$$

The presence of the $^{17}_8O$ atom cannot be detected chemically because too few atoms are formed in any particular experiment to give enough $^{17}_8O$ for analysis. Its presence is deduced from the fact that the equation must balance. There are nine protons and nine neutrons on the left-hand side of the equation, so there must be nine protons and nine neutrons on the right-hand side. If one proton, ^1_1H, is emitted separately, the remaining atom must contain eight protons and nine neutrons and so is $^{17}_8O$.

Not only must the number of protons and neutrons be the same on both sides of the equation, but also there must be equality of mass–energy. If rest-masses of the four particles are considered, we get for the masses involved in the reaction, using data from Table 28.2:

mass of He	=	4.002 60 u	mass of H	=	1.007 83 u
mass of N	=	14.003 07 u	mass of O	=	16.999 13 u
total before	=	18.005 67 u	total after	=	18.006 96 u

(The mass of nine electrons is included on both sides of the equation, though in practice the electrons will usually be lost, at least temporarily, in such a high-energy reaction.)

It can be seen that the rest-mass of the particles after the reaction is higher than it was before. This implies that the initial α particle must have had a high velocity. The mass associated with its kinetic energy must be at least

$$18.006\ 96\ u - 18.005\ 67\ u = 0.001\ 29\ u$$

This is equivalent to

$$(932\ \text{MeV/u}) \times 0.001\ 29\ u = 1.20\ \text{MeV}$$

which is also

$$1.20 \times 10^6 \times 1.6 \times 10^{-19}\ J = 1.92 \times 10^{-13}\ J$$

$\llcorner \alpha$ particle source

FIG 28.7 Occasionally an alpha particle travelling through air hits a nitrogen nucleus and causes a nuclear reaction. The forked track shows that this has happened.

If this is equated to the kinetic energy of the α particle, the minimum speed v it can have to cause the nuclear reaction is given by

$$\tfrac{1}{2}mv^2 = 1.92 \times 10^{\pm 13} \text{ J}$$

$$v^2 = \frac{2 \times 1.92 \times 10^{\pm 13} \text{ J}}{4 \times 1.66 \times 10^{\pm 27} \text{ kg}} = 5.78 \times 10^{\pm 13} \text{ m}^2 \text{ s}^{\pm 2}$$

$$v = 7.6 \times 10^6 \text{ m s}^{\pm 1}$$

Since the proton emerges with a high kinetic energy, this increases the necessary energy of the α particle.

A nuclear reaction such as this is said to be an *induced* nuclear ration. It cannot take place spontaneously, as it absorbs energy. It was just such a reaction that led to the discovery of the neutron by Chadwick in 1932. He bombarded a disc of beryllium, placed in a vacuum, with α particles from a polonium source and found an unusual, uncharged radiation coming from the beryllium (**Fig 28.8**). If a solid material containing many hydrogen atoms was placed in the path of the unusual radiation (paraffin wax was used), then the protons in the wax get knocked on by the radiation. This made it impossible for the unknown radiation to be gamma (γ) rays, as they have too little mass. Chadwick's deduction from the quantitative results of his experiment was that the unknown radiation consisted of uncharged particles of mass similar to that of the proton. He called these particles 'neutrons'. The equation for the reaction of the α particles and the beryllium becomes

$$^4_2\text{He} + {}^9_4\text{Be} \rightarrow {}^1_0\text{n} + {}^{12}_6\text{C}$$

Conservation of momentum and energy can be used to find the mass of the neutron. Question 28.8 indicates the way this was done.

Until the discovery of the neutron, high-energy α particles had been used to cause nuclear reactions. They were not particularly suitable particles for this, however. Alpha particles have a positive charge and therefore there is a strong repulsive force on them whenever they approach any other nucleus. This makes it difficult for them to penetrate the nucleus of another atom. The uncharged neutron made an ideal missile with which to penetrate the nucleus of atoms to examine their structure.

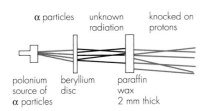

FIG 28.8 The experiment that resulted in the discovery of the neutron.

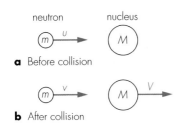

FIG 28.9 The collision for question 28.8.

QUESTION

28.8 A neutron of mass m travelling with velocity u hits a stationary nucleus of mass M head-on (**Fig 28.9**). After the collision, the velocity of the neutron is v and the velocity of the nucleus is V. The collision is an elastic collision.

a Write down equations showing conservation of kinetic energy and conservation of momentum for the collision.

b Show that eliminating v from these equations to find V in terms of u gives $V = 2mu/(M+m)$.

When the nucleus is a hydrogen nucleus of mass 1 u, V is found to be 3.3×10^7 m s^{-1}. When the nucleus is a nitrogen nucleus of mass 14 u, V is found to be 4.4×10^6 m s^{-1}.

c Substitute each of the values above into the equation given, and use the two resulting equations to eliminate u, and find the mass of the neutron. Use the mass of the neutron to find its velocity before the collision.

ANALYSIS

Nuclear reactions

In each of the reactions in Table 28.5, copy out and complete the nuclear equation (material to be completed/found is marked by question marks) and find the quantity of energy, E, necessary to balance the equation. Necessary data can be taken from Table 28.3.

TABLE 28.5 Reactions for the data analysis exercise

PARTICLE CAUSING REACTION	EQUATION	NOTES
alpha	$^{4}_{2}He + 7.7\ MeV + ^{14}_{7}N \rightarrow ^{1}_{1}H + ? + E$	The first induced nuclear reaction
proton	$^{1}_{1}H + ^{7}_{3}Li \rightarrow ^{4}_{2}He + ? + E$	The first experiment to show the possible release of a large quantity of energy
alpha	$^{9}_{4}Be + ^{4}_{2}He \rightarrow ^{1}_{0}n + \gamma + E$ (γ of 4.4 MeV)	Discovery of the neutron
deuteron	$^{2}_{1}H + ^{2}_{1}H \rightarrow ^{1}_{0}n + ? + E$	An example of fusion; fusion is the source of the Sun's energy
photon	$^{2}_{1}H + \gamma \rightarrow ^{1}_{1}H + ? + E$ (γ of 2.6 MeV)	–
neutron	$^{10}_{5}B + ^{1}_{0}n \rightarrow ^{?}_{?}Li + ^{?}_{?}He + E$	Control of a nuclear reactor
none	$^{238}_{92}U \rightarrow ^{234}_{90}Th + ? + E$	Radioactive decay
neutron	$^{235}_{92}U + ^{1}_{0}n \rightarrow ^{??}_{54}Xe + ^{90}_{38}Sr + 3^{1}_{0}n + E$	Fission; energy of around 200 MeV is emitted

From Rutherford's first induced nuclear reaction in 1919, the demand has been for higher and higher particle energies. Neutrons are ideal particles for causing nuclear reactions, but they cannot be accelerated to high speed in particle accelerators because they have no charge and so are unaffected by electric and magnetic fields. Some of the particle energies and accelerators used in the investigation of nuclear particles are given in Table 28.4. More detail of accelerators can be found on p. 539.

Nuclear reactions of considerable complexity are now known. Some important reactions in the development of nuclear physics are listed in the analysis exercise on nuclear reactions. There, brief details on different reactions are given, and you are asked to complete the equations and find the necessary energy to balance the equation. Note that any nuclear reaction that releases energy *could* take place spontaneously – but does not *necessarily* do so. A reaction that requires energy can only take place if the particle that causes it arrives with sufficient energy.

In all of these reactions, the energy required to balance the equation is measured in MeV. The corresponding figure in a chemical equation is usually of the order of a few electron-volts. It is for this reason that, whereas a conventional coal-burning power-station may burn a million tonnes of coal a year, a nuclear power-station requires only a few tonnes of fuel per year. It is worth remembering that each nuclear fission reaction releases of the order of a million times the energy that a single atomic (chemical) reaction releases.

TABLE 28.4 Some dates, energies and accelerators that have been significant in the history of nuclear particles

YEAR	ENERGY	INVESTIGATION
1917	8 MeV	alpha particles, from radioactive materials
1932	2 MeV	neutrons, after discovery by Chadwick
1932	2 MeV	protons, in accelerator of Cockcroft and Walton
1939	32 MeV	positive ions, in cyclotron of Lawrence and Livingston
1940	2.3 MeV	electrons, in betatron of Kerst
1950	900 MeV	ions, in synchrotron at Brookhaven
1954	5000 MeV	ions, in upper atmosphere
1976	300 000 MeV	protons, in super synchrotron at CERN, Geneva
1984	800 000 MeV	protons, in Fermilab synchrotron, USA
2005 (due)	1000 GeV	large hadron collider, CERN

 ## RADIOACTIVITY

Natural radioactivity has been stated to result in three types of radiation: alpha (α) and beta (β) particles and gamma (γ) rays. It has been seen that α particles are the least penetrating of the radiations but cause the greatest amount of ionisation. Any of these radiations is hazardous, and therefore unnecessary exposure to them should be avoided. It is not possible, however, to avoid radiation completely as it is part of our environment. Our own bodies are radioactive and contribute a large fraction of our dose of radiation; our houses and gardens are radioactive; the Sun is radioactive; and we get some very high-energy particles from outer space. Besides these sources of radioactivity, there are additional man-made sources. Fall-out from nuclear bombs still exists in the atmosphere; nuclear power-stations cause some radiation; and so do hospitals on account of some of the tests and treatments they use, and from their X-ray machines.

Whatever is written about radiation will find its critics. The problem is knowing how to draw a line between acceptable and unacceptable exposure. A physics textbook is not a suitable place to argue for or against any particular policy. It is nevertheless worth while stating some points about which there is almost complete agreement, so that precautions may be taken both within the laboratory and outside it. There is no particular order of importance in these ten items.

1 Any ingestion of radioactive material should be avoided. Radioactive particles lodged in the lungs, for example, are much more dangerous than if they were outside the body.
2 Hold any radioactive material well away from you and do not point radioactive sources at anyone.
3 When not in use, keep radioactive materials inside lead containers.
4 Reduce the time of contact to a minimum. Old watches with luminous dials are quite strongly radioactive, and people owning them wear them in contact with their wrists for a long time. Usually the backs are metal and so absorb α radiation.
5 Limit the amount of radiation you receive from X-rays and ultra-violet. There is a unit in a hospital in Brisbane, Australia, with a world-wide reputation for treating skin cancers. They recommend that people with fair skin should not sunbathe at all, as the risk of getting skin cancer is too high.

6 Keep a gentle draught through your house. Sealing up a room so that it is virtually airtight allows radon gas, which is radioactive, to collect in the room. The radon comes from the ground and from very slightly radioactive gypsum in the plaster on the walls. By itself this is no great strength, but you are likely to be in your house for a long time, and a weak exposure for a long time gives rise to increased risk.

7 A distinction cannot be made between natural and artificial radiation. An 8 MeV α particle from any source is the same.

8 Some radiation treatments in hospitals are extremely successful.

9 Some natural repair takes place within the body after radiation damage has occurred, so that a small rate of exposure for a long time may cause less damage than the same dose given at a high rate of exposure over a short time.

10 There are considerable variations in the background radiation level in different parts of most countries.

28.7 DETECTION OF RADIOACTIVITY

Photographic materials

A photographic plate was used to discover radioactivity, and photographic film is still used for detection of radiation. **Fig 28.10** shows the effect in photographic emulsion of high-energy particles. **Fig 28.10** shows an aluminium nucleus colliding with a nucleus in the photographic emulsion and producing six α particles, besides other fragments.

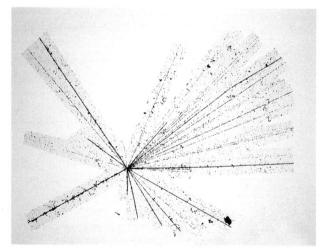

FIG 28.10 An aluminium nucleus collides at high speed with a nucleus in the photographic emulsion and produces, among other particles, a spray of six α particles.

Scintillation methods

Scintillation methods, too, have their modern equivalent. Researchers in the early part of the century spent many hours in darkened rooms counting tiny flashes of light from fluorescent screens. Now the same process is used with photomultiplier tubes in which the initial flash of light causes a few photoelectrons (see photoelectric effect, Chapter 29). These constitute a very tiny current, which can be amplified to give a pulse of current. The pulses can then be counted electronically.

Cloud chambers

A cloud chamber has also been mentioned. It depends for its operation on the fact that condensation of a liquid as dew requires something to start it forming. Dew readily forms on dust, for instance, to give fog. Condensation will also take place on ions formed in the air after a radioactive particle has passed through it. A diagram of the diffusion cloud chamber is drawn in **Fig 28.11**. The solid carbon dioxide cools the black surface to a low temperature and the ethanol vapour near it condenses on the ions caused by the passage of α particles. It is well worth while setting up this piece of apparatus if one is available, and you can get some solid carbon dioxide. You must illuminate the cloud chamber from the side in order to see the tracks, and you must be patient. It sometimes takes a few minutes after putting the lid on before the atmospheric conditions are right for the formation of the tracks. The black surface must be level. Take the Perspex lid off at some stage and you will see clouds of mist forming on the dirt in the air you have let in. Rubbing the Perspex lid creates a static charge on it, and this in turn creates an electric field, which enables the tracks to be seen more clearly. A photograph showing hundreds of these tracks is shown in**Fig 28.12**.

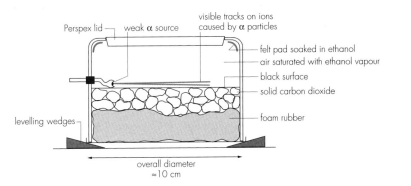

FIG 28.11 A diffusion cloud chamber.

FIG 28.12 Alpha particle tracks in a cloud chamber.

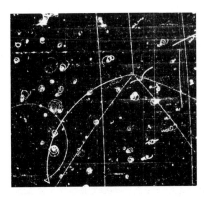

FIG 28.13 Bubble chamber photograph.

Bubble chambers

Bubble chambers are a modern development of the cloud chamber. These are used with particle accelerators and they can be very large (and very expensive). They use a tankful of liquid hydrogen instead of ethanol vapour. The hydrogen is exceedingly pure and, just before a burst of high-energy particles enter it, the pressure surrounding it is dropped so the hydrogen is above its boiling point. The ions formed by the passage of the nuclear particles act to cause centres on which the hydrogen boils. The tracks caused can be photographed. One such photograph is shown in **Fig 28.13**. One advantage of the bubble chamber is that, because it is at such a low temperature, very little thermal energy is possessed by any target atom, so all the effects caused must be due to the bombarding nuclei.

Geiger–Müller tube

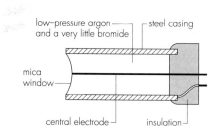

FIG 28.14 A Geiger–Müller tube.

The Geiger–Müller tube is a well known radiation detector, though it is being replaced commercially by other ionisation detectors and by solid-state detectors. Any detectors of ionisation can be used to detect α particles and β particles, and γ rays if it is sensitive enough. A Geiger–Müller tube is illustrated in **Fig 28.14**. The central electrode is between 400 V and 600 V positive with respect to the case. When an α particle enters the argon gas through the mica window, it causes some ionisation and a flow of charge takes place. The bromine present has the effect of preventing this flow becoming a continuous avalanche of charge. The pulse of current caused by the flow of charge can be amplified and the pulses counted electronically. The counter, which also provides the power supply, is called a *scaler*. If the rate at which pulses are produced is measured by the instrument, then this is called a *ratemeter*.

Solid-state detectors

Solid-state detectors are available that use semiconductor materials. A p–n junction is connected to a d.c. power supply so that it is not conducting. If some sufficiently energetic radiation is incident on it, charge separation occurs. Holes and electrons are produced. These cause a current to flow, which can be amplified and counted.

28.8 RADIOACTIVE DECAY

Radioactive decay is measured in becquerels: **one becquerel** (1 Bq) is a rate of decay of one nucleus per second. Any radioactive nucleus decays spontaneously. What causes it to change is not known. It is known that chemical reactions, temperature and pressure make no difference whatsoever to the likelihood of a radioactive nucleus undergoing radioactive decay. This is not really surprising when it is realised that these changes involve only the electrons within atoms and energies of the order of a few electron-volts. Even at 6000 K (the surface temperature of the Sun) the energy of thermal vibration is still only about one electron-volt (1 eV) whereas the binding energies of nuclei are of the order of 10 000 000 eV.

What is surprising is that some extremely stable nuclei should suddenly decay. Many uranium nuclei, for example, were formed with the Earth about 5000 million years ago. Ever since, they have remained on the Earth in atoms that have undergone all sorts of extremes of temperature, pressure and chemical change without altering at all. Yet if you have some uranium metal or uranium oxide in the laboratory, you will find that after all this time some of the nuclei suddenly, for no apparent reason, give off an α particle and hence decay into thorium. The radioactive decay equation is:

$$^{238}_{92}\text{U} \rightarrow \underset{\substack{\text{α particle}}}{^{4}_{2}\text{He}} + \underset{\substack{\text{energy mostly of} \\ \text{emitted α particle}}}{^{234}_{90}\text{Th} + 4.3\,\text{MeV}}$$

The only factor known to affect the rate at which α particles are produced by this nuclear reaction is the number of uranium-238 atoms present. We are forced therefore to use the laws of chance to analyse the decay pattern. There are several ways of dealing with this mathematically, so an actual example will be used.

EXAMPLE

The iron nuclide $^{59}_{26}$Fe is radioactive. A solution containing these radioactive iron atoms is used medically in the diagnosis of blood disorders. There is a 50:50 chance that any of the atoms will emit a β particle in a period of 46 days. What fraction of the original total remains after 184 days?

From the information given, we have that:

- on average, half of the original atoms decay during the first 46 days;

- on average, half of the atoms remaining after 46 days (i.e. after the first period of 46 days) decay during the second 46 days;

- on average, half of the atoms remaining after 92 days (i.e. after two periods of 46 days) decay during the third 46 days;

- on average, half of the atoms remaining after 138 days (i.e. after three periods of 46 days) decay during the fourth 46 days (i.e. up to 184 days).

This is shown in **Fig 28.15**.

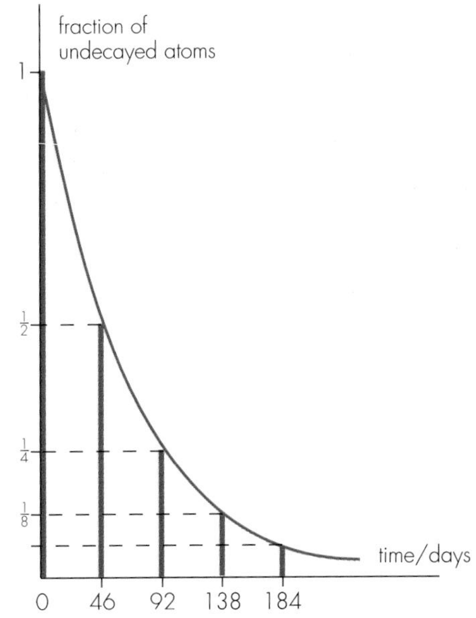

FIG 28.15 Radioactive decay, showing half-life.

The graph in **Fig 28.15** is clearly not a straight line. The average time it takes for half of the atoms of any nuclide to decay is called the **half-life** of the nuclide. Note that a half-life of 46 days does *not* mean that all of the nuclide has decayed after 2 × 46 days. In this case the fraction remaining after 184 days is 1/16 of the original.

QUESTIONS

28.9 Strontium-90 has a half-life of 28 years. If 32 g of strontium-90 is present in a sample at the start of a series of observations, how long will it be before there is only 1 g of strontium-90 remaining?

28.10 Show that the number of atoms present in 0.0632 g of $^{210}_{82}$Pb is 1.81×10^{20}. The half-life of $^{210}_{82}$Pb is 19 years. How many $^{210}_{82}$Pb atoms will be present after 95 years?

28.11 Plot a graph to show how the number of polonium-218 atoms in a sample changes with time, using the data in Table 28.6.

  a Find the half-life of polonium.

  b Check your answer by starting from some other point on your graph other than when the time is zero.

  c What relationship is there between the number of polonium-218 atoms present and the rate of decay of the atoms?

  d What would be the rate of decay when the number of polonium atoms present is 1.18×10^{12}?

28.12 Use the graph in **Fig 28.16** to find the fraction of $^{59}_{26}$Fe atoms remaining after: **a** 80 days, **b** 126 days and **c** 150 days.

TABLE 28.6 Data for question 28.11

TIME /s	NUMBER OF $^{218}_{84}$Po ATOMS /10^{15}	RATE OF DECAY /10^{13} Bq
0	4.38	1.64
50	3.63	1.36
100	3.01	1.13
150	2.50	0.935
200	2.07	0.775
250	1.72	0.643
300	1.42	0.533
350	1.18	0.422
400	0.98	0.366
450	0.91	0.304
500	0.67	0.252

The methods used for finding the activity of a nuclide have so far either assumed that an exact number of half-lives are being used or that a graph can be plotted. Another approach uses a log graph. The data from Example 28.3 are used this way in Example 28.4.

EXAMPLE 28.4

Using the data from Example 28.3, tabulated in Table 28.7, plot a graph to show the logarithm of the fraction of $^{59}_{26}$Fe atoms remaining against time. Use the graph to find the fraction remaining after 100 days, F_{100}. (In the table, lg F is the logarithm of F to base 10.)

TABLE 28.7 Data for Example 28.4

TIME /DAYS	FRACTION REMAINING (F)	lg F
0	1	0
46	1/2	−0.301
92	1/4	−0.602
138	1/8	−0.903
184	1/16	−1.204

Because each reading of F is the previous reading divided by 2, the value of lg F has to have lg 2 (0.301) subtracted from it to get the next reading. The graph must therefore be a straight line. It is shown in **Fig 28.16**.

This makes it easy to interpolate to find the value of x, which is the logarithm of the fraction remaining after 100 days.

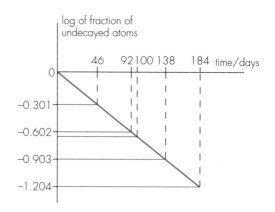

FIG 28.16 A logarithmic graph of radioactive decay (plotted in Example 28.4 and used in question 28.12).

Using similar triangles

$$\frac{184 \text{ days}}{\pm 1.204} = \frac{100 \text{ days}}{x}$$

$$x = \pm\frac{1.204 \times 100}{184} = \pm 0.654$$

But $x = $ lg F_{100}, giving

$$F_{100} = \text{antilog}(-0.654) = 10^{-0.654} = 0.22$$

The data in question 28.11, and the basic principle that the rate of radioactive decay depends only on the number of radioactive atoms present, show that

rate of disintegration ∝ number of atoms

or in calculus notation

$$\pm\frac{dN}{dt} \propto N$$

where N is the number of radioactive atoms present at time t and the minus sign indicates that N decreases with time. But the rate of disintegration is also known as the **activity** A, so

activity ∝ number of atoms
$$A \propto N$$

This therefore gives the equation

$$A = \frac{dN}{dt} = \pm\lambda N$$

where λ is a constant called the **decay constant**. Use of calculus is necessary to solve this equation, to get

$$N = N_0\,e^{-\lambda t}$$

where N_0 is the number of atoms present at time $t = 0$. Since $A \propto N$ this also gives

$$A_t = A_0\,e^{-\lambda t}$$

where A_0 is the activity when $t = 0$, and A_t is the activity at time t. The use of a calculator makes it easy to determine any decay pattern and to plot decay graphs even if you were unable to see how to get the equation from the initial proportion. Question 28.13 asks you to do this.

The half-life of a radioactive decay is given the symbol $t_{1/2}$. Calculation of the mathematical relationship between the half-life $t_{1/2}$ and the decay constant λ can

INVESTIGATION

Gamma ray absorption

Many schools and colleges have a very weak cobalt-60 source of gamma radiation. Use of such a source by sixth-form pupils may be allowed. You should keep the source in its lead container unless you are actually using it and you should handle it with tongs. Do not point it at anyone. Also, since radiation emerges in all directions from this source, position it so that you, and others in the laboratory, are all more than a metre away from it. You should also pay particular attention to any special rules that are applied to you, if you are allowed to do this investigation.

The arrangement of apparatus is shown in **Fig 28.17**. The detector may be a Geiger–Müller tube or a solid-state detector, and it must be connected to a suitable power supply and counter. The counter may be used as a ratemeter or as a scaler.

Start the investigation by finding the background count. Monitor the background count for at least 5 min before taking any other readings, and make sure that your cobalt-60 source is well out of the way in its lead container while you do this.

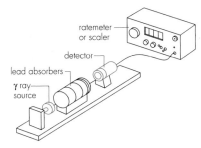

FIG 28.17 Experimental arrangement for the investigation.

Safety!

- *Never* point a radioactive source at yourself or anyone else.
- When not in use, keep it in its lead container.
- Follow all the rules applied to you.

Next, put the source into position in its holder. The distance between the source and the counter must be kept fixed, so arrange that there is just sufficient space for all the lead absorbers.

Take a reading of the count rate with no lead absorber in position, and record this value. Use the background count already measured to find the count rate, C, in the detector due to the gamma source alone.

Repeat the experiment with different thicknesses, x, of lead absorbers in position. Plot a graph of the natural logarithm of the count rate against the thickness of the lead. The graph will be a straight line if the absorption is given by the exponential equation

$$C = C_0\,e^{-\mu x}$$

Use your graph to find the value of μ.

An exponential law like this implies that a certain thickness of lead reduces the intensity of the radiation to a half. Twice this thickness reduces it to a quarter, three times to an eighth, etc. Find the thickness of lead that will reduce the intensity of gamma radiation to a half.

What thickness of lead is needed to reduce gamma ray intensity from that which gives a count rate of 3.7×10^{10} per second to one of 120 per second?

be done as follows. Assume that N_0 is the number of radioactive atoms present at time $t = 0$. Then after a time equal to one half-life has elapsed (i.e. when $t = t_{1/2}$), the number present will be $\frac{1}{2}N_0$. So substituting $N = \frac{1}{2}N_0$ and $t = t_{1/2}$ in the above equation gives

$$\frac{1}{2}N_0 = N_0\,e^{-\lambda t_{1/2}}$$
$$\frac{1}{2} = e^{-\lambda t_{1/2}}$$

Taking logarithms to base e ('ln') gives

$$\ln(0.5) = -0.6931 = -\lambda t_{1/2}$$
$$\lambda t_{1/2} = 0.6931$$

or

$$t_{1/2} = \frac{0.6931}{\lambda}$$

As expected, therefore, when the half-life $t_{1/2}$ is large, the decay constant λ is small; only a small fraction of atoms decay per unit time. When $t_{1/2}$ is small, a large fraction of the radioactive atoms decay per unit time, so λ is large. Note that since $t_{1/2}$ has the unit second, λ has the unit 'per second' (s^{-1}).

QUESTION

28.13 Plot the graph of $N = N_0\,e^{-\lambda t}$ for the range of values $t = 0$ to $t = 10\,s$ when $N_0 = 6.2 \times 10^{12}$ and $\lambda = 0.25\,s^{-1}$. Calculate the half-life of this decay?

28.9 USES OF RADIOACTIVE NUCLIDES

R adioactive nuclides are used in many completely different ways. In whatever way they are used, it is always necessary to take precautions so that the user is well protected from any harmful radiation. Five main categories of use are given here: as tracers; as penetrating radiation; for medical and biological use; for dating archaeological specimens; and as power sources. Each use will be treated separately, and the examples given are illustrations of some of the many uses now made of radioactivity.

All the uses depend on being able to obtain a specific radionuclide. These nuclides are manufactured by specialist firms. They are produced by irradiating different materials in the intense neutron radiation of a nuclear reactor. To make some radioactive magnesium atoms for use in the tracer experiment that is described later in this section, for example, requires that some magnesium sulphate is placed in a nuclear reactor for a time. The neutron flux penetrates the nucleus of the normal magnesium atoms and some neutrons can be absorbed, so creating radioactive magnesium atoms.

Generally speaking, materials are not made radioactive by being placed near a radioactive source of alpha, beta or gamma radiation. The lead case around a school cobalt-60 source does not become radioactive. This is because the absorption of the gamma radiation by the lead causes no change to the nuclei of the lead atoms.

Tracers

Use is made in tracer applications of the ability of detectors to measure extremely small concentrations of a radioactive nuclide. If, for example, a plant research centre wishes to investigate the ability of a tomato plant to take up magnesium through its roots, it is possible to water the plant with a fertiliser containing a known, extremely low, concentration of radioactive magnesium-28. If a detector is placed in contact with the leaves of the tomato plant, an increase in count rate soon after watering-in the fertiliser indicates that the magnesium has travelled through the plant to the leaves. This can happen within a time as short as 20 minutes.

Tracers are also used for detecting leaks in underground pipes and for measuring wear. Wear measured in a conventional way requires a long experiment until the object wearing away has lost appreciable mass or is appreciably smaller. If the object wearing away is radioactive, the minute quantities of matter worn away in the first few minutes of wear can be detected. This technique is used for measuring the wear of machinery.

Penetration of radiation

Gamma rays can be used in place of X-rays to photograph solid objects. To examine a piece of metal for internal cracks a gamma source is placed on one side of the metal and a piece of photographic film is placed on the other side. If a hidden crack exists then the gamma radiation will penetrate more easily and show up as a darker area on the film. This technique has the advantage that it does not need the high-voltage electrical sources required for alternative methods of crack detection.

The same technique can be used in many ways, the radioactive source being chosen so that it is suitable for the thickness of penetration required. The thickness of rolled sheets of paper or plastic can be controlled during manufacture by placing a beta source on top of the sheet and a detector beneath it. If the sheet is too thin, then the count rate will rise. This rise in count rate can be directly coupled to the rollers so that they are separated slightly.

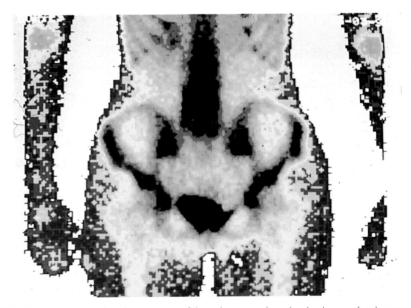

FIG 28.18 A gamma camera image using false colouring to show the distribution of radioactivity in a person's pelvic region.

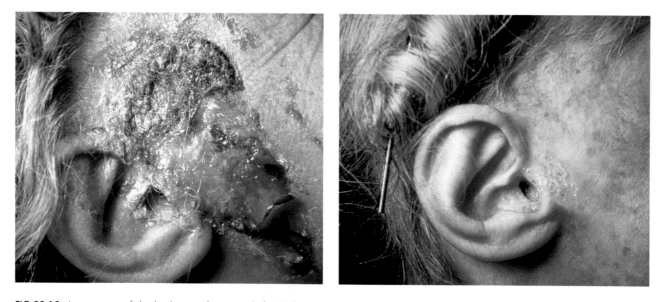

FIG 28.19 A carcinoma of skin by the ear of a patient before (left) and after (right) radiotherapy treatment.

FIG 28.20 Gamma camera is used in hospitals for imaging.

Medical and biological uses

When radiation is passed through living cells, it may do no harm at all. On the other hand, it can kill living cells or cause them to undergo a mutation, that is, to change the biological function of the cell in some way. It is immature cells and cells that are growing or dividing most rapidly that are most sensitive to radiation. The cells in mature insects divide very rarely, and for that reason some insects are able to survive extremely large doses of radiation.

Use is made of the sensitivity of rapidly dividing cells in the radiation treatment of cancer. Often cancer cells are growing rapidly and are therefore much more likely to be killed by a high dose of gamma radiation from a cobalt-60 source than are normal cells, which divide less frequently. **Fig 28.19** shows the dramatic effect that a course of gamma ray treatment can have on a skin cancer. The same is true of other cancers, but photographing them is not possible. **Fig 28.20** shows a gamma camera used for medical imaging. **Fig 28.18** shows an image produced by a gamma camera after a patient has been injected with a weak radioactive substance of short half life.

Mutations of living cells are usually harmful, but by irradiating plants it has proved possible to cause mutations that have led to improved varieties of many crops, such as wheat, peas and beans. New strains of plants have been produced that give larger yields and are more resistant to disease.

Archaeological dating

The atmosphere contains a small proportion of radioactive carbon-14. During a plant's life carbon dioxide is absorbed, so the plant too becomes slightly radioactive. The level of radioactivity of plants can be measured. When a tree is cut down and burnt, radioactive decay of any charcoal remaining begins. The longer ago the charcoal was formed, the lower is its radioactivity. This method of dating can now be used with an accuracy of better than to the nearest 100 years. By carrying out experiments on accurately dated events (at Pompeii, for example), it is possible to take into account variations in the level of atmospheric radiocarbon through the centuries.

Power sources

The obvious example here is the use of uranium-235 in a nuclear power-station (**Fig 28.21**). This was briefly referred to earlier. There are, however, other portable power sources that use radioactive materials. Some satellites contain radioactive materials that are decaying naturally and giving enough energy to remain hot throughout a long space flight. The hot material is used in conjunction with thermocouples to supply electricity. Some fire alarms contain a small amount of alpha-emitting substance. This keeps the air in the fire alarm slightly ionised, and any alteration in the level of ionisation, caused by smoke, can be detected and used to set off the alarm.

FIG 28.21 Sizewell nuclear power-station.

ANALYSIS

Decommissioning a nuclear reactor

A nuclear reactor throughout its life generates power from fission of uranium-235 by neutrons in a reaction of this type:

$$_0^1n + _{92}^{235}U \rightarrow _{38}^{90}Sr + _{54}^{143}Xe + 3_0^1n + 200\,MeV$$

The *fission products*, as the strontium (Sr) and xenon (Xe) are called, have very small mass and are mostly contained within the fuel cans. The neutrons, on the other hand, have high speed and are uncharged. They are able to escape from the fuel cans into the graphite (carbon) moderator, where they are slowed down. One of the three neutrons is needed to sustain the chain reaction, and the others are absorbed by the moderator, the fuel, the control rods and the structural materials of the reactor.

The plans for decommissioning nuclear reactors have to take into account the following information:

- A reactor may have a total mass of 300 000 tonnes, and 5% of this may be radioactive.

- The fuel elements can be removed in the normal way. This removes 99.99% of the radioactivity that was on-site when the reactor was working.

- The remaining radioactivity is mostly neutron-induced in the steel, concrete and graphite.

- Radioactivity is expressed in becquerels (Bq); 1 Bq is one disintegration per second.

- British law defines a radioactive substance as one with an activity greater than 400 Bq kg^{-1}.

- Garden soil contains the radioactive potassium-40 nuclide and frequently has an activity of 800 Bq kg^{-1}; coffee powder is even more radioactive at 1600 Bq kg^{-1}.

■ The activity of a source cannot be related directly to a dose of radiation. Some nuclides are more radiologically significant, as the dose rate depends not only on the number of radioactive disintegrations but also on the type of particles produced, their range and their energy.

■ There are about 2500 known nuclides. Of these, 79 that have a half-life longer than one year may be present in a reactor. Most of these 79 nuclides do not reach the activity of 400 Bq kg^{-1} to necessitate them being called radioactive.

■ **Table 28.8** shows the mass and the activity of 13 problem nuclides 10 years after the shut-down of a reactor. A dash indicates an insignificant amount of nuclide.

TABLE 28.8 The problem nuclides in the data analysis exercise*

NUCLIDE	HALF-LIFE /YEAR	CONCRETE (8000 TONNES) $/10^{12}$ Bq	GRAPHITE (2200 TONNES) $/10^{12}$ Bq	STEEL (3400 TONNES) $/10^{12}$ Bq	TOTAL ACTIVITY $/10^{12}$ Bq	PARTICLE
$^{3}_{1}H$	12	–	110	–	110	β
$^{14}_{6}C$	5 700	–	42	1.7	44	β
$^{36}_{17}Cl$	310 000	–	1.3	–	1.3	β
$^{41}_{20}Ca$	130 000	0.10	0.90	–	1.0	κ
$^{55}_{26}Fe$	2.7	0.82	4.0	2400	2400	κ
$^{60}_{27}Co$	5.2	0.28	11	680	690	γ
$^{59}_{28}Ni$	80 000	–	0.096	2.1	2.2	κ
$^{63}_{28}Ni$	92	–	17	230	250	β
$^{63}_{41}Nb$	20 000	–	–	0.013	0.013	β
$^{108}_{47}Ag$	130	–	0.0017	0.017	0.019	β
$^{151}_{62}Sm$	90	0.15	0.051	–	0.20	β
$^{152}_{63}Eu$	12	0.82	0.11	–	0.93	κ
$^{154}_{63}Eu$	16	0.10	0.61	–	0.71	β
Total		2.3	190	3300	3500	

*All values are given to two significant figures.
1 tonne = 1000 kg.
κ represents K capture – this is a mode of decay in which the nucleus captures one of the atom's orbiting electrons.

Answer the following questions using the data supplied.

a What is the activity in Bq kg^{-1} of each of the three main sections of the reactor?

b Why is it that over a period of 100 years the radioactive iron-55 in the reactor is less of a problem than the radioactive nickel?

c Why is it likely that the activity of the long half-life nuclides is low and the activity of the short half-life nuclides is high?

d How long will it be before the activity of the tritium (hydrogen-3) in the reactor is below 0.43×10^{12} Bq?

e What will be the activity of the tritium after 420 years?

f Using the equation
$$A_t = A_0 \, e^{-0.693t/t_{1/2}}$$
where $t_{1/2}$ is the half-life, A_t is the activity at time t, and A_0 is the activity at $t = 0$, find the activity of each isotope 110 years after ceasing to make use of the reactor for power generation. Take A_0 to be the values given and t to be 100 years, so
$$A_t = A_0 \, e^{-69.3/t_{1/2}}$$

g Which nuclides are the problem ones?

h Without making any further calculations, what would you expect the decay graph to look like?

SUMMARY

■ Atomic constituents:

Proton	charge $= +1$	mass $= 1.007\ 28$ u
Electron	charge $= -1$	mass $= 0.000\ 549$ u
Neutron	charge $= 0$	mass $= 1.008\ 67$ u

■ Nucleon: a proton or a neutron.

■ Nuclide: an atom of a particular specification.

■ Isotopes: nuclides with the same number of protons but with different numbers of neutrons.

■ Proton number (atomic number) Z: the number of protons in the nucleus of an atom.

■ Nucleon number (mass number) A: the number of protons plus neutrons in the nucleus of an atom.

■ Atomic mass unit (u): one-twelfth the mass of an atom of carbon-12.

■ Mass defect, Δm: the difference in mass between the mass of the constituent particles of an atom and the (smaller) mass of the whole atom.

■ Binding energy, ΔE: the energy equivalence of the mass defect, found using $\Delta E = \Delta mc^2$; it is the energy required to separate an atom into its constituent particles.

■ The properties of the three naturally occurring types of radioactivity (c is the speed of light in a vacuum):

		NATURE	TYPICAL SPEED	TYPICAL PENETRATION	CHARGE	IONISATION CAUSED
alpha	α	two protons plus two electrons	$0.1c$	6 cm of air	$+2e$	a great deal
beta	β	one electron	$0.9c$	few mm of aluminium	$-1e$	some
gamma	γ	electromagnetic radiation	c	several cm of lead	0	very little

■ Energy and mass are conserved in any nuclear reaction. The following ways of writing nuclear equations are equivalent:

$$^{10}_{5}\text{B} + ^{1}_{0}\text{n} \rightarrow ^{7}_{3}\text{Li} + ^{4}_{2}\text{He} \quad \text{or} \quad ^{10}_{5}\text{B}(\text{n},\alpha)^{7}_{3}\text{Li}$$

■ Radioactive decay is a random process. The rate of decay is proportional to the number of radioactive atoms present:

$$\frac{dN}{dt} \propto \pm N \quad or \quad \frac{dN}{dt} = A = \pm\lambda N$$

where λ is the decay constant. This gives an exponential decay equation:

$$N = N_0\,e^{-\lambda t}$$

For a half-life $t_{1/2}$, $N = \frac{1}{2}N_0$, which gives $t_{1/2}\lambda = 0.693$.

ASSESSMENT QUESTIONS

28.14 An α-source with an activity of 150 kBq is placed in a metal can as shown. A 100 V d.c. source and a 10^9 Ω resistor are connected in series to the can and the source. This arrangement is sometimes called an ionisation chamber.

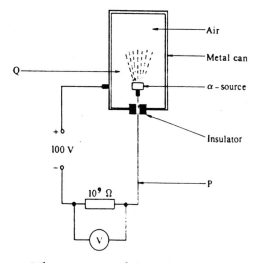

a What is meant in this case by *an activity of 150 kBq*?

b Describe how the nature of the electric current in the wire at P differs from that in the air at Q.

c A potential difference of 3.4 V is registered on the voltmeter.

  i Calculate the current in the wire at P. State any assumption you make.

  ii Calculate the corresponding number of ionisations occurring in the metal can every second. State any assumption you make.

d With the α-source removed from the metal can, the voltmeter still registers a potential difference of 0.2 V. Suggest two reasons why the current is not zero.

e The half-life of the α-source is known to be 1600 years. Calculate the decay constant and hence deduce the number of radioactive atoms in the source.

London 1996

28.15 a The fuel rods of a thermal nuclear reactor contain mostly uranium-238 and a small proportion of uranium-235.

  i Describe what happens to uranium-238 and to uranium-235 when they are being bombarded by neutrons inside a working nuclear reactor.

  ii Describe the controlled chain reaction by which energy is released from the fuel in a thermal nuclear reactor.

b i Describe the function of the moderator in a thermal nuclear reactor.

  ii State a material suitable for use as a moderator in a thermal nuclear reactor, giving a reason for your choice.

NEAB 1997

28.16 One mode of spontaneous nuclear change is called K capture. This happens when a nucleus captures an electron from the inner shell, the K-shell, of electrons surrounding the nucleus.

a When K capture occurs in iron, the iron nucleus $^{55}_{26}$Fe is changed into a manganese (Mn) nucleus. Copy and complete the nuclear equation for this change.

$$^{55}_{26}\text{Fe} + {}^{...}_{...}\text{e} \rightarrow {}^{...}_{...}\text{Mn}$$

b Copy and complete the table below for the nuclei in the equation in **a**.

nucleus	number of protons	number of neutrons
iron		
manganese		

c By reference to your answers in **b**, suggest what is the effect of electron capture on the nucleons.

OCR 1999

28.17 a Radon is an α-particle emitter.

  i Explain why radon gas is not very hazardous when outside the body but is particularly dangerous when inhaled.

  ii In some houses in Cornwall, unacceptably large amounts of radon are found because the houses are built on granite (which is a source of radon). Suggest one way of reducing the amount of radon in these houses.

b State and explain one use of a radioisotope emitting β-radiation.

c Radioactive tracers are sometimes used to locate leaks in underground water pipes.

  i State, with a reason, what type of emitter should be used as the tracer.

  ii Suggest a suitable value for the half-life of the tracer, explaining the reason for your choice.

Cambridge 1997

28.18 Boron has two isotopes represented by the nuclide symbols

$^{11}_{5}$B and $^{12}_{5}$B.

a In what way do nuclei of boron-11 and boron-12 differ?

b Calculate the number of atoms in a sample of boron-12 of mass 6.0 kg.

Cambridge 1997

28.19 The table below shows the binding energies, *BE*, per nucleon for some particles and nuclides which have low proton numbers.

particle or nuclide	^1_0n	^1_1H	^2_1H	^3_1H	^3_2He	^4_2He	^6_3Li	^7_3Li
BE per nucleon /MeV per nucleon	0	0	1.11	2.83	2.57	7.07	5.33	5.60

n ≡ neutron; H ≡ hydrogen; He ≡ helium; Li ≡ lithium.

a State what is meant by the *binding energy of a nucleus*.

b State the conclusion which can be drawn from the relatively high value for the binding energy per nucleon for helium-4.

c A possible reaction for use in a fusion reactor is one in which hydrogen-1 and hydrogen-2 fuse together to produce helium-3.

    i 1 eV = 1.6×10^{-19} J.

    Show that the energy released in each fusion reaction is 8.8×10^{-13} J.

    ii The speed of light, $c = 3.0 \times 10^8$ m s^{-1}.
Calculate the mass change when this reaction takes place.

    iii The Avogadro constant, $N_A = 6.0 \times 10^{23}$ mol^{-1}.
Calculate the mass of helium-3 that would be formed per second when producing a power output of 500 MW.

AEB 1997

28.20 The nuclear equation

$$^{235}_{92}U + ^1_0n \rightarrow ^{146}_{57}La + ^{87}_{35}Br + 3^1_0n + \text{energy}$$

represents a nuclear process which occurs in the core of a nuclear reactor.

a **i** What is the name given to this type of process?

    ii Explain how a chain reaction can be produced from this process.

b Describe briefly how a chain reaction is controlled in a nuclear reactor.

c The figure shows the variation with nucleon number (mass number) of binding energy per nucleon.

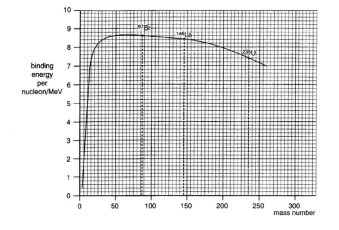

Use the curve to calculate a value for the energy which would be available from the fission of a $^{235}_{92}U$ nucleus into a $^{146}_{57}La$ nucleus and a $^{87}_{35}Br$ nucleus.
The power output from a typical nuclear power station is 2000 MW. If the efficiency of the power station is 25%, calculate a value for the number of fissions per second of uranium-235 nuclei in order to generate this power.

Cambridge 1997

28.21 Tritium, ^3_1T, decays by β^- emission. ^3_1T has a half-life of 12 years.

a Of which element is ^3_1T an isotope?

b Copy and complete the nuclear equation given below for this decay.

$$^3_1T \rightarrow X +$$

c Define the term *half-life*.

d An example of a thermonuclear fusion reaction is given by the equation below.

$$^2_1D + ^3_1T \rightarrow ^4_2He + ^1_0n + \Delta E$$

Mass of ^2_1D	=	2.0136 u
Mass of ^3_1T	=	3.0160 u
Mass of ^4_2He	=	4.0026 u
Mass of ^1_0n	=	1.0087 u

Calculate ΔE the energy released when one nucleus of helium is created.

London 1998

28.22 a Describe the relevance of binding energy to nuclei. In your essay, you should make reference to, and where appropriate include sketches, graphs, equations and examples illustrating the following:

    i the meaning of the term binding energy,

    ii the association between mass and energy,

    iii the variation of binding energy per nucleon with nucleon number,

    iv fission and fusion.

b When a high speed α-particle strikes a stationary nitrogen nucleus, it may cause a nuclear reaction in which an oxygen nucleus and a proton are formed, as shown by the following equation.

$$^{14}_7N + ^4_2He \rightarrow ^{17}_8O + ^1_1H$$

The masses of these nuclei are shown below.

Nitrogen-14	13.99922 u
Helium-4	4.00150 u
Oxygen-17	16.99473 u
Hydrogen-1	1.00728 u

Explain how mass and energy can be balanced in such a nuclear reaction. In your answer, you are expected to calculate the minimum speed of the α-particle for this reaction to take place.

Cambridge 1997

28.23 A space probe to one of the outer planets is to be powered by a radioactive source containing 2 kg of plutonium-238 of half-life 87 years (1 year $\approx 3.2 \times 10^7$ seconds).

   a Calculate the approximate number of plutonium atoms initially in the source.

   b Calculate the decay constant of plutonium-238.

   c Hence determine the initial activity of the source.

   d Unfortunately the rocket carrying the space probe explodes on take off and the probe's radioactive fuel is released into the Earth's atmosphere. How long will it take for the activity due to this release to fall to 10% of its initial value?

   e Suggest one reason why, in spite of this risk, radioactive power supplies are preferred to solar panels for space probes of this type.

   London 1997

28.24 Summarise the properties of α-, β-, and γ-radiation from natural radioactive sources by referring to their charge, mass, speed and penetrating properties. Use a table like the one below.

	α	β	γ
charge relative to the fundamental charge, e			
mass relative to the proton mass, m_p			
speed relative to the speed of light, c			
penetrating properties			

Cambridge 1997

28.25 **a** $^{12}_{6}C$ and $^{14}_{6}C$ are two isotopes of carbon.

   i State the number of electrons in a neutral atom of $^{14}_{6}C$.

   ii State the number of neutrons in a neutral atom of $^{14}_{6}C$.

   b $^{14}_{6}C$ decays by beta-minus emission. Copy and complete the nuclear equation below.

$$^{14}_{6}C \rightarrow \quad N+$$

   c Describe briefly how you would test whether $^{14}_{6}C$ decays *only* by beta emission.

   London 1995

28.26 A radioactive source in a school laboratory contains 8.46×10^{13} radioactive nuclei at a time when the activity of the source is 1.85×10^5 Bq. Calculate, for this source,

   a the decay constant,

   b the half-life,

   c the time taken for the activity to fall to 1.00×10^4 Bq.

   OCR 1999

28.27 A patient is to be given an injection of iodine-131 in an investigation of her blood volume.

Two identical samples of iodine-131 are prepared.

Sample A is injected into the patient while, at the same time, sample B is set on one side.

Each sample has an initial activity of 8.0×10^3 Bq.

Iodine-131 has a physical half-life of 8 days.

   a Calculate the activity of sample B after 24 days.

   b After 24 days the total activity of iodine-131 remaining in the patient's body is estimated to be 4.0×10^2 Bq.

   Calculate

   i the *effective* half-life of iodine-131 in the body of the patient.

   ii the *biological* half-life of iodine-131 in the body of the patient.

   NEAB 1998

28.28 The radioisotope $^{232}_{92}U$ emits alpha particles with an energy of 5.30 MeV and has a half life of 74.0 years. For a sample containing 1.30×10^{24} atoms calculate

   a the decay constant,

   b the activity,

   c the maximum theoretical power output.

   WJEC 1997

28.29 **a** The equation for radioactive decay is as follows:

$$A = A_0 \, e^{-\lambda t}$$

   i Identify the symbols A, A_0 and λ.

   ii Why is there a negative sign in this equation?

   b **i** Explain what is meant by the *half-life* of a radioactive substance.

   ii Sketch a graph to show how the number N of radioactive atoms present in a sample of a radioactive substance varies with time t, when the total time extends over at least three half-lives.

   Show how the half-life can be obtained from your graph.

   c Cobalt-60 (^{60}Co) is often used as a radioactive source in medical physics. It has a half-life of 5.25 years. How long after a new sample is delivered will the activity have decreased to one-eighth of its value when delivered?

   CCEA 1998

29

THE QUANTUM THEORY

LEARNING OBJECTIVES

At the end of this chapter you should be able to:

1. describe the photoelectric effect and understand why it cannot be explained by classical physics;

2. calculate a photon's energy and relate it to frequency and wavelength;

3. map energy levels and use them to calculate the wavelength of line spectra;

4. describe X-ray spectra;

5. outline the theory of wave–particle duality.

29.1 THE EMISSION OF LIGHT

In Theme 3 it was established by several experiments that light can be considered as a wave motion. The theory of the transmission of electromagnetic waves was established by Maxwell in 1865. His work is summarised by four equations relating such quantities as electric field strength, magnetic field strength, electric current, charge, the permittivity of free space ε_0 and the permeability of free space μ_0.

A deduction from Maxwell's theory is that an electromagnetic wave is a combination of two transverse waves that travel through a vacuum with speed c. One of these waves is an electric field wave and the other is a magnetic field wave at right-angles to it. In any electromagnetic wave, the changing magnetic field creates an electric field, and the changing electric field creates a magnetic field. The fields were shown in **Figs 11.35** and **11.36** for waves travelling in one dimension.

The value of c according to Maxwell is given by

$$c = \frac{1}{\sqrt{\mu_0 \varepsilon_0}}$$

The agreement between Maxwell's theory and the experimental value of the speed of light available at the time was very strong evidence for the fact that light is a form of electromagnetic radiation.

At the time Maxwell published his theory, the only parts of the electromagnetic spectrum that were known to exist were light, infra-red and ultra-violet. Predictions put forward by Maxwell concerning other electromagnetic radiations were not demonstrated practically until after his death. In 1887, as a direct consequence of Maxwell's theory, Hertz set about constructing electrical circuits to produce electromagnetic waves of a different, longer wavelength. The waves he produced came to be called *radio waves*. This is a good example of an important scientific development in which theory preceded practice. Radio waves were not discovered experimentally; they were deliberately produced as a result of Hertz having the necessary theoretical understanding. Since that time X-rays, gamma rays and microwaves have also been produced, and have been shown to be electromagnetic waves.

More recently, the definition of the metre has been based on the speed of light, so that now

$$c = 299\ 792\ 458\ \text{m s}^{-1}$$

exactly, by definition of the metre. Also

$$\mu_0 = 4\pi \times 10^{-7}\ \text{N A}^{-2}$$

exactly, by definition of the ampere. Combining these gives

$$\varepsilon_0 = \frac{1}{c^2 \mu_0} = \frac{1}{(299\ 792\ 458\ \text{m s}^{\pm 1})^2 \times (4\pi \times 10^{\pm 7}\ \text{N A}^{\pm 2})}$$

$$= 8.854\ 19 \times 10^{\pm 12}\ \text{C}^2\ \text{N}^{\pm 1}\ \text{m}^{\pm 2}$$

QUESTION

29.1 Show that $\text{C}^2\ \text{N}^{-1}\ \text{m}^{-2}$ is the unit of ε_0, the permittivity of free space, and show that $\mu_0 \varepsilon_0$ has the unit $\text{s}^2\ \text{m}^{-2}$.

While the transmission of light waves was explained very precisely by Maxwell, there was still a mystery concerning the source of light waves. According to Maxwell, electromagnetic waves are emitted by accelerating charges. All that was known about light production in the middle of the nineteenth century was knowledge gained experimentally from the emitted light. Nothing was known about atomic structure, so it was not possible to consider what charges were accelerating, but it was known that the emission spectra of glowing gases consisted of light of specific wavelengths in a very pronounced pattern. Before any progress could be made on the theory of light emission, however, the photoelectric effect was discovered, by Hertz, and this threw doubt on the validity of considering light as a wave motion.

29.2 THE PHOTOELECTRIC EFFECT

Hertz noticed in 1887 that a spark between two metal spheres could occur more readily if the spheres were illuminated with light from another spark. Light made the space around the spheres conduct electricity more readily. The experiment that follows is an experiment that was done at that time and is worth repeating. It uses only simple apparatus, but it is very instructive.

Hertz was not able to explain why the air in the space near his metal spheres became conducting when illuminated with the light from a spark. The radiation from the first spark was clearly assisting the formation of the second. It was also found that visible light from a bright source can initiate sparking provided the metals used for the spheres are sodium or potassium. This is not a recommended experiment because sodium and potassium are so highly reactive.

After the discovery of the electron in 1897, it was immediately suspected that the negatively charged particles that were emitted by a zinc plate, when illuminated by ultra-violet light, were electrons. By 1899 it had been confirmed that they were electrons by experiments that showed that they had the same specific charge

INVESTIGATION

Place a piece of clean zinc strip on the cap of a gold-leaf electroscope, as shown in **Fig 29.1**. The zinc can be cleaned by rubbing it with sandpaper. Charge the gold leaf so that it is negatively charged, and ensure that the gold leaf stays at a constant angle to the vertical. If the leaf does fall initially, it is probably because the air inside the electroscope is damp. It can be dried by standing it on a radiator for a while. Then bring an ultra-violet lamp near to the zinc strip. What is the effect on the charge stored on the gold leaf? How does the movement of the gold leaf vary with the intensity of illumination? (The intensity of illumination may be changed by moving the lamp closer to, or further from, the zinc strip.)

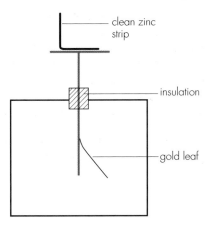

FIG 29.1 A gold-leaf electroscope.

Repeat the experiment using a positively charged electroscope. What charge does the ultra-violet light cause to be emitted from the zinc plate? The Sun can be used as the source of ultra-violet light if it happens to be sunny while you are doing this investigation. You must use direct sunlight, however, since ultra-violet radiation is absorbed by window glass.

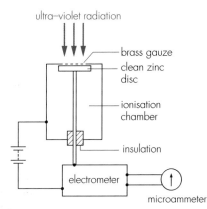

FIG 29.2 Demonstration of the photoelectric effect, using an electrometer.

This investigation can be extended if an electrometer is available. The apparatus is illustrated in **Fig 29.2**, and consists of a clean zinc disc held close to a brass gauze. The potential difference between these two items can be varied, and the flow of charge measured using the electrometer. The electrometer is a d.c. amplifier with a very high gain, so that it can detect very small currents. Investigate how the current varies with:

a the illumination, and
b the potential difference between the zinc plate and the gauze.

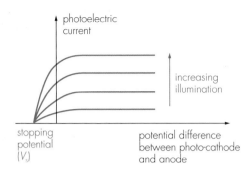

FIG 29.3 Typical results of the investigation into the photoelectric effect.

as the electron. This effect is shown by the investigation, and is called the *photoelectric effect*. The emitted electrons are referred to as *photoelectrons*.

The photoelectric effect produced some surprising results. It was the energy of the photoelectrons that was surprising. If graphs are plotted of photoelectric current against potential difference between the photo-cathode (the zinc in **Fig 29.2**) and the photo-anode (the brass), the result is as shown in **Fig 29.3**. As expected, a large photoelectric current is generated when there is greater illumination of the photo-cathode. Also as expected, increasing the potential difference between the electrodes cannot increase the photoelectron current beyond a certain maximum for any level of illumination. This is because the maximum current is limited by the rate at which photoelectrons are produced, and that is controlled by the illumination level, not the potential difference.

Stopping potential and threshold frequency

The totally unexpected result was that the energy of photoelectrons generated by dim illumination is the same as that generated by bright illumination. This is shown by the stopping potential. The **stopping potential** V_s is the reverse potential that is just sufficient to prevent photoelectrons from reaching the anode. If an electron is emitted with speed v, it will stop when it has changed all of its kinetic energy into potential energy. That is when

$$\text{kinetic energy} = \tfrac{1}{2}mv^2 = eV_s$$

where m is the mass and e the charge of the electron. Therefore V_s is directly proportional to the maximum kinetic energy of the photoelectrons, and **Fig 29.3** shows that the stopping potential does not depend at all on the brightness of the illumination. Dim illumination can produce photoelectrons of the same energy as bright illumination.

This result is surprising because it contradicts basic wave theory. The energy of any oscillating system is proportional to the square of the amplitude (see section 10.3). This implies that if one wave has an amplitude three times that of another wave, the energy of the first wave should be nine times that of the second wave.

Here we have a situation in which the number of photoelectrons produced depends on the intensity of illumination, but the maximum energy of these photoelectrons does not. It is for this reason that radio waves, however strong, do not affect photographic film. Likewise, red light does not affect photographic paper but blue light does.

The maximum energy of photoelectrons was found to depend on the frequency of the radiation causing them. This is shown in **Fig 29.4**. No photoelectrons are produced until the frequency is high enough: this frequency f_0 is called the **threshold frequency**. Further increase in the frequency results in a linear increase in the maximum kinetic energy of photoelectrons. The threshold frequency is such

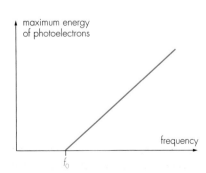

FIG 29.4 Threshold frequency.

that metals such as zinc will give photoelectrons only if the radiation used to produce them is in the ultra-violet or X-ray region of the electromagnetic spectrum. Sodium, potassium and caesium will give photoelectrons with visible light.

The quantum theory

Before the problem concerning the energy of photoelectrons became apparent, Max Planck had given a theory for the emission of radiation from a hot body. He had introduced the idea of discrete energy units. He suggested that, in much the same way that matter is not continuous but consists of a large number of tiny particles, radiation energy from a source of given frequency f is quantised. He further suggested that the basic quantum of energy from a source E is proportional to the frequency of the source. This gives

K $\quad E = hf$

where h is a constant of proportionality now called the *Planck constant*. The value of the Planck constant is found experimentally to be 6.625×10^{-34} J s.

Einstein's photoelectric equation

This concept was adopted by Einstein to give an explanation of the photoelectric effect in 1905. He assumed that radiation was quantised and that, to cause a photoelectron, a quantum of electromagnetic energy, called a **photon**, is absorbed by an atom in the surface of the cathode. Some of the energy absorbed by the atom releases an electron from it and from the surface of the metal. The rest of the energy becomes kinetic energy, some or all of which the electron may retain when it leaves the photo-cathode. Put as an equation this becomes:

K $\quad$ **energy of incoming photon** = **work done to remove electron from metal** + **kinetic energy of photoelectron**

If the photoelectron is emitted from the surface with no further interactions with other atoms, then it has the maximum kinetic energy corresponding to a speed v_{max}. For such an electron:

$$hf = \phi + \tfrac{1}{2}mv_{max}^2$$

where ϕ is called the **work function** energy of the metal. The threshold frequency f_0 is given when v_{max} is zero, so

$$hf_0 = \phi + 0$$

and so

$$hf = hf_0 + \tfrac{1}{2}mv_{max}^2$$

Since, as shown earlier,

$$\tfrac{1}{2}mv_{max}^2 = eV_s$$

the equation can be written as

$$hf = hf_0 + eV_s$$

This theory satisfactorily explains the photoelectric effect, but it does mean that when light interacts with matter it has to be considered rather differently from before. The wave nature of light is adequate to explain its transmission; the

EXAMPLE 29.1

When ultra-violet radiation of wavelength 319 nm falls on a metal surface, photoelectrons with a stopping potential of 0.22 V are produced. The same surface is then illuminated with ultra-violet radiation of wavelength 238 nm. Find: **a** the work function of the metal; **b** the threshold frequency for the metal; **c** the stopping potential when the wavelength is 238 nm; and **d** the maximum speed of photoelectrons for each wavelength. Assume:

$$e = 1.60 \times 10^{-19}\,\text{C} \qquad h = 6.63 \times 10^{-34}\,\text{J s}$$
$$c = 3.00 \times 10^8\,\text{m s}^{-1} \qquad m_e = 9.11 \times 10^{-31}\,\text{kg}$$

First the energy of one photon from each source of illumination must be found.

■ For $\lambda = 319$ nm:

$$f = \frac{c}{\lambda} = \frac{3.00 \times 10^8\,\text{m s}^{\pm 1}}{319\,\text{nm}} = 9.40 \times 10^{14}\,\text{Hz}$$

energy of one photon $= hf$

$$= (6.63 \times 10^{-34}\,\text{J s}) \times (9.40 \times 10^{14}\,\text{Hz})$$
$$= 6.23 \times 10^{-19}\,\text{J}$$

■ For $\lambda = 238$ nm:

$$f = \frac{c}{\lambda} = \frac{3.00 \times 10^8\,\text{m s}^{\pm 1}}{238\,\text{nm}} = 1.26 \times 10^{15}\,\text{Hz}$$

energy of one photon $= hf = 8.35 \times 10^{-19}\,\text{J}$

a For the 319 nm wavelength radiation, the stopping potential is 0.22 V. This means 0.22 joules per coulomb, so for a charge of 1.60×10^{-19} C the maximum kinetic energy of the electron is

$$(0.22\,\text{J C}^{-1}) \times (1.60 \times 10^{-19}\,\text{C}) = 0.35 \times 10^{-19}\,\text{J}$$

The work function is therefore

$$\phi = (6.23 \times 10^{-19}\,\text{J}) - (0.35 \times 10^{-19}\,\text{J})$$
$$= 5.88 \times 10^{-19}\,\text{J}$$

b The threshold frequency is the frequency for which the quantum of energy of the incoming photon equals the work function, so

$$hf_0 = \phi$$

giving

$$f_0 = \phi/h = (5.88 \times 10^{-19}\,\text{J})/(6.63 \times 10^{-34}\,\text{J s})$$
$$= 8.87 \times 10^{14}\,\text{Hz}$$

c For the radiation of wavelength 238 nm:

$$hf = \phi + eV_s$$

$$8.35 \times 10^{-19}\,\text{J} = 5.88 \times 10^{-19}\,\text{J} + eV_s$$

giving

$$V_s = \frac{(8.35 \times 10^{\pm 19}\,\text{J}) \pm (5.88 \times 10^{\pm 19}\,\text{J})}{(1.60 \times 10^{\pm 19}\,\text{C})} = 1.54\,\text{V}$$

d For the 319 nm wavelength, the maximum speed of photoelectrons caused by the photons is given by

$$\tfrac{1}{2} mv^2 = eV_s = 0.35 \times 10^{-19}\,\text{J}$$

giving

$$v = \sqrt{\frac{2 \times (0.35 \times 10^{\pm 19}\,\text{J})}{(9.11 \times 10^{\pm 31}\,\text{kg})}}$$
$$= 2.8 \times 10^5\,\text{m s}^{\pm 1} \quad \text{(to 2 sig. figs)}$$

Repeating for the 238 nm wavelength radiation gives

$$v = \sqrt{\frac{2 \times (1.54\,\text{V}) \times (1.6 \times 10^{\pm 19}\,\text{C})}{(9.11 \times 10^{\pm 31}\,\text{kg})}}$$
$$= 7.4 \times 10^5\,\text{m s}^{\pm 1} \quad \text{(to 2 sig. figs)}$$

It is easy to make arithmetical mistakes with problems such as these. Working can be simplified appreciably if energies are expressed in electron-volts. If this is done, it is still necessary to find the energy of a photon, but then it can be converted into electron-volts by dividing by the fundamental charge, 1.60×10^{-19} C. Working in the order of the encircled numbers produces the figures in Table 29.1. Note that a stopping potential of 0.22 V means that the photoelectrons have a maximum kinetic energy of 0.22 eV ①. After ② and ③ are calculated, ④ is obtained by direct subtraction and ⑤ must be the same as ④. This enables ⑥ to be obtained.

TABLE 29.1 An alternative approach to Example 29.1

WAVE-LENGTH /nm	PHOTON ENERGY /eV	MAXIMUM PHOTOELECTRON KINETIC ENERGY /eV	WORK FUNCTION /eV
319	3.90 ②	0.22 ①	3.68 ④
238	5.22 ③	1.54 ⑥	3.68 ⑤

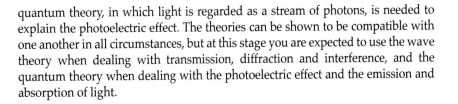

quantum theory, in which light is regarded as a stream of photons, is needed to explain the photoelectric effect. The theories can be shown to be compatible with one another in all circumstances, but at this stage you are expected to use the wave theory when dealing with transmission, diffraction and interference, and the quantum theory when dealing with the photoelectric effect and the emission and absorption of light.

QUESTIONS

29.2 a What are the frequency and the energy of a photon of red light of wavelength 700 nm?

 b What are the frequency and the energy of a photon of violet light of wavelength 400 nm?

29.3 a What is the threshold frequency of sodium of work function 2.3 eV?

 b What is the threshold frequency of caesium of work function 1.9 eV?

 c In which region of the electromagnetic spectrum do these frequencies lie?

29.4 What is the maximum kinetic energy of a photoelectron emitted from silver, of work function 4.7 eV, when illuminated with ultra-violet radiation of wavelength 122 nm?

29.5 The maximum energy for photoelectrons emitted from a metal with a work function of 2.9 eV is 8.7 eV.

 a What is the maximum energy of the incoming photons?

 b What is the maximum frequency of the incoming photons?

29.6 Assuming that numerical values are available, how can

 a the work function and

 b the Planck constant

 be obtained from the graph in **Fig 29.4**?

 c Explain why this graph always has the same gradient whatever metal is used for the photo-cathode.

 SPECTRA AND ENERGY LEVELS WITHIN ATOMS

In explaining the photoelectric effect, the quantum theory uses the idea that a photon of electromagnetic radiation has an energy E given by

$$E = hf$$

where h is the Planck constant and f is the frequency. This was quickly realised as being in some way connected with line spectra.

In section 27.1 it was stated that, when a gas is energised sufficiently, it emits visible light. If the light is examined using a spectrometer, the light is seen to consist of a few colours only. **Fig 29.5(a)** is a photograph of a hydrogen spectrum. The term 'line spectrum' indicates that certain specific wavelengths are present. The lines are created simply because the light entering the spectrometer passes through a narrow slit. Each colour shown in the photograph is really an image of the slit in the particular colour of the light seen. A continuous spectrum is obtained from a heated solid and covers a wide range of frequencies.

The difficulty of explaining line spectra stems again from Maxwell's theory of electromagnetic radiation. If the electrons in an atom are accelerating, then, according to Maxwell, they should be emitting radiation, and consequently losing energy. This would imply that the electrons would collapse into the nucleus of the atom and emit radiation of ever-increasing frequency while doing so. This clearly does not happen. Atoms do not collapse and the radiation produced has frequencies with definite values.

In 1913 Niels Bohr suggested that electrons within atoms could exist in stable energy states without emitting radiation. He suggested that the electrons within atoms could absorb energy in quanta of certain definite amounts when bombarded by other atoms or electrons during such processes as being heated. Once the electron has gained its quantum of energy, the atom is said to be in an **excited state**. Later, often no more than a microsecond later, the electron loses energy as a photon of electromagnetic radiation as it returns to its former, stable state. A line spectrum maps possible changes in energy levels within an atom.

The patterns of line spectra for different gases had been measured very accurately during the nineteenth century, and empirical equations had been found for their wavelengths. Any new theory therefore had to show agreement with the known patterns of wavelengths. Bohr succeeded in doing this for hydrogen to a high degree of accuracy, but later analysis showed that his theory was less successful with multi-electron atoms. The following example shows how the known wavelengths can be converted into energy levels for electrons within a hydrogen atom.

EXAMPLE 29.2

Some of the lines present in the hydrogen spectrum have wavelengths of 656, 486, 434, 410, 397, 389, ..., 365 nm. Some of these lines are visible and are shown in **Fig 29.5(a)**. The shorter wavelength lines are in the ultra-violet, and there they get very close to one another as a limit is reached at 365 nm. What energy levels are necessary within a hydrogen atom to allow these wavelengths to be produced?

First it is necessary to convert the wavelengths of the lines into their corresponding photon energies. This is done using $E = hf$ and since $c = f\lambda$ we get

$$E = \frac{hc}{\lambda}$$

The calculated values are given in Table 29.2.

TABLE 29.2 Wavelengths and corresponding photon energies for Example 29.2

WAVE-LENGTH /nm	PHOTON ENERGY /10^{-19} J	PHOTON ENERGY /eV
656	3.03	1.90
486	4.09	2.56
434	4.58	2.86
410	4.85	3.03
397	5.01	3.13
389	5.11	3.19
⋮		
365	5.45	3.41

Whenever energy is measured, it is important to state clearly the zero of potential energy being used. Within an atom, this is taken to be for an electron just released from the atom. If therefore a photon of 3.41 eV can be emitted from a hydrogen atom, there must be an energy level at −3.41 eV. That is, 3.41 eV lower than our arbitrary zero. If these levels are mapped for all the photons, we get the horizontal lines shown in **Fig 29.5(b)**, which is drawn to scale. Note that it has been assumed here that all the electrons fall to the *same* final energy level. There is a lower energy level within a hydrogen atom. This is at a value of −13.6 eV. When electrons fall to that level, the energy lost as a photon of electromagnetic radiation is much greater and the spectral line is short-wavelength ultra-violet.

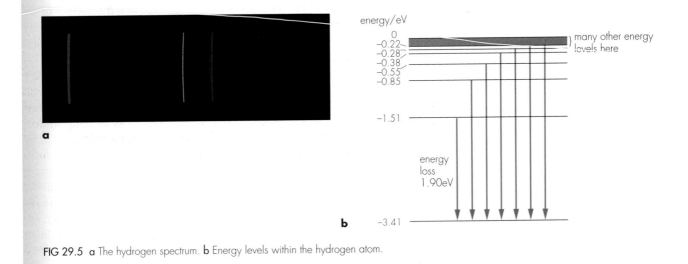

FIG 29.5 **a** The hydrogen spectrum. **b** Energy levels within the hydrogen atom.

29.7 Using values taken from Example 29.2, find the wavelengths of a series of spectral lines created when electrons fall (that is, lose energy) from different energy levels to the level at −13.6 eV.

29.8 In a helium–neon laser, electrons are forced from a low energy to a level of −4.031 eV in the helium. From there, they may fall back to a low energy level in the helium, emitting ultra-violet photons. It is also possible, when a helium atom collides with a neon atom, for electrons in neon atoms to be raised to a level of −4.026 eV by using the energy of a falling helium atom together with some thermal energy. The energy of helium atoms is continually pumping electrons into the −4.026 eV state. The electrons in the neon can be induced to fall to a level of −5.990 eV by photons of an energy equal to this energy fall. This produces the familiar red laser light.

   a Calculate the thermal energy required to push an electron into the −4.026 eV state.

   b What is the wavelength of the laser light produced?

X-rays were discovered by Röntgen in 1895. (In Germany, X-rays are still called Röntgen rays.) He was investigating cathode rays and found that, when his cathode rays had high energy, a radiation was produced that was very penetrating and caused a glow on a fluorescent screen.

Because of their ability to penetrate matter, they were used for medical investigations almost immediately after their discovery. The dangers associated with X-radiation were realised more slowly. Now there is pressure on doctors and dentists to use X-rays less. There used to be mass screening of the population to check people for tuberculosis of the lung. But this was stopped when it was realised that more harm was probably being done by the X-rays than good was being achieved by finding the lung disease. This was especially true when the incidence of tuberculosis of the lungs became small.

When X-rays are used now, the radiographers themselves have to be well screened from the X-rays, and the dose for the patient has to be kept to a minimum. This has been made possible by manufacturing photographic X-ray film that is of very high sensitivity. Figs 29.6 and 29.7 show a hospital X-ray machine and a typical X-ray photograph of soft tissue.

One reason for having thick glass on the front of a television set is to absorb the X-rays that are produced on the screen by the sudden slowing down of the electrons when they hit the screen. Although the level of radiation is very low from television screens, the dose received from them is built up over many hours. It is a bad idea for young children habitually to sit close to a television screen.

A modern X-ray tube is illustrated in **Fig 29.8**. If very penetrating, hard X-rays are required, then a high potential difference, typically 120 kV, is used between the cathode and the anode. X-ray tubes used by dentists typically use potential differences of 60 kV. A heated cathode produces electrons as in a cathode ray tube, and these can be focussed by the shape of the cathode onto an anode, which may need to be cooled by water or oil circulated through it if the power rating is high. As the electrons hit the anode, X-rays are produced. In practice, only about 1% of the power of the electron beam produces X-rays. The rest is wasted as heat.

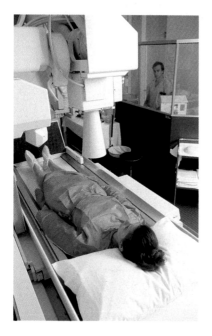

FIG 29.6 Hospital X-ray equipment has come a long way since Röntgen's initial discovery.

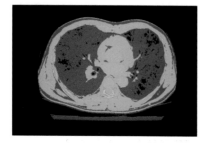

FIG 29.7 This false-colour computed X-ray tomography (CT) scan shows a section through a person's heart and lungs. The lungs are coloured purple with black flecks with the heart in shades of blue. A thoracic vertebra is coloured yellow. The aorta can be seen as a distinct blue circle.

QUESTIONS

29.9 An X-ray tube operates at 100 kV and the electrons through it constitute a current of 3.0 mA.

   a What is the power of the electron beam?

   b Why does the tube give better X-ray pictures if the electrons are brought to a point focus?

   c Assuming that most of the beam energy is wasted as heat, what problem arises if all the power of the electron beam is brought to a point focus?

29.10 a What is the energy of an electron if it is accelerated through a potential difference of 80 kV in an X-ray tube?

   b What is the maximum energy of a photon of X-rays from a tube working with this voltage?

   c What is the maximum frequency of the X-rays that the tube produces?

   d What is the wavelength of these X-rays?

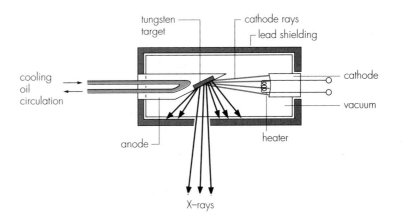

FIG 29.8 A modern X-ray tube.

Question 29.10 shows that the voltage used across an X-ray tube determines a maximum possible frequency, and hence a minimum possible wavelength, for the X-rays produced. Usually, however, when an electron strikes the target anode, its energy is not lost in creating a single X-ray photon. The electron can lose its energy in a series of encounters with target atoms or it may give several X-ray photons of lower energy. This analysis, by itself, would give rise to a continuous X-ray spectrum with a minimum possible wavelength. This continuous spectrum is found, but it is also found that there are some specific wavelengths that are much brighter.

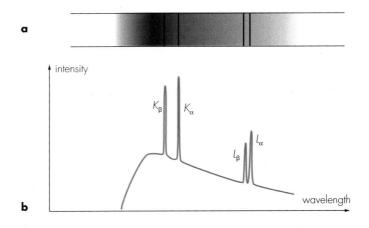

FIG 29.9 **a** An X-ray spectrum showing both a continuous spectrum and a superimposed line spectrum. **b** A graph showing how the intensity of X-rays varies with the wavelength.

The X-ray spectrum is both a continuous spectrum and a superimposed line spectrum, as shown in two ways in **Fig 29.9**. **Fig 29.9(a)** is a sketch of an X-ray spectrum and **Fig 29.9(b)** is a graph showing how its intensity varies with thewavelength. The wavelengths of X-ray line spectra were found to be determined by the element of the target. The photon energies of these lines are very high, so an electron within a target atom must have fallen into one of its lowest possible energy levels. A space will be available in one of these low levels only if the electron originally there has been knocked out by an incoming electron. The following analysis exercise shows how this is possible.

X-rays

Some of the energy levels within an atom of tungsten are given in **Fig 29.10**. The energies involved are given in keV, so are of the order of a thousand times greater than was the case with hydrogen. You may know that electrons closest to the nucleus in an atom are referred to as the K-shell electrons. These are the ones in the lowest energy state. When these electrons are removed on bombardment by electrons hitting the target, electrons from higher levels fall in to fill the gap. In doing so, they emit high-energy X-rays.

a Use the data on **Fig 29.10** and on the graph of **Fig 29.9(b)** to identify and find the photon energy of the K_α, K_β, L_α and L_β lines.
b What is the wavelength of each of these lines?
c At what wavelengths might you expect to find a K_γ and an L_γ line?
d The lines found in **c** are possible, but in practice their intensity is low. Why do you think this is so?
e What is the minimum potential difference that must be applied across the tube in order to get any K_α lines?

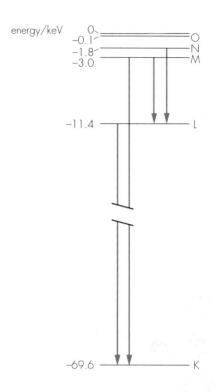

Fig 29.10 X-ray energy levels.

It was suggested in section 29.2 that electromagnetic radiation needed to be considered as a wave motion for transmission and as a stream of photons when it interacts with matter. The idea of both wave and particle approaches being applied successfully on an atomic level led de Broglie (pronounced 'de Broy') in 1924 to suggest that, if waves could behave as particles, perhaps particles could behave as waves.

De Broglie started with the following relationships for an electromagnetic wave, obtained from the quantum theory and Einstein's theory of relativity:

■ energy of a photon

$$E = hf$$

■ mass equivalence (m) of this energy (E)

$$E = mc^2$$

Equating these terms gives

$$hf = mc^2$$

Since $f = c/\lambda$, it follows that

$$hc/\lambda = mc^2$$

EXAMPLE 29.3

An electron is accelerated by a potential difference of 1500 V in a vacuum. What wavelength does such an electron possess?

We have

energy gained by electron = Ve
kinetic energy of electron = $\frac{1}{2}mv^2$

Equating these two equations and rearranging gives

$$v^2 = \frac{2Ve}{m}$$

$$v = \sqrt{\frac{2 \times (1500 \text{ V}) \times (1.6 \times 10^{\pm19} \text{ C})}{9.1 \times 10^{\pm31}\text{kg}}}$$

$$v = 2.3 \times 10^7 \text{ m s}^{\pm1}$$

Then we get

momentum of electron
$= mv$
$= (9.1 \times 10^{-31} \text{ kg}) \times (2.3 \times 10^7 \text{ m s}^{-1})$
$= 2.1 \times 10^{-23} \text{ N s}$

Finally, from De Broglie's equation $mv = h/\lambda$ we therefore get

$$\text{wavelength} = \frac{h}{\text{momentum}} = \frac{6.63 \times 10^{\pm34} \text{ J s}}{2.1 \times 10^{\pm23} \text{ N s}}$$

$$= 3.2 \times 10^{\pm11} \text{ m}$$

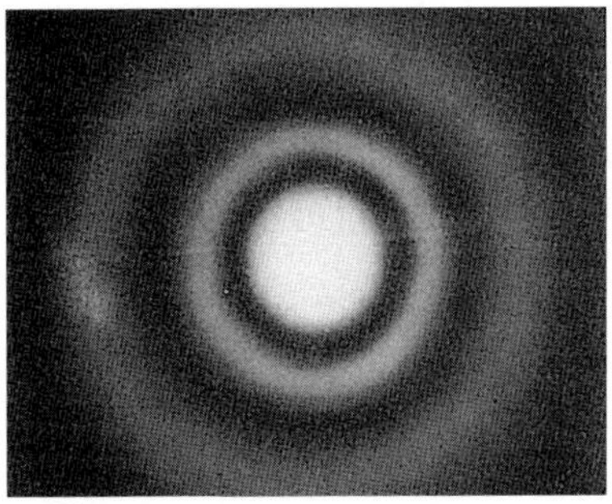

FIG 29.11 When electrons are diffracted by a thin layer of graphite atoms, this diffraction pattern is obtained.

So the momentum of a photon is given by

$mc = h/\lambda$

De Broglie suggested that the wavelength λ associated with particles of momentum mv should be given by the same relationship, namely

$mv = h/\lambda$

This equation can be tested for electrons. Example 29.3 shows how the wavelength of an electron can be found.

Example 29.3 shows that the wavelength of electrons at this speed is of the order of a tenth of the separation of atoms in solids. Davisson and Germer in 1926 used the planes of atoms in a crystal as a diffraction grating to test this theory. They found that electrons could be diffracted in the way de Broglie's theory suggested. An electron diffraction pattern is shown in **Fig 29.11**. The pattern is circular because, unlike with a diffraction grating, the atoms are not all aligned in the same direction.

QUESTIONS

29.11 What is the wavelength associated with electrons having been accelerated by a potential difference of 56 kV in an X-ray tube?

29.12 a Calculate the energy of a photon of light of wavelength 500 nm.

 b What is the momentum of this photon?

 c The power of a beam of this light is 4.0 W. How many photons pass a point in the beam per second?

 d What is the momentum of all the photons passing a point in the beam per second?

 e What force does the beam exert on an object in its path, assuming the light is absorbed by the object?

29.13 The first-order electron diffraction pattern occurred at an angle of 22° when electrons, having been accelerated in a vacuum by a potential difference of 350 V, were passed through a thin film of graphite. Assuming that you can use the same equation as for optical diffraction, $n\lambda = d \sin \theta$, find the separation of the atoms in the graphite.

SUMMARY

- Electromagnetic waves can be considered as streams of particles called photons.
- The energy of a photon is proportional to its frequency $E = hf$.
- The photoelectric effect is when a photon of incident light is able to cause an electron to be emitted from the surface of a metal. The energy of emitted electrons is found to depend on the frequency of the radiation and not on its intensity.
- Einstein's photoelectric equation:

$$\begin{array}{ccc} \text{energy of} & = \text{work done to remove} & + \quad \text{kinetic energy} \\ \text{incoming photon} & \text{electron from metal} & \text{of photoelectron} \end{array}$$

$$hf = \phi + \tfrac{1}{2}mv_{max}^2$$
$$hf = hf_0 + \tfrac{1}{2}mv_{max}^2$$
$$hf = hf_0 + eV_s$$

- The energy released when an electron falls from energy state E_2 to energy state E_1 equals the energy of the emitted photon of electromagnetic radiation:

$$E_2 - E_1 = hf$$

ASSESSMENT QUESTIONS

29.14 a A 60 W light bulb converts electrical energy to visible light with an efficiency of 8%. Calculate the visible light intensity 2 m away from the light bulb.
b The average energy of the photons emitted by the light bulb in the visible region is 2 eV. Calculate the number of these photons received per square metre per second at this distance from the light bulb.

London 1997

29.15 Monochromatic light of wavelength 450 nm falls on the cathode of a photocell at a rate of 25 µW. Only 10% of the photons produce an electron and all these electrons produce a current, I, in an external circuit. The following data is needed in this question.

the Planck constant $= 6.63 \times 10^{-34}$ J s
speed of light in vacuo $= 3.00 \times 10^8$ m s^{-1}
charge of an electron $= -1.60 \times 10^{-19}$ C

a Calculate
  i the energy of one photon in J,
  ii the number of photons falling on the cathode per second,
  iii the current, I.
b The same amount of energy per second falls on the cathode in the form of light of wavelength 600 nm.

The work function of the material of the cathode is 3.0×10^{-19} J.
Show that a photoelectric current will flow.

NEAB 1996

29.16 a **i** What is a *photon*?
  ii The output power in the beam of a certain helium–neon laser, which emits light of photon energy 3.1×10^{-19} J, is measured as 0.70 mW. The laser beam is pointed at a nearby wall. How many photons arrive at the wall in one second?
  iii Calculate the range of photon energies corresponding to the spectrum of visible light, which extends from a wavelength of 400 nm (violet light) to one of 700 nm (red light). Give your answers in electron volts (eV).
b Caesium has a work function of 1.90 eV.
  i Explain what is meant by *work function*. Suggest why different metals have different work functions.
  ii Calculate the maximum kinetic energy of the electrons emitted when a caesium surface is illuminated by violet light of wavelength

400 nm. (Make use of your answer to **a iii.**) Find also the speed of these electrons.

    iii Explain whether or not photoelectric emission would occur if the caesium surface were illuminated by red light of wavelength 700 nm. (Again make use of your answer to **a iii.**)

c Describe the principle of an experiment to measure the Planck constant. Your description should use the headings: apparatus, procedure, calculation of results.

CCEA 1998

29.17 a Describe the phenomenon of the photoelectric effect. In your answer, ensure that you cover the following points:
    i a description of the effect itself,
    ii the significance of the threshold frequency,
    iii the inability of the wave theory of light to explain the effect.

b Give an explanation of the photoelectric effect in terms of the particulate nature of electromagnetic radiation. In your answer, ensure that you cover the following points:
    i photons and photon energy,
    ii conservation of energy as applied to photo-electric emission,
    iii the effects of the frequency and of the intensity of the incident radiation.

c Describe the evidence provided by electron diffraction for the wave nature of particles. In your answer, you should consider the effect on the diffraction pattern of a change in the speed of the electrons and relate this to an expression for the de Broglie wavelength.

Cambridge 1997

29.18 a The following equation describes the release of electrons from a metal surface illuminated by electromagnetic radiation.

$$hf = \text{k.e.}_{max} + \phi$$

Explain briefly what you understand by each of the terms in the equation.

b Calculate the momentum p of an electron travelling in a vacuum at 5% of the speed of light.
    i What is the de Broglie wavelength of electrons travelling at this speed?
    ii Why are electrons of this wavelength useful for studying the structure of molecules?

London 1998

29.19 a Electrons can be liberated from a certain metal plate if the plate is illuminated with ultra violet radiation. If the ultra violet source is replaced by a filament lamp, no electrons are liberated.
Explain why electrons are liberated when an ultra violet source is used but not when a filament lamp is used.

b **i** Calculate the energy of a photon of ultra violet radiation of wavelength 110 nm.

    ii Calculate the momentum and the de Broglie wavelength of an electron of kinetic energy equal to the energy of a photon of wavelength 110 nm.

NEAB 1998

29.20 This question is about atomic spectra.
a Explain what is meant by the following italicised terms: *absorption* spectrum, *emission* spectrum, *ionisation energy* of an atom, *excitation energies* of an atom.
An atom has a single active electron. It has an ionisation energy of 5.9 eV and excitation energies from the ground state equal to 3.0 eV, 5.0 eV and 5.5 eV.
b Draw an energy level diagram for this electron in the atom, labelling the levels in eV.
c Calculate the longest wavelength which you would expect to observe in the absorption spectrum. Add to your diagram in part **b** an arrow clearly indicating the relevant transition.
d Without detailed calculation, identify the other lines which you would expect to be observed in the absorption spectrum of these atoms. Suggest briefly an experimental arrangement for observing this spectrum.
e Explain why you would expect there to be more lines observed in the emission spectrum than in the absorption spectrum.

O & C 1998

29.21 a Describe the principles of the photoelectric effect. Your answer should include
    i what is meant by a photon and how to calculate its energy,
    ii the significance of threshold frequency,
    iii the meaning of work function energy,
    iv the evidence for the particulate nature of electromagnetic radiation.

b The figure shows the variation with frequency f of the maximum kinetic energy E_{max} of photoelectrons emitted from sodium and zinc.

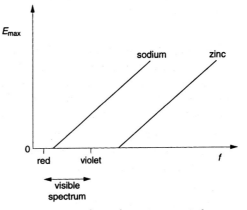

    i Compare the photoelectric behaviour of sodium and zinc.
    ii Suggest how the Planck constant may be obtained from the figure.

OCR 1999

29.22 A muon is a particle which has the *same charge* as an electron but its *mass* is 207 times the mass of an electron.

 a An unusual atom similar to hydrogen has been created, consisting of a muon orbiting a single proton. An energy level diagram for this atom is shown.

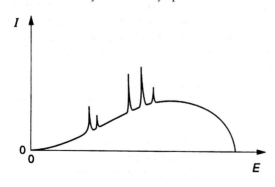

 0 eV ——————————————
 –312 eV ——————————————

 –703 eV ——————————————

 –2810 eV ————————————————— Ground state

 i State the ionisation energy of this atom.
 ii Calculate the maximum possible wavelength of a photon which, when absorbed, would be able to ionise this atom.
 iii To which part of the electromagnetic spectrum does this photon belong?
 b Calculate the de Broglie wavelength of a muon travelling at 11% of the speed of light.

London 1999

29.23 a Calculate the frequency of electromagnetic radiation emitted by a hydrogen atom which undergoes a transition between energy levels of -1.36×10^{-19} J and -5.45×10^{-19} J.
 The Planck constant, $h = 6.63 \times 10^{-34}$ J s.

 b A different emission from a hydrogen atom has a wavelength of 4.34×10^{-7} m.
 Calculate the angle of the first order diffraction maximum for this radiation when using a diffraction grating with a spacing of 2.00×10^{-6} m.

AEB 1998

29.24 The figure shows the variation with photon energy E of the intensity I of an X-ray spectrum.

 a Explain in detail the shape of this graph, with reference to the mechanisms of production of X-rays.
 b State and explain how, for an X-ray tube, the shape of the graph above may be changed.
 c Describe factors which affect the sharpness and the contrast of an X-ray image. Explain how the sharpness and the contrast may be improved.

OCR 1999

29.25 This question is about the wave-like properties of electrons.

 a State de Broglie's relationship in words.
 b **i** Show that the speed of electrons which have been accelerated from rest through a potential difference of 2.0 kV is equal to 2.6×10^7 m s^{-1}.
 ii Calculate the wavelength associated with a beam of these electrons.
 c What is electron diffraction? Suggest why electron diffraction is a useful tool for studying the arrangement of atoms in crystals, where the separation of the atoms is typically 0.3 nm.

O & C 1998

29.26 This question is about wave–particle duality.

 a The phenomena of diffraction and polarisation can be described only using a wave model. Describe briefly experiments to demonstrate these phenomena for electromagnetic waves.
 b Photoelectric emission can best be explained by the particle model.
 i What is photoelectric emission?
 ii For photoemission to occur with a given metal, the wavelength must be below a certain value. How does the particle model explain this?
 iii How does this model explain the relation between the maximum kinetic energy of the photoelectrons and the frequency of the incident radiation?
 c Particles such as electrons can exhibit wavelike properties.
 i Describe briefly an experiment to demonstrate electron diffraction. How are the results affected by increasing the energy of the electrons used in the experiment?
 ii An experimenter wishes to investigate the wave properties of electrons of wavelength 1.4×10^{-10} m.
 1 What is the momentum of these electrons?
 2 What is the kinetic energy of these electrons?
 3 What potential difference should be used to accelerate these electrons from rest to obtain this kinetic energy?

CCEA 1998

30
MICROCOSM

30.1 INTRODUCTION

The title to this chapter, 'Microcosm', needs some explanation. The other chapters in this book have titles that clearly define the subject matter of the chapter in a way that the title of this chapter does not. This title has been chosen in the first instance because this final chapter is more incomplete than any of the others. It deals with recent and current research and, because there are still many unknown aspects of cosmic and particle physics, the conclusion of the final chapter cannot yet be written.

One of the interesting features of physics is that strings of questions can always be posed until, at any particular time, the final question is unanswerable. For example:

■ 'What is matter made of?' can be answered by 'molecules'.
■ 'But what are molecules made of?' – answer 'atoms'.
■ 'And what are atoms made of?' – 'protons, neutrons and electrons'.
■ 'But what are these made of?' – …

Some answers will be given in this chapter, but much will be left unanswered.

A dictionary definition of 'microcosm' is a small scale of something cosmic. It is an extraordinary and fascinating aspect of physics that research into the ordered Universe, the cosmos, has many connections with research into sub-atomic particles. The Universe, with its diameter within an order of magnitude of 10^{26} metres, has been investigated by doing research using particles that are within an order of magnitude of 10^{-14} metres. The laws of physics that you have studied apply at *both* ends of this vast difference of scale, although some of the laws with which you are at present familiar have been found to be approximations when applied to extreme conditions. This chapter is intended to show you some of the links between the world of the very small inside the nucleus and the world of the very large, the Universe itself.

30.2 THE SIZE OF THE UNIVERSE

The Doppler effect

In Chapter 11 the pattern that waves make when they spread out from a source was considered. In question 11.5 you were asked to draw a diagram to show how the simple pattern of concentric circles changes when the source of the waves is moving. The answer to question 11.5(a) is shown in **Fig 30.1**. The diagram shows that the wavelength of the waves gets shorter in front of the source and that it gets longer behind the source.

Since the speed of a wave is unaltered by the movement of the source, it follows that the frequency increases in front of the source and decreases behind it. This effect is called the **Doppler effect** and it is a familiar effect if the source is one of sound waves. The frequency of the sound produced by the engine of a motor cycle being driven towards you is raised by the Doppler effect. As the motorcycle passes you, the frequency of the sound falls to a lower value. The familiar 'whoosh' is a change in frequency, as shown in **Fig 30.2**. The apparent frequency f' is related to the true frequency f by the equation

$$\frac{\text{apparent frequency}}{\text{true frequency}} = \frac{\text{speed of wave}}{\text{speed of wave} - \text{speed of source towards observer}}$$

or in symbol form

$$\frac{f'}{f} = \frac{c}{c \pm u}$$

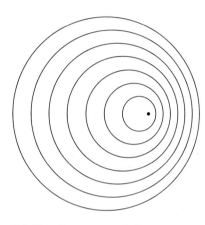

FIG 30.1 The pattern made by waves spreading out from a source moving towards the right.

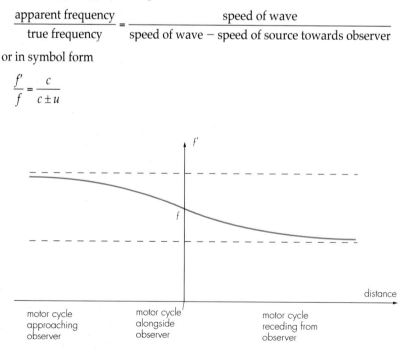

FIG 30.2 The graph shows how the frequency of the sound that a stationary observer hears varies as the motor cycle approaches and then recedes.

The importance of the Doppler effect in astronomy is that the effect is also observed with light waves. A star receding from the Earth produces light that we observe as having a longer wavelength than we would expect. There is an observable shift in its spectrum towards the red end, the lower-frequency end. This is shown in **Fig 30.3** and is called the **red shift**. It can be used to determine the speed of stars relative to the Earth. For any of the 200 billion or so stars in our own Galaxy, the red shift is very small because the speed of such a star is small in comparison with the speed of light. The equation above becomes

$$fc = f'c - f'u$$

giving

$$f'u = f'c - fc = (f' - f)c = \Delta fc$$

and for u small compared to c we can write

$$\Delta f = \frac{f'u}{c} \approx \frac{fu}{c}$$

where Δf is the change in frequency. In terms of wavelength, rather than frequency, this equation becomes

$$\Delta \lambda = \frac{\lambda u}{c}$$

Note that for a receding star Δf will be negative, indicating a decrease in frequency, and $\Delta \lambda$ will be positive, indicating an increase in wavelength.

QUESTIONS

30.1 A whistle of frequency 800 Hz is sounded on a train travelling at 50 m s^{-1}. What frequency will a person standing alongside the track hear when

  a the train is approaching the observer, and

  b the train is receding from the observer?

  [The speed of sound is 336 m s^{-1}.]

30.2 A racing car approaching a spectator has its engine producing a sound of frequency 200 Hz. A track-side observer hears a frequency of 240 Hz as the car approaches. What is the speed of the car? [The speed of sound is 336 m s^{-1}.]

30.3 The wavelength of the pale blue line in the hydrogen spectrum is 4.8627×10^{-7} m when measured in the laboratory. When the wavelength of the same spectral line is measured in light that is emitted from a star, the wavelength is found to be 4.8694×10^{-7} m. Calculate the speed of recession of the star. [The speed of light is 2.9979×10^{8} m s^{-1}.]

30.4 What fractional change takes place in the wavelength of light being emitted from a star that is receding from the Earth with a speed of a half the speed of light?

Two points to note

You should note the following two points about question 30.3.

First, although there is a red shift, the change in the colour of the light would *not* be detectable by eye, since the change in frequency is small.

Secondly, although all the data are given to five significant figures, the answer cannot reliably be quoted to more than two significant figures because the *change* in wavelength is only known to two significant figures.

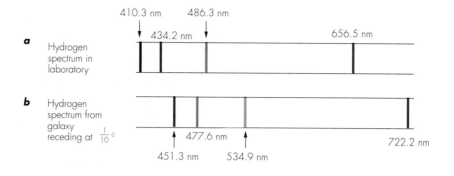

FIG 30.3 When the hydrogen spectrum is observed from a galaxy receding at a tenth of the speed of light, the whole spectrum becomes shifted towards the red end. Note that all the colours change correspondingly.

The Hubble constant H_0

Hubble collected data about a few stars from two different types of experiment: he had the information from their spectrum, which enabled him to calculate their speed of recession from the Earth; and he also found their distance from the Earth using parallax. He discovered that these two quantities are proportional to one another. This means that every galaxy is receding from every other galaxy and therefore that the Universe must be expanding. This discovery has been confirmed by subsequent research, and when applied to stars or galaxies becomes what is now known as **Hubble's law**:

$$v \propto d \quad \text{or} \quad v = H_0 d$$

where v is the speed of recession, d is the distance to the star or galaxy and H_0 is the Hubble constant.

This equation puts an absolute maximum on the present size of the Universe. Since the speed of light cannot be exceeded, the maximum value possible for the distance from the Earth to the edge of the Universe is given by $d_{max} = c/H_0$. The numerical value of H_0 is not known very accurately, but it is thought to be approximately 2.4×10^{-18} s^{-1}, so the distance from the Earth to the edge of the Universe is about 1.3×10^{26} m.

There is an interesting speculation concerning the expansion of the Universe: will it continue to expand for ever; or will it stop, reverse into a contraction, and lead to a so-called 'Big Crunch' and another 'Big Bang' (see section 30.5) some time in the distant future? At present, no one is confident about which of the graphs shown in **Fig 30.4** will prove to be correct, but it is thought that a Big Crunch will never occur, and the Universe will go on expanding for ever.

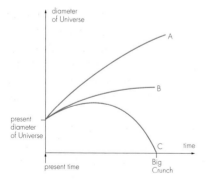

FIG 30.4 The Universe may continue to expand for ever (curve A), or may expand to a fixed diameter (curve B), or may stop expanding and contract back to produce a 'Big Crunch' (curve C).

The age of the Universe

Once the distance from the Earth to the edge of the Universe had been determined, it was easy in principle to calculate its age. Distant galaxies show red shifts that correspond to a recession speed from the Earth of $0.9c$. If these galaxies and our own Galaxy were formed at the Big Bang and have separated by a distance of up to 1.3×10^{26} m when travelling at 0.9 times the speed of light, then the time taken must be the age of the Universe from the Big Bang to the present. This method for calculating the age of the Universe gives its value as approximately 1.5×10^{10} years, or 15 billion years.

30.5 **a** Show that the unit of the Hubble constant is s⁻¹.

Using the value of the Hubble constant quoted, determine the speed of recession of:

b the nearest star to the Earth, Proxima Centauri, which is 4.2 light-years from the Earth;

c the centre of our Galaxy, the Milky Way, which is 2.6×10^4 light-years from the Earth;

d the Andromeda Galaxy, which is 1.5×10^6 light-years from the Earth;

e a quasar that is about 1.5×10^{10} light-years from the Earth.

[A light-year is a distance. It is the distance light travels in one year, i.e. $3.0 \times 10^8 \times 365 \times 24 \times 60 \times 60$ m.]

30.6 Express the Hubble constant in terms of kilometres per second per light-year.

30.7 What would happen to the red shift if the Universe stopped expanding and started to contract?

PARTICLE ACCELERATORS

In Chapter 27 details were given to enable you to calculate the deflection of particles when travelling in electric and magnetic fields. In Chapter 28 brief mention was made of some of the energies of particles that are used in high-energy research. All particle accelerators use electric fields to accelerate charged particles (most often protons and electrons). Most use magnetic fields as well to control the path that the particles take. The simplest accelerator uses a high voltage to accelerate the particles in a straight line.

One of the earliest particle accelerators was the *cyclotron*, which consists of two metal D-shaped cavities placed in a high-vacuum container and with a strong magnetic field acting vertically downwards (see **Figs 30.5** and **30.6**). When a proton, mass m, moves in the direction shown with speed v, it will travel in a circular orbit of radius r given by

force = mass × acceleration

that is, as before,

$$Bev = m \times \frac{v^2}{r}$$
$$v = \frac{Ber}{m}$$

The time t taken for the particle to complete a half-circle within one of the two Ds is therefore given by

$$t = \frac{\pi r}{v} = \frac{\pi m}{Be}$$

FIG 30.5 Photo of a cyclotron.

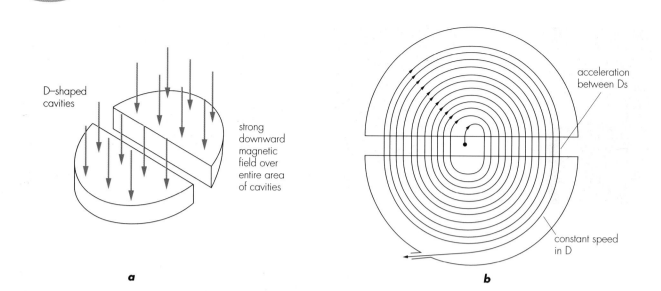

strong
downward
magnetic
field over
entire area
of cavities

acceleration
between Ds

constant speed
in D

a

b

FIG 30.6 a The arrangement of the Ds in a cyclotron. **b** The path of a proton. As it gains kinetic energy, it spirals outwards.

As the proton leaves a D, the potential difference between the Ds is arranged so that the D it is entering is negative with respect to the D it is leaving, so the proton gains energy as it moves across the gap from one D to the other. It then completes another half-circle, but this time at a higher speed. This process is repeated many thousands of times, with the proton travelling in a spiral (as shown in **Fig 30.6**) and picking up kinetic energy each time it moves across the gap between the two Ds. Inspection of the equation for *t* above shows that the time it takes for a proton to complete a half-circle is not dependent on its speed, so, if the potential difference between the two Ds is switched from positive to negative at the correct frequency, a continuous stream of high-energy protons can be produced.

A similar method is used in all accelerators, but today's high-energy machines produce particles that are so near to the speed of light that relativistic effects, which result in the particles increasing their mass as they approach the speed of light, are considerable and have to be taken into account.

FIG 30.7 The CERN accelerator. The large circle marks the path of the Large Electron-Positron Collider (LEP).

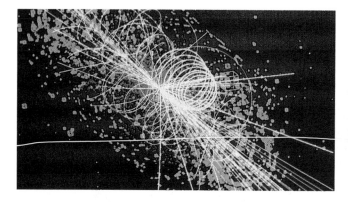

FIG 30.8 High energy decay showing sub-atomic particles.

FIG 30.9 Computer generated tracks superimposed on a view of a detector at the LEP collider.

In the CERN accelerator near Geneva, shown in Fig **30.7**, protons pass thousands of times around a doughnut-shaped evacuated chamber of radius 1100 m, increasing their energy on each revolution until they are travelling at over 99.999% of the speed of light. It is possible to arrange that protons and antiprotons travel in opposite directions around this system on parallel paths. When the two beams of particles have maximum energy, they are deflected so that they hit one another head-on in a particle detector. Photographs can be taken of the sub-proton particles that result from such high-energy collisions. An image is shown in **Fig 30.8**.

A new, even larger, accelerator of radius 4300 m is now in operation at CERN. It is the Large Electron–Positron (LEP) accelerator. The particles accelerated in this accelerator are electrons and positrons. Because of their much smaller mass than that of protons, their kinetic energy is not as high as that produced by the earlier accelerators, but the speed reached by the particles is even higher. The data provided by this accelerator are giving insight into the fundamental structure of matter and increasing our knowledge of the early Universe. A photograph taken using the LEP is shown in **Fig 30.9**.

QUESTION

30.8 The magnetic field in a cyclotron is 0.83 T. Protons are injected into it with a speed of 8.0×10^5 m s^{-1}. Calculate:

a the force that the magnetic field exerts on a proton;

b the acceleration of a proton;

c the radius of the circular path of the protons;

d the speed the protons have by the time the radius of their path is 0.25 m;

e the frequency of the potential difference that needs to be applied to the Ds;

f the increase in the kinetic energy of a proton during the time it is in the cyclotron;

g the gain in kinetic energy each time the proton crosses the gap between one D and the other, if the potential difference between them is always 500 V; and

h the number of times a proton completes a revolution.

[Mass of proton $= 1.67 \times 10^{-27}$ kg; charge on proton $= 1.60 \times 10^{-19}$ C.]

The problem

The proton, electron and neutron were thought to be the only three indivisible particles out of which the whole Universe is made. Those days have now gone. As a result of high-energy particle physics experiments with particle accelerators, hundreds of particles have been discovered with various different masses, charges and spin (angular momentum). These particles have been categorised into leptons, baryons and mesons. The **leptons** are light particles, such as the electron. The heavy particles are the **baryons**, and these include the proton and the neutron. Originally the **mesons** were taken as particles with intermediate masses, but since then many heavy mesons have been discovered. The distinction between baryons and mesons now is that *mesons* have spin of 0, 1, 2, 3, etc., whereas *baryons* have spin of $\frac{1}{2}, \frac{3}{2}, \frac{5}{2}$, etc.

In addition to these particles, all of them have an antiparticle. Any antimatter particle will annihilate its matter particle to produce energy. When a proton annihilates an antiproton, it produces a high-energy gamma-ray photon. Luckily for us there are not many antiprotons around on the Earth. All antiparticles have the same mass and spin as their particle, but have opposite charge. All particles and antiparticles except the electron, the proton and a particle of zero mass, called the neutrino, are unstable. In general they have half-lives that can be as short as 10^{-23} s. Even the neutron is unstable when isolated: it has a half-life of 636 s.

The multiplicity of particles has led to a search being made for a sub-structure of the familiar proton, neutron and electron. This has resulted in what is at present called the *standard model* in particle physics. It relates the particles themselves to the forces that they experience.

The four fundamental forces

Associated with mass is the idea of gravitational field. A gravitational field is defined as a region in which a mass experiences a force. A parallel situation occurs with charge. By definition, a charge experiences a force when it is in an electromagnetic field. These two forces, gravity and electromagnetic force, are fundamental forces and are the only two forces that exist outside of the nucleus of an atom.

We already know that within a nucleus there must be some other force or forces acting because the gravitational attraction of one proton for another in the nucleus is far smaller than the electromagnetic force of repulsion between protons. If only electromagnetic and gravitational forces existed, then a nucleus would be impossible. Suggesting *one* other force acting within the nucleus is unable to explain many features of the nucleus, but if *two* types of force act then a theory can be built up that is in agreement with experimental findings. These two forces are called the **weak force** and the **strong force**. They do not obey an inverse square law with distance but a law that makes them decrease very rapidly with distance so that they both act over only extremely small distances. The strong force acts only over a range of about 10^{-15} m and the weak force over about a thousandth of this distance.

Associated with these forces are particles that are affected by them. The particle that is affected by the strong force is the **quark**. Unfortunately it is not quite so easy to ascribe a specific particle to the weak force. The weak force is felt by both quarks and leptons, and is able to convert particles from one type to another.

Physicists are trying to rationalise these four forces further. It is at present an unfulfilled dream to provide a unified field theory covering all four forces.

Einstein tried, but did not succeed, and many physicists are still trying. There is some evidence that just after the Big Bang there was only one kind of force, but that somehow it divided into several parts to leave us with the four fundamental forces that we now have. A unification of the weak force with the electromagnetic force has been successfully proposed.

A solution

Certain particles do seem to be fundamental. Using very high-energy beams of particles to probe their interior, electrons and other leptons have proved to be devoid of any sub-structure to within 10^{-18} m of their centre. As a result of this type of experiment, Gell-Mann and Zweig put forward a quark theory, which has now been developed into the standard model. This stipulates that there are 12 fundamental particles and 12 corresponding antiparticles. They are listed in Table 30.1.

The antiquarks are indicated in writing as, for example, $\bar{u}$, and spoken as 'u-bar'. One speaks and writes of antineutrinos, and the antielectron is the positron, which was the first antiparticle to be discovered, in 1933, having been first predicted mathematically by Dirac.

There is no intention here to write a detailed analysis of how these particles can be combined to produce all known particles and reactions. As you can imagine, the task is a complex one, but it is appropriate to show how quarks combine to produce protons and neutrons. **Fig 30.10(a)** is not in any way intended to be a photograph of the inside of a proton; it is just a visual representation of the particles it contains. **Fig 30.10(b)** is the corresponding diagram for a neutron. You can calculate that the total charge on the proton is e and that on the neutron is zero.

This model of fundamental structure satisfies many of the requirements of experimental physics. There is however a real problem: no quark has ever been found experimentally. It is suggested that they are always confined within a baryon or a meson. Another way of putting this is to say that their half-life is so short (less than 10^{-25} s) when outside the nucleus that there is no instrument at present able to detect them.

TABLE 30.1 The fundamental particles*

QUARKS NAME	SYMBOL	CHARGE
up	u	$2e/3$
down	d	$-e/3$
strange	s	$-e/3$
charm	c	$2e/3$
bottom	b	$-e/3$
top	t	$2e/3$

LEPTONS NAME	CHARGE
electron	$-e$
electron-neutrino	0
muon	$-e$
muon-neutrino	0
tauon	$-e$
tauon-neutrino	0

* Plus 12 antiparticles that have the same mass but opposite charge.

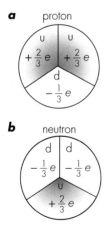

FIG 30.10 Representations of **a** a proton and **b** a neutron, showing the quarks that they contain.

QUESTIONS

30.9 Calculate the energy released when an electron annihilates a positron.

30.10 Baryons are always composed of three quarks. Antibaryons are composed of three antiquarks. Mesons are always a quark and an antiquark.

   a How many different baryons are there made of only up, down and strange quarks?

   b What is the structure of an antiproton and an antineutron?

   c What is the charge on a pion (π meson), which is composed of one u quark and one $\bar{d}$ antiquark?

   d What is the structure of, and the charge on, an antipion?

30.5 THE BIG BANG

Speculation about the formation of the observable Universe is not a new activity. In the early history of mankind, stories abound of the creation of the Earth – the sea and the sky and all the people, plants and animals on it. As astronomy has developed as a science, larger and more sophisticated telescopes and other instruments have increased our knowledge of the extent of the Universe, and many early theories concerning the formation of the Universe have been abandoned. This has only encouraged scientists to put forward new theories, which in most cases have been tested and found wanting.

The current approach to this problem begins by working backwards in time from the present. When this is done, the Universe gets smaller and smaller the further back in time one goes. This implies that at some stage it must have had an enormous density and temperatures reaching values of the order of 10^{13} K. Matter under these conditions is nothing like matter as we know it. The temperature and pressure are so great that we do not know whether the normal laws of physics apply, but we know enough to be certain that protons and neutrons would not have existed. This is where particle physics comes in. At the time when the temperature is hot enough, only fundamental particles and photons can have existed. Theory for further back in time than that has been worked out, and it is

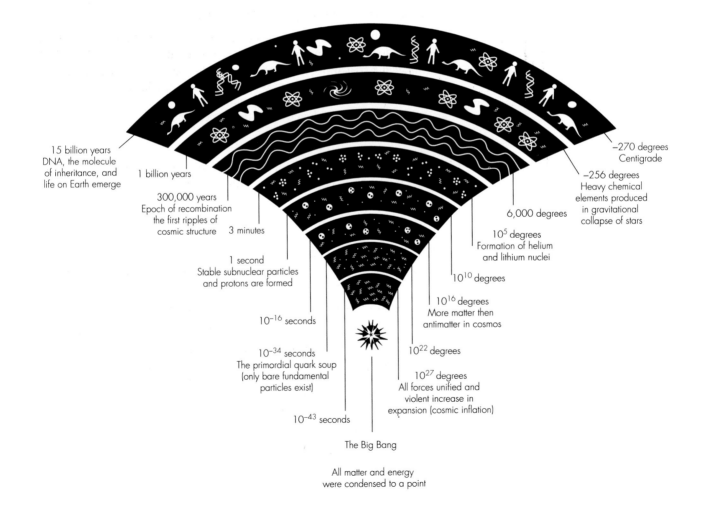

15 billion years
DNA, the molecule of inheritance, and life on Earth emerge

1 billion years

300,000 years
Epoch of recombination the first ripples of cosmic structure

3 minutes

1 second
Stable subnuclear particles and protons are formed

10^{-16} seconds

10^{-34} seconds
The primordial quark soup (only bare fundamental particles exist)

10^{-43} seconds

The Big Bang

All matter and energy were condensed to a point

10^{27} degrees
All forces unified and violent increase in expansion (cosmic inflation)

10^{22} degrees

10^{16} degrees
More matter then antimatter in cosmos

10^{10} degrees

10^5 degrees
Formation of helium and lithium nuclei

6,000 degrees

−256 degrees
Heavy chemical elements produced in gravitational collapse of stars

−270 degrees
Centigrade

FIG 30.11 A proposed time scale for the formation of the Universe (from *The Independent*, 1992).

believed that it is reliable to within a time less than a picosecond (10^{-12} s, a micro-microsecond) of the explosion called the Big Bang.

Now, calculations can be made with time going forward. If you start with a mix of radiation energy and fundamental particles, it is possible to put a time scale on to the formation of different constituent parts of matter as we know it. The time scale was shown on a chart in *The Independent* newspaper in 1992, and it is shown in **Fig 30.11**. The development followed this pattern. Treat the figures as orders of magnitude only.

Fig 30.11 gives only a very brief account of what is now a very full theoretical theory. There are certain features of the Big Bang theory that enable it to be tested practically, and this practical confirmation makes it more than just a theory. The following experimental evidence, some of which is rather sketchy, supports the theory:

■ The number of protons to neutrons in the Universe is very close to the 4:1 ratio predicted by the theory.
■ The expansion of the Universe is as predicted.
■ Particle accelerator collisions can reproduce conditions in which the energy of the collision is about the same as would have occurred at temperatures of 10^{15} K. These experiments help to support the theory.
■ The background microwave radiation from space is that which would come from molecules at a temperature of 2.5 K – just the temperature that the theory predicts.
■ The abundances of the light elements – hydrogen (and deuterium), helium and lithium – in the Universe agree well with the theory.

Present and future research will in many respects reinforce the theory, but it will probably also produce evidence that will seem to contradict it. There is no doubt that the formation of the Universe will continue to fascinate many people, both the researchers in the field and others, whatever theory is fashionable.

SUMMARY

■ The Doppler effect gives rise to an apparent increase in the wavelength of waves travelling from a wave source moving away from an observer, given by

$$\Delta\lambda = \frac{\lambda u}{c}$$

For light from a receding star, this gives a red shift.

■ Particle accelerators use electric and magnetic fields to control the path of, and to accelerate, charged particles.

■ Particles are categorised into leptons, mesons and baryons.

■ Twelve particles and their antiparticles are thought to be fundamental.

■ Working forward from the Big Bang, there is now fairly good agreement between the predictions of the theory and the observable Universe.

ASSESSMENT QUESTIONS

30.11 a i State the conditions in the Sun that allow nuclear fusion to take place.

  ii Explain why these conditions are necessary.

b The total mass of the protons in a star at its formation was 2.0×10^{30} kg. Energy is produced in the star by the fusion of protons. The net outcome of the fusion process is that 6.4 MeV are produced for each proton in the star. Assume that the star radiates energy at a constant rate of 3.9×10^{26} W during its lifetime.

  i Calculate the number of protons in the star at its formation.

  ii Estimate the lifetime of the star.

OCR 1999

30.12 A large asteroid recently passed the Earth at a distance similar to that of the distance between the Moon and Earth. A radar system was used to track its path. The frequency of the radar signal was 1.1×10^{10} Hz.

a At a certain point in the path of the asteroid, the time between the transmission of a radio pulse and the return of its echo was 2.9 s. Calculate the distance of the asteroid from Earth.

b The increase in frequency between the outgoing and returning radio signal was 6.6×10^{5} Hz.

  i Why did the return signal differ in frequency from the transmitted signal?

  ii Calculate the speed of the asteroid relative to the Earth at this point.

  iii Deduce whether the asteroid was moving away from or towards the Earth.

c In an attempt to estimate the radius of the asteroid, individual radar echoes were studied. The time between receiving the first part and last part of the echo of a particular pulse was 1.5 µs.

  i Assuming that the asteroid was spherical and that the duration of the outgoing pulse was much shorter than 1.5 µs, explain how this information leads to a value for the radius of the asteroid.

  ii Calculate the radius of the asteroid.

NEAB 1998

30.13 a Describe and interpret Hubble's redshift observations. In your account, you should describe what is meant by redshift, how it can be determined and the conclusions which can be reached.

b Using Newton's law of gravitation and making clear the other physical concepts involved, derive the expression

$$\rho_0 = \frac{3H_0{}^2}{8\pi G}$$

Explain clearly what each symbol in this expression represents and why it is important to continue trying to measure H_0 with improved accuracy.

UCLES 1997

30.14 a Hubble's law relates the speed of recession of a galaxy to its distance from Earth. Show how the maximum distance to the edge of the Universe may be obtained from this law.

b The Hubble constant is assumed to be between 50 km s^{-1} Mpc^{-1} and 100 km s^{-1} Mpc^{-1}. Hence calculate the maximum age of the Universe.*

NEAB 1998

30.15 a i Describe how the motion of galaxies varies with their distance from Earth.

  ii Hence explain why astronomers will never be able to record the presence of galaxies beyond a certain distance from Earth.

b Calculate the distance to a given galaxy if the speed of recession of the galaxy is 3.7×10^{7} m s^{-1}.

NEAB 1997

30.16 The diagram below represents an outline of a cyclotron used to accelerate protons to 10 MeV. The protons enter the cyclotron at the centre and spiral outwards.

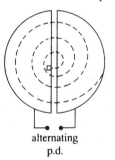

alternating
p.d.

a Explain why the speed of a proton increases each time it moves from one half of the cyclotron into the other half.

b The maximum radius of curvature of the path of the protons inside the cyclotron is equal to the radius, R, of the cyclotron. Show that the maximum speed of a proton is equal to

$$\frac{BQR}{m}$$

where B is the magnetic flux density of the magnetic field, Q is the charge of the proton and m is its mass. Neglect relativistic effects.

c Calculate the maximum speed of protons in a 1.20 m diameter cyclotron when the magnetic flux density is equal to 0.50 T. Hence determine the frequency of the alternating p.d. which must be applied to achieve this speed.

NEAB 1996

---

*1 Mpc = 1 megaparsec
1 parsec = 3,0857 × 10^{16} m

30.17 Below is a plan view of part of a cyclotron.

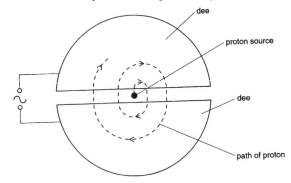

An alternating potential difference is connected between the dees. When a magnetic field is applied, the path of the protons is as shown.

a State the direction in which the magnetic field is applied. Explain your reasoning.

b Explain why
  i a potential difference between the dees is required,
  ii the potential difference must be alternating.

c Show that the time for the proton to travel a complete semicircle within one dee is independent of the speed of the proton and of the radius of its path.

OCR 1998

30.18 a State *two* properties of the strong force between nucleons.

b i Sketch a curve to show how the force between a proton and a neutron varies with separation for distances less than 5 fm. Give an approximate scale on the force axis. Label this curve 'n-p interaction'.

  ii On the same axes, sketch a curve to show how the force between two protons varies with separation. Label this curve 'p-p interaction'.

NEAB 1996

30.19 a State the quark composition of
  i a proton,
  ii a K^+ meson (i.e. a positive kaon).

b A meson consists of a quark and an antiquark.
  i State the charge, strangeness number and identity of the meson composed of an up antiquark, $\bar{u}$, and a down quark, d.
  ii In a collision with a proton, the up antiquark of the meson in **i** annihilates an up quark of the proton to release a high energy gamma photon. The remaining quarks form a baryon b, as shown in the equation.

$$(\bar{u} + d) + p \rightarrow \gamma + b$$

Identify the baryon formed and state its quark composition.

NEAB 1997

30.20 a i State the difference between a hadron and a lepton in terms of the type of force experienced by each particle.

  ii Give *one* example of a hadron and *one* example of a lepton.

  iii Hadrons are classified as either baryons or mesons. In terms of quark composition, explain the difference between a baryon and a meson.

b State the quark composition of a neutron.

NEAB 1996

30.21 A beta particle is emitted from an unstable nucleus when a neutron in the nucleus changes into a proton.

a i Complete the equation below which represents this process.

$$_0^1 n \rightarrow {}_1^1 p +$$

  ii State which particles in the process are baryons and which particles are leptons.

  iii What fundamental interaction or force has been involved in this process?

b i Describe the process given in part **a** in terms of changes in the quark composition of the baryons involved.

NEAB 1997

Appendix A: Answers to Questions

The following answers have been devised by the author. They are entirely the responsibility of the author and have neither been provided nor approved by the awarding bodies. These are suggested answers and may not constitute the only possible solutions.

Edexcel Foundation (London Examinations), the Assessment and Qualifications Alliance (covering AQA(AEB) and AQA(NEAB), Oxford, Cambridge and RSA Examinations Board, Northern Ireland Council for the Curriculum Examinations and Assessment, and the Welsh Joint Education Committee bear no responsibility for the example answers to questions taken from their past papers which are contained in this publication.

Answers to questions

In many of the examination questions physical constants are required in order to answer the question. Different examining boards have different policies regarding the supply of this information. In some cases the required data is given in, or at the end of, the question however some boards supply data booklets containing the information. If therefore you find it necessary to use additional information to that given in the question you'll find the required data in the appendix.

In these answers you will often find more significant figures quoted than are strictly justified. Advice is often given not to give answers to more significant figures than are stated in the question but if this is done with multi-answer questions then inaccuracies can be introduced by rounding up or down too soon. Also the numerical value of the answer affects the number of significant figures that should be given. 1 in 99 is a much smaller uncertainty than 1 in 11 although both 99 and 11 have two significant figures.

Here, therefore, I have attempted to give answers so that any unreliability is limited to the final digit.

Section questions

Chapter 1

1.1 s^{-1}, $kg\,m\,s^{-2}$, $kg\,m^{-1}\,s^{-2}$, $kg\,m^2\,s^{-2}$, $kg\,m^2\,s^{-3}$, $kg\,m^2\,s^{-3}\,A^{-1}$, $kg\,m^2\,s^{-3}\,A^{-2}$, $kg^{-1}\,m^{-2}\,s^4\,A^2$

1.3 (a) correct (b) incorrect (c) correct (d) incorrect (you do not need to know what moment of inertia is in order to tell that this equation is incorrect. Why not?) (e) correct

1.5 $kg^{-1}\,m^{-3}\,s^4\,A^2$

1.6 Mass is in joule per zipp^2, length is in joule per newton and time is in joule per newton zipp

1.8 Length, temperature interval (b) $kg\,m^2\,s^{-3}\,A^{-1}$ (c) vectors have magnitude and direction, scalars have only magnitude (d) scalar

1.9 Kilogram, coulomb and joule are units, the others are quantities:
(a) potential difference = energy/charge (or power/current)
(b) $kg\,m^2\,s^{-3}\,A^{-1}$ (c) homogenous equation since both sides have the base unit $kg\,m^2\,s^{-3}$

1.10 (a) velocity (b) current (c) power (d) force

1.11 m^2

1.12 $m^3\,kg^{-1}\,s^{-2}$

1.13 (a)(i) viscosity of fluid (ii) m^3, s, m^2, m (iii) $kg\,m^{-1}\,s^{-1}$
(b) $9700\,kg\,m^{-1}\,s^{-2}$ (Pa)

1.14 (a)(i) incorrect (ii) correct (iii) incorrect (iv) correct
(b) e.g. (i) 800 N (ii) 2000 W (iii) 400 000 J (iv) 500 Ω
(v) 500 nm

1.15 (a) kg s^{-1} (b) 49 cm s^{-1}

1.16 (b) 3.00×10^8 m s^{-1} (c) This is not a coincidence but indicates that light travels at this speed because it is part of the electromagnetic spectrum

Chapter 2

2.1 (a) 0.0013, 0.13% (b) 0.002, 0.2% (c) 0.0008, 0.08%

2.2 (a) +0.04% (b) +1.3% (c) −0.45% (d) +2%
(e) + 0.07%

2.3 636 mm, 3, so the final figure is the doubtful one, 635.7 mm has two doubtful figures

2.4 (68.9 ± 1.4) s, fractional uncertainty 0.02
This should perhaps be quoted as (69 ± 2) s

2.5 (a) 3.463 mm (b) 0.303 mm (c) sum 0.003, difference 0.03

2.6 (9.73 ± 0.13) m s^{-2} OR (9.7 ± 0.2) m s^{-2}

2.7 $(1.5 \pm 0.3) \times 10^{-3}$ N s m^{-2} Note that all lengths must be in metres and that 1 cm^3 = 10^{-6} m^3.

2.9 (a) (3.75 ± 0.05) m s^{-2} (b) 6.6 m s^{-1} (c) 18.6 m s^{-1}

2.10 Plot z^2 on the y-axis against f^2 on the x-axis
$R = (900 \pm 10)$ Ω, $L = (0.380 \pm 0.005)$(H)

2.11 Plot lg f on the y-axis against lg d on the x-axis. The gradient gives $n = -3$ so law is $F = k/d^3$

2.12 (a) 270 mm (b) 71 N m^{-1} (c) 2.43 N

2.13 (a) Use

$$T^2 = \frac{4\pi^2 l}{g} + \frac{4\pi^2 r}{g}$$

so plot T^2 against l. Gradient = 3.97, $g = 9.94$ m s^{-2}

(b) No – it affects the intercept on the graph but not the gradient

2.14 (b) Three forces, as the stand is touching two objects: a support force from the bench, a support force from the balance and its weight.
(c)(ii) Plot F against $1/x$ gradient = 3.44 giving $W = 45.3$ N and so $m_c = 4.6$ kg
(d)(ii) $d_0 = 40.4$ cm $a = 0.353$

Chapter 3

3.2 5.8 m s^{-1} in a direction S 31° W

3.3 70 m s^{-1} in the new direction

3.4 75.5 m s^{-1} from a direction N 4° W

3.5 (a) 2.6×10^{-6} rad ($= 1.5 \times 10^{-4}$ °)

(b) and (c) 2.6×10^{-3} m s^{-1} towards the centre of the Moon's orbit, i.e. towards the centre of the Earth

3.6 (a) 1080 J (b) zero (c) 830 J (d) −1080 J

3.7 (a) 0.023 N (b) zero (c) 0.0097 N

3.8 (a) Speed and kinetic energy are scalars, momentum is a vector
(b)(i) 68.7 N if P is vertical, 59.5 N if it is at right angles to bonnet
(ii) 68.7 N if P is vertical, 34.3 N if at right angles to bonnet

3.9 (b)(i) 25.6° (ii) 45.2 kN

3.10 (a) 6.9 N vertically upwards (b) 6.9 N

3.11 (ii) zero

3.12 (b)(i) 1713 J (ii) 165 J (iii) 15.5 m s^{-1} (iv) commences at 14 m s^{-1}, falls to 10 m s^{-1} and rises to 15.5 m s^{-1} at end

3.13 (a)(ii) compressive 157 N, shear (sideways, i.e. at right angles to compressive) 188 N
(b)(i) 322 N on 3.0×10^{-4} m^2 = 1.07×10^6 Pa

Chapter 4

4.1 (a) 31 m s^{-1} (b) 16 m s^{-1} (the ball really does move off at almost twice the speed of the club)

4.3 6.4 m s^{-2}

4.4 576 m

4.5 1.9 m s^{-2}

4.6 (a) 49 m s^{-1} in opposite direction
(b) 0.28 m in opposite direction

4.8 50 s, both travel 2000 m

4.9 1.43 s, 14 m s^{-1}

4.10 41 m

4.11 20 m s^{-1}

4.12 (a) 2.04 s (b) 4.08 s (c) 20.4 m (d) 91 m

4.13 (a) 68 m (b) 15.8 m s^{-1} (c) 68 m (d) 5.3 m s^{-1} upwards
(e) 18.5° (f) 16.7 m s^{-1}

4.14 48 000 N s due east

4.15 30 N s, 25 000 m s^{-2} both in opposite direction

4.16 3.1×10^{-20} N s

4.17 (a) 3.57×10^{29} N s (b) zero

4.18 765 N s in direction S 79° W

4.20 4 m s^{-2}, 33 m

4.21 (a) 1.59 m s^{-1} (b) 1.41 m s^{-2}

4.22 (a) 1.3 m s^{-2} (b) 9.3 m

4.23 (b)(i) 39.6 m s^{-1} (ii) 4.04 s (iii) 320 J

4.24 (a) 4.05 m s^{-1} (b)(i) 1.35 m s^{-1} (ii) 6.75 m s^{-1} upwards

4.25 (b) 0.158 m s^{-2}

4.26 (b) gradient of straight region = 0.17 (m s^{-1})

4.27 (a)(ii) 78 m s^{-1} at 59° to the vertical (b)(i) 2400 N

4.28 (a)(iii) 7.3(m s^{-1}) (iv) 10.4 m s^{-2}

4.29 (a)(i) 638 N (ii) 520 N (b) 8.0 m s^{-2}

4.30 (a) 2.07 s (b) 1.25 m

4.31 (a) 6.26 m s^{-1} (c) 16 N

4.32 (b)(i) 0.418 s (ii) 0.836 s (iii) 3.43 m
(c) range is the same for both and is 2.97 m (d) 45°

4.33 (a) 24.1 J (b) 24.5 m s^{-1}

Chapter 5

5.1 All three diagrams are the same with the tension in the string exactly equal and opposite to the weight of the bucket.

5.2 132 000 N

5.3 3.07×10^{17} m s^{-2}

5.4 61 000 N

5.5 (a) 1.5 m s^{-2} (b) 45 000 N (c) 55 000 N (d) 20 500 N

5.6 1.47 m s^{-2} downwards, 2.6 s

5.7 (a) 1800 N (b) 2.7

5.8 55 N

5.9 (c) 1.11 m s^{-2} upwards, 273 N downwards

5.10 (a) 0.61 m s^{-2} (b) 4900 N forwards
(c) 5960 N forwards

5.12 0.90 m s^{-2} forwards

5.16 (a) 0.296 m s^{-2} (b) 7.7 m s^{-1} (c) 41 400 N (d) 159 kW

5.17 (a) 9600 N (b) 11 800 N
(c) 15 200 N at an angle backwards from the vertical of 39°

5.18 (a) 6.15 m s^{-1} (b) 0.72 s (c) 2.72 s (d) 36 kW (e) 72 kJ

5.19 (a)(i) 1.66 m s^{-2} (ii)1330 N (iii) 400 m (iv) 13.3 m s^{-1}
(b)(i) drag = driving force (ii) 6.0 kW (c) 10.5 kW

Chapter 6

6.1 (a) 75 N (b) 350 N (c) 81 N (d) ≈ 100 N (e) 800 N

6.2 0.030 N m

6.3 $P = 1.4 \times 10^5$ N, support at B is zero, at
C = 5.6×10^5 N

6.4 3.6 N m anticlockwise

6.5 Force at A = 4.14×10^5 N, force at B = 4.36×10^5 N

6.6 6.0×10^7 N m anticlockwise

6.8 17.5 m

6.9 (b)(ii) 21 200 N

6.10 (a)(i) 200 N (ii) 300 N

6.11 (b) at A 113 N, at B 132 N

6.12 (a)(i) and (ii) down at A end and = 675 N
up at B end and = 1375 N
(b) 3.5 m from A

6.13 (b) 200 N

6.14 (b)(i) 30 mm (ii) 150 kPa (c) 20 mm

6.15 29.4 N

6.16 (b) 900 N (c) 7200 N

6.17 (b)(ii) at right hand edge of base
(iii) 0.10 N m, 0.70 N m, 3.0 N m (iv) 38 N

6.18 (a)(ii) 2700 N (iii) 2810 N

6.19 (a) 8.4 N m clockwise (b) 1.8 N m anticlockwise

6.20 (b) 100.44 g

Chapter 7

7.1 2400 N

7.2 7200 N

7.3 (a) 66 m s^{-1} (b) 3.1 N s

7.4 average total force = 8200 N, change in velocity = 29.4 m s^{-1}

7.5 (a) 2.14×10^{-20} N s (b) 2.14×10^{-12} N

7.6 5.7 m s^{-1} in direction van was travelling

7.7 0.63 m s^{-1}

7.8 aircraft slows down by 0.004 m s^{-1}

7.9 6500 J

7.10 (a) 2.94×10^6 J (b) because it will be counterbalanced by a weight which falls

7.11 6.9×10^4 J

7.12 1.6×10^{-15} J

7.13 10 J

7.14 0.027 J

7.15 12.5 J

7.16 2200 J

7.17 97 J

7.18   93 kW

7.19   (a) 40.8 kW (b) $P = F \times v$

7.20   (a) 6250 m (b) 375 MJ (c) 375 MJ

7.21   (a) 15 m s^{-1} (b) 43.4 m s^{-1}

7.22   16 000 N

7.23   8.1 N

7.24   (a) 603 J (b) 5.0 m s^{-1}

7.25   30.7 m

7.26   (a) 0.96 J (b) 3.6 m s^{-1}

7.27   (a) 7.6×10^{12} J (b) 166 MW (c) power is often produced at inconvenient times of the day, all the water would have to be released instantaneously at low water

7.28   3580 m s^{-1}

7.29 $1.5v$

7.30   (a) 0.156 m s^{-1} (b) 0.055 J

7.31   (a) 1.42 m s^{-1} (b) 60

7.32   of remaining nucleus $4E/226$, of alpha particle $222E/226$

7.33   (b)(i) 1.13 m s^{-1}

7.34   (a) 1080 W (b) m s^{-1} (c) 2.0 m s^{-1}

7.35   0.175 m

7.36   (b) 20.4 m (c) 27 J (d) 9.8 m s^{-2}

7.37   (a)(i) 4.9 m s^{-1} (ii) 0.097 N s (b)(i) 0.19 N s upwards (ii) 21 N (c) not elastic

7.38   (a) to the left (b)(i) $3u$ (ii) $t_2 = 3t_1$ (c) $v = 3u/2$ (d) an external force acted on the system when A hit the spring

7.39   (a)(i) 1.6 N s (ii) 54 N (b) 19 m

7.40   (a)(i) 15 kW (ii) 5.5 kW (b)(i) 460 W (ii) 630 N m

7.41   (a)(i) 320 N s (ii) 128 J (b)(ii) forwards (iii) 3.0 s (iv) 53 N (c) 160 J, 53 W

7.42   (b) 40 N (c)(ii) 288 W (d) 363 W

7.43   (c)(i) $u/2$ (ii) $mu^2/4$

7.44   (a) 25 m (b) 18 m s^{-2} (c) 810 kJ

7.45   (a) greater than weight (b) force of mallet on board, force of iron bar on board (c) 9.0 N s (d) 1.27 m s^{-1} (e) the weight of the block acts on the block as well as the force of the board on the block

7.46   (a) k.e. before = k.e. after = 132 kJ (b) k.e. now 99 kJ – some p.e. in springs (d) 66 000 N s

7.48   (a) 1.5 m s^{-1} (b) 47 kJ (c) 5900 W (d) 86%

7.49   (a)(i) 5.0 m s^{-2} (ii) 6000 N (iii) 540 kJ (b)(i) 36 kW (ii) 302 pence

7.50   (a)(ii) (1) 2.0 m s^{-2} (2) 0.96 N (3) 0.32 N upwards (b)(i) 2600 N (ii) (1) 2200 N (2) 44 kW (c)(ii) (1) 0.20 m s^{-1} (2) 9600 N s

7.51   (a)(i) 2000 N (ii) (1) 3700 N (2) 74 kW

Chapter 8

8.1   (a) 4π (b) $\frac{3}{4}\pi$ (c) $\frac{5}{2}\pi$ (d) 100π (e) $\frac{1}{18}\pi$ (f) 48π

8.2   0.105 rad s^{-1}, 5.2×10^{-3} m s^{-1}
1.745×10^{-3} rad s^{-1}, 8.7×10^{-5} m s^{-1}
1.454×10^{-4} rad s^{-1}, 5.8×10^{-6} m s^{-1}

8.3   (a) 1.99×10^{-7} rad s^{-1} (b) 7.27×10^{-5} rad s^{-1} (c) 2.6×10^{-6} rad s^{-1} (d) 2.6×10^{-6} rad s^{-1}

8.4 5.95×10^{-3} m s^{-2}

8.5   (a) 367 rad s^{-1} (b) 84 m s^{-1} (c) 3.1×10^4 m s^{-2}

8.6   (a) 1.18×10^4 m s^{-2} (b) proportional to r

8.7   340 rad s^{-1}

8.8   (b) 4.46 N

8.9 S_B = 1073 N upwards, S_T = 161 N upwards – so person must be held up either by a seat belt or by the cab not being upside down

8.10   (a) two forces, both downwards (b) 924 N

8.11   (a) 27 m s^{-2} (b) 2760 N

8.12   36°

8.13   18.3 cm

8.14   (d)(i) 13.3 m s^{-1} (ii) 6.6 m s^{-1} (iv) 75.6 N

8.15   (a)(i) 130 rad s^{-1} (ii) 950 m s^{-1} (iii) 0.095 N (b)(i) 650 rad s^{-1}

8.16   (a) 31 rad s^{-1} (c)(i) 0.62 m s^{-1} (ii) zero

8.18   (a)(i) 2.87 m s^{-1} (ii) 180 rad s^{-1} (iii) 5.2 m s^{-2} towards centre

8.19   (b)(iii) 36°

8.20   (a) 10 N m^{-1} (b) 0.67 m s^{-1} (c) force is 0.63 N

Chapter 9

9.1   (a) 1.0×10^{-42} N (b) 9.8 N (c) 3.6×10^{22} N

9.2   (a) 6×10^{-5} N (b) 20 N (c) 3×10^{-6} rad (= 1.7×10^{-4} degrees) (d) $M_e = mR^2 / x^2 \tan \theta$

9.3 3.4×10^8 m from the centre of the Earth

9.4   9.8 J kg^{-1}

9.5   (a) -6.26×10^7 J kg^{-1} (b) 1.28×10^7 m (c) 6.26×10^7 J kg^{-1}

14.9 6900 Ω in dark, 113 Ω in the light

14.10 (a)(i) ≈1000 Ω (ii) 100 Ω (iii) 20 Ω (iv) 250 kΩ
(b) 0.73 V

14.11 (a)(i) 301 Ω (ii) 297 Ω (iii) 293 Ω (iv) 179 Ω (b) when
V is small V^2 is very small and $R \approx 1/b$ (c) 130 Ω

14.12 (a) 2.8×10^{-8} Ω m (b) aluminium

14.13 (a) regard the coil as 190 circular turns (b) 15 Ω

14.14 (a) 8.4 V (b) 0.75 A

14.15 Answer in terms of the potential difference across
the internal resistance, not in terms of the motor
taking so much current that there is not enough for
the lights.

14.16 21.1 V, 0.56 Ω

14.17 5.2 V, 1.76 Ω

14.18 (a) 12 A (b) 2880 W (c) 4900 s

14.19 (a) 12.1 MJ (b) 2.4 Ω (c) 240 W

14.20 (a) 2.8 kW

14.21 £181.86

14.22 (a)(i) 320 C (ii) 2.00×10^{21} (b)(ii) 6.17×10^{-5} m s⁻¹

14.23 (c)(i) 1.02 V, 1.22 W (c)(ii) 7.53 m, 1.41 W,
4.4×10^{-4} m s⁻¹

14.24 (a) 2.4×10^{-2} m s⁻¹

14.25 (a) 529 Ω

14.26 (a) 240 V (b)(i) 4.0 A (ii) 2400 C (c) 5.5×10^{-8} m²

14.27 (a) 2120 Ω, 353 Ω (b)(i) the 25 W bulb

14.28 (b)(i) 1.4 Ω, 5.0 Ω (d)(i) 40 Ω (ii) 11 m
(iii) series 500 W, parallel 2000 W

14.29 (b) and (c) 155 Ω resistor

14.30 (a)(ii) 80 Ω

14.32 (a) 0.413 Ω (b) 0.409 A

14.33 (b) 0.0114 W (d)(i) 180 kC (ii) 1710 s

14.35 (a)(i) 450 C (ii) 9.0 V (iii) 7.2 V (iv) 4.0 Ω

14.36 (a)(i) 1.18 MW (ii) 7.5 MW; 3.8×10^8 kg h⁻¹
(b) 2.5 p per kWh

Chapter 15

15.1 a = 3.1 A, b = 3.2 A, c = 1.1 A, d = 4.3 A

15.2 170 μA, 64.08 mA

15.3 9 Ω; 1.0 W, 0.75 W, 2.25 W and 0.50 W to the internal
resistance

15.4 (a) 5.39 V (b) 0.61 V (c) 5.0 V (d) 0.33 mW, 100 mW
(e) 120 mW (f) 20 mW

15.5 (a) 1.5 A (b) C; -1.2 V, D; -7.0 V, E; 8.0 V (c) 11 V
(d) 1.6 Ω (e) 52.5 W, 16.5 W (f) 24.5 W, 2.7 W, 1.8 W,
40 W

15.6 (a) 80 kΩ (b) 20 kΩ (c) and (d) 36 Ω

15.7 42 kΩ in parallel (e.g. 39 kΩ + 1.8 kW + 1.2 kΩ)

15.8 If the fine control resistor has too little resistance its
effect is too small, so make the resistance of the fine
control resistor about a tenth that of the coarse
control resistance.

15.9 (a) 8.98 V (b) 0.69 V (c) 860 Ω

15.10 (a) 0.61 V (b) 95 kΩ

15.11 (a) 0.0375 Ω (b) 0.225 V

15.12 (a) 119 920 Ω (b) 1120 Ω (c) 40 Ω

15.13 (a)(i) 410 kJ (b)(i) 10 A

15.14 (b)(i) 6.7 V (ii) 3.0 V (iii) 0.30 A (iv) 6.7 V

15.15 (a)(i) 10 V (ii) 12 Ω (iii) 8.3 W (b)(i) 1.2 Ω (ii) 8.8 W
(iii) I = 1.36 A I_1 = I_2 = 0.68 A

15.16 (b) 1730 kJ (c)(i) 57.6 Ω (ii) 918 W (d)(i) two in
parallel, (ii) two in parallel with one in series,
(iii) two in series with one in parallel

15.17 (a)(i) 0.80 A (ii) 3.2 W (b) 0.80 A

15.18 (a)(i) 4.0 Ω (ii) 6.0 V (iii) 0.75 A (b)(i) 4.67 Ω
(ii) 0.86 A

15.20 (b)(i) 37.7 Ω (ii) 5.28 V

15.21 (b)(i) 4.0 Ω (ii) 8.0 Ω (iii) 3.0 Ω (iv) 1.0 A (c)(i) 4.8 kV
(ii) 4.6 kV

15.22 (c) 4.9 Ω

15.23 (b)(ii) 0.12 A

Chapter 16

16.1 (a) 9000 N (b) e.g. a horse (c) virtually impossible on
small objects

16.2 (a) 230 N (b) 1.37×10^{29} m s⁻²

16.3 5.8×10^{-9} N

16.4 (c) 1.37×10^6 C (d) 2.66×10^{-9} C m⁻² (Near a
conductor the value of the charge per unit area
always equals $E\epsilon_o$)

16.5 (a) zero (b) 1.6×10^{-23} N m

16.6 (a) +1400 V (b) (−)2.24 × 10⁻¹⁶ J (c) +2.24 × 10⁻¹⁶ J
(d) because of the negative charge on the electron
(e) 2.2×10^7 m s⁻¹

16.7 potential at A = −1300 V, at B = +2200 V, at C =
−1300 V, at D = −4800 V and at E = −1300 V

16.11 1.73×10^7 V m⁻¹

16.14 (a)(ii) 2.0×10^4 N C^{-1}, 0.12 N towards O'
(c)(i) 3.0 mm from $+ 1.0$ μC charge
(ii) 1.5×10^9 N C^{-1}

16.15 (b)(i) -120 kV (ii) -240 kV (iv) 400 kV m^{-1}
(v) 6.4×10^{-14} N

16.17 (c) 1500 V

Chapter 17

17.1 (a) 0.024 μF (b) capacitance is doubled

17.2 (a) 198 μC; 423 μC; 621 μC (b) 69 μF

17.3 (a) 0.2 μC on 0.010 μF, 1.0 μC on 0.050 μF, 1.2 μC on 0.030 μF (b) 60 V (c) 0.020 μF

17.4 (a) 6.0 mC, 18 mJ (b) 12.0 mC, 72 mJ

17.5 92.3 V

17.6 (a) 2.0 mC (b) 0.165 mC on each (c) 20 mJ by 100 μF, 1.36 mJ by 10 μF, 0.29 mJ by 47 μF (d) 20 V across 100 μF, 16.5 V across 10 μF, 3.5 V across 47 μF

17.7 maximum charge is 90 μC

17.8 maximum current (at $t = 0$) = 4.5 mA

17.9 (a) 4.5 mA (b) 90 μC (c) 20 ms (d) 20 ms (e) second (f) 0.62 after 1 time constant, 0.993 after 5 time constants

17.10 (a) 36.8% (b) 13.5% (c) 5.0% (d) 1.8% (e) 0.67%

17.13 (a) 6.0 s (b) 24 s (c) 8.6 s (d) time constant $\times$ ln 2 = half-life

17.14 (a)(i) 4.2 V (ii) 1.8 V (b)(i) 1.8×10^{-3} J (ii) 3.6×10^{-3} J

17.15 (a)(i) 0.018 μC (ii) 0.108 μJ

17.16 (c) 3.2 μA (e) 225 μJ

17.18 (a)(i) 6.0 mC (ii) 90 mJ (b)(i) charge (ii) p.d. (c) 60 mJ

17.19 (a) 54 μJ (b) 20.3 μJ (c) energy is lost as internal energy (heat) in R_2

17.21 (b)(i) 167 μF (ii) 1.5 mC (iii) p.d. across 100 μF = 1.5 V, p.d. across 200 μF = 7.5 V

17.22 (c) (reading from top of table) 18 μC, 54 μJ, 9 μC; 9 μC is supplied by the source at a p.d. of 6 V so 54 μJ is done by the source. The examiners accepted 27 μJ as the energy supplied to the capacitors.

17.23 (a) 500 μC (b) 1250 μJ

17.24 (a) false (b) false (c) true (d) false

17.25 (c)(i) 1.3 W (ii) 75 Ω

17.26 (a)(i) 2.0 mC (ii) 8.9 mJ (b)(i) 4.5 W (ii) 0.99 A (c) 18 W

Chapter 18

18.1 0.188 N

18.2 1.77×10^{-4} T

18.3 3.36×10^{-3} N

18.4 0.39 m

18.5 4.77×10^7 C kg^{-1}

18.6 (a) 0.40 mm s^{-1} (b) 3.1×10^{-24} N (c) 7.8×10^{22} m^{-1} (d) and (e) 0.24 N

18.7 (a) 43.6 m s^{-1} (c) positive

18.8 (a) 0.030 V m^{-1} (b) 0.15 m s^{-1} (c) 1.04×10^{23} m^{-3}

18.9 1.85×10^{-3} T

18.10 1.26×10^{-2} T

18.11 (a) 1.6 mT (b) 0.40 mT

18.12 (a) 8.5×10^{-5} N m^{-1} (c) twice the frequency of the a.c.

18.13 (a)(ii) down the page, 0.080 T (b)(iii) 3.82 A (c)(ii) 2.4×10^{-5} N

18.16 (b) 1.50×10^{-6} T (c) and (d) 4.50×10^{-6} N m^{-1} away from P (e) away from Q, 4.50×10^{-6} N m^{-1}

18.17 (a) N A^{-1} m^{-1} (b)(iii) 6.0 μT

18.18 (a) 1.41×10^{-5} T

18.19 (b) to the east (c) 410 A

18.20 (a)(iii) 6.0 A (iv) angle between current and field is zero so zero magnetic force (b)(iii) gradient 12.3 (A m^{-1}), $B = 3.19 \times 10^{-3}$ T

Chapter 19

19.1 0.45 m s^{-1}

19.2 0.37 V

19.3 (a) 0.50 A, 3.0 W (b) 0.43 W for friction, 0.091 W for heat (c) 1860 r.p.m.

19.4 Idealised graph is a straight line with zero speed when current is 6.0 A and maximum speed of 3000 r.p.m. when current is 0.50 A.

19.5 (c) 0.77 V

19.6 (b)(i) Expt. 1; oscillatory, Expt. 2; zero, Expt. 3; zero (ii) needle oscillates at 2 Hz; zero (oscillations too fast for needle response)

19.7 (b)(i) 3300 m^2 (ii) 0.133 V

19.10 (b)(ii) 3.0 s

19.11 (c) 8.0 V, 60 Hz

Chapter 20

20.2 (a) 4.0 mA (b) one leads the sum by $\pi/4$ rad, the other lags the sum by $\pi/4$ rad

20.3 (a) zero (b) total current of three phases is very small

20.4 (a) 6.00 A (b) 8.49 A (c) 339 V (d) 2880 W (e) 1440 W

20.5 (a) 4.0 A (b) 4.0 A (c) 3.16 A ($\sqrt{10}$ A) (d) C

20.6 $I = 0.28$ A when $t = 0$ ms, $I = -0.22$ A when $t = 2.0$ ms, $I = -0.28$ A when $t = 2.5$ ms, $I = 0.085$ A when $t = 4.0$ ms

20.8 $X_c = 480\ \Omega$, $C = 6.6\ \mu$F

20.9 $85\ \Omega$, 0.24 A

20.10 (a) 0.49 A (b) $V_R = 4.9$ V r.m.s, $V_C = 3.5$ V r.m.s (c) X_C now very small compared with R so $I \approx 0.6$ mA

20.11 7.7 μF

20.12 250 turns

20.13 (a) 1960 A (b) 4350 A (c) 0.45 (d) 3.0 MW

20.15 (a) the right hand one (b)(i) 240 V (ii) 0.40 A (iii) 96 W (c) 5.82 W

20.16 (a)(ii) 40 (iii) 8.3 mA; 333 mA

20.17 (a) 39.6 V (b) values in table; 0.41 on top line, 0.18 on bottom line

20.18 (b)(iv) 2750 V; 50 A, 5500 W (c)(i) 100 kW (83%), (ii) 21.9 kV

20.19 (a) 50 Hz (c) 25.6 Ω

20.20 (a)(i) 90 (ii) 0.45 A (iii) 57.6 W (b)(i) 17.0 V

Chapter 21

21.1 1.0 V to 5.6 V

21.2 720 Ω

21.3 600 Ω

21.4 (b) 0.40 V (c) 5.6 V

21.6 Z is a relay in all cases. In (a) Y is a thermistor, in (b) X is an LDR, in (c) X is a touch sensor, in (d) X is a thermistor and in (e) Y is a thermistor. The other components are resistors.

21.8 (a) 0.186 V (b) $I_{in} = I_f = 21.7\ \mu$A

21.9 Gain is 30 but output cannot be greater than 5.0 V or less than -5.0 V.

21.11 (b) The output changes to $+4.99$ V.

21.12 (b) A, 0.7 V, 46.5 mA; B; 10 V, zero; C; 6 V, 20 mA

21.13 Stage 2 factor = -39.2; $V_{OUT} = -19.6 V_{IN}$

21.14 (a) 4.5 V (b)(i) 3.18 V (ii) –9.0 V (iii) off (c) when the temperature rises to make the thermistor resistance equal to 1.2 kΩ the diode will come on. Only if the thermistor's resistance decreases linearly with temperature will this happen at 71 °C.

21.15 (a) 1.09 kΩ, 36 °C (b)(ii) 630 Ω

21.16 (a)(i) inverting (ii) -50

21.18 (c)(i) 3.0 mA, 1.5 mA (ii) -3.6 V

21.19 (a) -8.75 V (b) 4.38 mA

21.20 (b)(i) 0.121 mA, 0.064 mA (ii) -8.7 V

Chapter 22

22.1 (a) 13.6 (b) 13.6 g = 0.0136 kg

22.2 0.00081 m^3 = 810 cm^3

22.3 (a) 0.375 (b) 0.402

22.4 7.1×10^{-3} m s^{-1}

22.5 (a) 2.0×10^{-11} N (b) 8.7×10^{-22} J (c) 1.67×10^{-21} (d) 0.38 nm (e) 5.2×10^{-12} N

22.6 3.6×10^{-9} m

22.7 2.28×10^{-10} m if in cubical pattern; 2.56×10^{-10} m if atoms are close packed; separation of layers = 2.09×10^{-10} m

22.8 0.30 g s^{-1}

22.9 1.0×10^9 kg s^{-1} (a million tonnes a second!)

22.10 4.5×10^5 kg

22.11 (a) 0.94 N (b) 747 N

22.12 24.7 m

22.13 (a) zero (b) Antarctic ice is resting on land but Arctic ice is largely floating

22.15 (a) weight of aircraft is unchanged (b) v greater, ρ smaller

22.16 67 m s^{-1}

22.18 (a) 100.3 kPa

22.19 (a) 0.0191 N (b) 1.43 kg m^{-3}

22.20 (a) 2.05×10^5 kg (b) 3.51 MJ (c) 1.01×10^5 N (d) 0.52 m s^{-2}

22.21 Unit cell contains the equivalent of two atoms; $d = 4.3 \times 10^{-10}$ m (diameter of one atom = 3.4×10^{-10} m)

Chapter 23

23.1 2.1×10^5 Pa; 1.4×10^7 Pa

23.2 (a) area $= 7.8 \times 10^{-7}$ m^2 (b) 3.7×10^{-4}
(c) 4.1×10^{-7} Pa (d) 32 N

23.3 (a) 32 000 N (b) 22 m s^{-2}

23.4 (a)(i) 1.14×10^{11} Pa (ii) 1.11×10^8 Pa

23.5 (a) 4.21 mm (b) 0.0364 J (c) wire now plastic

23.6 (a) 10 N

23.7 (a) 1.0, 2.6, plastic

23.8 (a) 1.25×10^7 Pa (b) 6.25×10^{-5} J (c) 2.00×10^{11} Pa
(d) 2.45×10^{-3} J

23.9 (b) 4.0×10^{10} Pa

23.12 (a)(i) 0.019 (ii) 9.4 mm (b) 102 kN

23.14 (c)(i) 3.0 cm (ii) 196 N m^{-1}

23.15 (b)(i) 0.012 mm (ii) 2.1×10^{11} Pa (iii) 8.04×10^{-6} m^2
(iv) 1600 N

23.16 (a) Y (b) X (c) X

Chapter 24

24.1 69 °C

24.2 -56.3 °C

24.3 342.8 K

24.4 (d) 774 J

24.5 (b) 89 °C

24.8 (b) Use $Q/t = kA\Delta\theta/x$ where k is the thermal
conductivity; time $= 304\,000$ s $= 84$ hours

Chapter 25

25.1 391 °C

25.2 8.24×10^{-6} m^3

25.3 314 kPa, i.e. 214 kPa above atmospheric pressure

25.4 (a) 5.65×10^{-21} J
(b) Equilibrium occurs when the average energy of
the different molecules are equal.

25.5 (a) 500 m s^{-1} (b) 290 000 m^2 s^{-2} (c) 539 m s^{-1}

25.6 6.2×10^{-21} J

25.9 (b) 9.24×10^{-5} kg

25.10 (b)(i) 0.246 mol (ii) 298 K $= 25$ °C

25.13 (a) 9.1×10^5 Pa

25.15 (a) joule (c) 2.4×10^{11} m^{-3}

25.16 (c)(i) 4.7×10^{-26} kg (ii) 508 m s^{-1} (iii) 3600 J

Chapter 26

26.1 (a) 100 J (b) 160 J (c) Increase in internal energy is the
same. Since less work is done on the gas more heat
has had to be supplied to it.

26.2

A→B	-50	-70	20
B→C	25	25	0
C→D	100	140	-40
D→A	-75	-75	0

26.3 (b)(i) $PV =$ constant at constant temperature
(ii) $V \propto T$ at constant pressure
(iii) $P \propto T$ at constant volume
(c)(i) 26 200 J (ii) 18 700 J (iii) 5200 J {not zero, see (g)}
(d)(i) 0 (ii) -7500 J (e) -5200 J (f)(i) 18 700 J
(ii) 18 700 J (g) 5200 J of work are done by the gas
(h) infinity (i) Assuming a straight line, if the final
temperature is approximately 650 K the work done
by the gas is 5100 J and the decrease in internal
energy is 5100 J also – so no heat has been supplied
or taken away.

26.4 0.375

26.5 (a) 0.090 mol

(b) A→B	1480	0	1480
B→C	1880	2630	-750
C→D	-2520	0	-2520
D→A	-845	-845	0

(d) 179 kW (f) 5.8 g s^{-1}

26.6 20.4%

26.7 55.2% at 370 °C (643 K), 68.6% at 645 °C (918 K)

26.8 (b)(i) 8.30×10^5 J (ii) 1.07×10^7 J

26.10 (b) 2.0×10^8 J (c)(i) 1 kWh $= 3.6 \times 10^6$ J (ii) £67.00

26.12 (a) $\Delta U = 0$ (b) $\Delta W = 48$ J (c) $\Delta Q = -48$ J

26.13 8400 J

26.14 (a) 552 kW (c) £68 900 (d) £103 000

26.15 (b) 52 500 J

26.17 (b)(i) 38 400 W (ii) 90 000 kg (iii) 1.0×10^{-4} °C s^{-1}
(iv) 8.2 hour

26.19 (b) 18.0 °C (c)(i) 2.4×10^5 J kg^{-1} K^{-1}
(ii) 8.4×10^5 J kg^{-1}

26.20 (b)(i) $2P_o$ (ii) $3P_o$ (iii) $1.5P_o$

(d) work W is done on the gas, $\Delta U = 0$ since
temperature is constant so heat Q is extracted from
the gas. $W + Q = 0$

26.21 (a) B and C (b) 0.0197 mol (c) 610 K

26.22 (a) 105 kW (b) 0.70 (c) in claim 1 efficiency $= 0.76$ so
claim is unjustified; claim 2 justified

Chapter 27

27.1 3.2×10^{-17} J, 8.4×10^6 m s^{-1}

27.2 (a) 2.05×10^{15} m s^{-2} (b) 1.86×10^{-15} N
(c) 11 600 V m^{-1} (d) 163 V (e) 7.6×10^6 m s^{-1}

27.3 1.01×10^7 m s^{-1}

27.4 (a) amplitude doubled (b) twice the number of cycles
(c) half the number of cycles (d) one-eighth the
number of cycles (e) half the number of cycles (f)
twice the number of cycles (g) same number of
cycles in half the distance across the screen

27.5 (a) amplitude of 2.8 cm, 1 cycle occupies 2 cm
(b) 5 cycles occupy 1 cm

27.6 (a) 2.7×10^6 m s^{-1} (b) 3.0×10^{13} s^{-1}
(c) 1.13×10^6 m^{-1}

27.7 2.2×10^5 m s^{-1}

27.8 (a) 2.67×10^{-26} kg, 2.84×10^{-26} kg, 3.00×10^{-26} kg (b)
16.1, 17.1, 18.1 (c) a singly charged oxygen molecule

27.9 (a) 4.8×10^{-19} C, negative

27.10 (b) 4.56×10^{-10} kg s^{-1}

27.11 (a) 3.2×10^{-19} C (b)(i) 4.8×10^{-19} C

27.12 (a) gravitational (b)(i) 5.6×10^{-11} N C^{-1}
(c)(i) positive (ii) 3.3×10^{-26} kg
(d)(ii) 1.17×10^5 m s^{-1} (iii) 0.053 T (e)(i) increases by a
factor of $\sqrt{2}$ (ii) greater force exerted by magnetic
field than by electric field so particles' path is curved

27.14 (b)(i) 8.0×10^{-9} J (ii) use $E = mc^2$ to get 8.9×10^{-26} kg

Chapter 28

28.1 (a) 2.32×10^{-13} kg (b) 1.3×10^{-12} m^3

28.2 1.0×10^7 m s^{-1}

28.3 3.0×10^{-15} m

28.5 (a) 0.0256 u (b) 23.8 MeV

28.6 (a) 0.0155 u (b) 14.4 MeV (c) 14.4 MeV

28.7 (a) 0.0858 u (b) 80 MeV

28.8 (c) 1.0 u, 3.3×10^7 m s^{-1}

28.9 145 years

28.10 5.7×10^{18}

28.11 (a) (184 ± 2) s (c) rate of decay / number of atoms =
3.74×10^{-3} s^{-1} (d) 4.4×10^9 s^{-1}

28.12 (a) 0.30 (b) 0.15 (c) 0.104

28.13 2.8 s

28.14 (a) 150 000 disintegrations per second (c)(i) 3.4×10^{-9}
A (ii) 2.1×10^{10} (e) 1.37×10^{-11} s^{-1}, 2.1×10^{10}

28.16 (b) iron 26p, 29n, manganese 25p, 30n (c) a proton is
converted into a neutron

28.18 (b) 3.01×10^{26}

28.19 (c)(ii) 9.8×10^{-30} kg (iii) 2.8×10^{-6} kg s^{-1}

28.20 (c) energy from one fission $= 3.8 \times 10^{-11}$ J,
2.1×10^{20} s^{-1}

28.21 (a) hydrogen (d) 2.73×10^{-12} J

28.22 (b) increase in mass $= 2.14 \times 10^{-30}$ kg $\equiv 1.93 \times 10^{-13}$ J
speed of alpha particle at least 7.6×10^6 m s^{-1}

28.23 (a) 5.1×10^{24} (b) 2.5×10^{-10} s^{-1} (c) 1.3×10^{15}
(d) 9.2×10^9 s

28.24

	α	β	γ
	$+2(e)$	-1	0
	4(u)	small (1u / 1840)	0
	$\sim 0.1c$	$\sim 0.99c$	c
	stopped by paper	few mm Al	reduced by thick lead

28.25 (a)(i) 6 (ii) 8

28.26 (a) 2.19×10^{-9} s (b) 3.17×10^8 s (c) 1.33×10^9 s

28.27 (a) 1000 Bq (b)(i) 5.55 days (ii) 18 days

28.28 (a) 2.99×10^{-9} s^{-1} (b) 3.89×10^{15} Bq (c) 3300 W

28.29 (c) 15.8 years

Chapter 29

29.2 (a) 4.29×10^{14} Hz, 2.84×10^{-19} J (b) 7.50×10^{14} Hz,
4.97×10^{-19} J

29.3 (a) 5.55×10^{14} Hz (b) 4.59×10^{14} Hz (c) both in
visible spectrum

29.4 8.8×10^{-19} J

29.5 (a) 1.86×10^{-18} J (b) 2.80×10^{15} Hz

29.7 122 nm, 103 nm, 97.5 nm, 95.3 nm, 94.0 nm 93.3 nm,
92.9 nm .....

29.8 (a) 8.0×10^{-22} J (b) 633 nm

29.9 (a) 300 W

29.10 (a) 1.28×10^{-14} J (b) 1.28×10^{-14} J (c) 1.93×10^{19} Hz
(d) 0.0155 nm

29.11 5.2×10^{-12} m

29.12 (a) 3.96×10^{-19} J (b) 1.32×10^{-27} N s (c) 1.01×10^{19} s^{-1}
(d) 1.33×10^{-8} N (e) 1.33×10^{-8} N

29.13 2.1×10^{-10} m

29.14 (a) 0.096 W m^{-2} (b) 3.0×10^{17} s^{-1}

29.15 (a)(i) 4.42×10^{-19} J (ii) 5.66×10^{13} s^{-1}
(iii) 9.05×10^{-7} A (b) photon energy is 3.32×10^{-19} J
and this is greater than work function energy

29.16 (a)(ii) 2.3×10^{15} s^{-1} (iii) from violet at 3.11 eV to red
at 1.78 eV (b)(ii) 1.21 eV, 6.5×10^5 m s^{-1}

29.18 (b) 1.37×10^{-23} N s (i) 4.85×10^{-11} m

29.19 (b)(i) 1.81×10^{-18} J (ii) 1.82×10^{-24} N s, 3.6×10^{-10} m

29.20 (c) corresponds to 0.5 eV, $= 2.5 \times 10^{-6}$ m

29.22 (a)(i) 2810 eV (ii) 4.42×10^{-10} m (iii) X-rays
(b) 1.07×10^{-13} m

29.23 (a) 6.17×10^{14} Hz (b) 12.5°

29.25 (b)(ii) 2.8×10^{-11} m

29.26 (c)(ii) (1) 4.74×10^{-24} N s (2) 1.23×10^{-17} J (3) 77 V

Chapter 30

30.1 (a) 940 Hz (b) 696 Hz

30.2 56 m s^{-1}

30.3 4.12×10^5 m s^{-1} (Note that although 5 sig. figs. are
given in the question the answer is not more reliable
than to 2 significant figures since the change in
wavelength is only known to 2 sig. figs.)

30.4 $\frac{1}{2}$

30.5 (b) 0.095 m s^{-1} (c) 590 m s^{-1} (d) 3.4×10^4 m s^{-1}
(e) about the speed of light. The value comes to
3.4×10^8 m s^{-1} so one item of data given must
be incorrect.

30.6 2.3×10^{-5} kilometres per second per light-year

30.8 (a) 1.06×10^{-12} N (b) 6.35×10^{14} m s^{-2} (c) 0.99 mm
(d) 2.02×10^8 m s^{-1} (e) 1.28×10^8 Hz (f) 3.4×10^{-11} J
(g) 8.0×10^{-17} J (h) 213 000

30.9 1.64×10^{-13} J

30.11 (b)(i) 1.2×10^{57} (ii) 3.1×10^{18} s

30.12 (a) 4.4×10^8 m (b)(ii) 9000 m s^{-1} (iii) towards
(c)(ii) 230 m

30.14 6.2×10^{17} s

30.15 1.5×10^{25} m

30.16 (c) 5.7×10^7 m s^{-1}, 7.6×10^6 Hz

Appendix B: Useful Data

Physical quantities

QUANTITY	DEFINING EQUATION	UNIT	ABBREVIATION
mass	base unit	kilogram	kg
time	base unit	second	s
frequency	number of cycles/time	hertz	Hz
length	base unit	metre	m
angle	arc length/radius	radian	rad
angular velocity	angle/time	radian per second	rad s^{-1}
area	length × breadth	metre^2	m^2
volume	length × breadth × height	metre^3	m^3
density	mass/volume	kilogram per metre^3	kg m^{-3}
velocity	distance in given direction/time	metre per second	m s^{-1}
acceleration	increase in velocity/time	metre per second^2	m s^{-2}
momentum	mass × velocity	kilogram metre per second	kg m s^{-1}
force	mass × acceleration	newton	N
moment of force, torque	force × distance from axis	newton metre	N m
pressure, stress	force/area	pascal	Pa
strain	extension/original length	no unit	–
work, energy	force × distance in direction of force	joule	J
power	work/time	watt	W
gravitational field	gravitational force/mass	newton per kilogram	N kg^{-1}
gravitational potential	gravitational energy/mass	joule per kilogram	J kg^{-1}
thermodynamic temperature	base unit	kelvin	K
heat capacity	energy/temperature change	joule per kelvin	J K^{-1}
specific heat capacity	heat capacity/mass	joule per kelvin kilogram	$\text{J K}^{-1}\text{ kg}^{-1}$
specific latent heat	energy/mass	joule per kilogram	J kg^{-1}
electric current	base unit	ampere	A
electric charge	current × time	coulomb	C
electric field strength	force/charge	newton per coulomb	N C^{-1}
specific charge	charge/mass	coulomb per kilogram	C kg^{-1}
potential difference	energy/charge	volt	V
resistance	potential difference/current	ohm	Ω
resistivity	resistance × area/length	ohm metre	Ωm
capacity	charge/potential difference	farad	F
magnetic flux density	force/current × length	tesla	T
magnetic flux	magnetic flux density × area	weber	Wb
radioactive activity	counts/time	becquerel	Bq

Physical data

PHYSICAL QUANTITY	SYMBOL	NUMERICAL VALUE AND UNIT
speed of light in vacuum	c	2.998×10^8 m s^{-1}
Planck constant	h	6.626×10^{-34} J s
gravitational constant	G	6.672×10^{-11} N m^2 kg^{-2}
permeability of free space	μ_0	$4\pi \times 10^{-7}$ N A^{-2}
permittivity of free space	ε_0	8.854×10^{-12} F m^{-1}
Boltzmann constant	k	1.381×10^{-23} J K^{-1}
molar gas constant	R	8.314 J K^{-1} mol^{-1}
fundamental charge	e	1.602×10^{-19} C
specific charge of electron	$-e/m$	-1.759×10^{11} C kg^{-1}
electron rest-mass	m_e	9.110×10^{-31} kg
proton rest-mass	m_p	1.673×10^{-27} kg
neutron rest-mass	m_n	1.675×10^{-27} kg
Avogadro constant	N_A	6.022×10^{23} mol^{-1}
atomic mass unit	u	1.661×10^{-27} kg
electronvolt	eV	1.602×10^{-19} J
acceleration of free fall	g	9.81 m s^{-2}
standard temperature and pressure (s.t.p.)		
standard temperature, ice point, 0°C		273.15 K
standard atmospheric pressure		$1.013\ 25 \times 10^5$ Pa
density of air at s.t.p.	p	1.29 kg m^{-3}
molecular mass of air		$0.028\ 98$ kg mol^{-1}
molar heat capacity of air at constant pressure	$C_{p,\ m}$	29.1 J K^{-1} mol^{-1}
molar heat capacity of air at constant volume	$C_{v,\ m}$	20.8 J K^{-1} mol^{-1}
density of water	p	1000 kg m^{-3}
0° Celsius		273.15 K
triple point of water		0.01 °C $= 273.16$ K
specific heat capacity of water		4190 J kg^{-1} K^{-1}
mass of the earth		5.98×10^{24} kg
equatorial radius of the Earth		6.378×10^6 m
mean density of the Earth		5520 kg m^{-3}
period of rotation of Earth on its axis (≈1 day)		8.616×10^4 s
mean distance of Earth from Sun		1.50×10^{11} m
period of revolution of Earth (1 year)		3.16×10^7 s
mass of the Sun		1.99×10^{30} kg
radius of Sun		6.96×10^8 m
power output of Sun		3.9×10^{26} W
power intensity received at Earth (solar constant)		1.4 kW m^{-2}
mass of the Moon		7.35×10^{22} kg
radius of the Moon		1.74×10^6 m
mean distance of Moon from Earth		3.84×10^8 m
period of rotation of Moon about axis		
= period of revolution in orbit (1 lunar month, 27.3 days)		2.36×10^6 s
acceleration of free fall on the Moon		1.62 m s^{-2}

PREFIX	MULTIPLYING FACTOR	SYMBOL
atto	10^{-18}	a
femto	10^{-15}	f
pico	10^{-12}	p
nano	10^{-9}	n
micro	10^{-6}	μ
milli	10^{-3}	m
centi	10^{-2}	c
deci	10^{-1}	d

PREFIX	MULTIPLYING FACTOR	SYMBOL
deca	10	da
hecto	10^{2}	h
kilo	10^{3}	k
mega	10^{6}	M
giga	10^{9}	G
tera	10^{12}	T
peta	10^{15}	P
exa	10^{18}	E

Mathematical relationships

Algebra

$$a^2 = a \times a \qquad a^{\frac{1}{2}} = a^{1/2} = \sqrt{a} \qquad a^{-2} = \frac{1}{a^2}$$

$$a^{\pm\frac{1}{2}} = a^{\pm 1/2} = \frac{1}{\sqrt{a}}$$

If $a = 10^x$ $\lg a = x$

If $a = e^x$ $\ln a = x$

$\lg (ab) = \lg a + \lg b$ $\ln (ab) = \ln a + \ln b$

$\lg (a^n) = n \lg a$ $\ln (a^n) = n \ln a$

$\lg\left(\dfrac{a}{b}\right) = \lg a \pm \lg b$ $\ln\left(\dfrac{a}{b}\right) = \ln a \pm \ln b$

$(1+x)^n \approx 1 + nx$ if x is small

The two solutions to the equation $ax^2 + bx + c = 0$ are given

by $x = \dfrac{\pm b \pm \sqrt{b^2 \pm 4ac}}{2a}$

Trigonometry

$$\sin\theta = \frac{o}{h} \qquad \cos\theta = \frac{a}{h} \qquad \tan\theta = \frac{o}{a}$$

giving $a = h \cos\theta$ and $o = h \sin\theta$

In any right-angled triangle, Pythagoras's theorem gives
$o^2 + a^2 = h^2$

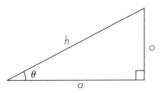

In the circle drawn the radius is 1 unit of length
the opposite side, PR, is sin θ units of length
the adjacent side, OR is cos θ units of length
and the arc length PQ is θ units of length

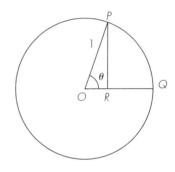

From the right-angled triangle we get $\sin^2\theta + \cos^2\theta = 1$
For small angles $\sin\theta \approx \theta \approx \tan\theta$, with angles measured in radians.

In any triangle

$$\frac{a}{\sin A} = \frac{b}{\sin B} = \frac{c}{\sin C}$$

and $a^2 = b^2 + c^2 \pm 2bc \cos A$

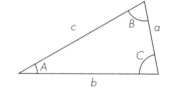

Geometry

Circumference of a circle of radius r	$= 2\pi r$
Area of a circle of radius r	$= \pi r^2$
Volume of a sphere of radius r	$= \frac{4}{3}\pi r^3$
Surface area of a sphere of radius r	$= 4\pi r^2$
Volume of a cyclinder of radius r and height h	$= \pi r^2 h$

For a straight-line graph $y = mx + c$ where m is the gradient of the graph and c is the intercept on the y axis

Calculus

The gradient of a graph is found by differentiating:

if $y = x^n$ the gradient, $\dfrac{dy}{dx}$ is given by $\quad \dfrac{dy}{dx} = nx^{n\pm1}$

if $y = \sin x$ $\qquad\qquad\qquad\qquad \dfrac{dy}{dx} = \cos x$

if $y = \cos x$ $\qquad\qquad\qquad\qquad \dfrac{dy}{dx} = \pm\sin x$

if $y = e^x$ $\qquad\qquad\qquad\qquad\; \dfrac{dy}{dx} = e^x$

The area beneath a graph can be found by integrating, which is the reverse process to differentiating:

if $y = x^n$ the integral, $\int y \, dx$ is given by $\int y \, dx = \dfrac{x^{n+1}}{n+1} + c$

if $y = \cos x$ $\qquad\qquad\qquad\quad \int y \, dx = \sin x + c$

if $y = \sin x$ $\qquad\qquad\qquad\quad \int y \, dx = \pm\cos x + c$

if $y = \dfrac{1}{x}$ $\qquad\qquad\qquad\quad \int y \, dx = \ln x + c$

if $y = e^x$ $\qquad\qquad\qquad\quad\; \int y \, dx = e^x + c$

$\qquad\qquad\qquad\qquad\qquad\quad$ where c is a constant

Appendix C: Guidelines for Data Analysis Exercises

Many answers are given in this section, but not all of them. You should treat the answers as checks that enable you to see if you are progressing in the right direction. In places hints are more appropriate than answers. You will need to use the numerical values of physical constants from time to time. These are given in the previous appendix. As with answers to questions, the final digit of an answer should be treated with caution. It is given to help with checking and to avoid introducing too many rounding up errors. It may not be justified by the accuracy of the data supplied.

Chapter 3

Forces in frameworks

Note that throughout this exercise the forces shown are the forces exerted by the struts on the joints. The diagram does not show the forces on the struts
(d) This will be a rectangle (e) 150 kN (f) 600 kN
(g) CD is in *tension*. It pulls on C to the right and it pulls on D to the left, so it is being stretched. Similarly JB, IC and DF are also in tension. IB and IH are two of nine in compression
(h) JI, DE, BC and HC

Chapter 4

Grand Prix facts and figures

(a) Graph is a series of positive and negative peaks with flat portions where a constant speed is maintained
(b) Graph goes upward continuously (the car never goes backward). The gradient is greatest when the car is going fastest
(c) Total track length is about 6000 m. If you hold a ruler at a speed of 62 m s^{-1} on the velocity–time graph, the area above the ruler *under* the graph is approximately equal to the area below the ruler *over* the graph, i.e. the average speed is given by (d)
(d) 6000 m/96.3 s = 2 62.3 m s^{-1}
(e) Acceleration is reaching a maximum or minimum of about 30 m s^{-2}, i.e. 3g. Values over this are impossible as the car would skid. If worn tyres are used or a slippery track exists then the value of the maximum acceleration will be less

Chapter 5

Passenger loading of a European Airbus

(a) 3000 km (b) 600 km (c) 75 000 kg (d) 1.88 m s^{-2}
(e) 141 000 N (f) 176 000 N (g) 60 000 kg (h) Range: 3000 km, 400 km, 800 km, 1200 km, 3000 km (i) 100

Chapter 7

Energy and momentum of balls

(a) 120 J, 165 J, 110 J, 16 J, 38 J, 75 J
(b) 6.2 N s, 11.8 N s, 3.2 N s, 1.4 N s, 1.6 N s, 3.0 N s
(c) Answers *must* be identical to (b)
(d) Answers also identical to (b)
(e) (i) 1.55 kg (ii) 0.25 kg (iii) 0.60 kg (f) 120 J
(g) Each mass is held by hand and so collisions are not between a single object and a ball
(h) Taking 'most nearly elastic' to mean the collision in which the velocity of separation is the highest proportion of the velocity of approach gives the answer – golf ball
(i) 4500 N, 1470 N, 2400 N, 1000 N, 520 N, 740 N

Chapter 8

The behaviour of aircraft tyres when rotating at high speeds

SPEED $v/\text{m s}^{-1}$	CLEARANCE $/\text{mm}$	ANGULAR VELOCITY $\omega/\text{rad s}^{-1}$	$r\omega^2$ $/\text{m s}^{-2}$	FORCE PER UNIT MASS $/\text{N kg}^{-1}$
70	21	156	11 000	11 000
80	25	178	14 200	14 200
90	29	200	18 000	18 000
100	34	222	22 200	22 200
110	42	244	26 900	26 900

(h) These data are suitable for tyres between 850 and 1000 mm. For a constant speed the acceleration of larger radii tyres is less than for small radii ($a = v^2/r$) but the mass is greater. These two factors partially cancel one another out

Chapter 9

Satellite motion

(a) 8.0 (b) 6.2 MJ kg^{-1} (found as minus half the potential energy or, more accurately, from $GM/2r$)
(c) approximately 60 MJ kg^{-1} (d) 1.8 × 10^{11} J (e) 3 × 10^{10} J; 4500 m s^{-1} (f) 480 MJ greater potential energy at greater distance (g) 6100 m s^{-1} (loss of 480 MJ of kinetic energy)

Chapter 10

Energy in SHM

(a) 2.45 cm (b) 20 s^{-1} (c) 0.52 m s^{-1} (d) 0.24 J
(e) ⑦ = ⑧ = ⑨ = 0.24 J; ① + ④ = ② + ⑤ = ③ + ⑥ = 0.24 J
(f) 0.012 J, 0.36 J, 1.19 J (g) as (f) (h) −0.36 J (i) 0.59 J
(j) 0.23 J, −0.95 J

Chapter 11

Seismic surveying

(b) SD$_8$ = 1.15 km, SXD$_8$ = 1.77 km, depth = 0.67 km
(d) $A = 2.6 × 10^{10}$ kg s^{-2} m^{-1} = 2.6 Pa
(f) The further seismometers give the same amplitude but will receive weaker vibrations; direct waves should show larger amplitude than reflected waves
(g) If a wave strikes the boundary at the critical angle, it can be refracted with an angle of refraction of 90°. Re-entry of this wave back to D$_8$ at the critical angle is possible, presumably as a result of irregularities in the boundary layer

Chapter 12

The camera

(a) 6.25 mm (c) $\sqrt{32}$ = 5.657 (e) (1/500) s; $f/16$
(g) 600–1200 mm

Chapter 13

Synthesisers

For Fig 13.17, (a) $3f_0$ and (b) $5f_0$

Chapter 14

Light-bulb construction

(a) $\frac{1}{4}$ A, 960 Ω (b) 0.0121 mm (c) A is long life, C is high temperature (d) and (e) see table

	A	B	C
Resistance/Ω	960	960	960
Diameter/mm	0.0129	0.0121	0.0113
Length/m	0.159	0.14	0.122
Surface area/m²	6.4×10^{-6}	5.3×10^{-6}	4.3×10^{-6}
Power lost per unit area/W m⁻²	0.93×10^{7}	1.13×10^{7}	1.4×10^{7}
temperature/°C	2420	2600	2730

(f) Cost for A, £10.10; cost for C, £11.60. Light energy from A, 3.9×10^{7} J; light energy from C, 5.2×10^{7} J. Cost of a unit of light energy from A/cost of unit from C = 1.2

(g) If you want less light, use a lower wattage bulb not a long-life one

Chapter 17

Charging a capacitor through a resistor

$V_C = 10$ V; $V_R = 0$ V; $I = 0$; $q = 0$; $Q = 4000$ μF

Chapter 22

Blood pressure

(a) 16 000 Pa, 10 700 Pa (b) 116 000 Pa, 110 700 Pa
(c) Extra pressure as a result of 1.2 m height between heart and feet = 12 500 Pa (d) 7300 Pa
(e) Normal pressure in brain is $(0.4 \times 1060 \times 9.8)$ Pa = 4200 Pa below systolic pressure, i.e. 11 800 Pa above atmospheric pressure.
When accelerating the drop in pressure across 0.4 m of body = $[0.4 \times 1060 \times (28 + 9.8)]$ Pa = 16 000 Pa below systolic pressure = 0

Chapter 24

Temperature scales

Absolute zero becomes -468 °F and the triple point 32 °F. Any other temperature on scale X becomes $T \times 500/273.16$ giving the boiling point of water as 683.02 °X or 215 °F. Scale F has an uncanny similarity to the Fahrenheit scale.

Chapter 25

Kinetic theory

Component of velocity in a direction towards the wall $= c \cos \theta$
Change in momentum at a collision $= 2mc \cos \theta$
Time between collisions with wall $= 2r \cos \theta/c$
Rate of change of momentum $= mc^2/r$
Pressure on wall due to one molecule $= mc^2/4\pi r^3$

Chapter 26

The behaviour of gases in a Turbofan jet engine

(a) 6290 mol (c) 664 K (d) 2 610 000 Pa (e) 96.3 MJ (f) 1.82 kg s⁻¹ (g) 157 tonnes (h) 386 K (j) 8 400 000 J (l) Increase in internal energy is always $nC_{v, m}\Delta T$ so the four values are 57.4 MJ, 96.3 MJ, -132.7 MJ and -21.0 MJ.
Efficiency $= (96.3 - 29.4)/96.3 = 0.69$ (m) 55 MW (n) 0.57

Chapter 27

Ships' radar

(a) 4.15 cm (b) 3615 (c) 0.200 ms (d) 60.0 km (e) just under 30 km (f) 10 cm in 0.2 ms so speed is 500 m s⁻¹ (g) 5.0 cm (h) 3.0 km (i) 37.5 m, 0.25 mm

Chapter 28

Nuclear reactions

The energies necessary to balance each of the equations are as follows:

6.5 MeV, 17.3 MeV, 1.3 MeV, 3.26 MeV, 0.39 MeV, 2.8 MeV, 4.3 MeV

Decommissioning a nuclear reactor

(a) 2.9×10^5 Bq kg^{-1}; 8.6×10^7 Bq kg^{-1}; 9.7×10^5 Bq kg^{-1}

(c) Long half-life materials have their long half-life because there is a low probability of any particular nucleus decaying per unit time

(d) 96 years (e) 3200 Bq

(f) All in 10^{12} Bq: H, 0.34; C, 43.5; Cl, 1.3; Ca, 0.9995; Fe, 1.7×10^{-8}; Co, 1.1×10^{-3}; Ni, 2.198; Ni, 117.7; Nb, 0.012 96; Ag, 0.011; Sm, 0.93; Eu, 2.9×10^{-3}; Eu, 9.3×10^{-3}

(g) Nickel-59, carbon-14, nickel-59, chlorine-36, calcium-41

(h) The graph will not be exponential; it falls fairly quickly but because of the long half-life isotopes then flattens out

Chapter 29

X-rays

(a) K_α 58.2 keV; K_β 66.6 keV; L_α 8.4 keV; L_β 9.6 keV

(b) K_α 2.13×10^{-11} m; K_β 1.86×10^{-11} m; L_α 1.48×10^{-10} m; L_β 1.29×10^{-10} m

(c) K_γ 1.83×10^{-11} m; L_γ 1.10×10^{-11} m

(e) While it is possible for an electron in the K-shell to be given exactly 58.2 keV to lift it to the L-shell, this is in practice unlikely. Inner electrons are normally dislodged from the atom and this requires a potential difference of at least 69.6 kV.

Index

Bold page references refer to an ilustration or figure.

Liverpool
Community
College